ELECTRIC FURNACE STEEL PRODUCTION

ELECTRIC FURNACE STEEL PRODUCTION

Edited by

Professor Dr Mont. Erwin Plöckinger

and

Dr-Ing. Otto Etterich

in collaboration with the Steelworks Committee of the VDEh, Association of German Iron and Steel Engineers

Translated by

E. B. Babler and **P. E. O. Babler**

A Wiley Heyden Publication

JOHN WILEY AND SONS

Chichester · New York · Brisbane · Toronto · Singapore

German edition: *Elektrostahl-Erzeugung*

Library of Congress Cataloging in Publication Data:

Elektrostahlerseugung. English.
Electric furnace steel production.

Translation of: Elektrostahlerzeugung.
'A Wiley Heyden publication.'
Includes bibliographical references and index.
1. Steel—Electrometallurgy.
I. Plöckinger, Erwin. II Etterich, Otto.
III. Verein Deutscher Eisehhüttenleute. Stahlwerksausschuss.
IV. Title.

TN745.E4313. 1984 669'.1424 83-25910
ISBN 0 471 90254 3

British Library Cataloguing in Publication Data:

Electric furnace steel production.
1. Electric furnaces. 2. Steel
I. Plöckinger, Erwin. II. Etterich, Otto.
III. Elektrostahl-Erzeugung. English.

ISBN 0 471 90254 3

Printed and bound in Great Britain.

Contents

Preface to the English Edition

The English-language version of the third edition of Stahleisen's book *Electric Furnace Steel Production*, which was decided on eighteen months after the appearance of the German edition, contains a few additions to include the latest advances in techniques. The literature reference lists have also been revised to include the latest important publications; precedence was given here to English-language works. The new literature references have been fitted logically but informally into the reading lists. Symbols in mathematical expressions have been changed to conform with British Standards.

In this book, translations of the names of organizations and titles of Literature references may not necessarily be the same as definitive translations.

THE EDITORS

Preface to the German Edition

The development in modern steelmaking processes briefly outlined in the Foreword by Hermann Schenck made the preparation of a completely new edition of Stahleisen's book, *Electric Furnace Steel Production*, necessary after the second edition went out of print. This new edition does justice to the increasing importance of electric steelmaking throughout the world; the process has grown more than ten-fold since the first edition, while total world crude-steel production has only trebled.

It was no longer possible for only one or two authors to produce the comprehensive material required to cover the many new specialist subjects, and we therefore endeavoured to find a larger number of contributors to cover the different areas. We acknowledge that a certain amount of overlap could not be avoided; however, this can be accepted for the sake of completeness of production of each chapter and section written by the various experts.

The authors were asked to produce brief and concise contributions because of the wide range of the subject. However, each chapter ends with a comprehensive list of literature references, so the reader can obtain further information on details of the subject from important publications. For the same reason, metallurgical fundamentals are not included; these can be found in many textbooks available.

Our thanks go to the forty-two contributors listed in the next section. They have been very understanding in preparing their manuscripts and agreed to condense and revise their work as necessary in order to enable a uniform presentation to be produced.

Our special thanks also go to the members of the Steelworks Committee of the VDEh (Association of German Iron and Steel Engineers) and their Secretary, Hans Wilhelm Kreutzer, for their valuable additions and suggestions which were a great help.

Last but not least we thank our friends Dr mont. Franz Sommer and Dr-Ing. Hans Pollack, whose first edition of this book produced in 1951 provided the foundation on which a compendium on this specialist field could be produced.

Düsseldorf, May 1979 *Erwin Plöckinger and Otto Etterich*

Foreword

Since the appearance of the second edition of this book in 1964 there have been significant advances in the technical and commercial development of electric furnace steel production. This upturn was occasioned mainly by the extensive reorganization and adaptation programme that steelworks were obliged to undertake with the introduction of steelmaking processes using pure oxygen. The increase in the demand for pig iron that this caused triggered a very great improvement in blast furnaces and associated techniques on which steelworks had concentrated until then. Steelworks using open-hearth furnaces and Thomas converters were unable, for economic reasons, to maintain their commanding position in the face of the advantages of the basic oxygen steelmaking process, and, almost without exception, had to give way to new technology.

This train of events made it necessary to rethink the problem of the metallic charge for steelmaking. The open-hearth furnace had always been able to adapt very well to fluctuations in the availability of scrap and pig iron. However, the proportion of cold scrap used in the basic oxygen process is comparatively limited, and increases in the amount of scrap available could not be absorbed by this process.

Electric-furnace operators saw the increase in the availability of scrap provide them with an opportunity to break away from their traditional area of special, high-grade steels and move into the realm of quality and mass-produced steels. They concentrated their efforts on creating a new generation of extensively automated ultra-high-power arc furnaces, the design of which had to cater for the newly tightened regulations for environmental protection and health and safety at work. In addition, they were able to take advantage of vacuum treatments and ladle-refining processes that had been developed at about the same time and which had a beneficial effect on the quality and cost of the steel. Lastly, improvements in the continuous casting process, and the expectation that a particularly suitable charge material could be created by using the direct-reduction process, both finally paved the way for a breakthrough for the electric arc furnace.

The sixteen chapters of this book discuss the latest state of technology clearly and exhaustively. The editors must be congratulated for bringing together this group of well-known professionals to provide expert coverage of the many aspects of the subject. The team completed its work so comprehensively that even special applications such as electric furnace melting for producing cast steel and grey and malleable cast iron are included. The extensive contents list and subject index, and the long table of literature references that complete each chapter, provide helpful guidance through the complex and very varied array of subjects in the book.

This reference work provides essential information about almost every prerequisite and condition needed for the successful operation of a modern electric furnace steelworks, including the principles of how to calculate its economic efficiency. It is an invaluable source of information for operators. The book's comprehensive exposition of intricate relationships is designed also to cater for the special wishes of equipment manufacturers and suppliers; it should arouse particular interest in those areas and provoke thoughts of new developments. The reader will gain an idea of the considerations which need to be entered into in order to make the best use of the essentially simple concept of arc melting to institute competitively operable innovations in steel production.

Aachen, May 1979 HERMANN SCHENCK

Contributors

Dr-Ing. Rudolf Baum, *Siegen-Geisweid, Krupp Stahlwerke Südwestfalen AG, Siegen 1*

Dr-Ing. Karl-Heinz Brokmeier, *Dortmund, Fried. Krupp GmbH, Krupp Industrie- u. Stahlbau, Duisburg 14*

Dr-Ing. Erwin Dötsch, *Holzen, Brown, Boveri & Co. AG, Dortmund*

Dipl.-Ing. Gerd Wilhelm Drees, *Bergkamen, Brown, Boveri & Co. AG, Dortmund*

Dipl.-Ing. Klaus Dziggel, *Dortmund 30, Hoesch Hüttenwerke AG, Westfalenhütte, Dortmund*

Dipl.-Ing. Hans Eßmann, *Schermbeck, Gutehoffnungshütte Sterkrade AG, Oberhausen*

Dr-Ing. Otto Etterich, *Düsseldorf, formerly Thyssen Edelstahlwerke AG, Werk Witten*

Professor Dr-Ing. Wilhelm Anton Fischer, *Ratingen, formerly Max Planck-Institut für Eisenforschung, Düsseldorf*

Dr-Ing. Jürgen Geiseler, *Tönisvorst 1, Forschungsgemeinschaft Eisenhüttenschlacken, Duisburg 14*

Ing. Hans Ulrich Haering, *Solingen, BFI Betriebstechnik GmbH, Düsseldorf*

Dipl.-Ing. Fritz Hegewaldt, *Herdecke, Brown, Boveri & Co. AG, Dortmund*

Dr-Ing. Rüdiger Heinke, *Siegen, Krupp Stahlwerke Südwestfalen AG, Siegen 1*

Dr Heimo Jäger, *Bruck/Mur, Vereinigte Edelstahlwerke (VEW), Kapfenberg*

Dipl.-Ing. Heinz Jung, *Völklingen, Stahlwerke Röchling-Burbach GmbH, Völklingen*

Ing. (grad) Helmut Kahnwald, *Tönisvorst, BFI Betriebstechnik GmbH, Düsseldorf*

Dr-Ing. Kurt Kegel, *Essen, Lehrbeauftragter TH Aachen and Karlsruhe University*

Dipl.-Ing. Walter Klein, *Staufenberg 1-Uschlag, VGT-Vereinigte Großalmerode Thonwerke, Großalmerode, formerly Didier-Werke AG, Duisburg*

Dr-Ing. Siegfried Köhle, *Hilden, Betriebsforschungsinstitut (BFI) VDEh-Institut für angewandte Forschung GmbH, Düsseldorf*

DIPL.-ING. HANS WILHELM KREUTZER, *Kaarst, Verein Deutscher Eisenhüttenleute (VDEh), Düsseldorf*

DR MONT. DIPL.-ING. GERD KÜHNELT, *Kapfenberg, Vereinigte Edelstahlwerke (VEW), Kapfenberg*

ING. (GRAD.) HORST LAUFER, *Herten, Fried. Krupp GmbH, Krupp Industrie- u. Stahlbau, Duisburg 14*

DIPL.-ING. HERMANN LÜNIG, *Moers, Thyssen Edelstahlwerke AG, Werk Witten*

PROFESSOR DR-ING. ALEKSANDER MAJDIČ, *Cologne, Forschungsinstitut der Feuerfest-Industrie, Bonn*

DIPL.-ING. DIRK MARCHAND, *Krefeld, BFI Betriebstechnik GmbH, Düsseldorf*

DIPL.-ING. JOACHIM MEYER ZU HERINGDORF, *Bochum, Fried. Krupp Hüttenwerke AG, Bochum*

PROFESSOR DR-ING. FRANZ NEUMANN, *Unna-Billmerich, Brown, Boveri & Co. AG, Dortmund*

DIPL.-ING. KARL-BERND NEVRIES, *Essen-Steele, RWE, Rheinisch Westfälisches Elektrizitätswerk, Essen*

DR-ING. HEINRICH OTTMAR, *Schwalbach, Röhrenwerke Bous/Saar GmbH, Schwalbach*

DIPL.-KAUFM. KARL PITTEL, *Köln, Verein Deutscher Eisenhüttenleute (VDEh), Düsseldorf*

DR MONT. DIPL.-ING. RUDOLF PLESSING, *Kapfenberg, formerly Vereinigte Edelstahlwerke AG (VEW), Kapfenberg*

PROFESSOR DR MONT. DIPL.-ING. CHEM. DIPL.-ING. MONT. ERWIN PLÖCKINGER, *Vienna, formerly Vereinigte Edelstahlwerke AG (VEW), Kapfenberg*

DR-ING. KLAUS POLTHIER, *Meerbusch, BFI Betriebstechnik GmbH, Düsseldorf*

DIPL.-ING. FRIEDHELM REGNITTER, *Schaffhausen, Stahlwerke Röchling-Burbach GmbH, Völklingen*

ING. (GRAD.) DIETER RUBENSDÖRFFER, *Wadgassen, Stahlwerke Röchling-Burbach GmbH, Völklingen*

DR-ING. JOSEF SCHIFFARTH, *Bocholt-Barlo, Foseco GmbH, Borken*

DR-ING. GÜNTER SCHMEIDUCH, *Schwalbach, Röhrenwerke Bous/Saar GmbH, Schwalbach*

DR-ING. ULRICH SIEGERS, *Schwalbach, Röhrenwerke Bous/Saar GmbH, Schwalbach*

DIPL.-ING. HEINZ SPERL, *Lintorf, Verein Deutscher Eisenhüttenleute (VDEh), Düsseldorf*

ING. (GRAD.) WOLFGANG STAWICKY, *Altena 8, Vereinigte Deutsche Metallwerke AG, Werdohl*

DR TECHN. DIPL.-ING. WOLFGANG THOMICH, *Meerbusch 2, Böhler AG, Düsseldorf*

PROFESSOR DR-ING. KLAUS TIMM, *Wentorf, Hochschule der Bundeswehr, Hamburg*

DIPL.-ING. JÜRGEN MACKEPRANG VOGEL, *Mülheim-Ruhr, Gesellschaft für elektrischen Straßenverkehr GmbH, Düsseldorf*

PROFESSOR DR-ING. MANFRED WAHLSTER, *Hanau 1, Leybold-Heraeus GmbH, Hanau*

DR-ING. LUDGER ZANGS, *Essen, Mannesmann Demag Metallgewinnung, Duisburg*

Electric Furnace Steel Production
Edited by E. Plöckinger and O. Etterich

1 Introduction

Otto Etterich, Düsseldorf

The many discoveries in the field of natural science since 1945 have led not only to a deepening of our knowledge but also to many new technological developments, the future significance of which cannot in many cases yet be appreciated. These developments have also affected the steel industry, where in about 1950 new processes led to a revolution in operational practice. Moreover, total world steel production has almost quadrupled since then: a rise which before then had taken the whole of the first half of this century.

Electric furnace steelmaking started at about the turn of this century. It gradually ousted the crucible steel process and took over the production of quality steels to an increasing extent, so that the terms electric furnace steel or electric steel became synonymous with high-grade steel. Newly developed processes have already overtaken this picture. Electric furnaces are also used for the production of plain carbon non-alloy steels in large quantities; on the other hand, high-grade steels are now made by other processes.

1.1 THE CONCEPT OF STEEL

The meaning of the word steel has undergone changes in the course of time. Euronorm 20 now provides the following standard definition: 'Steels are defined as those ferrous alloys which, in general, are suitable for hot forming. Except for certain high-chromium steels, they have a maximum carbon content of 2% by weight, which distinguishes them from cast iron.'

In the same standard the types of steel are grouped according to chemical composition and use. There are two groups, according to their chemical composition:

Non-alloy steels,
Alloy steels.

The groups are subdivided according to their characteristics in use:

basis steels,
quality steels,
special steels.

This rough classification cannot do justice to the many uses to which steel is put today as an indispensable material in all areas of our lives. An extraordinary number of steels are made for all the different mechanical, physical, and chemical applications. Their properties are adjusted for particular applications, not only by heat treatment but also to a great extent by the addition of suitable alloying elements.

The Steel and Iron List published by the Association of German Iron and Steel Engineers[1] mentions 20 alloying elements which are added to steel singly or in combinations of up to ten elements in different quantities. The list names 1800 standard German steels, 650 international standard steels, and 450 steels to foreign national standards as well as 6900 proprietary kinds. These numbers might give one the impression that all possibilities for the development of steel have been exhausted. This is not the case, however, because new demands are constantly made, and the steel industry tries to fulfil all the requirements by continuous development.[2]

1.2 DEVELOPMENTS IN THE MANUFACTURE OF CRUDE STEEL

Table 1 shows how total crude-steel production has developed since 1910; it is divided into total steel production and electric steel production, and is shown in percentage proportions for Germany, the USA and the whole world. Figure 1 shows how crude-steel production per head of population has developed in different countries and parts of the world. If we add crude-steel output to steel imports in any country, and then subtract steel exports, we obtain the quantity of steel supplied to the market in that country. From this we obtain the visible steel consumption per head of population, which has today reached a saturation value of 500–600 kg of crude steel per person per year in highly industrialized countries. European steel production already approaches these figures very closely.

Future steel production up to the year 2000 can be estimated from the expected increase in population and a growth in steel consumption of 5 kg per year per head of world population (corresponding to the average during 1965–75). This development can be seen in Figure 2, taken from population growth studies by UNO.[3] Under these conditions steel production will be 315 kg of crude steel per head of population in the year 2000; this is about 60% of the saturation value in today's industrialized nations. Other investigations have arrived at similar results.[4–6] The estimated output of electric furnace steel up to the year 2000 is based on a proportion of 30–40% of total crude-steel production in the world.

1.3 DEVELOPMENT OF STEEL PRODUCTION METHODS

After 1955, world production of open-hearth steel and of Bessemer steel declined rapidly with the rapid introduction of the basic oxygen steelmaking process and the steady rise of electric furnace steel production. Figure 3 shows developments since 1950.[7] The change in the proportions of the individual steelmaking processes in total world crude steel output has also altered the percentages of scrap used in these processes. This development is shown in Figure 4 from statistical data since 1930 and the estimated growth up to the year 2000.[8]

1.4 STEELMAKING PROCESS FLOW CHART

The essential production stages and flows in modern steelmaking can be seen in schematic representation in Figure 5. Starting from the raw materials (iron ore, scrap, and alloying element ores), crude steels of exactly defined compositions can be made either by deoxidizing pig iron in converters or from scrap or sponge iron in other melting systems. The multitude of pre- and post-treatment processes enable us to arrive at the same end-products with different combinations of processes.

High-tonnage steels can today be produced from scrap or sponge iron in ultra-high-power arc furnaces, or they can be made in blast furnaces and basic oxygen converters. High-alloy steel can be made from converter steel if, after tapping, this is treated in the ladle or in a special post-treatment vessel. The old idea that special steels are made in an electric furnace and high-tonnage steels by the open-hearth process has been superannuated by modern steelmaking. However, we are not saying that every possible process is always sensible. Local conditions, such as the raw-material and energy situation, the amounts to be produced within the production programme, and, above all, economy must determine the optimum method and the required plant layout. The single-programme steelworks with its specific structure must be considered separately from the multi-programme works, which has to fulfil many diverse needs. Today, high-tonnage steelworks, which are equipped for secondary refining, are all able to produce both high-grade and special steels. But this makes sense only if the molten special steel is then passed into a special finishing plant. Special and high-grade steels require specific treatment in finishing and after-care which ends only at the customer. Apart from a few exceptions, hot-forming of special steels in a plant geared for the highest rates of large-tonnage production makes little sense.

1.5 SPHERES OF APPLICATION FOR ELECTRIC FURNACES

Electric furnaces can very easily be adapted for steelmaking under local conditions, as long as the necessary power is available. This can now be generated from solid, liquid, or gaseous fuel, hydro-power, or nuclear energy. Moreover, in arc furnaces, liquid or gaseous fuel can be directly added through

Table 1. Output of crude steel and of electric furnace steel in Germany and the USA, compared with world production. (After data from the Statistische Bundesamt, Aussenstelle Düsseldorf, Eisen und Stahlstatistik and the Statistical Handbook of the British Steel Corporation, Statistical Services, 1969)

Year	Germany[a] Ingots and steel castings			USA			World		
	Crude-steel production (millions of tonnes)	Electric furnace steel production (millions of tonnes)	Proportion of electric furnace steel production (%)	Crude-steel production (millions of tonnes)	Electric furnace steel production (millions of tonnes)	Proportion of electric furnace steel production (%)	Crude-steel production (millions of tonnes)	Electric furnace steel production (millions of tonnes)[b]	Proportion of electric furnace steel production (%)
1910	13.699	0.036	0.26	28.838	0.052	0.18	66.100	0.121	0.18
1911	15.019	0.061	0.41	25.989	0.028	0.11	66.300	0.118	0.18
1912	17.302	0.074	0.43	33.926	0.014	0.04	82.500	0.127	0.15
1913	18.935	0.089	0.47	34.053	0.021	0.06	85.900	0.170	0.20
1914	14.946	0.088	0.59	25.437	0.016	0.06	65.800	0.150	0.23
1915	13.258	0.132	1.00	33.983	0.047	0.14	69.111	0.310	0.45
1916	16.183	0.190	1.17	45.313	0.128	0.28	86.158	0.552	0.64
1917	16.587	0.220	1.13	47.689	0.243	0.51	88.788	0.818	0.92
1918	14.980	0.240	1.60	46.789	0.410	0.88	81.841	1.134	1.39
1919	7.847	0.082	1.04	36.658	0.277	0.76	61.400	0.698	0.87
1920	9.278	0.088	0.95	44.678	0.353	0.79	74.700	0.650	0.87
1921	9.997	0.079	0.79	20.847	0.086	0.41	46.000	0.380	0.83
1922	11.714	0.105	0.90	37.404	0.194	0.52	70.700	0.550	0.78
1923	6.305	0.079	1.25	46.972	0.284	0.60	80.400	0.700	0.87
1924	9.835	0.082	0.83	39.482	0.230	0.58	80.400	0.700	0.87
1925	12.243	0.128	1.05	47.105	0.341	0.72	92.100	0.920	1.00
1926	12.342	0.073	0.59	49.986	0.330	0.66	95.000	0.900	0.95
1927	16.311	0.162	0.99	45.654	0.377	0.83	102.000	1.000	0.98
1928	14.517	0.157	1.08	52.369	0.461	0.88	109.900	1.300	1.18
1929	16.246	0.164	1.42	57.336	0.541	0.94	120.900	1.600	1.32
1930	11.538	0.117	1.01	44.350	0.312	0.75	95.100	1.170	1.23
1931	8.292	0.108	1.30	26.361	0.239	0.90	69.700	1.230	1.76
1932	5.770	0.087	1.51	13.890	0.144	1.04	49.800	1.300	2.61
1933	7.612	0.135	1.77	23.714	0.305	1.29	68.100	1.400	2.06
1934	11.916	0.210	1.69	27.099	0.355	1.31	82.500	1.700	2.06
1935	16.447	0.329	2.00	35.134	0.530	1.51	99.700	2.200	2.21
1936	19.208	0.459	2.39	49.588	0.715	1.44	124.197	2.750	2.21
1937	19.840	0.678	3.42	52.788	0.827	1.57	135.400	3.000	2.22
1938	23.329	0.974	4.17	29.218	0.513	1.76	109.800	5.100	4.64
1939	23.733	1.121	4.72	47.899	0.933	2.09	137.300	4.500[c]	3.27
1940	21.540	1.446	6.70	60.767	1.542	2.54	141.500	6.000[c]	4.24
1941	26.574	1.790	6.74	75.152	2.602	3.46	154.200	7.500[c]	4.86
1942	27.096	1.967	7.26	78.048	3.605	4.62	154.000	9.700[c]	6.30
1943	30.603	1.973	6.45	80.592	4.163	5.17	161.500	10.200[c]	6.32

Year	Germany[a] Ingots and steel castings: Crude-steel production (millions of tonnes)	Electric furnace steel production (millions of tonnes)	Proportion of electric furnace steel production (%)	USA: Crude-steel production (millions of tonnes)	Electric furnace steel production (millions of tonnes)	Proportion of electric furnace steel production (%)	World: Crude-steel production (millions of tonnes)	Electric furnace steel production (millions of tonnes)[b]	Proportion of electric furnace steel production (%)
1944	28.916	2.008	6.94	81.323	3.844	4.13	155.500	9.500[c]	6.11
1945	1.200[c]	—	—	72.305	3.125	5.19	114.100	5.700[c]	4.99
1946	2.842	0.032	1.13	60.422	2.325	3.85	111.600	5.600[c]	5.02
1947	3.060	0.071	2.32	77.016	3.436	4.46	136.100	7.000[c]	5.14
1948	5.561	0.159	2.86	80.415	4.517	5.62	155.800	—	—
1949	9.156	0.204	2.23	72.531	3.432	4.73	161.600	8.500[c]	5.20
1950	12.121	0.334	2.76	90.392	5.748	6.06	191.800	12.496	6.51
1951	13.506	0.413	3.05	99.074	6.377	6.44	214.400	12.580	5.9
1952	15.805	0.509	3.22	87.766	6.069	6.91	214.700	13.604	6.1
1953	15.420	0.522	3.39	101.251	6.500	6.42	234.600	13.401	5.7
1954	17.434	0.677	3.88	80.115	4.853	6.06	223.400	12.222	5.5
1955	21.336	0.924	4.33	106.173	7.462	7.03	270.000	19.642	7.3
1956	23.189	1.153	4.97	104.525	8.168	7.81	283.400	22.216	7.8
1957	24.507	1.348	5.50	102.253	7.663	7.49	293.400	23.249	7.9
1958	22.785	1.533	6.73	77.342	5.943	7.68	274.100	24.431	8.9
1959	29.435	1.805	6.13	87.070	7.618	8.75	307.600	29.458	9.6
1960	34.100	2.174	6.36	91.920	7.481	8.14	347.900	34.560	9.9
1961	33.458	2.365	7.07	90.453	7.736	8.55	352.300	37.832	10.7
1962	32.563	2.567	7.88	91.171	8.047	8.83	361.800	39.099	10.8
1963	31.597	2.647	8.38	101.477	9.751	9.61	388.800	44.526	11.5
1964	37.339	2.998	8.03	117.993	11.320	9.59	438.100	50.778	11.6
1965	36.821	3.137	8.50	122.490	12.280	10.03	458.900	54.356	11.8
1966	35.316	3.090	8.75	124.700	13.488	11.01	474.900	58.015	12.2
1967	36.744	3.108	8.46	118.020	13.689	11.60	498.500	62.900	12.6
1968	41.159	3.684	8.12	121.900	15.253	12.51	530.800	69.870	13.2
1969	45.316	4.146	9.15	131.175	18.263	13.92	576.000	76.696	13.3
1970	45.041	4.436	9.85	122.120	19.155	15.69	599.400	85.187	14.2
1971	40.313	4.030	10.00	111.780	19.761	17.69	584.400	85.811	14.7
1972	43.705	4.480	10.25	123.530	22.593	18.29	632.100	91.719	14.5
1973	49.521	5.156	10.41	139.950	25.489	18.21	698.700	108.507	15.5
1974	53.232	5.748	10.80	135.235	26.008	19.23	709.800	114.290	16.1
1975	40.415	5.076	12.56	105.8	20.6	19.5	645.000	108.000	16.7
1976	42.415	5.260	12.40	116.1	22.3	19.2	676.000	120.000	17.8
1977	38.985	5.060	12.98	113.7	24.8	21.8	675.000	124.000	18.4
1978	41.253	5.968	14.47	124.4	29.2	23.5	717.000	137.000	19.1
1979	46.040	6.428	13.96	123.3	30.4	24.7	748.0[c]	150.0[c]	20.2

[a]Territory at that time.
[b]Without those countries whose production cannot be classified according to process.
[c]Estimated.

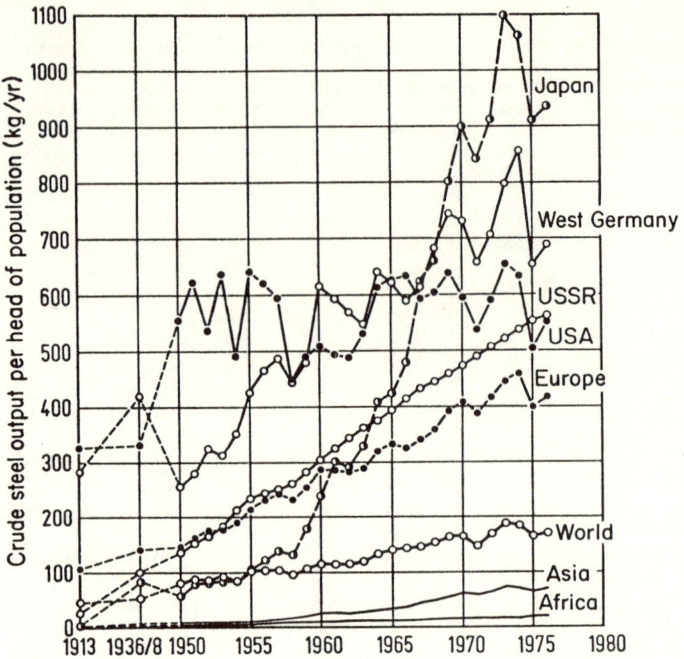

Figure 1. Growth of crude-steel output per head of population (Germany before 1945: territory at the time). (From Statistisches Bundesamt, Assuenstelle Düsseldorf, Eisen und Stahl Statistik)

fuel/oxygen burners in order to boost melting rates or to overcome power-supply shortages.

With regard to metallic charges, the highest possible utilization of alloying elements is achieved in electric furnaces when processing low- or high-alloy steel scrap and by using ferro-alloys.

The electric furnace, with its high thermal efficiency, is eminently suitable for melting down scrap. The iron content in scrap has already been reduced once from iron ore, and the reduction energy which it contains must not be wasted. Bearing in mind the loss in electric power generated, the electric arc furnace still uses about 30% less energy for the same process than the blast furnace and the basic oxygen steelmaking plant.[9] Moreover, in the electric furnace, scrap can be replaced up to 100% by unreduced iron ore (sponge iron). Figure 6 illustrates the versatility of the electric furnace for varying amounts of scrap, sponge iron, and pig iron. Electric furnaces account for 51% of the total span of the three basic raw materials (scrap, pig iron, and sponge iron), while open-hearth furnaces take about 17% and basic oxygen converters take about 9%. The maximum amount of sponge iron or scrap a basic oxygen

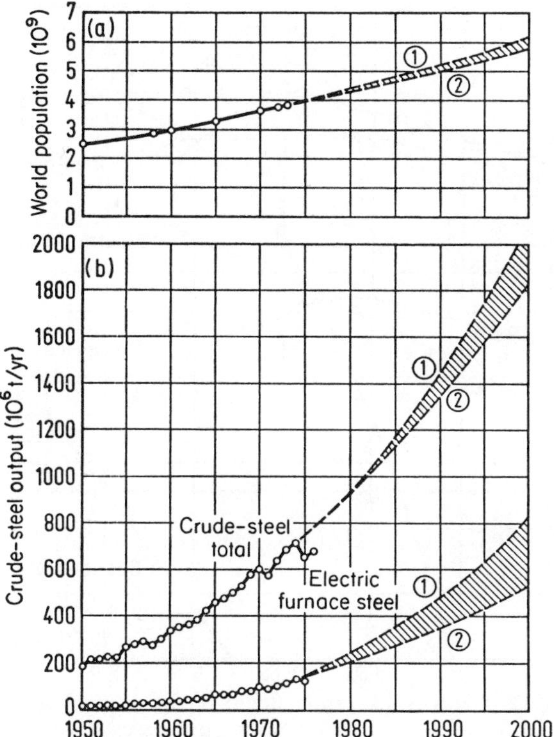

Figure 2. Estimated steel output up to the year 2000. Estimated world population growth from *UNO World Population Prospects* with high (1) and low (2) growth rates. Rise of world crude-steel output when assuming an average growth rate of 5 kg crude steel per head of population. Assumed share of electric furnace steel output 30 or 40%

converter can process is only about 30% of the charge. The growing availability of sponge iron increases the economic significance of the electric furnace. This is of great importance in the developing countries, which have no coking coal and very little scrap, but often have cheap means for making sponge iron. In these countries, moreover, scrap is generally far from plentiful.

All electric furnaces are extremely flexible and can be adapted to changing circumstances and production levels. They can be operated as continuous melters or on melt/refine cycles; their electric power can be limited and controlled to suit the demand capacity of the works complex, and in the case of arc furnaces electric power can be combined with oxygen-fuel burners as the occasion demands. Even the weight of individual melts can be varied within relatively wide limits; this is helpful when making special steels, which are often

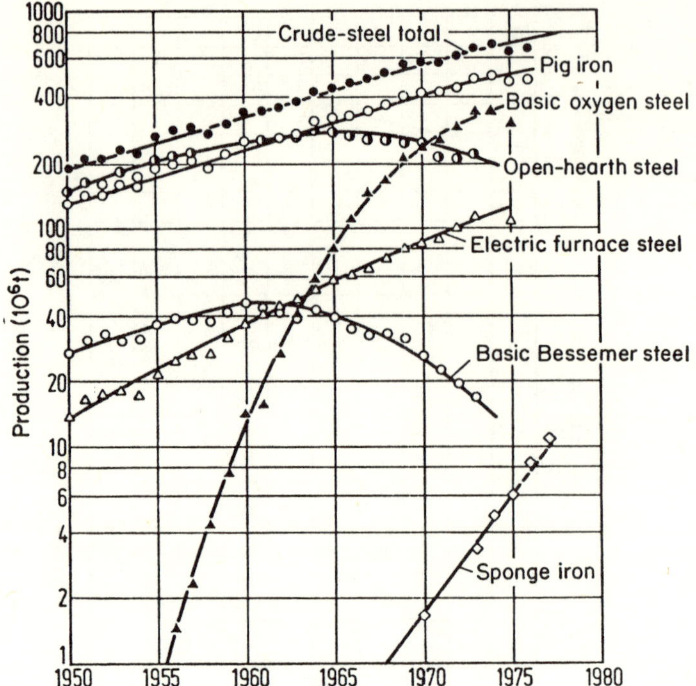

Figure 3. Growth of crude-steel, pig-iron, and sponge-iron production and of steelmaking processes since 1950[7]

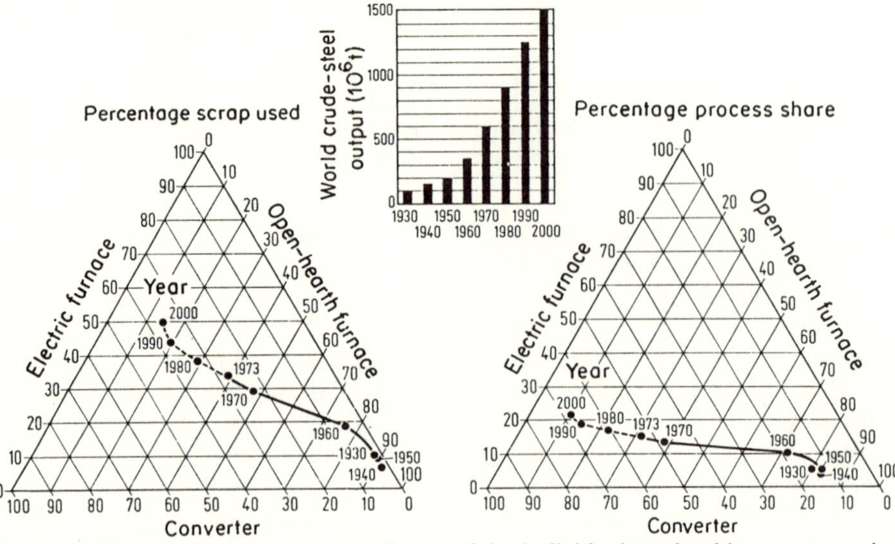

Figure 4. Changes in the percentage shares of the individual steelmaking processes in the total world crude-steel production and the scrap involved. Statistical data since 1930 and estimated growth until the year 2000[8]

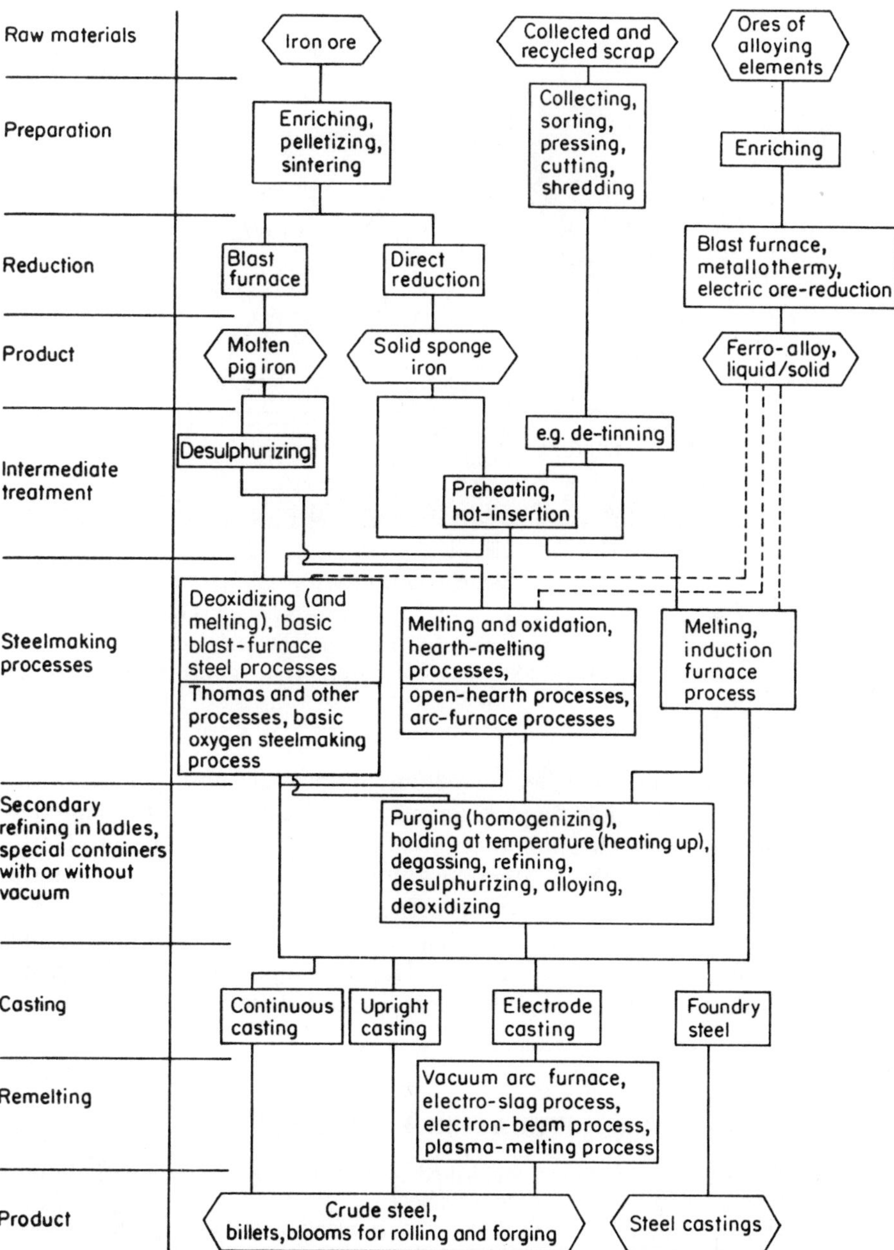

Figure 5. Flow chart for iron- and steelmaking processes

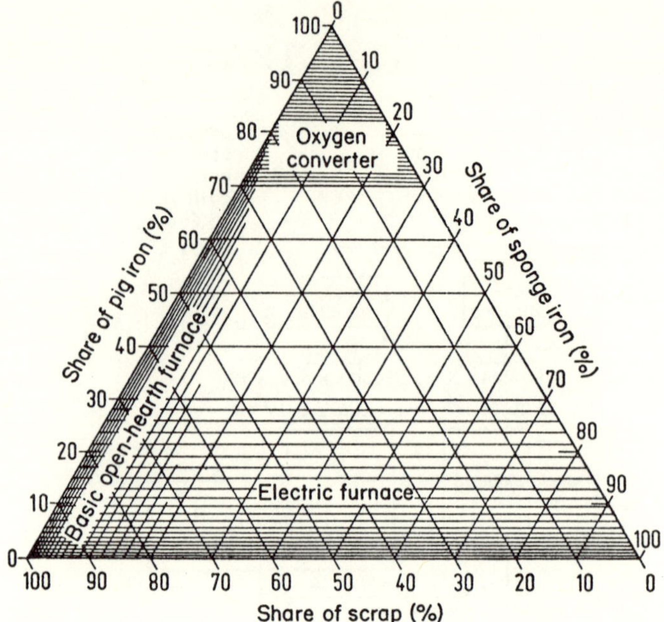

Figure 6. Ternary diagram with regions for the application of the oxygen steelmaking process, the open-hearth, and the electric process for making steel from pig iron, sponge iron, and scrap

wanted in smaller amounts. Electric melting furnaces are unsurpassed for the execution of varied production programmes as they can be readily adapted to the raw materials available and the quality requirements of the end-product.

Comparision of specific building costs for steelworks shows that electric steelworks using ultra-high-power furnaces in the 1–3 million tonnes per year class occupy a favourable position. The high processing costs are mainly determined by power and electrode costs. Hence the operating costs determine the competitiveness of electric steelworks compared with basic oxygen steel plants.[10]

1.6 LITERATURE REFERENCES

1. Stahl-Eisen-Liste. 6th edition, published by VDEh, the Association of German Iron and Steel Engineers. Düsseldorf, 1977.
2. Jäniche, W., Material steel — Properties, application and considerations about further development (German). *Stahl u. Eisen* **96** (1976), 1207–19.
3. *UNO-World Population Prospects 1970–2000*. Yearbook United Nations, New York, 1973. Publications Sales, 72, XIII, 4.
4. Kootz, T., K. Lenzmann, and A. Altgeld, Gegenüberstellung der Stahlgewin-

nungsverfahren aus technischer und wirtschaftlicher Sicht mit Betrachtungen über die weitere Zukunft. *Thyssenforsch.* **5** (1973), 59–77.

5. Möllenkamp, F. W., E. Markworth, and H. Tschabitscher, Design and layout of modern high-power arc furnace plant for the production of tonnage steel (German). *Elektrowärme Internat.* **29** (1971), 535–48.
6. Rumberger, M., The worldwide supply and the demand for coking coal up to 1985 (German). *Stahl u. Eisen* **97** (1977), 705–13.
7. Kalla, U., and R. Steffen, Direct reduction: progress and plans. *Iron Steel Internat.* **50** (1977), 307–20.
8. Amelung, E., and H. W. Kreutzer, Entwicklung der Schrottversorgung und ihre Bedeutung für die Stahlindustrie. *Stahl u. Eisen* **98** (1978), 715–17.
9. Barnes, R., The current state of iron and steel technology (German). *Stahl u. Eisen* **94** (1974), 1077–84.
10. *Baukosten von Stahlwerken und betriebliche Verarbeitungskosten der Stahlwerksverfahren.* H. R. Schenck, K. Consemüller, H. J. Zimmermann, E. Quadflieg, and P. D. Busch, Düsseldorf, 1970 (Stahleisen-Einzeldarstellungen technisch-betriebswirtschaftlicher Zusammenhänge).

Astier, J., Possibilités de pénétration de l'énergie électrique en sidérurgie (French). *Rev. Métallurg.* **74** (1977), Nos. 8/9, 493–503.

Gale, W. K. V., Steelmaking in Britain: 50 years of progress. *Iron & Steel intern.* **50** (1977), No. 5, 283, 285, 289, 291–2, 295–6.

'New' EF technology stretches steelmaking capacity with low capital investment. *Metal Producing* **15** (1977), No. 10, 37–42.

Worldwide steel shortage can be a threat by 1985. *Metal Producing* **16** (1978), No. 1, 52–4.

Chandler, Harry E., A look at American steel technology from the user point of view. *Metal Progr.* **114** (1978), No. 5, 40–61.

Stelzer, Reiner, Hans-Klaus Wapler, and Gert Wiethoff, Der Strukturwandel der Erzeugungsanlagen. *Stahl u. Eisen* **99** (1979), No. 9, 450–58 (Aussch. Anlagentechn. 580).

Kishida, Toshio, Derzeitiger Stand und zukünftige Entwicklung der Stahlerzeugung im Lichtbogenofen. *Denki-Seiko* **50** (1979), No. 1, 15–22.

Oxygen and electricity have revolutionized American steelmaking in the last 20 years. *Metal Producing* **17** (1979), No. 6, 52–7.

Fetters, Karl L., The future of the iron and steel industry. *Iron & Steelmaker* **6** (1979), No. 6, 7–13.

Kuhl, R. J., Capital investment and steel industry, Growth in Developing Countries. *Steel Times* **207** (1979), No. 6 (International Supplement), 28, 32, 34–5, 38–9.

Holzgruber, Wolfgang, Wandlungen in der Verfahrenstechnik bei der Herstellung von Qualitäts- und Edelstählen. *Železarski Zbornik* **13** (1979), No. 2, 45–61.

Harvey, D. S., Recycling of wastes produced during stainless steelmaking (German). In *Abfall- und Reststoffe der Eisen- und Stahlindustrie*. Kommission der Europäischen Gemeinschaften. Luxemburg, 1978, pp. 111–37.

Stein Callenfels, G. W. van, The changing pattern of technology in large-scale steelmaking. *Ironmaking & Steelmaking* **6** (1979), No. 4, 145–55.

Analysis of energy demand in steelmaking. *Ind. Heat.* **46** (1979), No. 7, 24, 26.

Umene, Eiji, The steel industry in transition. In *Proceedings of the Third International Iron & Steel Congress*, 16–20 April 1978, Chicago, Ill., Metals Park, Ohio, 1979. 40–47.

Iron and steel — state of the art. *Ind. Heat.* **46** (1979), No. 10, 6–8, 10.

Szekely, Julian, The role of innovative steelmaking technologies. *Iron and Steelmaker* **6** (1979), No. 12, 25–30.
Graf, Hans, Rudolf Baum, Gerd Kröncke, and Hans Wilhelm Kreutzer, Actual questions of the use of electric heat in the iron and steel industry (German). *Stahl u. Eisen* **100** (1980) No. 10, 509–17 (Stahlwerksaussch. 1037).

Electric Furnace Steel Production
Edited by E. Plöckinger and O. Etterich

2 *Historical Development*

Erwin Plöckinger, Kapfenberg

2.1 GROWTH TENDENCIES IN ELECTRIC STEEL PRODUCTION

The first attempts at using electric power for melting metal go back as far as the beginning of the nineteenth century, long before the invention of the electric generator by Werner Siemens (1867), and could not therefore achieve practical importance at first. With the development of electrical power generation and the availability of adequate power supplies towards the end of the nineteenth century, the experiments were taken up again on a wide scale. This led to a large number of proposals which resulted in the first tangible success in practical application at the start of the present century.

Without entering into details of historical developments leading to the introduction of steelmaking in electric furnaces (treated extensively in section I of the second edition of this book[1]) we shall try here broadly to describe the course of development which resulted from the interaction of progress and knowledge in electrical technology, metallurgy, and steel quality requirements. All these factors contributed about equally, so that today electric steelmaking in the broadest sense accounts for an appreciable proportion of all steel production.

The heat required — for melting the metal charge, or for holding at temperature, superheating, carrying out certain metallurgical reactions, or melting slag-forming additions — can be generated in various ways by electric power. The type of energy supply, and therefore of heating system, determines the type of equipment, and influences the nature of the refractory lining of the furnace interior and the possibilities for refining. A widely varying range of electrically heated melting systems are used side-by-side these days because certain types of furnace are specifically suited to attaining the qualities required of different finished steels.

The types of furnace now in use can be classified into the following groups, according to the way electrical energy is used to produce heat:

(1) Arc furnaces;

(2) Induction furnaces;
(3) Resistance furnaces;
(4) Electron beam furnaces;
(5) Plasma furnaces.

With the exception of electron beam furnaces, which can only be used for vacuum melting, all the other furnace types can be used either in air at atmospheric pressure, or under reduced pressure (vacuum), with or without a protective gaseous atmosphere, or at excess pressure. Further subdivision from a metallurgical point of view arises according to the possibility — or necessity — of melting on an acid, basic, or neutral lining, or using a melting process which takes place in a water-cooled vessel on a lining formed from the charge materials themselves.

All these furnace types are now in industrial use for primary melting, but to very different extents. The limits to the possibilities of refining which characterize each type of furnace, and economic considerations, too, have increasingly led to supplementing the primary melting devices (not only the electric furnaces) with separate after-treatments for refining. This means the melting process — from charging to pouring and solidification — is subdivided into stages which can be performed in turn in different, specially adapted units. A variety of special electrically heated furnaces or reaction vessels is necessary[2,3] for these secondary refining processes; these units are usually quite different from primary melting furnaces in construction and operation. We can distinguish here between ladle systems and remelting refining systems.

Reaction vessels for ladle refining are heated, where necessary, by

Electric arc;
Induction; or
Graphite rod resistance elements.

The design of these systems is similar to that of the corresponding arc, induction, and resistance primary melting furnaces.

Electric remelting furnaces are today indispensable for works making special steels. As with electric primary melting furnaces, we distinguish between different types of furnace according to the method of heating:

Vacuum arc furnaces;
Electro-slag remelting furnaces (resistance furnaces);
Electron beam remelting furnaces;
Plasma remelting furnaces.

All these types of furnace involve melting the solid charge in a water-cooled mould, where the remelted steel solidifies continuously into an ingot, generally with a directional grain structure. Electro-slag remelting furnaces and plasma remelting furnaces can operate in an ordinary atmosphere or in a high- or low-pressure protective gaseous atmosphere.

Finally, it should be pointed out that electric arc heating and especially resistance heating are used to control ingot solidification, using the principle of the electro-slag remelting furnace but without remelting.[4]

The development and present usage of the above types of electric furnaces and electrically heated reaction vessels will now be briefly described.

2.2 ARC FURNACES

It was the directly heated arc furnace, based on the Héroult furnace, which gained most ground after the First World War. Here heating is effected by arcing between graphite electrodes and the metallic charge or melt. Indirect arc furnaces (arc-radiation furnaces) were never of any practical importance, and are now no longer used in steelworks.[1]

Development from the original single- or two-phase system to the three-phase furnace now universally used for steel production took place in parallel with the development of efficient and adequate three-phase power supply networks. In this way large modern arc furnaces of up to 360 t capacity with transformer ratings of 160 MVA were developed from the early directly heated 500–1500 kg furnaces which used a modest 250–300 kVA at a maximum of 110 V.[5–9]

The rapid progress of arc furnaces to their dominant status in electric steelmaking (they now account for over 90% of total world electric furnace steel-melt output) was largely made possible by progress in the technology of refractory linings associated with the metallurgical requirements for refining and reduction of steel in addition to advances in electrical engineering and furnace design. While furnaces have developed there has been constant revision in melting practice in line with the latest state of knowledge of physical and chemical relationships and reaction kinetics; this process is still going on.

Arc furnaces were originally intended to be used not only for melting a solid metallic charge but also for oxidation, alloying, and refining, so that the steel could be made in the furnace in a single operation. The result was the two-slag process, which for many years was basic practice in arc furnace operation; this comprises oxidation under an oxidizing slag, followed by refining (deoxidizing and alloying) under a reducing and refining slag.

Arc furnaces always originally had acid linings, and accordingly melting practice was similar to the old crucible steel process. The advantage was the low hydrogen content of the steel produced and a relatively high alloy recovery when making high-alloy steels from alloy charges. However, the impossibility of dephosphorizing and desulphurizing with an acid lining led to the use of relatively pure charge materials or liquid charges from other melting units.

Efforts were therefore made to go over increasingly to basic linings and to adjust furnace operation and melting practice to suit the lining, the raw materials available, and the steel qualities required. With increasing control over the basic melt process and with improved lining technology and better basic linings, it soon became practicable to make basic electric furnace steel of equivalent and, in most cases, superior quality.

Acid-lined arc furnaces are today used practically only for producing steel castings; rolling and forging steels are made almost exclusively in basic lined arc furnaces.

The advent of new technologies, primarily the use of gaseous oxygen for oxidizing and of vacuum technique for degassing in addition to decarburizing and deoxidizing, and last but not least the need for economy and competiveness with other steelmaking processes, have all had a decisive effect and brought about changes in melting practice in basic arc furnaces.

The growing use of the single slag melting process (similar to practice in basic open-hearth furnaces or basic oxygen furnaces), the shortening of melting time with the help of oil–oxygen burners, working with gaseous oxygen instead of ore for oxidizing, and finally the addition of secondary refining processes forced the arc furnace increasingly into the role of a melting and oxidizing furnace and in many cases relieved it of refining work completely. This enables full use to be made of high transformer outputs for modern ultra-high-power (UHP) furnaces for almost the entire power-on time.

The transfer of metallurgical refining to the ladle has increased furnace output for given furnace capacity and considerably improved thermal efficiency.

In their development, arc furnaces have acquired induction stirring coils to intensify bath motion; the charging process (including preheating of scrap) and the handling of furnace additions have been mechanized, and parts of the refractory lining have been replaced by water-cooled elements.

The development of efficient automatic analytical apparatus, which minimizes the time taken for analysis, and the use of computers to calculate the alloying process have resulted not only in reduction of the time required to process a charge but have also increased the accuracy with which the desired chemical composition of the steel can be achieved while at the same time economizing on alloying elements.

Small arc furnaces of below 2–3 t charge capacity are at an economic disadvantage compared with other electric furnaces such as induction furnaces. The small single-phase arc furnaces for slag melting are an exception. These are used for secondary processing in ladle refining as well as for remelting processes and heating ingot tops.

Finally, it is intended to replace graphite electrodes by permanent metal electrodes to prevent carbon contamination of the melt, for example during the production of nickel-based alloys and alloys with the lowest possible carbon content. Arc furnaces of this type equipped with rotating water-cooled electrodes are already in service;[10] they operate mainly in a protective gaseous atmosphere or in vacuum.

2.3 INDUCTION FURNACES

The development of induction furnaces for steelmaking started in 1900 with the low-frequency channel furnace introduced by Kjellin. As with arc furnace development, there were many different proposals for electrical and mechanical design which led to the development of special types of furnace for different applications.[1]

The first and simplest solution was based on the Kjellin furnace, in which the melting hearth consisted of a circular channel running round the primary coil. The Röchling–Rodenhauser furnace developed from this had heating channels separated from the working hearth to increase melting capacity and assist refining. These channel furnaces, working at mains frequencies and connected direct to the mains, are now only of historical interest. They were employed with a molten charge supplied in duplex operations, mainly for refining open-hearth steel.

Modern induction-heated mixers for hot metal and foundry iron and various types of holding furnaces now in operation were developed from these furnaces. Separation of the furnace bath melting channel has basically been retained, and the melting channel, together with the inductor, is now usually an interchangeable unit, connected to the bath by a flange.

Mains-frequency crucible induction furnaces arrived in 1935 as a notable improvement.[1] Like the mains-frequency channel furnace, they have the advantage of being operable direct from 50 or 60 Hz mains. They are used today not only as melting furnaces, but also as superheating and holding furnaces, especially for grey iron in foundries with charges up to 60 tonnes.

Mains frequency crucible furnaces have only limited suitability for use with 100% scrap charges; however, they can be operated economically on a continuous basis with a high output using a 'hot heel' of metal in the furnace.

The crucible lining is usually acid, which precludes refining work. However, little refining work is carried out in basic-lined crucible furnaces, which now have successful linings for bath weights up to about 9 t.

The most common induction furnaces used for steelmaking are medium-frequency crucible furnaces (coreless induction furnaces, formerly called high-frequency crucible furnaces).[1] The first successful trials to develop this type of furnace go back to before 1930, when the high frequency was generated for the first time by rotary converters, instead of by the old spark gap method.

Unlike 50 Hz mains-frequency furnaces, medium-frequency crucible furnaces are well suited to melting cold charges of any piece size. The characteristic motion of the bath depends on the power input and frequency; it increases with increasing power input but declines with rising frequency. Two types have been developed, depending on charge weight and power: small furnaces operating at 200–500 V and 500–10 000 Hz, and large furnaces for 2000–3300 V at 150–300 Hz. Static frequency-changers have replaced rotary converters for both types of furnace.

Crucible linings are still made mainly from acid refractory compound and the furnaces are almost entirely used for remelting, without any refining; they are also used for making steel castings and high grades of grey and malleable cast iron. Even with basic linings, these furnaces are usually employed only in one type of remelting process, for which pure and high-grade charges are again necessary.

Medium-frequency crucible furnaces, through the avoidance of carburization, are particularly suitable for melting high-alloy scrap with a low carbon content (austenite, nickel-based alloys). However, this application has now been replaced by processing highly alloyed scrap, with a high alloy recovery, in basic arc furnaces, partly combined with secondary refining.

On the other hand, they are widely used as vacuum-induction furnaces with a bottom outlet, first developed by W. Rohn,[1] further developed into tilting furnaces, for primary melting of high-alloy iron or nickel-based materials. Such furnaces are available up to 30-t capacity. A very recent development is to combine vacuum-induction furnaces with additional plasma heating[11] or appended secondary refining units with electron-beam heating.[12]

The use of medium-frequency crucible furnaces[13] in pressurized form, for instance for producing nitrogen alloy steels, is still confined to experimental applications.

2.4 RESISTANCE FURNACES

The principle of resistance heating used in indirect resistance furnaces is based on passing an electric current through a resistive material, usually graphite, which is thus heated and gives up its heat to the charge to be melted in a crucible or melting hearth. Tamman resistance furnaces of this type were used for many years in laboratories;[1] however, most have been replaced by small medium-frequency furnaces which are more suitable.

In about 1936 a new directly heated resistance furnace was introduced into steelmaking practice; this was the graphite rod furnace developed by Messrs Otto Junker.[1] Furnaces of this type, in which a certain amount of refining work can be done with a reactive slag, have a capacity of up to 2 tonnes and were built for steel and grey-cast iron foundries. However, they have been almost completely overtaken by induction crucible furnaces, and now have no practical importance in the production of high-grade steel. On the other hand, graphite-rod heating is gaining increasing importance for degassing vessels and for handling molten steel.

In contrast to indirectly heated resistance furnaces, direct-resistance heating has become very important in the last ten years for secondary refining in remelting processes and elsewhere. This will be studied more closely in section 2.8.

2.5 ELECTRON-BEAM FURNACES

Electron-beam furnaces are of no importance for primary electric steelmaking, and costs make it improbable that they will be used in this field in the near future. However, the use of electron-beam heating is a remarkable innovation of the last few years for oxidizing and refining reactions combined with the melting of a solid charge in a vacuum-induction furnace. This became known as the AIRCO[12] or CRP process;[14] it is a continuous process for making ultra-pure steels, usually high in chromium, where electron-beam heating can remove almost all harmful gases and accompanying elements in the vicinity of the beam. The output of these furnaces is up to 7 tonnes/h.

2.6 PLASMA FURNACES

The endeavour to achieve a decisive improvement in the purity of steels without vacuum, and to increase alloy recovery, has led to melting furnaces being developed to enable them to operate in an inert atmosphere. One result of this development is the plasma furnace first mentioned for steelmaking in 1962.[15]

The first 140-kg experimental plasma furnace was constructed like an arc furnace, only the graphite electrodes were replaced by a plasma burner, which descended vertically through the furnace roof towards the bath. Plasma furnaces were developed to the extent that they now operate with melt weights of 10–30 t.[16] The plasma gas is invariably argon, because it constitutes a neutral atmosphere. The development of an efficient plasma burner was essential for this; it uses direct current and needs a bottom electrode. For furnaces above about 5-t melt capacity it is advantageous to have several plasma burners mounted radially, through the walls. More recent developments are aimed at eliminating the water-cooled bottom electrode by using a.c. plasma burners.[17]

From the output figures available, it appears that plasma furnaces will be able to compete with ultra-high-power arc furnaces.

2.7 ELECTRICALLY HEATED VESSELS FOR LADLE REFINING

The transfer of individual refining operations from the melting furnace to the ladle, or into a separate reaction vessel, necessitated extra heating to make up for the heat losses and to adjust the temperature for casting. This is especially true for small melts or for prolonged treatment. Electric arc and induction heating are used for this today, but graphite rod resistance heating is used for circulation degassing or partial degassing.

Arc-heated ladles and reaction vessels, operating on a principle first practically used by C. W. Finkl,[18] work like three-phase arc furnaces with graphite electrodes; they can be operated at reduced pressure or in a protective atmosphere. A limited amount of refining slag work is also possible, as in an arc furnace. The steel bath is homogenized by purging with argon.

The ASEA–SKF process[19] for ladle treatment of molten steel before casting uses induction stirring, but the bath motion here is different from that in induction crucible furnaces. The primary coil is mounted in a static frame, into which the ladle is transferred for the duration of the treatment. Heating is effected as with three-phase arc furnaces, at atmospheric pressure. It is also possible to use this process for degassing steel under vacuum and for effective desulphurizing, deoxidation, and separation of non-metallic impurities.

Both these processes are well developed and are used for producing high-purity, special steels with very low gas contents.

2.8 ELECTRIC FURNACES FOR REMELTING PROCESSES

2.8.1 Vacuum Arc Furnaces

Metal melting in a vacuum arc furnace with a consumable electrode was first practised by V. Bolton[1] in 1905 for melting and degassing of tantalum; it was then used about forty years later for making titanium, zirconium, and molybdenum. The favourable experience gained led in the 1950s to the use of this process for remelting steel.[20] In 1962, world-installed remelting capacity had already reached 200 000 t per year.

All vacuum arc furnaces work with a d.c. arc, where the cathode is the consumable electrode. Fundamentally they can be used for electro-slag remelting, which will be discussed later, but then the metallurgical disadvantages of direct current have to be accepted.

During the past ten years the simpler and cheaper electro-slag remelting (ESR) process has greatly reduced the installation of new arc furnaces for special steelmaking. These can be replaced by the ESR process where reduction of gas in steel is less important than improving purity and primary recrystallization.

The largest vacuum arc furnaces in operation today produce ingots of 1500 mm diameter at a maximum weight of 52 t.

2.8.2 Electro-slag Remelting Furnaces

The furnaces used for electro-slag remelting (ESR) actually belong to the group called directly heated resistance furnaces, because the heat is generated by a current passing through a conductive slag, which acts as a resistance, between the consumable electrode and the melt. The latter is in a water-cooled mould and solidifies continuously into a 'growing' ingot.

The first remelting furnaces (1935) developed by Hopkins[21] were based on research by Kellog and were used at first for producing ingots weight up to 4 t. After the Second World War the ESR process was further developed, mostly in the USSR,[22] and rapid progress was made in the 1960s in Western Europe,[23] where the first large plant was started in 1967 for 1000-mm diameter ingots weighing up to 22 t. Today, the largest ingot which can be produced by this process has a diameter of 2.3 m and can weigh over 100 t. The installed remelting capacity has already overtaken that of vacuum arc furnaces.

ESR furnaces have been developed for special uses in which they can work in a protective atmosphere at pressures ranging from a few mbar up to 25 bar.[2]

2.8.3 Electron-beam Remelting Furnaces

The development of electron-beam remelting systems began around 1955 using various design principles.[24] All these have the common feature that one or more electron beams melt the steel from a consumable electrode as well as heating a bath of molten steel in a water-cooled mould where the melt then solidifies. Electron-beam remelting furnaces are now built with high-power beams up to 1700 kW and can remelt ingots with diameters of up to 1000 mm, weighing up to 15 t. They are used for making ultra-pure steels and alloys.

2.8.4 Plasma Remelting Furnaces

The application of plasma burners to remelting furnaces was developed over the past ten years.[25] The results from different furnace designs and the possibility of working with plasma gases at various pressures point to interesting possibilities for refining iron- and nickel-based alloys. Specific industrial applications cannot yet be estimated.

2.9 LITERATURE REFERENCES

1. Sommer, F., and E. Plöckinger, *Elektrostahlerzeugung*. 2nd edition. Düsseldorf, 1964 (Stahleisen-Bücher. Bd. 8).
2. Plöckinger, E., Die Sekundärmetallurgie des Stahles. *Berg- u. hüttenm. Mh.* **121** (1976), 340–9.
3. Barraclough, K. C., The newer specialist steelmaking and steel refining processes. *J. Iron Steel Inst.* **207** (1969), 826–36.
4. Plöckinger, E., Die Entwicklung und Anwendung des B.E.S.T.-Verfahrens für die Herstellung großer Schmiedeblöcke. *Berg- u. hüttenm. Mh.* **122** (1977), 466–73.
5. Roesch, K., On the evolution of the arc furnace (German). *Elektrowärme* **26** (1968), 166–70.

6. The development of electric arc steelmaking. *Iron Steel Internat.* **48** (1975), 451, 453–4.
7. Etterich, O., 70 years steelmaking in the arc furnace (German). *Stahl u. Eisen* **95** (1975), 1028–31.
8. Baldwin, R. L., Electric furnace steelmaking. *Iron & Steelmaker* **3** (1976), No. 11, 31–3.
9. Köstler, H. J., Der Beginn der Elektrostahlerzeugung in Österreich. *Berg- u. hüttenm. Mh.* shortly to be published.
10. Lowry, J. D., and M. P. Schlienger, The anatomy of a non-consumable arc system. In 3rd International symposium on electroslag and other special melting technology, *Symposium Proceedings*. P. III, 8–10 June, 1971, Pittsburgh, pp. 83–99.
11. Asada, Ch., and T. Adachi, Plasmainduktionsschmelzen. *Neue Hütte* **16** (1971), 657–60.
12. Hunt, Ch. d'A., and C. V. Harrison, Airco's facility for steel refining and casting with induction furnaces and electron beams. *Iron Steel Eng.* **48** (1971), No. 8, 85–8.
13. Kubisch, Ch., Druckstickstoffstähle — eine neue Gruppe von Edelstählen. *Berg- u. hüttenm. Mh.* **116** (1971), 84–8.
14. Matejka, W. A., Kontinuierliches Elektronenstrahl-Frischverfahren zur Herstellung von nichtrostendem Stahl. In *Proceedings of the 4th International Conference on Vacuum Metallurgy*. Tokyo, 4–8 June, 1973. Publ. by the Iron and Steel Institute of Japan, Tokyo, 1974, pp. 151–5.
15. McCullough, R. J., A new concept in melting metals. *J. Metals* **11** (1962), 907–11.
16. Fiedler, H., Zum Stand der Plasmaschmelztechnik in der DDR. *Neue Hütte* **21** (1976), 69–72.
17. Brockmeier, K.-H., Entwicklungstendenzen bei der Erzeugung von Edelstahl, insbesondere unter Einsatz von Plasmaöfen. *Chem.-Ing.-Techn.* **46** (1974), 135–8.
18. Finkl, C. W., Vacuum arc degassing process at A. Finkl & Sons Company. *J. Iron Steel Inst.* **209** (1971) 33–6.
19. Tiberg, M., T. Buhre and H. Herlitz, Das ASEA-SKF-Verfahren zum Feinen von Stahl. *ASEA-Z.* **11** (1966), 47–50.
20. Sperner, F., Schmelzen von Stahl im Vakuumlichtbogenofen. *Berg- u. hüttenm. Mh.* **106** (1961), 407–16.
21. McKeen, W. A., L. G. Joseph and D. M. Spehar, Melting alloys by the Hopkins process. *Metal Progr.* **82** (1962), No. 3, 86–9.
22. Latas, J. V., and B. I. Medowar, *Das Elektroschlacke-Umschmelzen*, Moscow, 1970, pp. 38–75.
23. Duckworth, W. E., G. Hoyle, *Electroslag Refining*. London, 1969.
24. Ardenne, M. v., S. Schiller and U. Heisig, Stand und Entwicklungstendenzen der Elektronenstrahltechnologie. *Technik* **21** (1966), 307–13.
25. Rosner, H.-O., Plasmaverfahren in der Schmelzmetallurgie. *Stahl u. Eisen* **95** (1975), 61–4.

Electric Furnace Steel Production
Edited by E. Plöckinger and O. Etterich

3 Energy, Raw Materials, and Consumables

RUDOLF BAUM, GEISWEID

About 1300 kg of raw and auxiliary materials are required to melt one tonne of structural steel in an electric furnace. These are distributed as shown in the table below, taken from a statistical study made in 1968.[1]

Material	Amount per tonne of crude steel kg
Scrap or metal	1100.0
Ferro-alloys	55.0
Electrodes	5.5
Lime	45.0
Dolomite	21.0
Carbon supply	15.0
Fluorspar	5.0
Refractories (furnace)	15.0
Ladle ramming compound Ladle bricks	8.0
Shutter material Casting-pit lining bricks	8.0
Insulating plates	4.5
Casting flux and covering agent; glaze	3.0
Moulds	15.0
Total	1300.0

For higher-quality alloys, scrap is partly replaced by ferro-alloys. Chill-casting by the electric steel process results in 300 kg of residual materials, which are either recycled or have to be dumped. In addition, energy is required in the form of electricity for melting and auxiliaries, as well as oil, gas, and oxygen; these are all consumed in the process and generate waste heat. Compressed air and cooling water are also needed; the latter is usually recycled. The evaporation losses of between 3% and 6% have to be made up with fresh water.

Special attention should be given to residual materials for dumping, because as a rule they are relatively light and need a large protected transit storage space close to the melting site.

It is essential that all the materials should flow freely, so a computer is used nowadays for movement and stock control. The following figures illustrate average quantities required to produce 50 000 t of crude steel per month. As can be seen above, the output figure is multiplied by 1.3 to derive the total quantity of input material required: 65 000 t. If all this is brought in wagons carrying an average load of 25 t each, then 2600 wagons need to enter the steelworks each month; that is about 100 wagons for every working day. All this requires space for the tracks and facilities for movement and handling.

These raw materials, consumables, and energy sources will be described in the sections that follow. Attention will be paid to availability, production, physical and chemical attributes, and transportability as well as to applications in steelworks and the possibilities of substitution.

3.1 ELECTRIC POWER

Jürgen Mackeprang Vogel and *Karl-Bernd Nevries, Essen*

A reliable energy supply is a basic prerequisite for an efficient economy and for satisfying the fundamental needs of the people in an industrialized country. The energy supply must be adequate, available, and inexpensive to keep the country internationally competitive.[1]

If this general statement is related to electric power, then a retrospect over the past thirty years of economic history in the Federal Republic of Germany shows that steady growth has always been accompanied by a marked increase in the demand for electricity. In spite of some scatter in individual years there is a decided trend with high continuity. This growth rate dropped from about 11–12% per annum in the years of reconstruction to 7–8% today, and there are no signs that the trend is going to break in the next few years. Such an assumption would disregard the fact that power consumption and its changes are among the many interdependent factors of our complex economic system.

The most important criterion in estimating future electric power needs is the relationship between power consumption and Gross National Product (GNP). Comparison of the annual rates of change in these economic factors over many years and in many countries shows a very high measure of correlation. An increase in demand for electric power always goes together with an increase in GNP, and the rate of change in electric power consumption exceeds the rate of change in GNP by an average of 3%.

Thus the link between economic growth and the demand for electric power is clear. If we expect even a moderate economic growth rate then this automatically includes the expectation of a higher rate of power-generation, even considering improvements in energy-saving techniques.[2]

3.1.1 Power-generation Development and Energy Demand

Oil will soon become too costly and valuable to be used as a fuel for generating electricity; the same will apply to natural gas. We will then only have coal for conventional power stations and uranium for nuclear power. In West Germany, the demand for brown coal (lignite) cannot be appreciably increased. From the measured known reserves we know that pit coal will continue to fuel a considerable proportion of electricity generated. In 1976 the primary energy sources used in West Germany for electric power generation were distributed as follows:[2]

Hydro-power	4%
Nuclear energy	7%
Brown coal	29%
Pit coal	29%
Fuel oil	10%
Natural gas	17%
Others	4%
	100%

The most important statistics about the growth of electricity consumption and fuel consumption for electric power generation in West Germany are shown in Tables 1 and 2, which are based on combined findings of leading economic institutions for the past few years. The figures for pit coal have meanwhile been increased at the expense of oil and gas because of the ten-year agreement between the coal-producers and the electricity supply industry.

Table 3 shows the installed electric power generating capacity and utilization in West Germany. Electric heating is being used more and more; this will lead to better load distribution because many heating processes can be geared to work at off-peak times and at a lower tariff.

When considering the prospects for possible future developments we must remember the present worldwide political sensitivity of the subject of electricity-generation, and we must stress that with the present state of technology, if electricity can be superseded at all this could only be done in

Table 1. Development of electricity consumption in West Germany

	kWh	1970	1975	1985
Industry	10^9	112.5	125.5	226
	%	(56.5)	(49.6)	(48.1)
Transport	10^9	7.9	8.9	14.5
	%	(4.0)	(3.5)	(3.1)
Domestic and small consumers	10^9	78.8	118.6	229.5
	%	(39.5)	(46.9)	(48.8)
Total consumption	10^9	199.2	253.0	470
	%	(100.0)	(100.0)	(100.0)
Consumption during power generation, distribution losses, assessment, and statistical differences	10^9	51.2	56.6	95.0
Gross power consumption	10^9	250.4	309.6	565.0

Table 2. Fuel consumed in electric power-generation in West Germany

	Actual 1975 (%)		1985 (%)	
	in Mt equivalent pit coal units[a]			
Pit coal	24.4	(24.8)	32	(18)
Brown coal (lignite)	30.6	(31.2)	34	(19)
Fuel oil	8.6	(8.8)	12	(7)
Natural gas	17.5	(17.8)	23	(13)
Nuclear energy	7.1	(7.2)	62	(35)
Hydro-power	5.2	(5.3)	6	(4)
Others	4.8	(4.9)	7	(4)
	98.2	(100.0)	176	(100)

[a]1t equivalent pit coal units = 29.31 GJ = 8.14 MWh.

many applications if we accept technological retrogression. Any substitute for electric power in industry would be limited to peripheral applications.

This also applies in principle to district heating, which would additionally be limited to areas of high thermal density. Solar energy and environmental heat (heat pumps) will in the long run be able to cover only a small proportion of the overall requirement.[1]

3.1.2 Expansion and Operation of Supply Networks

When trying to get an idea of all the technical and economic progress in mains

Table 3. State and development of power-station output in West Germany from 1970 to 1976

	Peak power (MW)	Utilization time (h/a)
1970		
All power stations:	50 833	4 780
of which public	33 701	4 825
industry	16 249	4 640
German State Railways	833	5 210
1975		
All power stations	74 356	4 060
of which public	57 582	4 190
industry	15 684	3 698
German State Railways	1 090	4 495
1976		
All power stations	81 798	
of which public	64 943	
industry	15 493	
German State Railways	1 362	

distribution, we must remember that while demand in the Federal Republic of Germany grew by a factor of about 2.4 in 25 years, the length of the supply system only grew by a factor of somewhat over 2. The total system length of 400 000 km in 1950 increased to 855 000 km by 1975.

The enlargement of power stations, the interchange of electricity inside West Germany and the EEC and especially with the Alpine countries, as well as the appearance of high-consumption centres in the Federal Republic of Germany have all caused great structural changes in the high-voltage and ultra-high-voltage networks.

Increasing costs and the endeavour by the electricity suppliers to keep the tariff stable as far as possible have forced the early adoption of stringent rationalization in the operation of supply networks. This led to far-reaching automation and widespread adoption of information technology.

Ever-increasing automation in mains network operations made the use of computers essential. Among other things, the computers help to sort, reduce, and process incoming information. Many metropolitan electrical authorities have computers directly coupled to electronic remote control. This development will certainly continue.[3]

3.1.3 Organization and Utilization of Electricity Supply Networks

High- and medium-voltage transmission systems have been extended so as to provide the most reliable and economical electricity supplies possible. This has been particularly helped by standardizing and modernizing all network components and by rationalizing operations. With this in mind, essential components such as transformers, circuit-breakers, and converters had to be made rapidly replaceable. Rational stock-keeping and limitation to as few voltages as possible became imperative. In the last twenty years this led to a cut in the many prevailing medium range voltages to two only: 20 and 10 kV. Other voltages are disappearing fast.

The most important voltage networks in the Federal Republic of Germany are as follows.

High-voltage networks Networks of 380 kV act as a closely meshed extra-high-voltage interconnection grid as well as for linking power sources with consumers and for smoothing out temporary peaks and regional disparities. They also optimize power-station utilization and provide reserves to deal with fluctuations, thus safeguarding the now-customary reliability of electricity supplies. Such networks have gradually taken over these functions from the 220 kV networks over the last twenty years. They are either supplied direct by large generators via 380 kV transformers or from the 220 kV network via 220/380 kV transformers that link the two high-tension networks. For a few years now, 110 kV networks have been supplied via 380/110 kV transformers when it is necessary and possible to provide a concentrated power supply for the 110 kV network without using the 220 kV intermediate voltage.

For about fifty years, the 220 kV networks have provided the functions of transmitting and interconnection networks; they still serve today — together with the 380 kV grid — for transregional transmission. Large electric-power consumers, like the chemical and steel industries, are already supplied directly at this level. This is especially true for large electrolysis plants and electric steelworks with ultra-high-power arc furnaces.

Electricity is supplied from the power stations via transformers working at 220 kV generator voltage, or via 380/220 kV transformers, and is stepped down either to 110 kV via 220/110 kV transformers or via other transformers from 220 kV down to medium-voltage for industrial mains networks.

Networks of 110 kV handle most distribution today; industrial works consuming up to about 100 MW are supplied from them direct. When it is not possible to provide 110 kV overhead transmission lines because of routing problems, 110 kV cables are used to transmit electricity to closely built-up areas.

Medium-voltage networks These are for electricity distribution in regional and local areas; the most important ones operate at 20 and 10 kV.

Networks of 20 kV are mainly used in rural areas and small towns, while 10 kV networks are used in large towns. However, for economic reasons the two voltage levels are not used side by side.

3.1.4 Special Problems with Power Supplies and Network Disturbances

When planning an efficient power supply for a new plant, the main aim is to provide sufficient and economical power with proper safeguards. Industrial management should have a clear idea of its maximum demands and of the possibilities of transferring varying loads to times that make best use of different tariffs. The solution of these problems is one of the fundamental tasks for power planning in industry; however, this is not the place to consider it in greater depth. Special problems arise with devices which can feed back interference into the network. Among these devices there are:

(1) Appliances which draw quickly changing or surge currents, thus causing mains-voltage fluctuations.
(2) Appliances which draw non-sinusoidal currents, producing a distorted mains voltage curve and causing harmonics (such as 150 Hz, 250 Hz, 350 Hz).
(3) Appliances which superimpose frequencies between the harmonics on to the mains voltage.
(4) Appliances which impose an unbalanced load on the three power-supply feeders, thus causing the three mains-phase conductor voltages to become asymmetrical.

Expert treatment of these special problems requires close co-operation between the manufacturer and operator of any such devices planned and the power authority. The latter can only decide under what conditions a new plant can be connected when the technical data and operating conditions are known.

Industrial and trade-consumer devices which cause mains disturbances requiring correction at source are listed below.

3.1.4.1 Devices which cause voltage disturbances

(1) Arc furnaces.
(2) Welding appliances.
(3) Motors that start or reverse frequently.

3.1.4.2 Devices which produce harmonics

(1) Rectifiers for electrolysis, etc.
(2) Drives controlled by rectifiers.

(3) Lighting installations controlled by rectifiers.
(4) Converters for induction or dielectric heating.
(5) Mains-frequency induction furnaces.
(6) Arc furnaces.
(7) Resistance-welding machines with ignition control.

3.1.4.3 Devices which superimpose interharmonic frequencies on the mains

This group includes all devices which are powered by direct converters operating at frequencies below 50 Hz, and especially:

(1) ESR (electro-slag refining) plants.
(2) Slow-running drives, such as for cement mills.
(3) Resistance-welding machines controlled by coded pulses.

3.1.4.4 Devices which cause asymmetrical voltages

This group includes all consumer units which load the three supply-phase feeders unequally, and especially all units that operate on two phases only, e.g.:

(1) Mains-frequency induction furnaces.
(2) Resistance-welding machines and arc-welding devices of any kind, where the welding current does not load the three phases equally.
(3) Arc furnaces, especially during the melt-down period.
(4) Tinning furnaces.

The corrective actions for interference comprise the following:

(1) Collection of all data about every device which can feed interference into the mains. These data should be obtained from the manufacturers by the intending user. In planning new systems or the extension of installations it is advisable to have an early meeting between the manufacturer, the user, and the supply authority; this will save time and trouble later.
(2) Determining all data concerning and affecting the electrical power supply. The supply authority will investigate any phenomena that could affect the mains.
(3) Assessment of the magnitude of all interference, based on data from (1) and (2), by the supply authority.
(4) Assessment of the interference by the supply authority, having particular regard to possible mains-circuit variants and, if applicable, the existing load imposed by any plant already installed.
(5) Discussion of all necessary preventive measures to avoid interference in the supply.

(6) Execution of the measures.
(7) Checking the efficacy of the measures.

The data required under point (1) must include the rated current, rated power, and rated voltage for the following:

Arc-furnace installations The number of furnaces, the short-circuit power of each furnace, the nature of the charges to be melted (e.g. coarse or fine scrap, pellets, etc.), staggered or random operation of multi-furnace installations and operating times, for instance for continuous working.

Rectifier plant Pulse count of the plant, active and reactive power under variable load, especially for drives, and the possibility of working with a reduced number of pulses.

Resistance-welding machines Number of machines; for each machine: maximum welding power (see VDE 0545), span of time during which the switch-on time and connection time are varied, whether spark-ignition control is used, whether machine power can be switched over, and whether there are any interlocks.

Direct-frequency converters The operating frequency or operating frequency range.

A guide for assessing interference effects on the mains is given in the VDEW (Federation of German Power Stations) publication[4] entitled *Principles of the Assessment of Mains Interference*. The relevant supply authority assesses whether the disturbance anticipated in the public supply can be tolerated, having regard to any interference already present. The possibilities for effective reduction of mains interference are discussed between the manufacturer, the user, and the electricity supply authority. Basically, the following corrective actions have to be considered:

In the mains
(1) Increasing the short-circuit capacity of the mains supply at the connection point by increasing the transformer output or by adding more supply circuits.
(2) Feeding the new plant from a higher voltage level with adequate short-circuit capacity.

In the plant
(1) Using circuits with reduced interference, for example higher pulse rectifiers;
(2) Changing the operating schedule, e.g. by staggering the operation of several arc furnaces;

(3) Using interlocks which avoid causing more than one surge at a time, e.g. when using several resistance-welding machines;
(4) Installation of dynamic compensation to avoid fluctuations in reactive power;
(5) Using filter circuits to absorb harmonics;
(6) Balancing circuits to correct asymmetrical loads in the three mains feeders.

When deciding on countermeasures it must be remembered that mains interference not only affects other mains-users but will affect the plant causing the disturbance as well. This is especially true where the interfering device is not supplied via a separate transformer. Systems with power electronics are particularly susceptible to interference or damage when a certain interference level is exceeded. In the absence of other agreements between customer and power-electronics manufacturer, power electronics equipment in West Germany should comply with VDE 0160/part 2/10.75.3.5[5] concerning electrical operating conditions, or the local ruling standard elsewhere. If these conditions are exceeded, the manufacturer and the user should agree on specific remedies. It is therefore necessary to look at the whole interference problem (public mains supply and consumer circuit) in order to arrive at an optimum solution.

3.1.5 Costs, Prices, Sales

The growth of electricity prices in Western Germany during the last twenty-five years is characterized by a long period of stability between 1953 and 1971 (Table 4). Such long-lasting stability was possible only because the state electricity supply authorities used rationalization to the fullest extent by changing over to ever-larger generating units and employing the latest advances in power-generation and distribution. In this way the rise in pit coal prices could at least partly be compensated by using less coal per kWh. From 1950 to 1974 specific equivalent pit coal consumption was reduced from 580 g/kWh to about 333 g/kWh. Up to 1960 the savings were remarkable, but they steadily became less so when the limits of technological possibility had been reached. Distribution losses, too, were reduced in a similar manner by partly renewing the distribution networks and also by increasing distribution voltages; the losses, which were over 10% in 1950, are now only about 6%. More savings were made by increasing the meter-reading and accounting periods.

Since the beginning of the 1970s costs have changed in a fundamentally different manner. In the 1960s it was still possible for the supply undertaking to limit price increases largely by rationalization, making use of technology, and by decreasing costs through increasing system size, as well as by raising distribution efficiency and expanding sales. However, these possibilities have now, since 1970, been almost exhausted. On the other hand, expenses keep on

Table 4. Total income from supply of electricity (only from public power stations)

Year	Special customers			Tariff customers		
	Proportion (%)	Income (DM × 10^6)	Average price (Pf/kWh)	Proportion (%)	Income (DM × 10^6)	Average price (Pf/kWh)
1950[a]	70.5	924	5.39	29.5	1 250	17.43
1960[a]	69.6	3 921	7.30	30.4	3 960	16.06
1970[b]	59.9	6 784	6.88	40.1	8 172	12.36
1975[b]	59.4	10 912	8.13	40.6	13 087	14.29

Notes: [a]Federal Republic of Germany less West Berlin and Saar.
[b]Federal territory including West Berlin and Saar.

rising because of the high outlay required for environmental protection, which has often increased the cost of a new power station by over 10%. Only by electricity price-increases can the economic basis for further developments and a reliable power supply be maintained.

3.2 INDUSTRIAL OXYGEN, NITROGEN, AND ARGON, AND FUEL FOR ADDITIONAL ENERGY SUPPLIES

Rudolf Baum, Geisweid

Today the furnace is the preferred medium for rapidly melting down scrap and additional materials. In addition to electrical power, fuel oil (kerosene or petroleum) or natural gas is needed to be burned with oxygen to assist the rapid melting of scrap close to the vessel wall between the electrodes.

Argon and nitrogen are used for purging and for blowing solids into the melt during ladle post-treatments.

We shall now discuss industrial oxygen, nitrogen, and argon, as well as oil, petroleum, kerosene, and natural gas in more detail.

3.2.1 Physical and Chemical Data on Gaseous Oxygen, Nitrogen, and Argon

Air contains the following gases at the percentages stated:

	(% by volume)	(% by wt)
Nitrogen	78.08	75.46
Oxygen	20.95	23.19
Rare gases	0.94	1.30
Carbon dioxide	0.03	0.05

Table 1 contains the most important physical and chemical data on air products.[2] It can be seen that the constituents can only be separated out below −183°C or 90.15 K.

3.2.2 Availability of and Requirement for Air-separation Products

At present, total world oxygen-extraction from air amounts to 45 million tonnes per year. This corresponds to 0.009% of all the oxygen produced annually by photosynthesis. The ratios of oxygen consumed by the metal and chemical industries, by breathing, and by primary energy combustion are now 1:20:300.[3] If primary energy consumption rises further as forecast, then it is reckoned that up too 7% of all photosynthesized oxygen could be required for combustion.

Table 1. Physical and chemical data for air-separation products

		Oxygen	Nitrogen	Argon
Atomic number		8	7	18
Atomic weight		15.994	14.0067	39.948
Melting-point	(°C)	−218.8	−210.0	−189.37
	(K)	54.35	63.15	83.78
Standard density	(kg/m^3)	1.42895	1.25046	1.7837
Boiling-point	(°C)	−183.0	−195.81	−185.87
	(K)	90.15	77.34	87.28
Critical temperature	(°C)	−118.38	−146.9	−122.4
	(K)	154.77	126.25	150.75
Critical density	(kg/m^3)	0.430	0.311	0.531
Critical pressure	(MN/m^2)	5.0798	3.3833	4.8641
Triple-point temperature	(°C)	−218.80	−210.0	−189.38
	(K)	54.35	63.15	83.77
Triple-point pressure	(kN/m^2)	0.15199	12.53231	68.75436

In the 1950s, plants with capacities of 500–5000 m^3/h were adequate. Today the needs of the chemical and metal industries are 30 000–60 000 m^3/h; production is met by the Linde–Frenkl low-pressure process. The energy required to produce oxygen at normal pressure and ambient temperature is 0.45 kWh/m^3, while liquid-oxygen production by the high-pressure process would need 1.3 kWh for the same amount of oxygen.

3.2.3 Industrial Breakdown of Air

All the gas components of air,[4–6] such as nitrogen, oxygen, argon, krypton, and xenon, are separated in low-temperature air breakdown plants. Users of the gases should be aware that argon requires very high analytical expenses compared with oxygen and nitrogen, because 'pure' argon must not contain more than a few parts per million (ppm) of oxygen.

3.2.4 Physical and Chemical Data on Fuel

Table 2[7–18] presents the most important data about fuels used in steelworks for drying and heating ladles, distribution channels, etc., or for supplying

Table 2. Physical and chemical data for selected fuels

	Unit	Natural gas[b] Holland	Fuel oil EL DIN 1603	Fuel oil S DIN 51 603	Kerosene ASTM D 1655	Petroleum	Propane
Carbon content	%	65.1[c7]	85.5[16]	86[13]		85.7[14]	81.8[13]
H_2 content	%	21.9[c7]	13.5[16]	11[13]		14.3[14]	18.2[13]
S content	%	0.02–0.25[8]	0.55[13]	2[13]	0.04[15]	0.025–0.05[14]	
Free water content	%		<0.10[13]	0.1–0.3[16]	0.0030[15]	0.030[14]	
Density liquid (15°C)	kg/m^3		860[13]	900–1200[13]	550–830[12]	800–820[13]	492.8[a11]–585[8]
gaseous		0.78[13]–0.829[17]			2.10–2.30[12]		2.01
Viscosity	cSt	12.8[13]	6/20°C[13]	150/50°C[13]	max. 15/−39°C[12]	7.9/−34.4°C	2/−50°C[11]
Freezing-point	°C	ca. −183[d9]	−10[13]	0–38[16]	−49[8]	−20–−70[8]	−190[8]
Flashpoint	°C	450–650[e18]	70–120[13]	120–140[13]	43–66[12]	21–55[13]	
Boiling-point	°C	ca. −162[d9]	100–360[16]	300[13]	180–300[12]	130[8]	−44[10]
Calorific value Hu	kJ/kg	43 400[6]	42 700[13]	39 770[13]	43 580[15]–46 000[12]	40 800[13]	46 360[13]
Ho	kJ/kg	55 900[6]	45 400[13]	42 280[13]		42 900[13]	50 340[13]
Theoretical air amount	m^3/kg	10.14[17]	11.2[13]	10.6[13]	12[12]	11.41[14]	12[13]
Waste gases							
Damp	m^3/kg	11.39[17]	11.8[13]	11.4[13]	12[13]	12.17[14]	13[13]
Dry	m^3/kg	9.28[17]	10.2[13]	10.0[13]			11[13]

[a]At saturation pressure (25°C).
[b]m^3 in normal condition.
[c]Converted from methane.
[d]Derived from the main constituents of methane.
[e]Ignition temperature at 5–15 vol.%.

additional energy for the furnace itself. Propane–oxygen burners are invaluable for heating distribution channels. For economic reasons, drying and heating of ladles is effected by means of natural gas/compressed air burners.

In Japan kerosene–oxygen burners are used for assisting the melting down of scrap. In trials with fuel oil, corrosion damage caused by sulphur was noticeable. In Europe, petroleum is used; its properties resemble those of kerosene, but petroleum contains more free water. Positive results have been achieved in trials with natural gas–oxygen burners,[19] but the lower calorific value and the poorer transfer of heat into scrap must be compensated by burner design.

3.2.5 Transportation and Storage of Air-separation Products, Gaseous Fuel, and Crude-Oil Derivatives

The arrangements for transportation and storage depend on the maximum requirements. As demand rises, storage costs rise, too, while the associated transportation costs (per unit transported) drop appreciably.[20] The specific weights of the gases are low, and so large volumes and high pressures are required for economical transportation. The transportation of liquid fuel is no problem. Tanker-lorries are used, and their contents are pumped into static tanks. For safety reasons, static tanks have to be surrounded by catchment basins so that the entire contents can be contained in the event of spillage.

Gas cylinders are the most expensive form of storage. Their sizes are 2, 5, 10, and 27 l for laboratory and similar uses, while 40 and 50 l cylinders are available for small industrial users. The older, 40 l cylinder contains 8 kg of oxygen at 150 bar, equivalent to 6 m^3 at normal temperature and pressure (NTP); it weights 70 kg. The newer 50 l cylinder of chrome—molybdenum steel contains 13 kg of oxygen at 200 bar (equivalent to 10 m^3 at NTP); it weighs only 63 kg. There is also a 50 l cylinder for nitrogen, compressed air, or argon, containing the equivalent of 15 m^3 gas at 300 bar; however, it is still in the development stage. Acetylene and propane are stored in 40 or 50 l cylinders containing 6.3 or 10 kg, respectively.

For gas-users consuming more than 300 m^3 per month there are packs of cylinders, 10 to 30 in a unit, with a central pressure-reducing valve. These packs are suitable for connection to local ring mains, with taps where required. The design of such ring mains is subject to stringent safety regulations and will not be discussed here.[21]

The distance between the air-separation plant and the consumer is critical for selection of the most economical means of transportation for large supplies. For example, for a consumer of 10 000 m^3/h at a distance of up to 38 km a direct pipeline is still economical. For longer distances or smaller quantities it is cheaper, although production costs more, to use liquid-gas transporters (because of their lower weight) and storage in cold evaporators. Liquid gas is sent out in insulated tankers and is pumped into storage on arrival with hardly any loss. Road-transporters for liquid gas have capacities of 6000–24 000 l; rail-tankers go up to 30 000 l. A cold evaporator scheme for liquid, deeply cooled gases is shown in Figure 1. The gas is evaporated by heat and is conducted at ambient temperatures to its user. Normally ambient temperature provides the heat of evaporation.

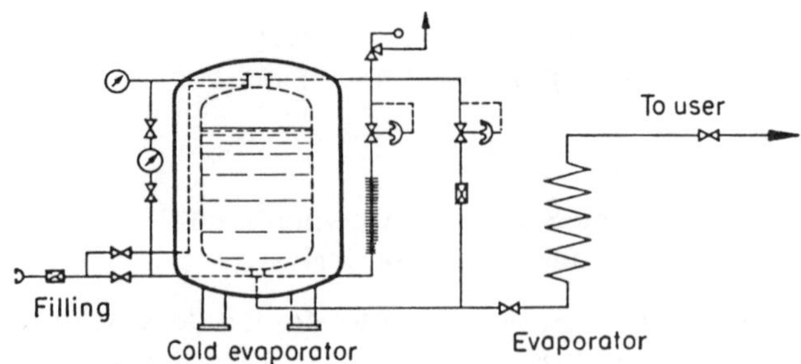

Figure 1. Scheme of a cold-evaporator installation for liquid, deeply cooled gases[20]

If there are several large consumers in a district, then pipelines are the most rational form of transport for gases. They constitute a single medium for storage and transportation, and there are no empty containers to transport. In the Ruhr, twelve large steelworks and chemical plants are supplied from an air-separation plant via a pipeline, 300 mm in maximum diameter, 191 km long, at a pressure under 30–40 bar. Up to 160 000 m^3 of oxygen per hour are moved in this way.

3.3 THE CHARGE MATERIALS: SCRAP, PIG IRON, AND SPONGE IRON

Hans Wilhelm Kreutzer, Düsseldorf

The process of making steel from ore or scrap is shown schematically in Figure 1. The 'indirect' way of making steel from iron ore is via ore-reduction in a blast furnace. However, the molten high-carbon pig iron is 'over-reduced' in the blast furnace and has to be refined in a converter straight away to reduce the carbon content. It would not make sense to transfer this process, which needs no external heat supply, to an electric furnace. Small amounts of solid pig iron are sometimes charged direct into an arc furnace, but only for metallurgical and quality reasons.

The main task of the electric furnace, with its high-energy concentration, is melting solid metallic charges of scrap, sponge iron, ferro-alloys, and the required additives.

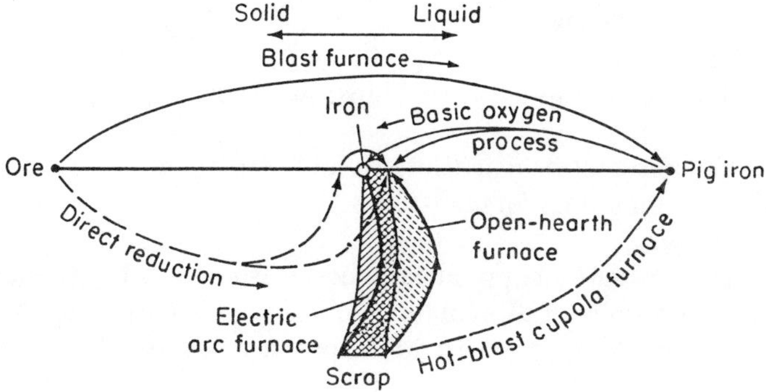

Figure 1. Flow chart of steelmaking process

3.3.1 Scrap

Scrap is a collective term for all production waste and for steel and cast iron and consumer goods which are no longer fit for use. Scrap is the most important home-produced raw material for steelmaking in countries without their own source of iron, or others, such as West Germany, where it is uneconomical to process domestic ores.

3.3.1.1 Characteristics of scrap

Scrap can be classified[1] according to:

(1) Recycling time and place of production: new scrap with a short recycling time and old scrap (old iron) with a longer recycling time;
(2) Origin: production and processing scrap as well as collected and old scrap;
(3) Use: blast-furnace scrap, steelworks scrap, foundry scrap, and scrap for the chemical industry;
(4) Deliberations of the scrap consumer: own scrap, bought-in scrap, recycling scrap such as broken castings (mould castings, engineering castings, commercial castings, pottery).

The following grading results from the scrap-consumers' point of view concerning the use of scrap in steelworks:

(a) Own scrap from iron and steelmaking and foundries
Recycling scrap and foundry waste (bloom scrap, casting risers, sculls, etc.).
Casting scrap (moulds, bottom-pouring plates).
Own scrap in steelworks from rolled or forged products (blue scrap).
Bright scrap from further processing (such as pipe scrap).

(b) From purchased scrap
New scrap
Sheet-metal waste, stamping scrap, chips, swarf, etc.
Old scrap
Scrapped steel and cast-iron articles.
Broken scrap from industrial plant.
Ship-breaking scrap.
Collected scrap, worn-out appliances (cars, washing machines, etc.).
Scrap from refuse-incineration plants.
Scrap from scrap-processors (shredders, presses, shears).

Lists of types of scrap are maintained in the industrialized countries to help users buy the right scrap. These lists, which differ in each country, are continually updated to accord with scrap-handling practice and the physical specifications for recycling scrap which are negotiated between scrap merchants and users. Earlier West German scrap type-lists and those of other countries were aimed primarily at the use of Thomas converters, open-hearth, and electric furnaces, and the size-classifications were geared to existing scrap chutes, charging boxes and baskets.[2] The growth in the sizes of basic oxygen furnaces and arc furnaces enabled scrap of ever larger piece-size to be charged, thus saving extensive and costly breaking down.

The scrap list in Table 1 should be regarded only as a fleeting example. It relates only to unalloyed carbon steel scrap. One particular class of scrap is used as a guide for price-fixing, and the other types are traded at higher or lower prices. Similar lists apply to scrap castings and to steel-foundry scrap.

3.3.1.2 Reduced scrap-availability

The need to use scrap in steelmaking arose only with increasing production of crude steel. In 1880 about 1400 kg of pig iron was used to make 1 tonne of crude steel, but this dropped in 1925 to about 840 kg pig iron per tonne of crude steel.[3] Today a highly industrialized country has to reckon with crude-steel production of 700–800 kg per inhabitant and visible steel consumption of 500–600 kg per inhabitant, with an average requirement of 700 kg of pig iron and sponge iron and 400 kg of scrap per tonne of crude steel; scrap imports and exports are about equal. Table 2 shows this for West Germany for 1970–7.

Total scrap consumption — as already described — comprises the steel industry's own scrap, new scrap from the further treatment of steel, and old scrap. The proportion of steelmaking scrap is steadily reducing. Continuous casting of steel and improvements in hot and cold deformation methods have led to more efficient use of materials. The same is also true for further processing of steel.

Table 1. List of scrap types in West Germany

Types of scrap General rules
The types of scrap defined relate to carbon-steel scrap only. This must be absolutely clean and free from:
Any constituents which are harmful in steelmaking (see also Commercial Regulations for the Supply of Unalloyed Iron and Steel Scrap dated 31 May 1976, Item 8c). Enclosed, hollow items are excluded
Non-ferrous metals, alloys, castings, and all non-metallic materials are counted as harmful
In addition, all scrap containing parts of old vehicles, motors, oil-filled gear-boxes, chips, sculls, sinter, and slag will not be accepted
(0) Old scrap, at least 3 mm thick, maximum sizes 1.50 × 0.50 × 0.50 m
(1) Old scrap, at least 6 mm thick, maximum sizes 1.50 × 0.50 × 0.50 m
(2) New scrap, at least 3 mm thick, maximum sizes 0.60 × 0.50 × 0.50 m and new scrap at least 5 mm thick, maximum sizes 1.50 × 0.50 × 0.50 m
(3) Heavy industrial dismantling scrap (containing no pipes), minimum 8 mm thick, sizes 1.50 × 0.50 × 0.50 m
(4) Shredded scrap
(5) Steel swarf, without cast iron, free machining, and woolly swarf, to be handled by lifting magnet
(6) Packs of new light scrap
(7) Packs of old black scrap
(8) New scrap, under 3 mm thick, over 0.60 m long, to be handled by magnet, suitable for pressing

New scrap (which has never been used) is very pure and has a defined chemical composition. It is especially valuable for making steels which have to comply with stringent specifications for quality and consistency. Declining availability of new scrap is a drawback unless the quality of old scrap is continuously improved.

The diminishing supply of production and processing scrap is compensated by a steady rise in the supply of old scrap.[4,5] This is why the total amount of scrap used in the last ten years has remained about constant, in spite of far-reaching technological developments. This is illustrated in Figure 2, especially for the period after 1960.

Many studies have been effected to ascertain the time taken for steel to be recycled as old scrap and in what proportion.[6–9] These investigations, conducted for different uses (steel buildings, ship-building, the motor industry, and the consumer goods industry), are, however, overlaid by economic trends in the steel industry. All sources of scrap are exploited during a boom when scrap prices are high, but not during a recession, when the price drops and there is a danger that scrap will be dumped and lost forever.

Table 2. Specific consumption (kg/t) of material for crude steel ingot-making by the basic oxygen process (LD), open-hearth furnaces (OH), electric furnaces (El), and total steel production (ALL) in the Federal Republic of Germany

	1970				1975				1977			
	LD	OH	El	ALL	LD	OH	El	ALL	LD	OH	El	ALL
Pig iron and blast furnace ferro-alloys	889	360	37	688	885	338	28	698	877	353	25	708
Scrap	223	699	1013	406	226	717	989	394	237	706	1013	389
Pellets and sponge iron	1	3	7	3	2	5	63	9	2	1	57	9
Ore and welding slag	25	90	33	36	34	97	33	39	33	84	4	37
Ferro-alloys	4	10	53	10	7	11	44	12	7	12	58	13
Total	1142	1162	1143	1143	1154	1168	1157	1151	1156	1156	1157	1156

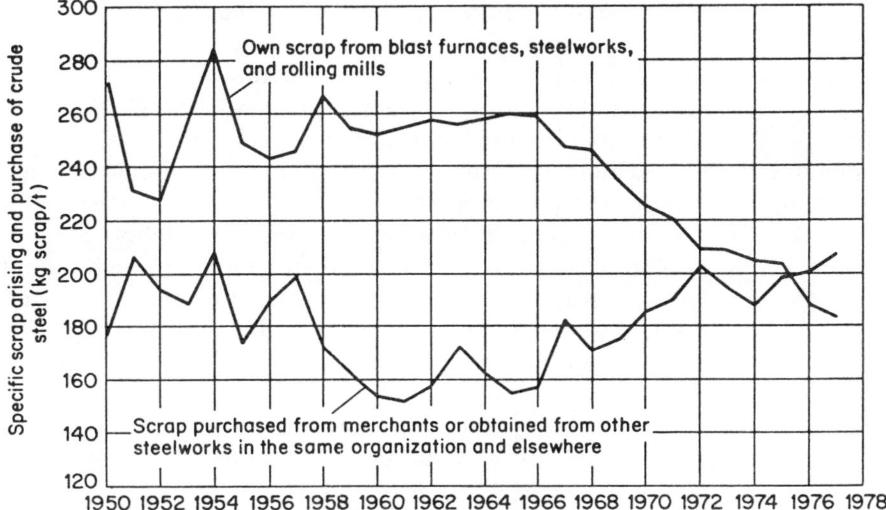

Figure 2. Own scrap and bought scrap for crude-steel production (for ingots, continuous casting, and casting steel)

3.3.1.3 Harmful and undesirable ingredients in scrap

There are now so many alloyed and non-alloy steels that it is difficult to collect correctly sorted scrap and recycle it. All the various types of steel must be correctly identified *in situ*, and proper sorting should take place when dismantling large objects (bridges, hangars, ships).

Different steels are present in consumer goods combined in many ways with other materials. Admixtures and coatings, which may be harmless in steel, can increase the amount of slag, attack the refractory lining more severely, and lead to an increase in the emission of harmful substances; this is economically and environmentally undesirable. For example, synthetic resins, especially PVC, fluorinated compounds, and sulphur oxides below the dewpoint can cause corrosion in the gas-purification plants.[2,10]

The types of scrap listed in Table 1 apply only to unalloyed carbon steel scrap. There are many unalloyed steels, especially large quantities of unalloyed deep-drawing steels, where higher proportions of alloying elements are harmful. During melting down under strongly oxidizing conditions in both basic oxygen furnaces and arc furnaces, elements such as aluminium, titanium, silicon, and vanadium, as well as small amounts of manganese and chromium, are oxidized and scorified. When scrap is remelted without oxidation in a basic-lined induction crucible furnace there is much less scorification.

Alloys and highly alloyed scrap, containing chromium, nickel, tungsten, or molybdenum, must be collected separately and accurately channelled into making alloy steels. The substantially higher prices that high-alloy steels command in the scrap trade are a strong inducement for separate collection.

This inducement does not apply to scrap contaminated with sulphur, copper, tin, zinc, or lead, and sometimes also antimony or arsenic. These elements are harmful, and attention should be paid to the following points:

Sulphur is introduced into scrap from oil, grease, paint, and sulphur-bearing, free-cutting steels. With the exception of free-cutting steels, the presence of sulphur can be greatly reduced by thorough sorting. Desulphurization is possible in the arc furnace but is very limited in single-slag high-output operation; it is possible to only a very small extent in the induction furnace. It is now possible to make fully killed steels with extremely low sulphur contents by using newer secondary desulphurizing processes, even when scrap is melted in an arc furnace under black slag.

Zinc and lead originate mainly from anti-corrosion protective coatings, so that particularly old, collected scrap could be relatively highly contaminated with these elements.

During melting down, the zinc in the scrap evaporates (boiling-point 907°C) and will be wholly or partly oxidized, according to the oxygen content of the waste gases emitted from the furnace. These days, the zinc oxide level in exhaust gases must be kept very low; this means more expense for scrubbing.

In contrast, lead boils at 1700°C but its oxides boil at only 1475°C, i.e. below the steel-bath temperature.[2] Any lead that is not oxidized and combined with the slag settles on the bottom of the hearth because of its high density. Lead penetrates the refractory lining and can cause serious damage and steel breakout. The solubility of lead in the steel-melt increases with rising temperature and reaches concentrations of up to 0.025%; this will certainly cause trouble with hot deformation of high-alloy steels.

Copper and tin are particularly harmful.[11] Their solubility in molten steel is unlimited. The tin comes from tin plate and solder in the scrap; copper is found because it is widely used in electrical engineering, plumbing, and refrigeration. Copper and tin have a lower affinity for oxygen than has iron, and they cannot be scorified. Copper – tin alloys segregate at the grain boundaries during solidification and cause red brittleness in hot-forming, even at low temperatures from 850 to 1100°C. Cold ductility is also affected. Copper and tin concentrations in non-alloy scrap must be limited to 0.15% and 0.020%, respectively, so that the scrap can be used without any precautions. Antimony and arsenic are also harmful, like tin.

Swarf needs special attention, as it is often contaminated with harmful and unwanted constituents and its recycling value is reduced. Swarf is therefore shown in scrap lists (Table 1) at reduced prices. Swarf should always be fresh, broken, and degreased when recycled. It should be separately collected and delivered according to its chemical composition. Great care should be exercised at the collection-point to keep steel swarf separated from brass and copper. The price of non-ferrous swarf is more than ten times higher than that of steel swarf, and should be sufficient inducement for this.

3.3.1.4 Scrap-preparation

Scrap can be prepared chemically or physically. Machine preparation can be done by oxyacetylene cutting, or with mechanical or hydraulic shears, presses, breakers, briquetting, drop-hammers, and by shredding or milling. De-oiling is done by centrifuging or burning.[2]

Scrap shears The development of scrap shears makes it increasingly possible to do away with oxyacetylene cutting for reducing scrap to chargeable size. Hydraulic shears operated by one man today have a cutting force of 800–1200 t. Developments in scrap-preparation have now led to shears with cutting strengths of up to 2000 t and shear blades up to 2.5 m long, so that whole waggons, large parts of dismantled ships, and other heavy scrap can be cut immediately without preparation.

Scrap presses Fully automatic hydraulic presses are used to compress light, bulky scrap into packs measuring up to 60 × 60 × 150 cm. The method preferred today for preparing new loose scrap and light production scrap, such as from the car and consumer goods industries, for electric-furnace charges is to compress it into packs. Care must be taken, however, to ensure that the composition is uniform and that no non-metallic contaminants are present.[12]

Shredders Despite the increasing quantity and proportion of light scrap (see Tables 1 and 3) and the increasing demands of the scrap consumers, the quantity and proportion of packs of ferrous and non-ferrous scrap are steadily declining in the Federal Republic of Germany. Shredded scrap is on the increase because shredders have largely replaced presses for processing old scrap. In 1976, twenty-four shredder systems with various powers between 500 and 4000 hp were in operation in West Germany, and more were planned.[2,13].

A shredding plant consists of the shredder itself, plus a dust-removal plant and separating and magnetic drums, together with conveyor belts for carrying and sorting the steel scrap, non-ferrous scrap, and stones (Figure 3). An old car, complete with engine, axles, transmission, seats, and tyres can be offered up to a large shredder without any preparation. In the shredder, which is actually a hammer mill, rotary hammers operating against an anvil edge tear small pieces off the car body. Powerful suction is used to separate non-metallic materials from metals in the shredder, conveyors, and drums. Next follows magnetic separation of ferrous from non-ferrous scrap and from contaminants such as wood, rubber, and plastic. However, the ferrous output may still contain non-metallic impurities and non-ferrous metals if these are too strongly

Table 3. Figures for scrap originating in West Germany

	1965		1973		1975		1977	
	(t × 10^6)	(%)	(t × 10^6)	(%)	(t × 10^6)	(%)	(t × 10^6)	(%)
Light scrap	1.761	21.2	2.496	21.9	2.037	22.1	2.041	20.8
light loose scrap	0.448	5.4	0.691	6.1	0.536	5.8	0.686	7.0
packs of ferrous scrap	1.095	13.2	1.305	11.5	0.984	10.7	0.858	8.7
non-ferrous packs	0.218	2.6	0.142	1.2	0.073	0.8	0.068	0.7
shredded scrap	—	—	0.358	3.1	0.444	4.8	0.429	4.4
Swarf	1.277	15.4	1.590	14.0	1.314	14.3	1.237	12.6
clean blue swarf								
cast iron and steel chips								
Heavy scrap	4.342	52.3	6.448	56.7	5.177	56.3	5.653	57.6
steel scrap								
cupola furnace scrap								
Other scrap	0.932	11.1	0.843	7.4	0.675	7.3	0.883	9.0
cast iron scrap								
alloy scrap								
other assorted scrap								
Total	8.312	100.0	11.377	100.0	9.203	100.0	9.814	100.0

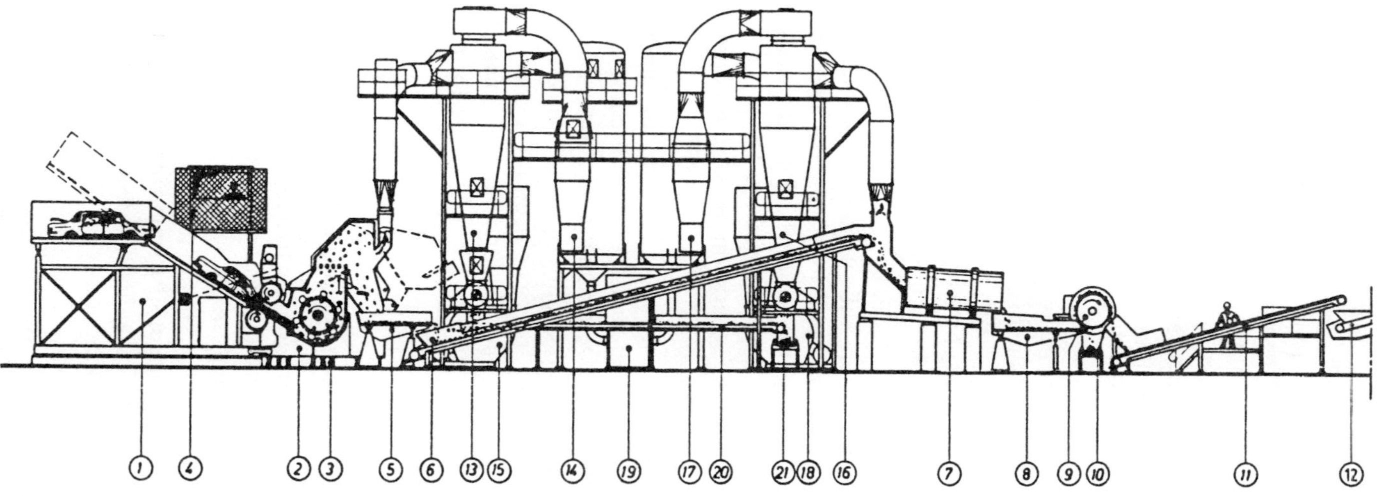

Figure 3. Sectional view of a shredder plant

1 approach chute with tilting tray;
2 shredder, hydraulically operated;
3 vibration damping;
4 control cabin;
5 vibrating conveyor;
6 belt conveyor;
7 separating drum;
8 vibrating conveyor;
9 magnetic separation;
10 non-ferrous sorting belt;
11 steel sorting belt;
12 steel ejection belt.

Shredder dust removal:

13 separator 1;
14 wet scrubber 1;
15 fan 1.

Separator drum dust removal:

16 separator 2;
17 wet scrubber 2;
18 fan 2;
19 water preparation for wet scrubbers 1 and 2;
20 dust collection belt;
21 dust ejection belt (to container)

bonded to the steel or hammered in and cannot be separated by the magnets. Manual sorting is still essential where very high purity is demanded.

A 1000–4000 hp shredder working single shifts produces about 2000–8000 t per month at about 1 t/m^3 piled weight. One tonne of input material contains about 75% clean, shredded scrap, 24% refuse, and 1% non-ferrous metals. The composition of the shredded scrap naturally depends on the input material. This is especially true for the percentages of copper, tin, and zinc and of chromium, nickel, and other steel-improvers.[14]

Much cleaner scrap in smaller pieces, with a piled weight of 2–3 t/m^3, can be obtained from cryogenic shredders in which the scrap is cooled in liquid nitrogen down to at least −196°C. Although, due to cold enbrittlement of the scrap, the shredder requires only 25% of the driving power otherwise needed, this process could not establish itself on account of the high price of nitrogen, of which about 450 l are required per tonne of scrap.

Scrap mills Another development is the scrap mill, which is a framework carrying a mill-enclosure with vertical walls. In the enclosure there is a breaker on a vertical shaft with two bearings. The input is scrap which has previously been cut up with shears. The output is similar to shredded scrap. The process is again followed by air-blast sifting and magnetic separation, as with shredders.

Chemical preparation As distinct from physical preparation by mechanical breaking up, or magnetic separation, chemical scrap separation consists of dissolving harmful ingredients from the scrap. De-tinning of tin plate waste by the Goldschmidt process has been practised and steadily improved since 1891; however, it is used only for new tin-plate scrap.[2] Specifications demanding very low tin contents in scrap will allow only incinerator scrap to be used in future, when we shall have better chemical, metallurgical, or chloride-forming processes for de-tinning.

3.3.2 Pig Iron and Cast-iron Fragments

There will only be a few cases when it will be economical to add solid pig iron to an electric furnace, unless this is done for metallurgical reasons to provide a carbon carrier, or to dilute undesirable elements in the scrap. Pig iron is only used in an electric-arc furnace in special cases.[15] On the other hand, it can be advantageous to add granulated pig iron continuously to control the degree of oxidation during refining or melting, when shredded scrap or sponge iron with a low degree of metallization are also fed in continuously.

Table 4 shows the chemical composition of various types of pig iron and foundry iron,[16] because some steelworks use fragments of moulds, bottom-pouring plates, and cast-iron chips in charges.

Table 4. Chemical composition of different types of pig iron

	Chemical composition (%)				
	Si	Mn	P	S	c_{tot}
For steel production:					
pig iron for steel	max. 1.0	0.4–1.0	0.08–0.25	max. 0.04	3.5–4.5
Thomas pig iron	max. 1.0	0.4–0.8	1.5 –2.2	max. 0.08	3.2–3.8
For castings:					
hematite (standard)	2.00–2.50	0.7–1.0	max. 0.12	max. 0.04	3.8–4.2
	2.50–3.00	0.7–1.0	max. 0.12	max. 0.04	3.7–4.1
foundry pig iron (standard)	2.00–2.50	0.6–0.9	0.5–0.7	max. 0.04	3.6–4.1
	2.50–3.00	0.6–0.9	0.5–0.7	max. 0.04	3.6–4.0
nodular cast-iron	0.1 –2.5	max. 0.15	max. 0.060	max. 0.010	2.5–4.3

3.3.3 Sponge Iron

Sponge iron is a product of granulated or lump or pelletized iron ore, which is reduced in the solid phase together with gaseous or solid reducing agents. As reduction takes place in the solid phase, the iron is not separated from the gangue of the ore. The sponge iron thus still contains gangue, which will only be separated out as liquid slag during melting. If the sponge iron is to be charged into an electric furnace, it should come from an ore low in gangue and high in metallization in order to keep down the amount of slag and with it to minimize energy consumption.[17]

3.3.3.1 Production

In 1976 the Commission of the EEC sponsored a bibliographical study of the direct reduction of iron ore.[18] The development of the direct-reduction process

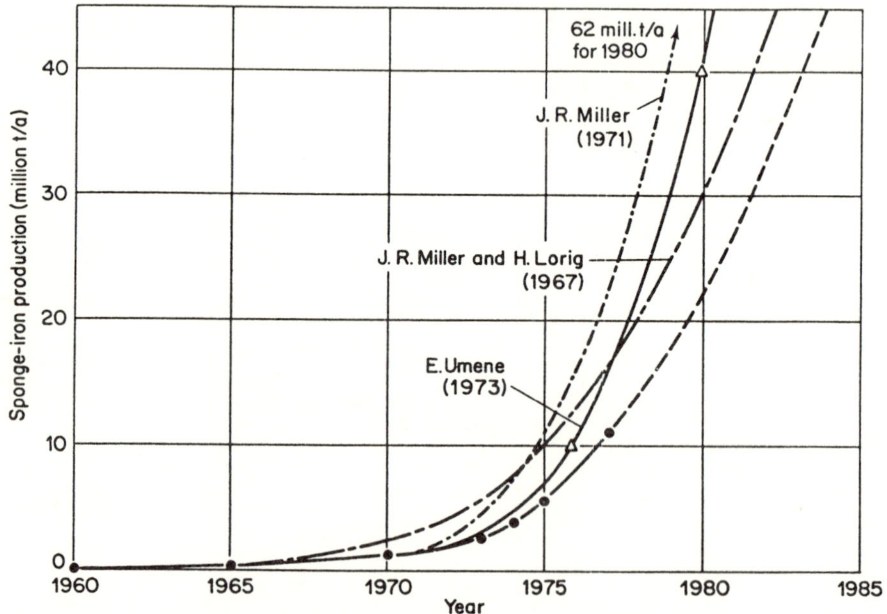

Figure 4. Estimated and actual development of sponge-iron production[19]

was assessed in this study,[18,19] (Figure 4), which also critically examined natural gas and coal reserves and the availability of iron ore for direct reduction. The assessment found fifteen methods of great technical importance, eleven of limited importance, and twenty-eight of no technical importance.

The processes can be divided into four groups according to the type of

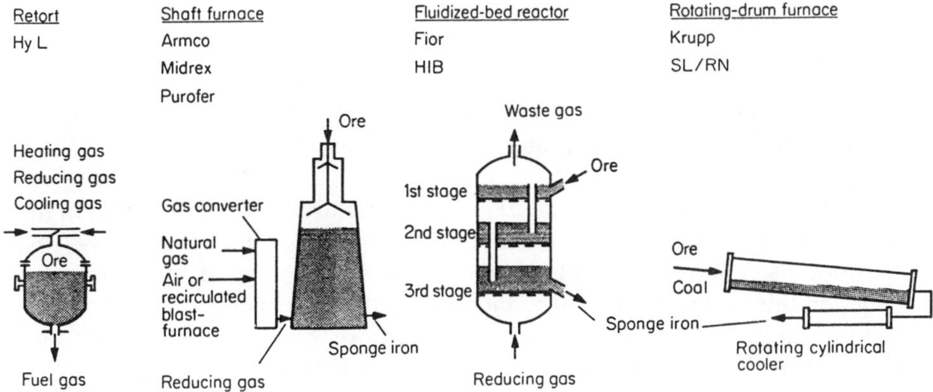

Figure 5. Schematic representation of different sponge-iron processes[19]

reaction vessel: retort, shaft furnace, fluidized bed reactor, and rotating drum furnace, as shown schematically in Figure 5.[19–26] The first three methods generally use natural gas for reduction. A reducing gas can be prepared from oil-or coal-gasification in the conventional manner or by making use of process gas from high-temperature reactors. The reducing gases must be free from sulphur, or the catalysts will be poisoned. Sulphuretted hydrogen must be removed. The nitrogen content must also be taken into account in processes which use gas recirculation.

The rotating-drum furnace processes use solid carbon agents. At the start of the development of these processes, coal such as anthracite and breeze was mainly used because of the low gas content. However, with the present state of development virtually all types of coal can now be used as long as the ash melting point is high enough.

3.3.3.2 Characteristics of sponge iron

Sponge iron suitable for charging into an electric furnace should have a degree of metallization of at least 80%. Reduction products with a degree of metallization under 80% should be classified as pre-reduced ore.

Sponge iron (depending on its ore base) is generally characterized by only small percentages of undesirable ingredients. The overall iron content of sponge iron is in most cases 85% or more, with up to 2.5% carbon. The remainder is mostly unreduced gangue and oxygen combined with iron. Sponge iron has an apparent density of ≤ 4 g/cm^3. The density will be higher if it is briquetted after reduction. The piled density depends on the nature of the ore and the degree of metallization. However, one can assume an average piled density of 1.75 t/m^3.

The most important points about the composition of sponge iron are:

Total iron content;
Metallic iron content;
Degree of metallization (ratio of metallic iron to total iron × 100 (%));
Type of gangue content and of other elements such as carbon, sulphur, phosphorus, silica, alumina, lime, and magnesia.

The metallic iron content, the degree of metallization, and the carbon and sulphur contents are to a great extent affected by the production process. Only the amounts of phosphorus, silica, lime, and magnesia depend on the ore used.

Depending on the method of reduction and the type of charge required, iron may be finely divided, with a grain size of not more than 4.5 mm, or it may be in pellets of 6–15 mm, or larger pieces of up to 25 mm or briquettes. Lump-size and shape significantly affect behaviour in handling.

Table 5. Reference data of physical properties of sponge iron

Shape	Bulk density	Pure density	Apparent density	Porosity
	t/m^3	g/cm^3	g/cm^3	%
Lump	1.9	7.0	~3.7	50
Pellet	1.7	7.0	~3.3	55
Briquette	2.2	7.0	~5.3	20

Reference data about physical properties are presented in Table 5. The average thermal conductivity

$$\frac{\lambda}{\rho \cdot c_p} \; [cm^2 \cdot s^{-1}]$$

as measured in argon at 760 torr is about:

800°C	900°C	1000°C	1100°C	1200°C	1300°C
0.012	0.015	0.018	0.021	0.024	0.027

The mechanical strength of sponge iron depends on the method of preparation and use. Typical values for abrasion <0.5 mm could be 6%, and 90% for tumbler strength >3.15 mm.

The tendency of sponge iron to reoxidize depends on conditions in use and the method of production. Typical values for free exposure to the weather are shown in Table 6. Resistance to reoxidation is increased by suitable inactivation treatment or by briquetting. Lumps of sponge iron can be ignited

Table 6. Reference data for reoxidation of a pile of sponge iron in free storage (without post-treatment)

Location	Degree of metallization after *n* months in store						
	$n = 0$	1	2	3	5	7	9
On surface	93	72	58	52	46	43	41
0.1 m down	93	90	86	82	75	68	62
0.4 m down	93	93	93	92	91	91	90

at around 140°C or in pellet form at about 270°C. When produced in a rotary furnace, the ignition temperature may be 100–150°C higher because of surface consolidation in production.

3.3.3.3 Use of sponge iron

In highly industrialized countries where steel is made conventionally in oxygen converters, open-hearth furnaces, and arc furnaces, the use of sponge iron depends very much on economic conditions. In times of high prosperity, when scrap is scarce and expensive, sponge iron has to be used to fill the 'scrap gap'.[27] During a recession, scrap prices fall and it becomes more economical to use scrap unless qualitative considerations make it essential to use sponge iron.

3.4 ALLOYS

Otto Etterich, Düsseldorf

Over the last twenty-five years the use of alloying elements has increased more than proportionally to the rise in total steel production. This is due to the more than commensurable increase in the use of special high-grade steels, especially the rust- and acid-resisting types. Growing demands for better steel properties have led to the development of new high- and micro-alloyed steels.

New metallurgical understanding, the treatment of steel after tapping in secondary processes, and a pronounced leaning to economy led to changes in the use of the different alloying ingredients. An example here is the use of highly carburized ferro-chromium (carburé) in place of low- or medium-carburized ferro-chromium types (*suraffiné* and *affiné*). Similar changes can be seen with nickel and molybdenum. Today the main source ingredient no longer is nickel metal but ferro-nickel and nickel oxides in the most diverse physical forms, and molybdenum oxide instead of ferro-molybdenum.

3.4.1 Physical Character of Alloy-carriers

The change away from employing electric-arc furnaces solely for making special types of high-grade steel and their transformation into fast-melting machines, used in direct conjunction with highly developed refining techniques carried out in secondary installations, also brought about changes in the requirements for the physical character of the alloy-carriers. Their external form must make handling easy and enable transfer into furnaces, ladles, vacuum installations, etc. to be effected as rapidly and as free from loss as is possible.

Fine-grain materials can be blown into the melt. Alloy in lump form is moved direct to charging points in modern steelworks via conveyors and bunkers with automatic weighing. These automatic handling-devices have dust-extraction systems and make very high demands on particle-specification. Only the smallest amounts of undersized particles and dust can be tolerated. The nominal particle-size range must be appropriate to the purpose and solubility of the alloying element, to ensure that alloy is distributed equally throughout the melt and dissolved in a given time.[1] The dust content and proportion of undersize particles present must also be kept at a minimum so as to avoid uncontrolled losses in transport, dust-extractors, and in the waste gases from vacuum installations. Apart from material losses, which can also occur through slagging, the chemical composition of the steel may also be affected. Moreover, experience shows that undersize particles can mean higher levels of contaminants in the steel (flakes, loss of purity) and the quality of the steel may be affected.

When crushing ferro-alloys, a higher proportion of dust and undersize particles is unavoidable with smaller grain-specifications. Therefore, finer grain sizes (below 80 mm) should not be specified for ferro-silicon, ferro-manganese, and ferro-chromium, unless absolutely essential for processing reasons. Alloys charged in buckets or fed directly into larger induction furnaces can have fragment sizes up to 200 mm.

Powdered or very fine grain alloy-carriers can be blown into the melt or be briquetted or canned. These methods are already in industrial use for the supply of nickel or molybdenum oxides and for concentrates of other alloying elements. Alloy-carriers in granular or pellet form can also be used for all kinds of charges; however, aluminium is also usually used in ingot form.[2]

3.4.2 The Chemical Composition of Alloying Elements and their Form for Delivery

The origin of alloying elements for steel is confined largely to particular geographical regions (Table 1).[3] These elements are usually extracted from the

Table 1. Reference data on important steel-alloying elements. (World reserves for 1977 in thousands of tonnes and concentrations of reserves in various countries)

Metal	Proved and probable reserves	Concentration of reserves by countries in 1977		Consumption				
		3-nation share (%)	Countries and their percentage share	World 1971 ($t \times 10^3$)	West Germany 1973	USA 1973	West Germany 1976	USA 1976
					crude steel production (%)			
Manganese	1 764 300	87.8	South Africa 43.6, USSR 38.5, Gabon 5.7	9200	0.64	0.70	0.665	0.62
Chromium (chromite)	2 841 000	96.4	South Africa 73.9, Rhodesia 19.7, USSR 2.8	1700	0.243	0.175	0.280	0.19
Nickel	68 000	58.9	New Caledonia 22.7, Cuba 21.5, Canada 14.7	517	0.068	0.065	0.048	
Molybdenum	8 761	75.4	USA 40.0, Chile 24.0 USSR 11.4	74	0.0146	0.0150	0.0177	0.025
Tungsten	2 033	76.2	Peoples Republic of China 52.4, USSR 11.9, Canada 11.9	31.6	0.0034	0.0012	0.0022	
Vanadium	9 698	95.0	USSR 74.9, South Africa 18.7, Chile 1.4	16.4	0.0052	0.0050	0.0053	0.068
Cobalt	2 451	57.6	Zaire 27.8, New Caledonia 15.6, Zambia 14.2	21.8	0.0015	0.0012	0.0016	
Niobium	10 820	89.7	Brazil 75.8, Canada 7.6, USSR 6.3		0.0016	0.0010	0.0022	0.001

Data from the Federal German Institute for Geo-Sciences and Raw Materials.
Consumption figures for the world, the Federal Republic of Germany, and the USA from data by C.W. Sames in the US Bureau of Mines bulletin *Mineral Facts and Problems*.

ores and their concentrates and converted into ferro-alloys via reduction processes which are often relatively costly.[4] In many cases world reserves of alloying metals are limited, and every means should be adopted in all stages of production, application, and recovery for recycling to ensure that utilization is maximized. The chemical reaction behaviour during steelmaking has particularly to be watched, to keep slagging losses to a minimum.[5]

Agreement to standardize the chemical composition and physical form of alloying elements has been concluded between producers and consumers in the alloy committee of DIN (German Industrial Standards). Every alloying element is also standardized with a view to economy in use. The ISO (International Organization for Standardization) tries to lay down worldwide standards for delivery conditions, sampling, and analysis, as well as alloy composition and external form.

The following survey of the most common alloying elements is based on the DIN standards, literature, and manufacturers' data. Information about the purpose and use of individual alloying elements and about techniques for utilizing other alloying elements that are rarely used can be found in the literature.[6–8]

3.4.2.1 Aluminium

Table 2. Composition of pure commercial aluminium to DIN 1712 and 1725[9]

Designation	Code-name	Content of permitted companion elements (% wt) Max. total	Si	Fe	Ti	Cu	Zn	Al
Pure aluminium	Al 99.9H	0.1	0.050	0.035	0.006	0.005	0.04	
Primary aluminium	Al 99.8H	0.2	0.15	0.15	0.03	0.01	0.06	
	Al 99.7H	0.3	0.20	0.25	0.03	0.01	0.06	
	Al 99.5H	0.5	0.30	0.40	0.03	0.02	0.07	
	Al 99 H	1.0	0.5	0.6	0.03	0.02	0.08	
Remelting aluminium	URAL I/41I	—			Traces	4.5	0.7	>92

Delivered as 1 kg and 5 kg bars in various forms such as breeze and granulate.

3.4.2.2 Boron

Table 3. Boron alloys to DIN 17567, January 1970[9]

Code-name	Material No.*	Chemical composition (max. % wt)							
		B	Al	Si	C	Mn	P	S	Co
FeB 16	0.4816	15–18	4.0	1.0	0.10	0.50	0.005	0.001	0.005
FeB 18	0.4818	18–20	2.0	2.0	0.10	0.50	0.005	0.001	0.005
FeB 12c FeB 17c	0.4812 0.4817	10–14 14–19	0.50	4.0	2.0	0.50	0.005	0.1	0.005

*From the type Nos in DIN 17007, sheet 3, still in draft at time of writing.

Table 4. Composition of complex boron alloys[8]

Type	Chemical composition (% wt)				
	B	Al	Si	C	Remainder
Manganese–boron	7.5			3	Mn
Nickel–boron	15–18	1	1.5		Ni
Cobolt–boron	15–18	1			Co

Delivered in lump, granulate, or powdered form.

3.4.2.3 Cerium

Composition of a typical misch alloy:[10] Cerium 45–55%; lanthanum 20–25%; neodymium 15–20%; praseodymium 5–8%, silicon 0.15%; iron 0.1%; aluminium 0.08%; carbon 0.08%; magnesium 0.02%; remainder: other rare-earth metals.

Form of delivery: profiled bars of 1 and 5 kg, bar cuttings 15 mm.

3.4.2.4 Chromium

Table 5. Composition of ferro-chromium, ferro-chromium-silicon, and chromium from DIN 17565, December 1968[9]

Name		Code-name	Material no.[a]	Chemical composition (wt %) Cr	C	Si	P max.	S max.	N
Ferro-chromium suraffiné	Ferro-chromium low C	FeCr70C01 FeCr70C02 FeCr70C04 FeCr70C06 FeCr70C08 FeCr70C10 FeCr70C50	0.4010 0.4011 0.4012 0.4013 0.4015 0.4016 0.4017	65 to 75[b]	max. 0.010 max. 0.020 max. 0.040 max. 0.060 max. 0.080 max. 0.10 max. 0.50	max. 1.5[c]	0.030	0.010	max. 0.050
	Ferro-chromium low C, containing nitrogen	FeCr70NC10	0.4020	60 to 72[b]	max. 0.10	max. 1.5[c]	0.030	0.010	2.5 to 4.0[d]
Ferro-chromium affiné	Ferro chromium medium C	FeCr70C1 FeCr70C1,5 FeCr70C2 FeCr70C4	0.4030 0.4032 0.4036 0.4038	65 to 75[b]	0.5 to 1.0 over 1.0 to 1.5 over 1.5 to 2.0 over 2.0 to 4.0	max. 1.5[c]	0.030	0.050	—

carburé	Ferro-chromium high C	FeCr70C5 FeCr70C8	0.4052 0.4056	60 to 72[b] 60 to 72[b]	4.0 to 6.0 over 6.0 to 10.0	max. 1.5 max. 1.5	0.030 0.030	0.050 0.030	— —
	Ferro-chromium high C, containing silicon	FeCr70C6Si	0.4058	60 to 72[b]	4.0 to 8.0	1.5 to 10	0.030	0.030	—
Silicon chromium	Ferro-chromium silicon	FeCr40Si FeCr60Si	0.4070 0.4072	40 to 45[b] 55 to 65[b]	max. 0.050[c] max. 0.050[c]	45 to 35[b] 25 to 20[b]	0.020 0.020	0.010 0.010	— —
Chromium metal	Chromium metal (alumino-thermic)	Cr99	0.4090	99.0 to 99.3	max. 0.030	Fe: max. 0.20 Al: 0.10 to 0.30 Si: max 0.10	0.020	0.035	—
	Electrolytic chromium	Cr99.9	0.4099	min. 99.9	Contents of other constituents on demand				

[a]After DIN 17007, sheet 3 (in draft at time of writing).
[b]The variation in any one lot must not exceed ± 2 wt %.
[c]Lower maxima can be negotiated.
[d]The contents quoted apply for melted FeCr types. Sintered types with N contents from 5 to 8 wt % can be supplied by agreement.
Note: In addition to the above analysis data: Common iron-rich chromium ores required a decrease of the lower Cr limit to 50% in an ISO revision.
Form of delivery: see section 3.4.4.

3.4.2.5 Cobalt

Table 6. Composition of commercial cobalt alloys[4]

Name	Proposed code-name	Chemical composition (%) Co	Fe	Ni (max.)	Cu (max.)	Mn (max.)	S (max.)	C (max.)	Ca (max.)	N	O	H cm³/100 g	Density (g/cm³)	Melting range (°C)	Remarks
Ferro-cobalt	FeCo 85	80–90	18–8	0.8	0.02	0.05	0.01	0.02	—	—	—	—	~8.4	1450–1490	—
Cobalt metal	Co 97	95–99	<1.5	2.5	0.02	0.02	0.01	0.01	—	—	—	—	~8.6	1450–1490	Partly granules, rondelles
Pure cobalt	Co 99	>99	<0.2	0.6	0.04	—	0.04	0.06	—	—	~0.03	~7	~8.8	1470–1490	Partly granules, rondelles
Cobalt powder	Co 990	>99	<0.05	0.03	0.001	0.015	—	—	0.015	—	0.8	—	~8.7	1450–1490	Cobalt powder
Electrolytic cobalt	Co 99.5	>99.5	<0.07	0.3	0.003	0.015	0.005	0.01	—	~0.005[a]	~0.001[a]	~2[a]		1490–1495	Zn <0.002 Si <0.003

[a]Degassed in vacuum.
Note: Cobalt 97 and Co99 come as granules, cubes, and rondelles in the trade; they are the types mainly used for alloy steelmaking.

3.4.2.6 Copper

Copper is used only in small amounts for steel alloys, and then as pure copper scrap, electrolytic copper containing 99.8% Cu, or copper-bearing scrap steel.

3.4.2.7 Manganese

Table 7. Composition of ferro-manganese, ferro-manganese–silicon, and manganese after DIN 17 564, Decer

Name			Code-name	Material number[a]	Chemical composition by wt % Mn[b]	C	Si	P max.	S max.	N	
Ferro-manganese	suraffiné	Ferro-manganese, low C	FeMn85C01	0.3185	80–92	0.05–0.50[c]	1.0–1.5[c]	0.25	0.030	—	—
		Ferro-manganese, low C, low P	FeMn85C01P015	0.3186	80–92	0.05–0.50[c]	1.0–0.50[c]	0.15	0.030	—	—
		Ferro-manganese low C, with N	FeMn85NC01	0.3187	80–92	0.05–0.50[c]	1.0–1.5[c]	0.25	0.030	2.0–2.5[d]	—
		Ferro-manganese, low C, low P, with N	FeMn85NC01P015	0.3188	80–92	0.05–0.50[c]	1.0–1.5[c]	0.15	0.030	2.0–2.5[d]	—
	affiné	Ferro-manganese, medium C	FeMn80C1	0.3180	75[e]–85	above0.5–2.0[e]	0.5–1.5[c]	0.25[e]	0.030	—	—
		Ferro-manganese, . medium C, with N	FeMn85NC1	0.3181	80–90	above0.5–2.0[c]	0.5–1.5[c]	0.25[e]	0.030	1.0–2.0[d]	
	carburé	Ferro-manganese, high C	FeMn75C7	0.3175	75[e]–80	6.0–8.0	max. 1.5[f]	0.35	0.030	—	—
		Ferro-manganese, high C, low P	FeMn75C7P015	0.3177	75[e]–80	6.0–8.0	max. 1.5[f]	0.15	0.030	—	—
Ferro-manganese–silicon			FeMn65Si	0.3160	58–72	0.1–0.5	23–35	0.20	0.010	—	—
			FeMn70Si	0.3162	65–75	0.5–2.0	15–25	0.20	0.010	—	—
Manganese		Electrolytic manganese	Mn99.9	0.3199	min. 99.9	max. 0.010	max. 0.01	0.010	0.040	—	max. 0.010
		Manganese metal	Mn97	0.3197	95–93	max. 0.030	max. 0.50	0.050	0.008	—	3.0–1.5
		Manganese with N (sintered)	Mn90N	0.3190	88–90	max. 0.050	max. 0.50	0.050	0.010	6.0–7.0	≈2

[a]Corresponding to DIN 17 007, sheet 3 (draft).
[b]The percentage range in one lot must not exceed ± 2 wt %.
[c]If a maximum is wanted in any range, to be agreed on ordering.
[d]The quoted contents apply for melted FeMn types; sintered types with N 5–8 % wt can be supplied by agreement.
[e]Lower contents are subject to negotiation.
[f]A silicon content of 0.5 wt % max. by agreement, when ordering.
The material number is then 0.3276 for Fe Mn C7 and 0.3278 for Fe Mn C7 PO 15.
Forms of delivery: see section 3.4.4.

3.4.2.8 Molybdenum

Table 8. Composition of ferro-molybdenum after DIN 17 561, December 1965[9]

Code-name	Material number[a]	Mo[b]	Chemical composition by wt % Si	C	S	P	Cu
			Maximum				
FeMo70	0.4270	60–75	1.0	0.10	0.10	0.10	0.50
FeMo62	0.4262	58–65	2.0	0.5	0.10	0.10	1.0

[a]A standard is in preparation for material numbers of the main group O. The material numbers quoted are proposed.
[b]The percentage range in one lot must not vary by more than ± 2 % wt.
Forms of delivery: see section 3.4.4.

Table 9. Molybdenum metal ('technical')

Type	Mo min.	C max.	SiO_2 max.	Composition wt % Mn	S max.	Cu max.	P max.	O max.	Fe max.
Molybdenum metal 'technical'	98	0.15	0.4	0.10	0.03	0.02	0.05	1.5	0.2

Forms of delivery: sintered, powder or compressed.

Table 10. Molybdenum oxide[11]

Type	Mo	Composition in % wt C as pitch	Cu max.	S max.	P max.
Molybdenum oxide in cans	min. 57		0.15	0.10	0.05
Molybdenum oxide in briquettes	about 53	about 11	0.15	0.10	0.05

Forms of delivery: molybdenum oxide in 10 kg cans; briquettes bonded with pitch, each with 1.25 kg molybdenum, where 4 briquettes (= 5 kg Mo) are packed in one carton.

3.4.2.9 Nickel

Table 11. Characteristic compositions of nickel[12]

Type	Chemical composition by wt %						
	Ni	Co	Cu	C	S	Fe	Other
Nickel pellets	99.97	0.0005	0.001	0.015	0.0015	0.02	
Electrolytic nickel	99.92	0.06	0.005	0.01	0.001	0.002	
Nickel granulate	95.0	1.3	0.4	0.02	0.003	0.4	1.1 Oxygen
Nickel granules	91.0	0.5	0.2	0.55	0.06	1.65	5.5 Silicon

Note: The very different production processes account for the various compositions, shapes and forms delivered:
Nickel pellets: spherical pellets, mostly 5–10 mm diameter
Electrolytic nickel: cut pieces, about 100 × 100 × 12 mm
Nickel granulate: mostly 0.2–0.4 mm
Nickel granules: mostly 0.5–6.5 mm.

Table 12. Typical composition of sintered nickel oxide[12]

Type	Chemical composition in % wt				
	Ni	Co	Cu	S	Fe
Sintered nickel oxide	76.00	1.00	0.75	0.006	0.30

Note: Nickel and nickel oxide are usually supplied in tins and steel barrels, according to the product, but then always uniformly, with pure nickel contents of 10, 22.5, 15, and 250 kg.
Supplied as granulate: 0.2–0.4 mm.

Table 13. Composition of ferro-nickel, after DIN 17 568, January 1970[9]

Code-name	Material number[a]	Chemical composition by wt %					
		Ni + Co[b,c]	C	Cr	Si	P	S
				Maximum			
FeNi25	0.4425	20 to 30	0.030	0.10	0.05	0.030	0.040
FeNi25C	0.4427	20 to 28	2.0	2.0	4.0	0.040	0.040
FeNi25C5	0.4429	20 to 28	2.0	2.0	4.0	0.040	0.30
FeNi55	0.4455	50 to 60	0.05	0.050	1.0	0.020	0.010

[a]Conforming to material number in DIN 17 007, sheet 3 (draft).
[b]Cobalt content is 1.0 wt % max.
[c]The tolerance within one lot is usually 1.0 wt % absolute.
Form of supply: Ferro-nickel is cast into pigs, the pieces weighing 12–40 kg.

3.4.2.10 Niobium–tantalum

Table 14. Composition of ferro-niobium to DIN 17 569, draft March 1977[9]

Code-name	Material number[a]	Nb[b]	Ta	Chemical composition by wt % Al max.	Si max.	Ti max.	C max.	S max.	P max.	Sn max.	Co max.
FeNb63			max. 0.5								
FeNb63	Ta1	58 to 68	1.0	2.5	4.0	2.5	0.2	0.10	0.10	0.15	0.05
FeNb63	Ta2		2.0								
FeNb65		63 to 68	max. 0.5	1.0	2.5	0.4	0.15	0.05	0.10	0.10	0.05
FeNb60	Ta8	58 to 63	max. 8.0	2.5	1.5	1.5	0.15	0.10	0.10	0.20	0.05

[a]Material numbers laid down only in the final standard.
[b]The tolerance within one lot is ± 2 % Nb.
Note: In ferro-silicon–tantalum alloys, use is shifting towards higher niobium content types, because they suffer lower melting losses.
Form of supply: see section 3.4.4.

3.4.2.11 Phosphorus

Table 15. Normal phosphorus alloy

Type	Chemical composition by wt % P	Mn	Si max.	Ti	Cr	V
Normal phosphorus alloy	20–30	0.4–0.7	10	1–2.5	0.3–0.5	0.5–1.0

Form of supply: In pieces 6–80 mm and 80–250 mm, and granulated up to 6 mm.

3.4.2.12 Sulphur

Sulphur is now usually introduced into steel as elementary sulphur. Its use as iron sulphide and roasted pyrites has been greatly reduced for quality reasons.

3.4.2.13 Silicon

Table 16. Composition of ferro-silicon and silicon to DIN 17 560, December 1965[9]

Code-name	Material number[a]	Chemical composition by wt %				
		Si[b]	Al	P	S	C[c]
				Maximum		
FeSi10	0.3310	8.0 to 15.0	max. 0.8	0.15	0.04	2.5
FeSi25	0.3325	20.0 to 30.0	max. 0.8	0.08	0.04	0.8
FeSi45	0.3345	42.0 to 48.0	max. 1.5	0.05	0.04	0.2
FeSi75	0.3375	73.0 to 79.0	1.0 to 2.0	0.05	0.04	0.1
FeSi75-Al1	0.3376	73.0 to 79.0	max. 1.0	0.05	0.04	0.1
FeSi90	0.3390	87.0 to 95.0	1.0 to 2.5	0.04	0.04	0.1
FeSi90-Al1	0.3391	87.0 to 95.0	max. 1.0	0.04	0.04	0.1
			Al	Fe	Ca	
Si 98	0.3398	97 to 99	max. 1.5[d]	max. 0.5	max. 0.35[d]	

[a]A standard for material numbers in the main group O is in preparation. The material numbers quoted are proposed.
[b]Tolerances within one lot must not exceed ± 2 wt %.
[c]The values for carbon are for information, but are not binding for acceptance.
[d]Lower maxima are subject to agreement.
Form of supply: see section 3.4.4.

3.4.2.14 Nitrogen

Nitrogen is introduced into the melt mostly through nitrogen-bearing ferro-alloys. These can be found under section 3.4.2.4 for chromium (Table 5) and section 3.4.2.7 for manganese (Table 7).

3.4.2.15 Titanium

Table 17. Composition of ferro-titanium from DIN 17 566, December 1968[9]

Code	Material number[a]	Ti[b]	Chemical composition by wt % Total-Al max.	Si max.	Mn max.	C max.	P max.	S max.
FeTi30	0.4530	28–32	4.5	4.0	1.5	0.10	0.050	0.060
FeTi40	0.4540	36–40	6.0	4.5	1.5	0.10	0.10	0.060
FeTi50	0.4550	46–50	7.5	4.0	1.0	0.10	0.10	0.060
FeTi70	0.4570	65–75	2.0	0.20	1.0	0.20	0.040	0.030
FeTi70VB Vacuum-treated	0.4571	65–75	0.50	0.10	0.20	0.20	0.030	0.030

[a]From DIN 17 007, sheet 3, (in draft)
[b]The tolerance within one lot must not exceed ± 2%.
Form of supply: see section 3.4.4.

3.4.2.16 Vanadium

Table 18. Composition of ferro-vanadium from DIN 17 563, December 1965[9]

Code-name	Material number[a]	V[b]	Chemical composition by wt % Al	Si	C	S	P	As	Cu
					Maximum				
FeV60	0.4706	50–65	2.0	1.5	0.15	0.05	0.06	0.06	0.10
FeV80	0.4708	78–82	1.5	1.5	0.15	0.05	0.06	0.06	0.10

[a]A standard for material numbers in main group O is in preparation. The material numbers quoted are proposed.
[b]Tolerances within one lot must not exceed ± 2 % by wt.
Form of supply: see section 3.4.4.

3.4.2.17 Tungsten

Table 19. Composition of ferro-tungsten to DIN 17 562, December 1965[9]

Code-name	Material number[a]	W[b]	Si	C	Mn	Cu	S	P	As	Sb	Sn	Σ As + Sb + Sn
		Chemical composition by wt %										
								Maximum				
FeW80	0.4380	75–85	0.6	1.0	0.6	0.2	0.05	0.05	0.10	0.08	0.10	0.20

[a] A standard for material numbers in main group O is in preparation. The material numbers quoted are proposed.
[b] Tolerances within one lot must not exceed ± 2 % by wt.
Form of supply: see section 3.4.4.

3.4.2.18 Zirconium

Usual alloy in steelworks.[10]

Table 20. Ferro-zirconium–silicon

Type	Chemical composition by wt %				
	Zr	Si	Ca	Al	Fe
Fe Zr Si	35–42	47–52	0.5–0.8	1.2–1.7	Remainder

Form of supply: 10–15-mm pieces or granulated 0 to 2 mm.

3.4.3 Alloys for Deoxidation and Refining

The aim of deoxidation is to purify molten steel by adding elements which react with the impurities, mainly with oxygen, but also with sulphur, and to segregate them from the steel bath, or to convert them into less harmful compounds. Silicon, manganese, aluminium, calcium, magnesium, boron, zirconium, and cerium (misch metal) form the bases for a series of such alloys.

Table 21. Composition of calcium–silicon to DIN 17 580, December 1968[9]

Name	Code-name	Material number[a]	Chemical composition by % wt Ca	Ca + Si min.	C max.	Al max.	P max.	S max.
Calcium–silicon	CaSi	0.3650	29–33	90	1.20	1.80	0.070	0.060
Calcium–silicon low carbon	CaSiC50	0.3655	29–33	90	0.50[b]	1.80[b]	0.070	0.050[b]

[a]Corresponding to DIN 17 007, sheet 3 (draft).
[b]Lower maxima are subject to negotiation.
Form of supply: In lumps or cominuted.

Table 22. Reference values for the composition of deoxidation alloys, preferred on calcium–silicon base[13]

Type	Chemical composition by wt %									
	Ca	Si	C max.	Mn	P max.	S max.	Al	Ti	Fe	Others
Alcasil 6	29–31	52–58	0.3		0.03	0.05	5–8	0.1–0.2	4–6	
Alcasil 10	25–31	52–58	0.3		0.03	0.05	9–12	0.1–0.2	4–6	
Alcasil 20	19–27	47–53	0.3		0.03	0.05	18–22	0.1–0.2	4–6	
Alcasil 30	18–22	40–45	0.3		0.03	0.05	28–34	0.1–0.2	4–6	
Alcasil 35	15–20	35–40	0.3		0.03	0.05	34–40	0.1–0.2	4–6	
Alcasil 55	9–13	25–36	0.3		0.03	0.05	50–56	0.1–0.2	4–6	
Mn–CaSi	18–23	49–55	1.0	14–20	0.05	0.03		<0.2	8–10	
Mn–CaSi–Al	18–23	50–56	1.0	8–10	0.05	0.02	2.7–3.7	<0.2	8–10	
Mn–CaSi–Al 1	22–29	48–53	1.0	6–11	0.05	0.02	4–8	<0.2	Remainder	
Mn–CaSi–Al 2	13–15	40–42	1.0	10–12	0.05	0.02	10–12	<0.2	Remainder	
Ba–CaSi	13–18	46–54	1.0					<0.1	18–20	Ba 11–18
Ba–CaSi–Al	10–16	34–43					13–21			Ba 9–12
Mg–CaSi 5	24–29	48–56	1.0						4–6	Mg 5–6
Mg–CaSi 10	25–30	52–56	1.0						4–6	Mg 8–12
Mg–CaSi 15	26–30	52–57	1.0						4–6	Mg 13–17
Mg–CaSi 20	23–27	51–53	1.0						4–6	Mg 17–18
Mg-CaSi 30	16–20	40–45	1.0						4–6	Mg 28–32
Ti–CaSi	17–22	43–53	1.0					14–24	10–15	
Cercal S 1	12–20	52–54	1.0			0.03		<0.1	10–12	SE 15–21
Cercal S 2	10–14	46–48	1.0			0.03		<0.1	10–12	SE 24–28
Cercal Al 1	5–7	20–25	1.0			0.03	36–40	<0.1	2–4	SE 25–27
Cercal Al 2	10–14		1.0			0.03	55–63	<0.1	2–4	SE 23–26
Cercal Al 3		15–20	1.0			0.3	45–50	<0.1	2–4	SE 20–22

Supplied: lumps and commercial-size granules.

Table 23. Composition of the usual alloys containing magnesium, cerium, and zirconium[10]

Type	Chemical composition by % wt								
	Mg ca	Ce ca	Zr ca	Si ca	Ca ca	Al	Mn	C max	Fe
Cer–MM–Silicide	—	39		30–35	—			0.10	Remainder
FeSiMg 5	5.5	1		46	1–2			—	Remainder
FeSiMg 10	10	1		46	1–2			—	Remainder
CaSiMg	10	—		53	25–30			—	Remainder
FeMg 10	10	—			—			0.03	90
FeMg 15	15	—			—			0.03	85
FeZr			85	—	—	—	—		15
ZL 80			1.5	80	2.5	max. 1.5	—		Remainder
ZL 80-S			1.5	80	2.5	2–3	—		Remainder
SMZ			6	60–65	0.6–1	1.2–1.6	5–7		Remainder

Supplied: Granulated and partly pulverized.

Table 24. Composition of nickel-based deoxidation alloys[12]

Type	Chemical composition by % wt				
	Mg	Ca	Si	C	Fe	Ni
Incocal 10[a]		5.4	0.6	0.2	0.15	Remainder
Incomag 1 LC[b]	15		0.2	0.1	0.22	Remainder
Incomag 3 LC[a]	4.5			0.1	—	Remainder

Supplied: [a]Pigs of 1.5 and 7 kg.
[b]In pieces 20 to 55 mm.

3.4.4 Particle-size Classes of Standardized Ferro-alloys (West German Recommendations)

Table 25. Ferro-silicon to DIN 17 560[9]

Particle class	Particle size range mm	Oversize particles max. % wt	Undersized particles max. % wt Total	Fines (dust) <3.15 mm
1	75–200		15	4
2	35–100		15	4
3	10–75		15	4
4	3.15–35	10[a]	8	
5	3.15–10		10	
6	3.15–6.3		10	
7	max. 3.15		—	

[a]Oversize particles max. 10% of the weight supplied. No piece must exceed the 1.5-fold upper limit of the specified particle size in two or three directions.

Table 26. Ferro-manganese and silicon–manganese to DIN 17 564,[9] ferro-chromium and silicon–chromium to DIN 17 565[9]

Particle class	Particle size range mm	Oversize particles max. % wt	Undersize particles max. % wt Total	Fines (dust) <3.15 mm
1	25–200		10	5
2	10–100		10	5
3	3.15–50	10[a]	5	
4	3.15–25		5	
5	max. 3.15		—	

[a]See note to Table 25.

Table 27. Ferro-molybdenum to DIN 17 561, ferro-tungsten to DIN 17 562, ferro-vanadium to DIN 17 563, ferro-titanium to DIN 17 566, and ferro-niobium to DIN 17 569

Particle class	Particle size range mm	Oversize particles max. % wt	Undersize particles max. % wt FeW	FeV	FeMo	FeNb	FeTi
1	3.15–100		3	3	3	5	10
2	3.15–50	10[a]	5	5	5	10	10
3	3.15–25		5	8	5	10	10
4	max. 3.15		—	—	—	—	—

[a]See note to Table 25.

3.4.5 Summary of National and International Standards for Ferro-alloys

Table 28

Country Name of standard	West Germany DIN	Czechoslovakia CSN	France AFNOR NF	Japan JIS	USA ASTM	USSR GOST	ISO DIS
FeSi	17 560	42 2205/06	A-13-010	G 2302/1969	A 100-69	1415-70	5445
FeMo	17 561	42 2250		G 2307/1969	A 132-74	4759-69	5452
FeW	17 562	42 2260		G 2306-1969	A 144-73	17 293-71	5450
FeV	17 563	42 22551		G 2308-1969	A 102-73	4760x-49	5451
FeMn		42 2278	A-15-020	G 2301	A 99-66	4755-70	5446
	17 564	42 2228					
FeSiMn		42 2232	A-13-030	G 2304	A 701	4756-70	5447
FeCr		42 2240/2244	A-13-040	G 2303	A 101	4757-67	5448
	17 565			G 2317			
FeSiCr		42 2248	A-13-050	G 2315	A 482	11 861-66	5449
FeTi	17 566			G 2309-1969	A 324-73	4761-67	5454
FeB	17 567			G 2318	A 323	17 848-69	
FeNi	17 568			G 2316			
FeNb	17 569			G 2319-1969	A 550-73	16 773-71	5453

3.5 ELECTRODES AND RESISTANCE RODS

Kurt Kegel, Essen

H. Davy and J. G. Children first experimented with electric arcs between carbon electrodes around 1801–15. The electrode material was developed systematically, although it was originally intended for other uses. In 1842, R. Bunsen described the production of 'galvanic carbon' from coal and from 'baking' pit coal by a heating process, and at the same time recommended improvement of the material by densification. Infiltration with a sugar solution, followed by annealing, resulted in greater strength and density, and, in addition, the sugared carbon was converted into graphite by the high arc-temperature. In 1879 William Siemens built the first usable electric-arc melting furnace with a graphite crucible and a compressed carbon electrode. Starting in 1900, Héroult had considerable success in refining iron with his arc furnace. This increased the demand for carbon and graphite electrodes. The development of industrial graphite electrodes was also boosted by Atcheson's process of converting carbon into graphite in a resistance furnace. The principle of this process is still being practised today. It was used in Germany from 1916 onwards to make arc-furnace electrodes.

3.5.1 Raw Materials

Natural graphite occurs in South and North Korea, Austria, the USSR, and West Germany. It has a dihexagonal–bipyramidal lattice structure with a hexagonal stratified lattice grid.[1,2] The Mohs hardness is 1 and the density 2.26 g/cm^3. Graphite sublimes at 3650°C due to its low vapour pressure; it does not melt under normal conditions and oxidizes in air above 500–600°C. Its electric and thermal conductivity depend on direction; both are favourable in the direction of the strata. Natural graphite is hardly ever used for electrodes because its strength is poor compared with synthetic graphite or artificial carbon. Therefore electrodes are made from coal-tar and pitch products and from petroleum coke.

3.5.2 The Manufacture of Carbon and Graphite Electrodes for Arc Furnaces

Graphite has many different uses in industry because of its unique combination of physical and chemical properties. Graphite is a good electrical conductor and a superior lining and crucible material in the most varied media; it has a high thermal conductivity and sublimation temperature, a low elastic modulus, very good thermal shock resistance and hot strength, and can be machined easily.[3,4]

The production of electrodes with well-defined properties is mainly based on two raw materials. These are petroleum coke (an oil-refinery product) and coal-tar pitch. Anthracite and pitch coke are also used for special groups of materials. Thus industrially produced 'synthetic' graphite is a family of several materials with different properties. Every component essentially contains carbon of a different structure. The type and origin of the carbon affect grain-orientation, the size and number of pores, and the extent of graphitization. Accordingly, different properties result which are adapted for the use intended.[1,2,5] Specific qualities have been developed for graphite electrodes and resistance rods for electric furnace steel production.

An electrode is a two-component system: the mechanical part of the electrode and the graphite nipple.

The process for manufacturing synthetic graphite is shown schematically in Figure 1. The basic process is always adhered to, whatever the size of the finished product, but certain variations in materials and processing are made, according to the purpose of the final article and its associated properties. Electrodes are made in four stages. The starting material, crude petroleum coke, is calcined at temperatures up to 1300°C.

The lumpy coke product resulting is then crushed and ground and then graded according to grain size. Part is ground down into dust. The dry mix is then collected and coal-tar pitch is added in a heated mixing machine to produce the 'green' mix. This can be rammed, jar-rammed, and formed by flexible-plunger moulding or in a worm-extruder, extruding press, forging press, or isostatic press. Graphite electrodes are usually given their shape by an extruding press. Nowadays, graphite electrodes of larger diameter are also jar-ram-moulded, usually under vacuum. Electrode properties, especially anisotropy, can be affected by the compression process used. Electrodes made by the extrusion press have low electrical resistance, high thermal conductivity, and a low coefficient of expansion in the axial direction; however, these properties are relatively poor in the radial direction. With electrodes produced by jar-ram-moulding these properties are only slightly anisotropic. The cylindrial moulding, which is still ductile, is then cooled down to room temperature, when it is firm enough to be transported with ease. In this condition the electrode is called 'green'.

The green electrodes shaped in this way are now inserted into an annular kiln and carefully supported to avoid distortion during baking. The baking schedule is exactly specified with regard to heating up, steady heating, and cooling time. During heating up, the temperature range up to 500°C is critical, because the body resoftens in this region; the bonding media — tar, pitch, and perhaps synthetic resin — turn into coke, while gases are liberated which cause porosity; at the same time newly formed coke interlinks strongly with the original coke particles. The maximum temperature during this baking process is about 1300°C, and this stage takes about three weeks. The outcome is a hard, porous, brittle but stable synthetic carbon body, with well-defined properties; it can be used in this state as a carbon electrode in reducing furnaces or as an anode in aluminium electrolysis.

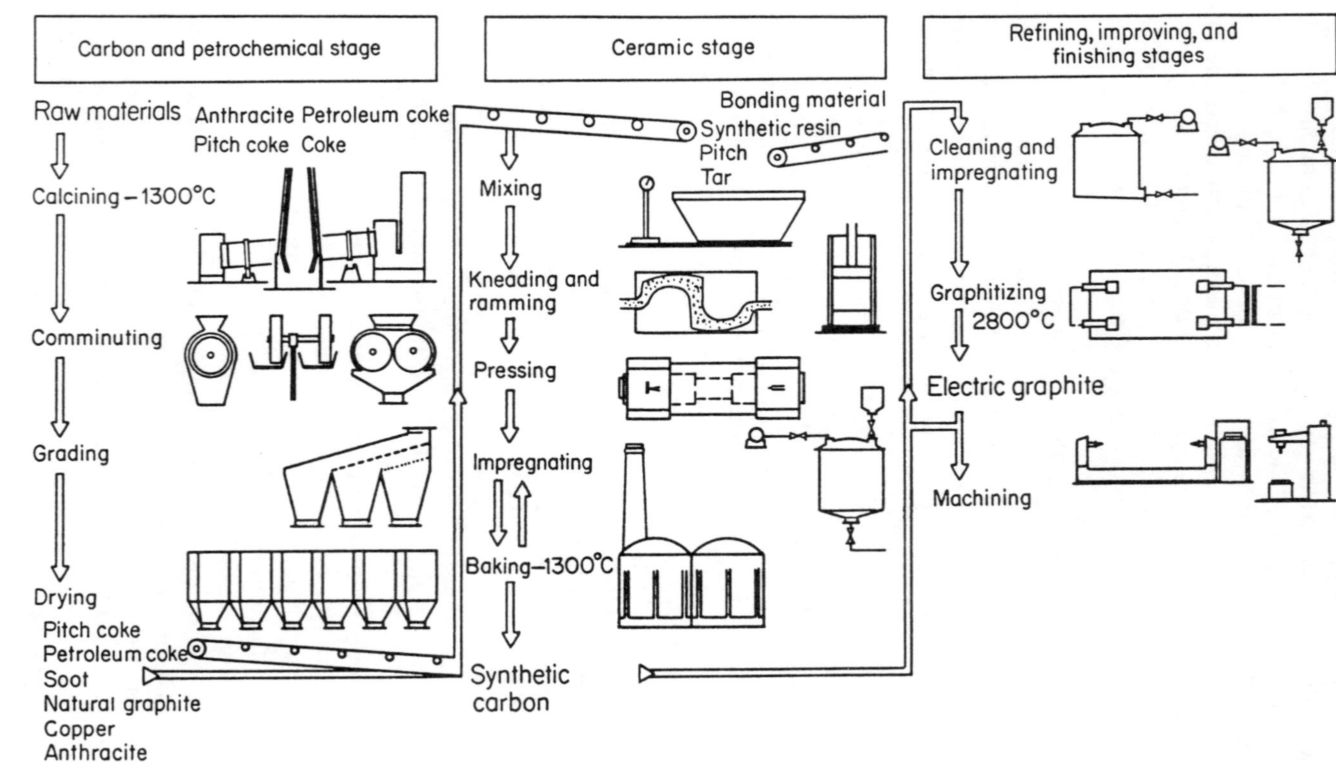

Figure 1. Flow chart depicting the production of synthetic carbon and electrode graphite

The strength and electrical properties necessary for the most highly stressed graphite electrodes for arc furnaces can be attained and improved by impregnation, i.e. filling the pores and voids with bonding agents. If necessary, this can be done before converting the synthetic carbon product into graphite.

This stage — graphitizing — is performed in special resistance furnaces at 2700–3000°C. The conversion into graphite begins at 2200°C; the crystal growth and arrangement are predetermined by the internal structure of the 'synthetic' carbon. The process depends on the interaction of time and temperature. The graphitizing process takes about 2 to 3 weeks, during which eight to ten times more energy is consumed in making a tonne of graphite than is used to produce 1 tonne of electric-furnace steel.

The graphite electrode is stronger than its predecessor the carbon electrode, and has a lower resistivity. Machining is carried out with conventional machine tools.

3.5.3 Forms of Delivery and Technical Properties of Graphite Electrodes

The sizes and tolerances for common, commercially available graphite electrodes are laid down and standardized internationally in IEC publication 239 (DIN 48 601 in the Federal Republic of Germany). Working Group 14 of IEC Technical Committee 27 have revised IEC Recommendation 239 (Graphite Electrodes) and brought it up to date in the light of the development of the ultra-high-power arc furnace over the last ten years. Table 1 illustrates the last compromise proposal which was due to be ratified in 1981. The weight of a 600 mm diameter moulded electrode 2100 mm long is about 1000 kg.[5–9] The main quality classes satisfy the requirements of steelworks operations. In tolerance class S the electrode diameter takes into account the heavy loading between the contact jaws and the electrode at maximum specific power.

Double-tapered graphite nipples with three- and four-start threads and pitch-locking pins or tablets, or polythene bags packed with tough cement, are used for joining electrode sections. Working Group 14 for electrodes, mentioned above, approved nipples type TP3 and TP4 for electrodes of 250–550 mm nominal diameter; they only approved type TP4 nipples (four threads per 25.4 mm) for nominal diameters of 600, 650, and 700 mm.

In arc furnaces the electrodes convey the energy to the charge to be melted. To do this the graphite or carbon electrodes must have the following properties: strength, elongation, elasticity, thermal expansion, and, in a certain respect, the right dimensions. Data on these properties can be found in Table 2. Electrodes must have the greatest possible strength, also at high temperatures, and low resistivity. Good thermal conductivity is advantageous. All physical values depend upon temperature.

Table 1. Diameter and length of graphite electrodes

Nominal diameter	Specified diameter	Actual diameter				Flat point dimension B	Length		Remarks
		Tolerance class R		Tolerance class S[a]			Specified value	Permissable deviation	
		max.	min.	max.	min.				
(mm)	(mm)	(mm)	(mm)	(mm)	(mm)	(mm)	(mm)	(mm)	
75	76	78	73	76.5	75.5	72	1000, 1200, 1500	+ 50 −100	
100	102	103	98	102.5	101.5	97	1000, 1200, 1500	+ 50 −100	Trend continues
130	130	132	127	130.5	129.5	126	1000, 1200, 1500	+ 50 −100	
150	152	154	149	152.5	151.5	146	1200, 1500, 1800	+ 75 −100	
175	178	179	174	178.5	177.5	171	1200, 1500, 1800	+ 75 −100	
200	203	205	200	203.5	202.5	197	1200, 1500, 1800	+ 75 −100	
225	229	230	225	229.5	228.5	222	1200, 1500, 1800	+ 75 −100	

250	254	256	251	254.5	253.5	248	1200, 1500, 1800	+ 75 −100
300	305	307	302	305.5	304.5	299	1500, 1800, 2100	+ 75 −100
350	356	357	352	356.5	355.5	349	1500, 1800, 2100	+ 75 −100
400	406	408	403	406.5	405.5	400	1500, 1800, 2100	+ 75 −100
450	457	460	454	457.5	456.5	452	1500, 1800, 2100 1800, 2100, 2400	+ 75, −100 +100, −150
500	508	511	505	508.5	507.5	503	1800, 2100, 2400	+100 −150
550	559	562	556	559.5	558.5	554	1800, 2100, 2400	+100 −150
600	610	613	607	610.5	609.5	605	2100, 2400, 2700	+150 −150
650	660	663	657	660.5	659.5	655	2100, 2400, 2700	+150 −150
700	711	714	708	711.5	710.5	705	2100, 2400, 2700	+150 −150

[a] In some parts of Europe the tolerances are less.

Table 2. Technical data for electrode graphite

Quantity	Unit	Average values
Resistivity ρ	$\Omega mm^2/m$	5–12
Compressive strength σ_c	N/mm^2	10–35
Bending strength σ_b	N/mm^2	4–20
Tensile strength σ_t	N/mm^2	2–8
Density, apparent γ	t/m^3	1.55....1.75
Density, true γ_t	t/m^3	2.20....2.25
Porosity P	%	20–30
Elastic modulus E	MN/mm^2	3–12
Brinell Hardness (10/100HB)	daN/mm^2	4–10
Thermal conductivity λ	W/m·K	120–250
Coefficient of thermal expansion 20…500°C α	l/K	
in axial direction		1.2–2.5
perpendicular to axial direction		2.2–2.8
Thermal shock resistance		Very good
Specific heat capacity c	Wh/kg·K	
at 50°C		0.22
at 1500°C		0.55
Softening temperature	°C	None
Melting-point	°C	None
Sublimation temperature	°C	3560
Total content of ash-forming impurities A	%	max. 0.5

Figure 2 shows how thermal conductivity, mean specific heat, mechanical strength, elastic modulus, and thermal expansion coefficient vary with temperature.

3.5.4 Testing of Electrodes and Nipples

Consistent equality cannot be assured only by testing the end-product. The raw materials have to be tested on receipt to check their constituents and grain structure. Above all, the sulphur content requires special treatment in production. Density, bond strength, and the percentage of volatile ingredients are determined with sample mouldings. Then follows a check of the size of pre-baked electrodes; any that are oversize are rejected. After graphitizing, the electrode surface is turned on a lathe. The diameter must lie within relatively narrow tolerances, but length is less important. Special attention must be given to the threads of the nipples and sockets. Finally, the electrodes are crack-tested and checked for resistivity. Non-destructive resistivity tests are made in several places to check consistency and to establish an average resistivity value, since resistance contributes to losses in service.[8–10]

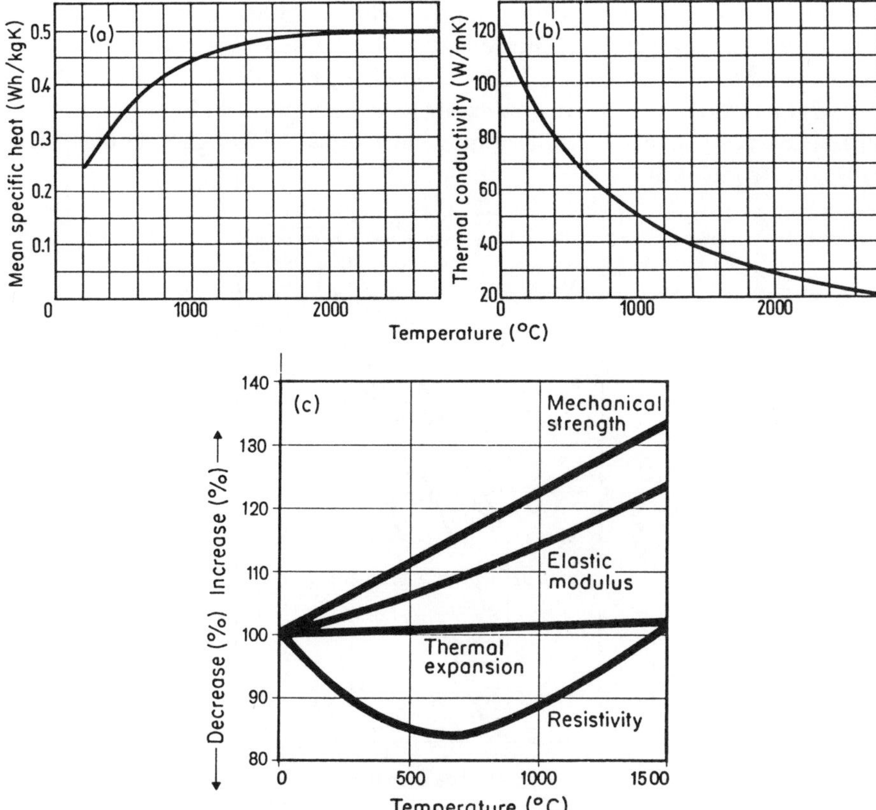

Figure 2. Average specific heat (a), thermal conductivity (b), and various mechanical quantities (c) for graphite in relation to temperature

3.5.5 Energy Rating and Current-carrying Capacity

The current-carrying capacity of an electrical conductor is determined by current density, power, cooling conditions, and the material itself. The current density of graphite is very much less than that of metals. Any unnecessary heating of the electrodes caused by high resistivity must be avoided. The cooling conditions for the electrodes in an arc furnace are far from ideal. The arc temperature is 4500–6600°C, and the electrodes are heated by heat conduction. The temperature of the electrode tip is far above 2000°C;[11] in fact, temperatures of 2000°C have been measured in an arc furnace 80 cm from the electrode tip. The current is not evenly distributed over the electrode cross-section because of skin effect and proximity effect. It has been found[12] that in a 500 mm electrode, skin effect causes a local increase of about 20% in resistance and reactance. This raises the question of whether it makes sense to use large-diameter electrodes, though the nominal resistance will be lower. For a 500 mm electrode the current density will be 25.5 A/cm^2 at a current of 50 kA. In reality, local current density will be much higher through current displacement (see section 4.1.1.2). In practice, load diagrams are drawn for various electrodes, which are applicable to any arc furnace. Figure 3 shows the current-carrying capacity of graphite electrodes.[12,13] Higher current loads cause electrode wear.

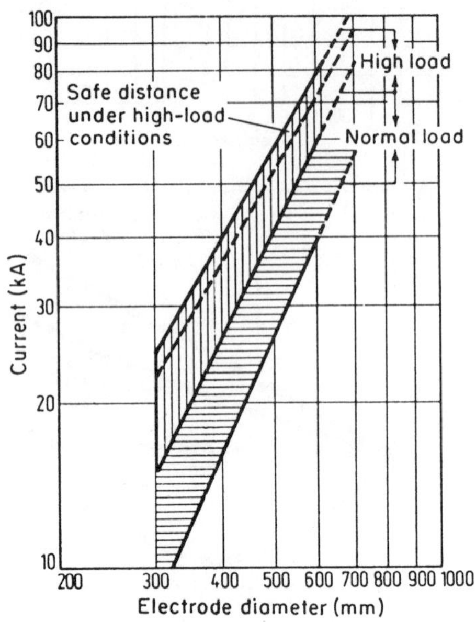

Figure 3. Current-carrying capacity of graphites electrodes under conditions of high and normal loads

3.5.6 Nippling

Nippling is a method of connecting two pieces of electrode with a tapered or cylindrical nipple; cylindircal nipples are rare nowadays. This connection must fulfil two important conditions: it must be mechanically stable and must not cause any additional resistance. A sound mechanical connection is achieved by joining the carefully fitted parts of lower electrode, nipple, and upper electrode, after thoroughly cleaning the threads and faces with dry oil-free compressed air, and then tightening up. Care must be taken to ensure that the three parts to be connected are all precisely vertically aligned and screwed together, so that the seating is correct and the current density is even (see Figure 4C). If there is the least amount of binding when the parts are being joined, then they must be unscrewed again, and the obstacle removed. When

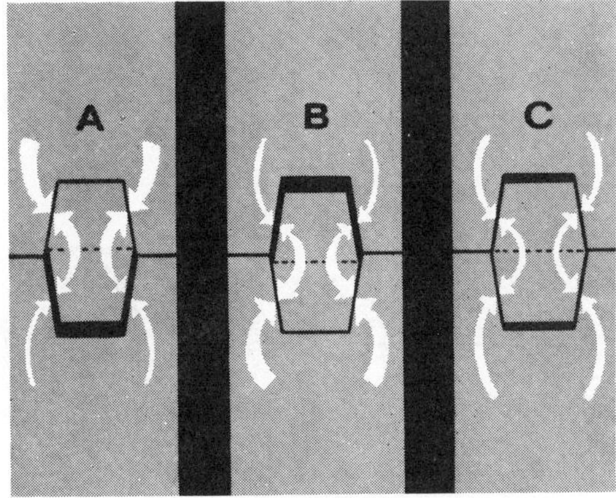

Figure 4. Special phenomena with nippled electrodes. (Works photo by Union Carbide)

Table 3. Controlled torque for wrenches or mechanized equipment (Union Carbide)

Electrode diameter (mm)	Torque (Nm)	Electrode diameter (mm)	Torque (Nm)
75	22–28	400	850–950
100	45–55	450	1050–1450
150	100–125	500	1300–1800
200	200–260	550	1700–2400
250	350–500	600	2100–2800
300	500–700	650	2400–3400
350	700–950	700	2800–4000

tightening up, the correct torque, which depends on the diameter, has to be applied with appropriate tools (Table 3, Figure 5).[8,9] A direct mechanical connection is usually not adequate. It is advisable to lock the parts together indirectly using tablets, studs, or cushion inserts. These jointing aids are made of pitch and melt at about 110°C; they either expand in volume so that the nipple and electrodes 'grow together' or they coke at above 350°C and thus form a permanent joint. This method of connection should, if possible, be used on cold electrodes; otherwise it has to be done fast enough to prevent premature coking in hot electrodes.

Nippling can be carried out in the pit beside the furnace or on the furnace itself. Both methods have advantages and disadvantages. When nippling is performed on top of the furnace, there will be less thermal shock to the electrodes; on the other hand, nippling beside the furnace can be done with more care and

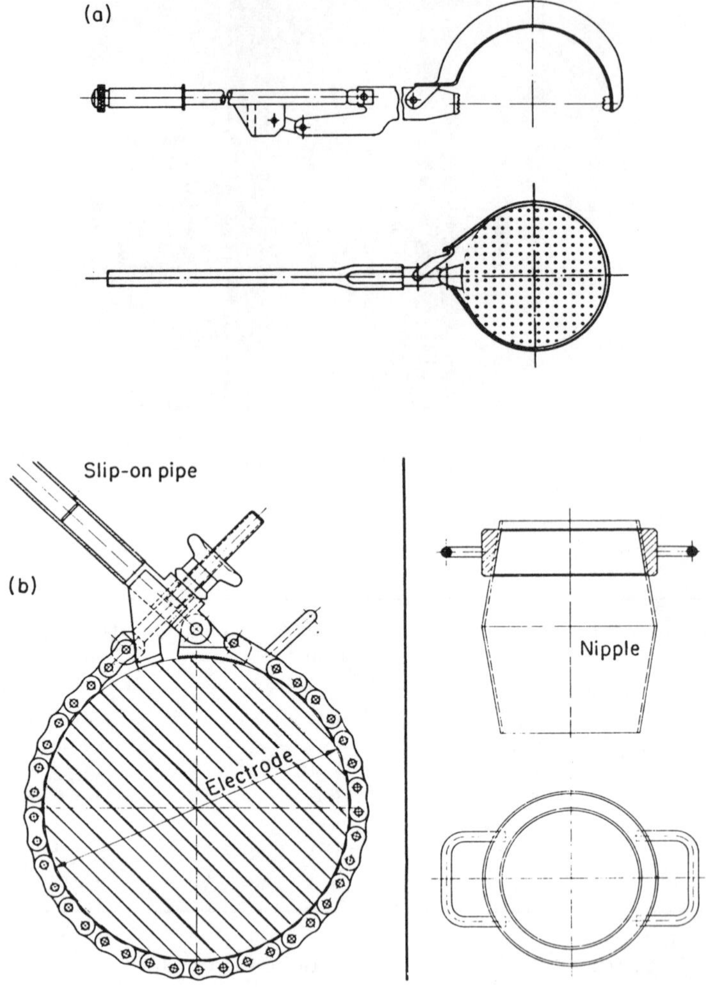

Figure 5. Tools for screwing electrode sections together. (a) Adjustable torque wrench, electrode spanner (suitable for pneumatic operation) (drawings by SIGRI); (b) electrtode wrench; nipple carrying ring for electrodes (drawings by C. Conradty)

attention. Figure 6 illustrates tools which enable the electrode parts to be handled gently.

3.5.7 Special Treatment of Electrodes

Hollow electrodes These are ordinary electrodes, drilled out so that the arc starts in the bore. In this way we get higher local temperatures than with solid

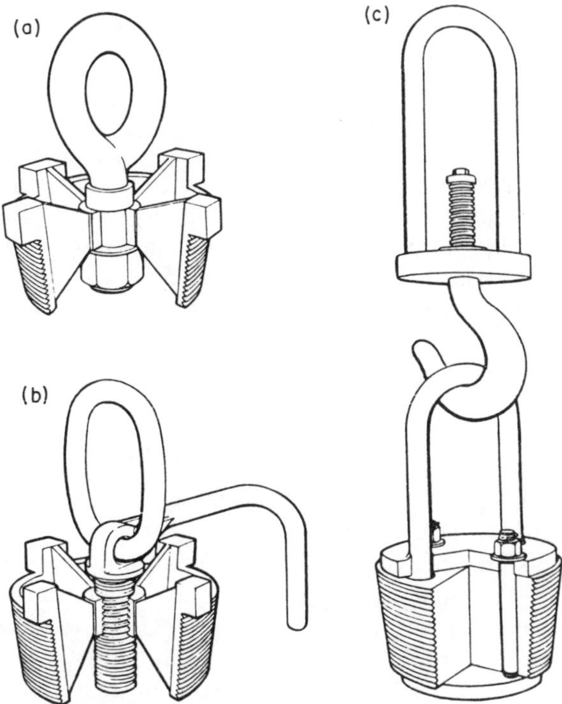

Figure 6. Tools and crane gear for connecting and transporting electrodes (drawings by Union Carbide)

electrodes, which results in more intense electron emission, thus stabilizing the arc. Hollow electrodes help to protect the refractory wall-lining; one disadvantage is a higher specific wear-rate.[12,14]

Doped electrodes During the production of these electrodes additives are mixed into the graphite which reduce the electron work-function, producing a more stable arc. However, there is one condition which considerably limits the choice of material: the dopant must withstand the highest electrode temperatures. A high degree of very short-lived success resulted with alkali-impregnated electrodes, using thorium oxide (which is unfortunately radioactive) and with additions of titanium oxide. The titanium oxide results were the most promising; however, it was not possible during the trials to solve the mechanical problems. Further experiments are necessary, probably with other materials or different dosages.[13,15]

Coated electrodes To reduce wear on the sides, electrodes are coated with an electrically conductive material. Such electrodes are already used in many steelworks, but they require special contacts and dust-protection.[8]

3.5.8 Graphite Rod Heating

There are a few processes where graphite rods are used as radiation heating elements. This applies to small melting furnaces and vacuum installations which are heated by thermal radiation. The heat source is the graphite rod, heated by its own resistance (Joule's Law). It gives up its energy by radiation (Stefan–Boltzmann Law). Today such graphite rods are used at temperatures up to 2200°C. The heating rods are self-protected by an envelope of carbon dioxide, and therefore last a long time.[8,9]

3.5.9 Permanent Electrodes

There have been frequent attempts to develop permanent electrodes for arc furnaces. A permanent electrode consists of a water-cooled copper tube, which is the electrode stem, carrying an expendable tip of graphite, tungsten, or a similar material. A stable arc requires very high temperatures to maintain a sufficient flow of electrons. About 1800–2000°C are required for tungsten; however, this leads to rapid oxidation. Graphite oxidizes more slowly under similar conditions, so it is seldom possible to use any material but graphite. The graphite tip on the water-cooled stem wears rapidly, so continuous operation is not possible. On the other hand, a permanent electrode with a copper tip is satisfactory for continuous operation in a melting furnace under a protective atmosphere or vacuum.[16,17]

The huge rise in the price of crude oil, which is the main source of raw materials for graphite electrodes, together with the accompanying rise in energy costs, has recently revived and spurred development work for the carbon-electrode industry. This work is extended to enable alternative raw materials and energy-saving manufacturing processes to be adopted and consumption in arc-furnace operation to be cut. Almost all the development work still continuing is aimed at the eventual production of a compound electrode consisting of a metallic upper part with a consumable tip screwed on the end; the tip would be made from graphite or a composite material suitable for extreme temperature operation.[18]

3.6 REFRACTORY MATERIALS

Aleksander Majdič, Bonn

By international agreement, refractory materials are defined[1–6] as ceramic products whose refractoriness, i.e. pyrometric cone softening-point, is not less than 1500°C (ISO/R 836—1968). Certain ceramic products do not comply with this definition because their refractoriness is less than 1500°C; however, they have all the right characteristics otherwise. Such products are now increasingly called scale- or heat-resistant. Refractory and heat-resistant products may be bricks that are shaped and fired, or unfired (chemically bonded), or, more rarely, melted and cast; otherwise they are amorphous cements and compounds. Unshaped products are shaped and fired at the place where they are to be used, and they then attain the requisite properties for their use. The internationally defined limit of 45 vol.% total porosity distinguishes sealants (<45 vol.%) from heat-lagging materials.

Refractory materials for secondary refining furnaces, and for parts in contact with molten metals and slag, are usually exposed in use to a multitude of cumulative stresses. It is only possible to select an appropriate refractory lining if the stresses inherent in a particular application and the properties of the refractory are known as accurately as possible.[7–11]

It is impossible to optimize every property. A compromise has to be found between properties which are unfavourable in the quest for long lining-life. The characteristic data are measured by various methods, mostly standardized.[12,13]

The life of a refractory lining is limited, on the one hand, by corrosion, erosion, and softening, and on the other, by cracking, which leads to spalling or loss of adhesion of the ceramic material. The effects of vapour, slag, and the melt can produce pronounced zoning and thereby change considerably the original characteristics of a refractory. The design of furnaces influences the life of refractory linings more and more.

3.6.1 Characteristic Properties

Usually the properties of a refractory material can be derived from its structure, although the connections are not always clearly recognizable. The material properties depend on the type and percentage of mineral components, and on their size, shape, and arrangement, and their porous structure. Data about the properties of refractory materials are shown in Table 1.

The chemical and mineral composition form the basis for classification and for fundamental assessment of behaviour in operation, e.g. resistance to slag attack. With reasonably pure base materials, the effects of chemical composition are usually less important than the mechanical properties, but with demands for higher quality even very small amounts of foreign oxides can greatly influence the result.

For bricks in the silica–fireclay–corundum range, the *refractoriness* (pyrometric cone value) is, for a given Al_2O_3 content, an approximate guide to fluxes present. Pyrometric cone values are of no practical importance for basic and other highly refractory products.

Table 1. Reference data on chemical and physical properties and behaviour of refractory materials

Refractory type	Chemical composition (% wt) SiO_2	Al_2O_3	TiO_2	Fe_2O_3	CaO	MgO	Cr_2O_3	Na_2O+K_2O	Density (g/cm^3)	Apparent density (g/cm^3)
Ceramically bonded quartz products	>99	0.2	0.2						2.20	1.8–1.9
Silica bricks	>95	0.3–2	0–2	0.5–3	1.5–3.5			0.05–0.3	2.30–2.38	1.7–1.95
Acid fireclay bricks		20–30		<2.5	0.5–1.0	0.5–1.0		2–3.5	2.50–2.65	1.9–2.3
Fireclay bricks		30–45	1.2–5	0.6–3	0.5–1.0	0.5–1.0		0.5–2.5	2.55–2.75	1.8–2.45
Sillimanite bricks		60–65	<3	<1.5				<1	3.05	2.3–2.5
Mullite bricks		70–75	<1.5	<1				<1.2	~3.2	~2.4
Bauxite bricks		80–85	<3.5	<2.5				<0.5	~3.6	2.6–2.8
Corundum bricks (80–85% Al_2O_3)		80–85	<2.7	<1.5				<1.3	3.5–3.7	2.7–2.9
Corundum bricks (>95% Al_2O_3)		>95	<0.2	<0.5				<0.5	3.96	2.9–3.1
Chrome–alumina bricks		80–93					7–20		~4.0	~3.2
Magnesite bricks	0.2–6	0.1–5		0.2–6	0.8–3.9	85–98			3.5–3.8	2.85–3.1
Magnesite–chrome bricks	<5	<11		5–12	<3	55–80	6–20		3.7–3.85	2.8–3.1
Chrome–magnesite bricks	<5	<22		7–15	<2.5	25–55	15–35		3.8–4.0	2.8–3.2
Chromite bricks	<7	<28		9–18	<2	<25	>25		3.8–4.1	3.0–3.3
Forsterite bricks	<40	<8		<8	<1.5	50–60	<1.5		3.3–3.4	2.6–2.85
Sintered dolomite bricks	<1.5	<1		<1	<61	>36			3.38	2.6–2.9
Fused lime bricks	0.3–1	0.3–2		0.3	96–97.5	0.2–1			3.30	2.87
Fused magnesite–chrome cast bricks	2.4	7.2		13	1.3	56	20		3.9	3.15
Zircon–silicate bricks (62–5% ZrO_2)	33–34	0.5–1							4.60	3.6–3.8
Silicon carbide bricks (90% SiC)									3.1	2.5–2.8
Carbon bricks (>90% C)									1.9	1.5–1.7
Graphite bricks (99% C)									2.2	1.5–1.7
Light refractory silica bricks	>93									0.8–1.25
Light fireclay bricks		30–45							Corresponding to dense bricks	0.5–1.3
Light aluminous bricks		60–99							~2.6	0.8–1.6
Aluminous silicate fibre materials		~45								0.05–2.2

[a]Typical values in oxidizing atmosphere with no corrosive components; heated on one side.

Physical characteristics							Reaction to chemical attack			
Open pores (vol. %)	Compressive strength cold (N/mm^2)	Heat resistance in compression (°C)	Thermal expansion at 1000°C (%)	Mean specific heat 20–1000°C (J/(kg·K))	Thermal shock resistance	Maximum temperature limit in use[a] (°C)	Basic slags	Acid slags	Molten metal	Others
14–18	50–80	1400	<0.01		Very good above 1000°C	1400 (up to 1650 after devitrification)	Severely attacked by basic slag. Good resistance to alkaline vapour	Good resistance	Good	
17–24	20–70	1620–1700	1.0–1.5			1650				
14–25	>35	>1250	~0.9			1200–1450	Attacked	Mildly attacked	Not to be recommended for use inside furnaces over 1300 °C	
12–29	15–85	1300–1500	0.4–0.6	1.05–1.15		1200–1450	Attacked	Mildly attacked		
18–23	>40	>1550	0.55			1550–1700	Satisfactory resistance; more severely attacked as basicity increases	Good resistance	Good resistance	
19–22	>40	>1600	0.55		Good	Up to 1750				
22–24	>40	>1450	0.6			1500–1600				
16–23	>50	>1600	0.6			1750				
18–23	>40	>1700	0.8			1800				
~18	~70	>1700	0.8				Highly resistant to CaO–FeO–SiO_2 slags			
13–18	40–115	1560–1700	1.2–1.3	1.15–1.34	Poor to moderate	1450–1850				
16–23	>20	≥1500	~1.0	1.04–1.12	Moderate to poor	1600–1750	Very good resistance	Attacked		
16–23	>15	≥1500	~0.9	1.04–1.12	moderate to poor	1600–1750				
17–23	>30	≥1500	0.8	0.93	Low		Good against basic and acid slags		Good resistance	Sensitive to hydration
15–23	>25	≥1600	1.1	1.05	Moderate	~1600	Little attack	Heavily attacked		
14–25	30–80	>1700	1.35	1.00	Good	1500–1700	Good resistance	Slightly attacked		
13	60	>1700	1.35	0.94	Good		Good resistance	Slightly attacked		
6–14	40	>1700	~0.9	~1.0	Poor		Little slag attack because of dense structure			
17–22	20–100	>1500	0.4	0.75	Good	Up to 1650	Good resistance			
15–22	60–80	>1700	0.55	1.1	Very good	1600	Attacked by FeO, CaO and MgO			
13–16	30–100	No softening	~0.25	>1.5	Good	~600 (2000 in reducing atmosphere)	Good to excellent resistance if no oxidizing components present		Not wetted	Liable to oxidation
13–16	27	No softening		>1.5	Good					
	2.5–6					1500				
Over 45%	2–15		Somewhat less than dense bricks	Similar to dense bricks	Fair	1250–1400	Not generally resistant to slag attack because of high porosity. Prone to wear			
	2–10					1500–1800				
					Very good	~1260				

The *apparent density* (weight per unit volume) of a product (for non-shaped products after firing at selected temperatures) is decisive for calculating the quantities required and for economic assessments. The apparent density is affected by the density of the base material and the porosity of the product.

Porosity has a considerable effect on properties resulting from the mineral and chemical structure; every product is affected in the same direction. Cold compressive strength generally decreases with increasing porosity, while permeability and liability to slag attack increase; on the other hand, thermal lagging effect is usually improved.

The *cold compressive* and *cold bending strength* generally give a good indication of how bricks are made (moulding pressure and firing) and about the regularity of structure to be expected in a given batch. However, these properties yield little information on the behaviour of the product in operation and at high temperatures.

The *dimensional accuracy* of the bricks is the result of their manufacturing process, including the firing. In general, if there is much shrinkage during drying (e.g. plastic production) or during the firing process (e.g. with sintered alumina products), less accuracy must be expected. If the dimensional accuracy is inadequate with normal production methods, then the manufacturer or the user must take additional steps, such as grinding or grading.

Thermal expansion is virtually constant for the type of material. It is determined mainly by the base material of the refractory, for example by silica modifications (SiO_2) and periclase (MgO). Whether expansion is reversible or permanent is important for determining the width of the expansion gaps. The expansion gap-width is usually a proportion (0.5–0.8) of the expansion measured in the laboratory, but depends on design and furnace operating conditions. On the first heating of unshaped refractories there is a mixture of reversible expansion and irreversible expansion and shrinkage. Even low compression stresses can markedly affect the expansion/shrinkage behaviour.

Volume stability, i.e. secondary expansion and shrinkage without external load (under own weight), are usually tested with the product heated on all sides; however, in practice, refractory materials in brickwork are usually heated from one side only. Thus it could happen that, under the influence of the cooler side, the shrinkage will not only be less than expected from tests with heating from all sides, but in service some materials will actually grow in volume. *Secondary shrinkage* takes place over a long period which cannot be predicted from laboratory testing. We must bear in mind that in *secondary expansion* due to conversions or reactions, appreciably smaller values are obtained when the temperature rises slowly or when the material is under load. The operating temperature is often set in service by secondary shrinkage of 1–1.5%.

Softening under pressure, which is measured under constant heating and load conditions, is governed by various factors. These are the crystal structure of the refractory matrix, the amount of material, the viscosity, the wetting behaviour of the glassy phase, and, to a considerable extent, by porosity and the arrangement of the pores. It is therefore clear that the firing system and the nature of the furnace gases (oxidizing, reducing) will also exert a strong influence. These different influences underline the fact that softening behaviour cannot be studied in isolation from other properties, and that it can vary markedly in service.

Load-flow (creep) is softening of the refractory under constant temperature and load sustained for a long period, 50 h, for example. The rate of flow decreases with time. There are exponential relationships between flow, sagging, time of observation, load, and test temperature.

Hot strength is the most important thermomechanical characteristic in practice; it is affected by the distribution, size, and nature of the crystals, the size and composition of the melts, the nature of the bonds, and the structure. Hot strength thus consists of complex integrated effects; behaviour in relation to temperature varies according to how the load is applied (bending, compression, or torsion). The rate of loading at a given temperature also influences the start of deformation or flow. Hot bending and hot-compression strengths have become increasingly important for the assessment and selection of a refractory lining. Of special importance for crack- and tear-formation is the temperature at which the material behaviour changes from brittle–elastic to quasi-viscous (glass-point). We try to understand fatigue and flow processes that lead to stress relief by measuring *stress–strain behaviour* in relation to temperature.

Thermal fatigue results from the sensitivity of materials to repeated heating and cooling or to temperature fluctuations. It depends on the physical and mechanical properties of the material (the coefficient of thermal expansion is especially significant) and on its pore structure and grain structure. The resistance of non-porous materials to thermal fatigue is generally low; on the other hand, highly porous materials (light refractory bricks) are often sensitive to temperature fluctuations. Bricks with good thermal fatigue-resistance are mechanically weaker because of their higher porosity and their particular grain structure. Sensitivity to heating or liability to thermal fatigue also depend on brick shape and loading and increase rapidly with brick size. The original thermal fatigue characteristics of a material are modified to such an extent by operational conditions, and especially slagging, that fatigue data have only limited value in predicting behaviour in new applications.

Contact reactions can occur between bricks of different compositions; this is, however, usually affected by slags and the furnace atmosphere.

Various methods are used to test the resistance of refractories to slag attack; the method chosen depends on the process used. It should be remembered that most short-duration laboratory tests are performed in a homogeneous temperature field; they therefore cannot simulate changes in the material over extended periods and under the influence of temperature differences. The disadvantage of laboratory tests intended to simulate real conditions is that each test is only meaningful for one particular case, and the difference in behaviour between small experimental furnaces and large works furnaces cannot always be foreseen.

Thermal conductivity is important for energy housekeeping in refining systems. It should be low, as a rule, so that heat losses through conduction to the exterior are kept to a minimum. If necessary, a two-layer refractory wall must be built, using insulating materials. In some cases high-thermal-conductivity products (graphite, silicon carbide) are used to reduce the temperature and wear on the hot side of the lining to a minimum. Moreover, the thermal conductivity depends not only on the chemical composition but also greatly on pore volume, texture (direction of graphite lamellae), and on the prevailing atmosphere. At lower temperatures, porous materials have a lower thermal conductivity than when they are hotter. The most important thermal conductivity values of refractory bricks are shown in Figure 1. The thermal conductivity of unshaped products is generally lower than with fired bricks of the same type.

The *electrical resistance* of refractories is important for the insulation of electric furnaces. It is very high at room temperature, but drops with rising temperature. Figure 2 shows reference data on the electrical resistance of the most types of brick.

3.6.2 The Various Types of Refractory Product

3.6.2.1 Silica bricks (>93% SiO_2)

Silica bricks are made of crushed quartzites, sandstone, flint, etc. bonded in almost every case with slaked lime. The various quartz raw materials differ in conversion behaviour; this affects the properties of the finished product. Calcined silica products consist predominantly of crystalline SiO_2 which occurs in the modifications cristobalite, tridymite, and residual quartz. The liquid fraction increases distinctly only above 1650°C, which means that these bricks are highly refractory. On account of different, sometimes abrupt, changes in the SiO_2 modifications, silica products are liable to damage during heating up and cooling below 600°C. If the modification change is incomplete (residual quartz), after-expansion will result above 1300°C, the extent of which depends not only on the raw material and amount of residual quartz, but also on heating-up rate and on the compression loading.

3.6.2.2 Fireclay bricks (10–45% Al_2O_3)

The majority of all refractories are still made from fireclay products. They are made from fireclay–clay mixtures, first moistened with water, then moulded or

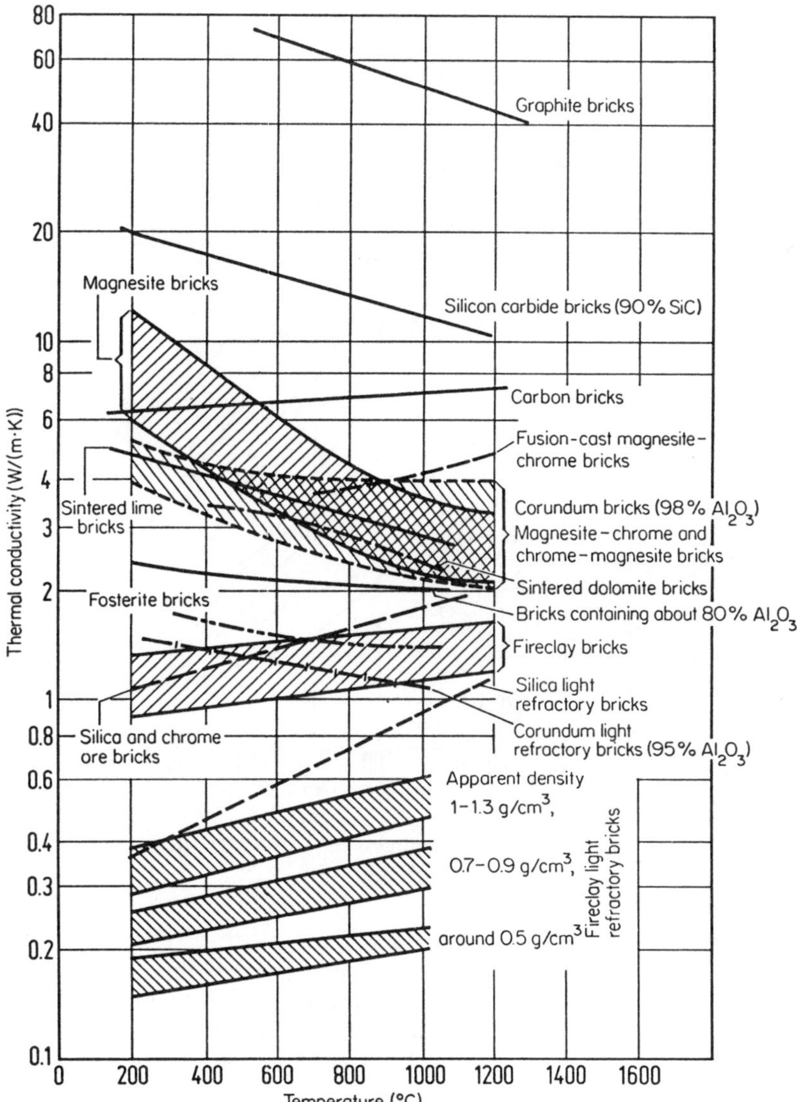

Figure 1. Thermal conductivity of refractory bricks in air, in relation to temperature

tamped in a plastic, semi-dry, crumbly, or dry condition. The properties of fireclay bricks can be varied through changes in the ratio of clay to fireclay, the grain structure and the firing temperature. They are composed of mullite, glass (potassium alumino silicate), cristobalite, and residual quartz. The glass content is between 25% and 60%, depending mainly on the alkali in the raw

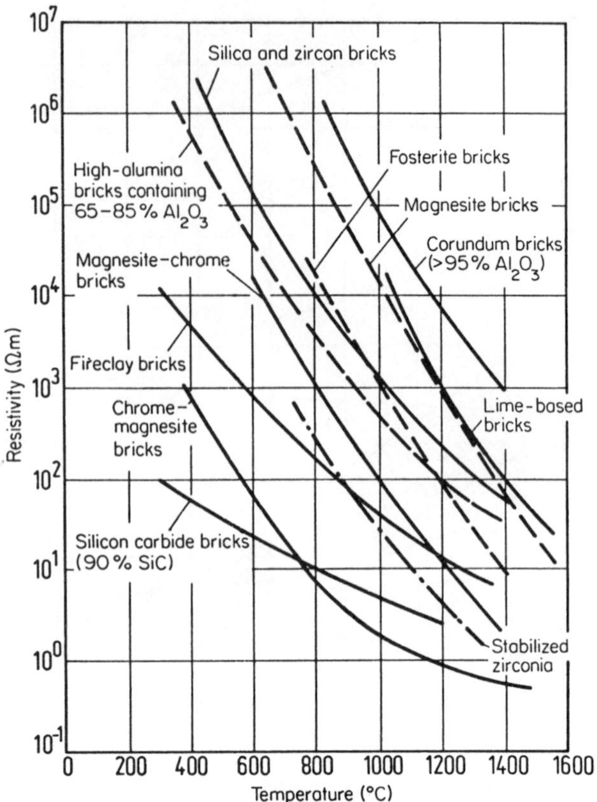

Figure 2. Resistivity-changes of refractory products with temperature

materials. This is why fireclay bricks heated on all sides already begin to soften under load at 900–1100°C. A higher Al_2O_3 content, low flux content, and high firing temperature all help the material to withstand purely thermal stresses. In the case of slagging and erosion, the significance of alumina and flux content has been found to vary from case to case; low porosity is always desirable in case of corrosion.

3.6.2.3 High-alumina bricks (>45% Al_2O_3)

The alumina content of fireclay bricks is limited by the composition of kaolin to about 45% Al_2O_3. Higher percentages of alumina are obtained by using higher-alumina, granular constituents such as calcined bauxite, sillimanite, mullite, or corundum, which are usually bonded with clay or with special cements, such as mixtures of clay, fine sintered alumina, or phosphates. The properties of high-alumina bricks of similar chemical composition vary widely, according to the raw material used, the grain structure, cement, and firing conditions. Very pure, dense high-alumina bricks can be made from synthetic mullite and pure corundum, with excellent high-temperature properties.

3.6.2.4 Magnesite bricks (>80% MgO)

The magnesia raw materials (sintered magnesite, magnesia from sea-water) and their end-products contain various amounts of Fe_2O_3, CaO, SiO_2, and Al_2O_3. The quantities and proportions of minor constituents determine the mineral structure and with it the high-temperature properties and the behaviour with slag. The ratio of CaO to Al_2O_3 (C/S) is of particular importance, as it characterizes the silicates which are present in addition to periclase (MgO), spinel (MgO Al_2O_3), and ferritic magnesia. Of these silicates, fosterite (2MgO SiO_2) and di-calcium silicate (C_2S) have high melting-points, and monticellite (CMS) and merwinite (C_2MS_2) together with calcium ferrite (C_2F) — which may be present — have low melting-points. Even small amounts in a melt (5–10%) cause early softening, because the melt coats the roundish periclase particles so that they can slide past one another. Magnesite bricks have high thermal expansion (about 2% at 1400°C), which means they also have low thermal shock-resistance; this can be improved by special grain-coarsening processes and additions of Al_2O_3 and Cr_2O_3. They are sensitive to moisture and steam below 300°C, but resist attack by basic slags and metal oxides. High-temperature fired, low-iron magnesite bricks (Fe_2O_3 about 0.5%) deserve special mention here.

3.6.2.5 Magnesite–chrome, chrome–magnesite (25–80% MgO), and chrome ore bricks (<25% MgO, >25% Cr_2O_3)

Basic chrome–magnesite products contain not only the mineral phases found in magnesite bricks, but also complex Cr–Al–Fe spinels. As with magnesite bricks, the mineral structure is primarily determined by the C/S ratio. The properties of bricks consisting of sintered magnesite and chrome ore cannot be determined from the two main constituents, because they form a non-homogeneous structure which depends on composition and grain-size. Various granulations have made possible successful production of a series of magnesite–chrome ore bricks with a thermal shock-resistance in some cases close to that of fireclay bricks. By reason of their high oxide content, a frequently changing atmosphere can cause growth and embrittlement of these bricks. With high firing temperatures of about 1750°C, or by using fixed mixtures of magnesite and chrome ore, a structure is obtained with a high proportion of bonded periclase/periclase and periclase/chrome spinels. Compared with normally fired, silicate bonded, magnesite–chrome bricks, this type of brick has a much better high-temperature strength, better volume-uniformity, and much less liability to soften. Fired products are often replaced by chemically bonded bricks and particularly internally reinforced bricks, clad with steel sheet. Chrome ore bricks have acquired only limited use. They are very durable when exposed to acid slags, but are specially sensitive to iron-oxide attack above 1500°C (considerable growth). Because of their neutral behaviour, they are used as buffer bricks between chemically different refractories.

3.6.2.6 Sintered dolomite bricks (about 35% MgO, about 60% CaO)

Sintered dolomite products do not keep well in store because they are inclined to hydrate. This can be countered by dipping in tar and by tempering tar or pitch-bonded dolomite bricks. In many cases, dolomite and magnesite products can be substituted for one another.

3.6.2.7 Lime products (>95% CaO)

Products made from molten lime (crystal lime) are important for vacuum induction-melting of low-oxygen steel because of their high purity and low vapour pressure.

3.6.2.8 Tar- and pitch-bonded bricks and tar-impregnated fire bricks

Bricks of this type in the magnesite–dolomite range are mainly used in oxygen-blown converters and in electric arc furnaces. In addition, fireclay and high-alumina bricks are impregnated with tar or pitch after vacuum treatment. The main characteristic of these bricks is that, when heated in the absence of air and/or when in use, a combined structure of refractory oxide components and carbon is formed, which appreciably improves resistance to slag attack. About 5% of carbon is separated by cracking.

3.6.2.9 Fosterite bricks (about 52% MgO and about 38% SiO_2)

The raw materials for these are olivine and dunite, a related rock. The physical properties are close to those of magnesite bricks. Certain properties can be improved by additions of MgO and Cr_2O_3. They are used in the metal-extracting and glass industries.

3.6.2.10 Zircon-bearing products

Products based on zircon ($ZrSiO_4$) and stabilized ZrO_2 are generally distinguished by good stability against slag and molten glass.

3.6.2.11 Carbon- and graphite-bearing products

Pure carbon products cannot be melted in a technical sense. They have high thermal conductivity, are highly resistant to slag and metal, and have a good thermal shock-resistance. However, they oxidize above about 600°C. Combined products, fundamentally of oxides and a large proportion of graphite (for example, magnesia–graphite bricks with 10–26% graphite), were developed for the side-walls of electric arc furnaces. Graphite–fireclay materials are mainly used for casting steel; increasing use is being made of graphite–alumina and partly of very complex high-graphite products, mostly compounds, for chutes and run-outs.

3.6.2.12 Silicon carbide bricks

These products have great hot-strength and high thermal shock resistance, high thermal conductivity, and good wear-resistance. After the formation of a protective layer of silica they resist oxidation up to about 1600°C. Their resistance to infiltration and attack by non-ferrous metals and silicate slags is good to excellent.

3.6.2.13 Fusion-cast bricks

These are mullite, corundum, corundum/zircon, corundum/mullite, magnesite/chrome, and chrome/corundum bricks. Compared with ceramically produced bricks, the main difference is in the intercrystalline bonding and interlocking of the mainly large crystals and that there are no open pores. They are therefore very resistant to attack by slags and melts, but their application is limited by poor thermal shock resistance.

3.6.2.14 Quartz products

The characteristic properties of quartz products are high purity, extremely low thermal expansion, and unusually high resistance to thermal shock and fatigue. Quartz products lose their thermal fatigue resistance by devitrification to cristobalte about 1100°C. Ceramically bonded quartz products are used for applications such as nozzles for continuous casting.

3.6.2.15 Light refractory bricks and insulating bricks

Light refractory bricks are classified in groups according to the recorded temperatures at which their linear dimensions shrink by no more than 2% in a standard 12-h test. Apparent density is an additional criterion. Strength, wear-resistance, and compressibility may also be important, depending on where they are to be used. The same applies to softening under load, because some low stress in the brick wall cannot be ruled out. Pre-fired materials are used in the production of most light refractory bricks. Special manufacturing processes are involved, such as foaming the ceramic mixture, which is normally fine-grained, and using expanded materials (bloating clay, hollow corundum pellets).

When light refractory bricks are used for backing linings, care should be taken to avoid a heat build-up on the hot side of the bricks because of their effectiveness in insulation. The temperature-drop across the inner lining will be much flatter than across ordinary bricks, and there is a danger that overheating might occur combined with greater softening and perhaps melting, together with slag attack. Use of commercial mortar can compensate part of this heat-lagging effect because of the higher thermal conductivity in the mortared gaps.

3.6.2.16 Ceramic-fibre materials

Ceramic fibres, normally with a maximum service temperature of 1260°C, are being increasingly used loose, or as pliable fabrics or mats, plates, or papers (permeable to air and partly to liquids), for insulation of particularly irregular parts, and where accessibility is poor.

3.6.2.17 Unshaped refractory and heat-resistant products

These are pre-prepared mixtures of one or more refractory components and one or several cements. They are either supplied ready for application or have to be mixed with a suitable liquid.[11] Processed unshaped products are graded according to the chemical and minerological structure of the main constituent and the type of cement used. Types of bonding used are hydraulic, ceramic, chemico-mineral, or organo-mineral and organic. These are subdivided, according to application, into gap-filling (mortars), surface coatings, and materials for monolithic construction and repairs (ramming, plastic compound, refractory concrete, and injection paste).

The VDEh (Association of German Iron and Steel Engineers) and the refractory industry have together prepared a twelve-digit code to improve the classification of unshaped products.[14] The use of unshaped refractory products offers the user quick delivery, simple storage, short application times, and thinner walls because of low thermal conductivity. Unlike bricks that have definite fixed properties, these materials only attain the required properties in use. Due to the temperature-drop, a distinct falling-off in properties occurs in the wall; this makes the lining yield particularly well to accommodate changing stresses. The development of strength in conjunction with temperature is of fundamental importance. When comparing the properties of different products one has to bear in mind that these depend decidedly on the way test-pieces are prepared.[12]

3.6.3 Brick Sizes, Packing, and Storage

It is in the interests of both producers and users of refractory bricks that brick format should be standardized as far as possible. Standard formats have been agreed[15,16] for fireplaces and for parts of different furnaces and transport ladles for the refining industry.

A standard flat, two-way wooden pallet (800 × 1200 mm) has been introduced and standard stacking plans have been laid down for a series of brick sizes; this simplifies transportation and storage.[17] When binding stacks of bricks with steel tape, corners and edges are protected with cardboard. Corrugated sheet is being increasingly used for packing the pallets. Light refractory bricks which are not very strong are often packed in cartons that are easy to handle. Unshaped products are usually transported and stored in water tight bags or metal drums.

3.6.4 Specifications: Acceptance Tests

The groups described above are classified in specification groups according to application and technical characteristics, and increasingly with a view to statistical assessment.

The European Federation of Refractory Producers (PRE) have drawn up recommendations for acceptance tests for refractory products.[18,19] Standards are available for acceptance tests for refractory products.[20]

On economic and statistical grounds, the tests should involve no more than three important qualities that should, if possible, be characteristic for the particular application.

3.7 ADDITIVES, SLAG-FORMERS, CARBURIZING AGENTS, AND CASTING AIDS

Klaus Dziggel, Dortmund

This section deals with consumable steelworks materials which are the sources of most of the left-over materials that have to be disposed of. Nearly all of these have a relatively low density and occupy a lot of storage space. They should be kept as dry as possible in store according to their use.

3.7.1 Lime

Lime is used in steelworks in the shape of limestone (i.e. untreated lime), mixtures of burnt lime and fluxes (for example, for desulphurization), and as lime for melting or crystal lime).

Limestone consists mainly of $CaCO_3$ and should contain the smallest amounts of undesirable elements, especially phosphorus and sulphur, and only small amounts of SiO_2. Table 1 shows the typical composition of limestone from two different sources. Chip-sizes between 18 and 65 mm are used in steelworks. Limestone can be stored indefinitely.

Table 1. Limestone analyses[1]

	1 Carrara (%)	2 Rhineland (%)
CaO	55.60	54.50
MgO	0.58	0.87
SiO_2	0.14	0.70
Al_2O_3	0.04	0.33
Fe_2O_3	0.01	0.10
SO_3	—	0.03
P_2O_5	—	—
Na_2O	—	0.01
K_2O	0.01	0.12
CO_2	43.60	42.80
H_2O	—	—

When limestone is used for refining, endothermic de-acidification, i.e. separation of CO_2 on heating, must be considered. During melting down, the CO_2 emerging has a refining effect on the metallic charge.

Burnt lime can be used in place of limestone if higher power consumption or oxidation during melting is to be avoided.

When limestone is burned with added coke in a shaft furnace, hard-burnt lime results. The quality of this depends on the chemical composition of the limestone, its grain, the degree of calcination, and the amount and nature of the fuel residue. The percentage of impurities increases with decreasing grain-size (see Figure 1).

Compared with hard-burnt lime, soft-burnt lime has a higher porosity, and therefore a larger internal surface area, which speeds its solubility in slag. For this reason, it is now preferred to hard-burnt lime as an additive agent during oxidizing and refining in all steelmaking processes.[2–4] Table 2 presents typical values for the physical properties of three types of lime and Table 3 shows typical chemical compositions. When gas or oil is used instead of coke, the impurity percentage in soft-burnt lime is less than in the hard-burnt variety produced in a shaft furnace. The CO_2 content of the former can be lowered.

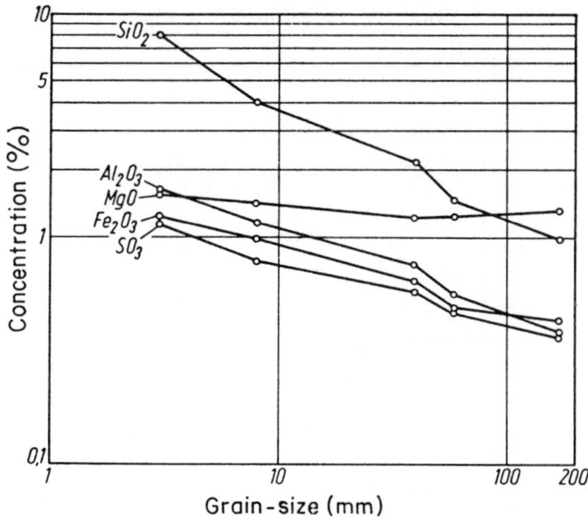

Figure 1. Impurity content in burnt lime in relation to grain-size[1]

Table 2. Guide to selected values for lime of different calcination grades[1]

		Soft-burnt	Medium-burnt	Hard-burnt
Density	(g/cm^3)	3.35	3.35	3.35
Apparent density	(g/cm^3)	1.5–1.8	1.8–2.2	>2.2
Total porosity	(%)	46–55	34–46	<34
Specific surface area after BET	(m^2/g)	>1.0	0.3–1.0	<0.3
Reaction behaviour				
(a) Wet quench curve R-value	(°C/mm)	>20	2–20	<2
(b) Rough grain titration 5-min value	(ml 4n HCl)	>350	150–350	<150

Burnt lime is supplied in grain sizes of 3–50, 10–40, and 40–200 mm. During transportation and storage in bunkers, hydration and mechanical action of lump lime can produce a great deal of abrasion.

Fine lime (particle sizes 0–1 or 0–3 mm), ground from soft-burnt lime, is increasingly used in steelworks because it can be blown into the furnace with air or oxygen. It is transported and stored in silo containers. These are emptied by dosimeter wheels, bucket wheels, worms, conveyors, or pneumatic discharge devices. If compressed air is used, it must be as dry as possible, and the air must be discharged from the silo through dust filters.

Table 3. Quality-development of steelworks lime[1]

Year		1940	1950	1960
CaO	(%)	78–85	82–93	86–94
MgO	(%)	1–5	1–5	1–3
SiO_2	(%)	1–5	1–4	1–3
$Al_2O_3 + Fe_2O_3$	(%)	1–4	1–4	1–3
S	(%)	0.1–0.3	0.07–0.3	0.05–0.15
CO_2	(%)	5–10	3–8	2–6
Year		1972		
		Normally burnt	Soft-burnt	
CaO	(%)	89–94	94–97	
MgO	(%)	1–3	1–2.5	
SiO_2	(%)	1–3	1–2.5	
$Al_2O_3 + Fe_2O_3$	(%)	1–3	1–2.5	
S	(%)	0.05–0.10	0.005–0.05	
CO_2	(%)	2–5	1–3	

Mixtures of lime with iron oxides, in the form of ore, scale, filtered dust, red mud (also with felspar), bauxite, and colemanite, are used in loose, briquetted, or pre-melted form for ladle-refining processes, such as dephosphorizing or desulphurizing, carried out during tapping (see Table 4).[5–12]

Table 4. Mixtures for post-desulphurizing and post-dephosphorizing steel[1]

No.	CaO (%)	CaF_2 (%)	Fe_2O_3 (%)	$Al_2O_3 + SiO_2 + MgO + CO_2$, etc. (%)
1	72.4	19.6	—	8.0
2	67.7	24.1	—	8.2
3	59.2	31.8	—	9.0
4	56.4	8.8	29.2	5.6
5	51.7	13.2	27.6	7.5
6	65.2	26.2	—	8.6

Nos 1–3 and 6 for desulphurizing.
Nos 4 and 5 for dephosphorizing.

Melt or crystal lime can only be produced in arc furnaces, because of the high temperature required — above 2600°C. It solidifies into large high-density crystals and is thus less liable to absorb water than is burnt lime. Rammed or moulded into bricks, it is used to line melting vessels and ladles where the lowest oxygen sulphur contents are required in the melt.

There are many natural deposits of limestone ($CaCO_3$) combined with large amounts of magnesium carbonate ($MgCO_3$). Burnt lime produced from these is more difficult to melt, and this is why the MgO content should not exceed 10%. This product, called dolime in America, has a somewhat smaller desulphurizing effect, but has the advantage of higher wear-resistance in the slagging zone in refining vessels.

3.7.2 Fluorspar

Fluorspar is natural calcium fluoride; for steelworks use it should not contain more than 5% SiO_2, 10% $CaCO_3$, and certainly not more than 0.2% sulphur. The slag-liquefaction effect is governed by the CaF_2 content.

It should be used as little as possible because it gives rise to fluorine compounds (environmental protection) and attacks refractory linings severely. However, there is no effective substitute for the melting of refining slags.

3.7.3 Bauxite and Other Slag-liquefiers

These are preferred as liquefying additives for refining slags. Bauxite consists of the aluminous hydrates diaspor (AlOOH) and gibbsite ($Al(OH)_3$) and weathered silica-rich rock, mixed with Fe_2O_3 and TiO_2. For steelworks use, the normal chemical composition is 40–60% Al_2O_3, 15–30% Fe_2O_3, 5% SiO_2, and 3–5% TiO_2.

Other slag-liquefiers used in steelworks are quartz sand, which contains 90–98% SiO_2, crushed calcium fireclay from old casting pit bricks, slag from aluminothermic production of titanium, and broken glass; these are preferred slag-reformers for acid-melting processes.

3.7.4 Ore for Use in Steelworks

Ore is still usually used in steelworks to enrich the melting slag with oxygen for the oxidizing process using gaseous oxygen. Ore is used in lumps or pelletized and, less frequently, as fine ore. Recycled roll-sinter and scale have also been used. With the latter one must be wary of undesirable alloy ingredients.

3.7.5 Carburizing Media

Coke, anthracite, crushed or ground electrodes, electrographite, petroleum coke, pitch coke, and complex carbon compounds such as silicon carbide all seem suitable for carburizing steel melts. The first ones named are contaminated with sulphur and incombustible materials and are therefore usually charged together with scrap. For this reason, only complex carbon-carriers are used for fine adjustment of the specified carbon content in steels.

Electrode fragments and briquetted mixtures of steel swarf and coke are well suited for adding carbon to molten steel covered by a thick layer of slag because these carbon-carriers are dense enough to sink through the slag into the melt.

3.7.6 Casting Aids

The task of casting aids is twofold: to protect the inner faces of moulds and thus give steel ingots a smooth, non-porous surface and to absorb non-metallic impurities liberated from the steel. For many years, lacquers with the most diverse compositions, molasses, and carbon-compound vapour, formed by non-stoichiometric combustion of acetylene or other hydrocarbons on the inner surfaces of moulds, have all been used for this purpose.

3.7.6.1 Casting fluxes

Casting fluxes were developed with the aim of depositing low-viscosity, high-wetting slag on the walls of moulds, etc. in order to facilitate wetting by the rising steel and to absorb non-metallic compounds which separate from the steel. Originally, cryolite (Na_3AlF_6) and finely ground fluorspar were used for this, but they could not be widely used because of their toxicity. Their place was taken by flue dust, from pit coal-fired power stations, which met the requirements relatively well. The average composition of this kind of casting flux is 14% C, 3% CO_2, 0.2% H_2O, 4% Na_2O, 3% K_2O, 2% CaO, 40% SiO_2, 22% Al_2O_3, 7% Fe_2O_3, 2% MgO, and 1% TiO_2. This shows that the main slag-liquefiers are alkaline oxides, Fe_2O_3 and Al_2O_3, while carbon provides the insulation effect.

Fluctuation in the chemical composition and physical structure of flue dust used as casting flux led to the development of synthetic fluxes for the rising cast, which are available in consistent quality. Their characteristics can be optimized and adapted to suit process requirements and the specific properties and quality requirements for the various types of steel. They consist mainly of pre-melted calcium–aluminium silicates, such as blast-furnace slag, synthetic wollastonite, synthetic diopside, nepheline–syenite, and other refining slags. Aluminous silicates and alkaline aluminous silicates such as spherulite, etc. are used as expanders. Alkali carbonates, fluorides, and borates are used as fluxes. Moreover, carbon-bearing, insulating slag-forming substances are sometimes added, but cannot be used for casting ELC (extra-low-carbon) steels because of possible surface carburizing. Table 5 lists the compositions of three typical synthetic casting fluxes with their most important physical and chemical properties.

3.7.6.2 Continuous-casting fluxes

These are modified ingot-casting fluxes in which the quality and quantity of the

Table 5. Typical chemical compositions and physical data for three synthetic casting fluxes

Type	Chemical composition (% wt)												Physical properties			
	C	CO_2	H_2O	Na_2O	K_2O	CaO	SiO_2	Al_2O_3	MgO	Fe_2O_3	TiO_2	MnO	F	Softening temperature (°C)	Hemisphere temperature (°C)	Flow-point (°C)	Density (kg/m^3)
1		2	0.5	4	2	28	42	12	7	0.5	0.1	0.5		1000	1240	1280	500
2	3	3	0.5	5	2	27	40	10	6	0.5	0.1	0.5		1050	1250	1265	550
3	4	4	0.5	4	3	34	33	8	4	0.5	0.1	0.5	6	1080	1220	1235	750

flux component have usually been altered in such a way that the rate of slag-formation corresponds to the cast cross-section and machine requirements. The chemical analyses and physical properties of present-day synthetic continuous casting fluxes are shown in Table 6. Although the CaO to SiO_2 ratios lie within a wide scatter-band, the four fluxes have an almost uniform melting character. However, their behaviour in the mould is affected by the widely varying melt-down rate. This depends on the mineral structure and grain-size of the slag base and the flux; thus behaviour cannot be assessed from chemical composition alone nor solely from melting characteristics. Only knowledge of viscosity in the 1200–1500°C temperature range and of melt-down rate will allow definite conclusions to be made about flux suitability; special test methods have been developed for this.

3.7.6.3 Covering agents for ingots

Covering agents can be divided into the following groups, depending on application and effect: insulating, low-, medium-, and highly exothermic covering powders, and welding fluxes. Detailed particulars about the raw materials used, chemical composition, bulk weights, applications, advantages, and drawbacks may be found in Table 7.

Recently, covering powder has been replaced by insulating or exothermic cover-plates, with or without provision for expansion. The advantage of these plates is easy handling, but the snag is that each ingot format requires a special plate, which means that the right-size plates have to be kept available in store.

3.7.7 Slags for Ladle-refining and Remelting Processes

These effective refining slags are used mainly for dephosphorizing, desulphurizing, deoxidizing, and for the removal of non-metallic inclusions. They are prepared from pre-dried mixtures of ingredients or are pre-melted in an arc furnace.[13] Table 8 lists a number of typical slag compositions used for dephosphorizing and desulphurizing in the ladle (Perrin process) and for electro-slag remelting (ESR).

In the ESR (electro-slag refining or remelting) process, slags are also used with additions of MgO and TiO_2 for special applications. In all cases a certain electrical conductivity besides viscosity and wetability are important for the refining process.[14]

Table 6. Typical chemical compositions and physical data for four synthetic casting fluxes

Type	Chemical composition (wt %)								Physical properties			
	CaO	Al_2O_3	SiO_2	MgO	Fe_2O_3	Na_2O	K_2O	F	Softening temperature (°C)	Hemisphere temperature (°C)	Flow-point (°C)	Density (kg/m^3)
1	37	3	40	1	<1	4	<1	7	1080	1160	1175	850
2	33	7	25	1	<1	7	<1	6	1100	1150	1165	1100
3	24	10	32	1	<1	9	2	7	1100	1150	1160	720
4	36	4	33	4	<1	6	<1	7	1090	1140	1145	650

Table 7. Data and assessments for ingot-covering agents, for both rising or downhill casting. (Chemical composition in percentage weight. Bulk weight in kilograms per litre)

Type	1	2	3	4	5	6	7	8
Characteristic	Insulation		Low	Medium Exothermic	High	High	Welding flux	
Al	—	—	8–14	14–18	18–22	22–6	27–30	—
SiO_2/Si	~40	~45	~15	~14	~25	~15	~6Si	~25Si
TiO_2	< 0.5	~ 1	~ 0.5	~ 0.5	~ 1	~ 0.5	< 0.1	~ 0.5
Al_2O_3	~20	~15	~35	~35	~25	~30	~ 7	~ 8
FeO	—	—	—	—	—	—	~20	—
Fe_2O_3	~ 5	~ 6	~ 5	~ 5	~ 5	~ 1	~25	< 1
MnO	< 0.5	< 0.2	< 0.5	< 0.5	< 0.5	< 0.5	< 0.1	~25
CaO/Ca	~ 3	~ 2	~ 2	~ 2	~ 2	~ 2	~10Ca	~25Ca
MgO	~ 2	~25	~ 5	~ 5	~ 5	~ 5	~ 0.1	~ 0.5
$Na_2O + K_2O$	~ 5	~ 5	~ 8	~ 6	~ 8	~ 8	~ 1	~10
C_{tot}	15–20	< 1	~10	~ 8	~ 8	~ 8	—	—
Cl	—	—	< 5	< 5	—	< 5	—	—
F			< 1	< 1	< 1	< 1		~ 5
Bulk weight	0.5–0.7	0.1–0.2	0.6–1.0	0.6–1.0	0.4–0.8	0.4–0.8	1.4–1.8	1.3–1.6

Notes for Table 7

Type 1: Raw material basis: flue dust; application; ingots and slabs over about 5 t, blooms; for all plain and medium-alloy steels. Can be used as ladle cover. Advantages and disadvantages: good insulation generally because of low inclination to form slag, dust clouds when handled.

Type 2: Raw material basis: specially treated minerals; application: ingots and slabs over about 2 t, blooms; for all plain and medium-alloy steels, ball-bearing and spring steels. Can be used as ladle cover. Advantages and disadvantages: practically no dust, vitrifies in presence of casting fluxes with strong tendency to vitrify; therefore loses insulating properties.

Type 3: Raw material basis: dust from aluminium ball mills, oxygen-carriers, organic and/or light inorganic materials. Application: ingots and slabs over about 5 t; for all plain and medium-alloy steels. Advantages and disadvantages: little smoke in exothermic reaction, moderately stable cover, large top volume needed for ingots and slabs.

Type 4: Raw material basis: dust from aluminium ball mills, partly with metallic aluminium. Oxygen-carrier, light organic and inorganic materials, expansion agent sometimes added. Application: as 3. Advantages and disadvantages: light smoke if exothermic reaction is controlled, stable cover; for good cover the covering agent must match the casting flux.

Type 5: Raw material basis: as for 4. Application: ingots and slabs over about 3 t, preferably for alloy steels, ferritic chromium and austenitic chrome–nickel steels. Advantages and disadvantages: generally good cover, better top yield for given hood conditions; smoke development as for 4.

Type 6: Raw material basis: as for 5. Application: ingots and slabs over about 2 t, preferably for alloy steels, ferritic and austenitic chrome–nickel steels. Advantages and disadvantages as for 5.

Type 7 and 8: Raw material basis: metallic aluminium and/or ferrosilicon or calciumsilicon, oxygen-carriers, effective catalysts added. Application: ingots and slabs, mainly under 3 t, for ferritic chrome– and austenitic chrome–nickel steels, high-carbon steels, such as ball bearing-, high-speed-, and ledeburitic steels. Advantages and disadvantages: almost parallel ingot tops; bad thermal efficiency making little use of free enthalpy, steel overheats in hot top; formation of 'reguli' with inclusions in the ingots, and steel flows from the hood in between the mould and the already solidified ingot wall (apron-formation). More insulating powder must be added at the end of the exothermic reaction.

Table 8. Chemical compositions and amounts required of pre-melted slags

Application type		Chemical composition (% wt)					Amount used in percentage of steel weight
		CaO	Fe-oxides	CaF_2	Al_2O_3	SiO_2	
Dephosphorizing	1	50–60	20–30	12			
	2	50–60	20–30		25		2–3
Desulphurizing	1	53	<1		42	<5	
(Perrin)	2	45			35–38	2–3	4
	1	33		33	33		
	2	0–20		40–80	0–35		
ESR	3	40		20	40		
	4			70	30		
	5	10–30		30–50	20–40		

3.8 DISCARD MATERIALS

Jürgen Geiseler, Rheinhausen

In electric furnace steelmaking, in addition to end-products (ingots, continuous castings, and steel castings), other substances are also produced which have to be disposed of unless they can be recycled. These materials can be classified according to value as follows:[1]

(1) By-products which can be sold profitably outside the works;
(2) Recycling materials which can be re-used in the works and can earn a credit;
(3) Residues which can be re-used inside or outside the works without gain or loss;
(4) Discard materials, dumping of which has to be paid for.

It was said in the introduction to Chapter 3 that about 1300 kg of raw materials and auxiliaries are required to produce one tonne of steel in an electric furnace. A one-tonne unfinished ingot results in about:

85 kg production scrap
150 kg slag
15 kg dust
50 kg gravel

3.8.1 Production Scrap

Steelworks production scrap arising from slagging-off, tapping, transferring between vessels, secondary treatments, and casting is almost all recycled and re-used in the works where it originates. It can naturally also be sold as a by-product outside.

3.8.2 Slag

Today, arc furnaces are used for making all kinds of steel, from crude, high-tonnage steels to highly alloyed special steels, usually in combination with secondary treatment in ladles, or under vacuum, or in convertors. Different amounts of slag of widely varying chemical composition arise at different locations in the steelworks, depending on the charge, the process, and production programme. This is illustrated below in three examples.

3.8.2.1 Black slags, rich in iron oxide

Refining slags arising in the production of plain or low-alloy steels are broadly similar in composition to slags from other steel-production systems using a similar programme (Table 1). The percentage of CaO in refining slags from electric arc furnaces is lower than in slags from basic oxygen-converters. However, these slags contain more MgO because of the more vigorous reaction with the refractory lining and any patching compound used (Figure 1). The increasing use of water-cooled furnace walls and roofs will probably lower the MgO content in slags, as there will be less lining to attack and this will not be compensated by intensive use of refractory spraying mixtures.

The discarded oxidizing slags from basic oxygen-converters are increasingly used as building-materials for roads, canals, etc.[2–5] The volume-stability of these slags is of particular importance here. The free CaO and MgO content must be limited because the volume grows on hydration of these phases. In slags containing higher percentages of MgO (>3%) this volume-increase only ends after several years. Higher MgO contents in black arc-furnace slags make their sale for these purposes difficult.

3.8.2.2 Reduction and desulphurizing slags

In the two-slag process (Table 1), either the oxidizing slag is drawn off and a lime-based slag is added with fluxes, or the heavy metal oxide-rich oxidizing slag is reduced and, with additional fluxes, converted into lime-based slag. These slags are characterized by low percentages of iron oxides and other heavy metal oxides. When cooling below 500°C they usually decompose on account of dicalcium silicate conversion. It is hoped that introduction of other ions into the crystal lattice[7] will have a stabilizing effect and overcome the dust problem in transportation.

3.8.2.3 Slags from high-chromium steelmaking

High-chromium low-carbon steels are made by melting a charge of scrap and ferro-alloys in an arc furnace, followed by oxidation in a vacuum–oxygen-decarburizing process (VOD) or argon–oxygen-decarburization (AOD). In the arc furnace/VOD method the chromium oxide-rich slag is reduced with ferro-silicon after melting in the arc furnace. This slag is tapped and drawn into a ladle, and the melt is transferred into the VOD plant without slag. Lime and flux are added and, after oxidation and slag reaction, the slag mentioned in Table 1 is produced. In the arc furnace AOD process the melt-down slag from the arc furnace is transferred to the AOD converter and then reduced. In these two different processes for making high-chromium low-carbon steels, different quantities of slag are produced at different times, but they can be very similar in composition.[8,9]

Black, oxide-bearing slag can be used in building for less important applications and therefore is a by-product; however, up to now no use has been found for white reduction slag or decomposition slag. Because these slags are

Table 1. Examples of the range of chemical composition of slags from various steel-production processes

Steel-production process	LD	Open hearth		Electric arc furnaces					
Steel quality	Unalloyed/ low alloy	Unalloyed/ low alloy	Unalloyed/ low alloy	Low/medium alloy		With high Cr content			
Number of slags	1	1	1	2		2		2	
Process variant						VOD		AOD	
Type	Oxidizing slag	Oxidizing slag	Oxidizing slag	Oxidizing slag	Refining slag	Reducing and refining slag	Desulphurizing slag	Reducing and refining slag	Desulphurizing slag
Fe_{tot} [a]	15–24	10–19	10–19	10–19	max. 2	1–3	max. 1	1–3	max. 1
CaO	40–50	35–45	35–45	35–45	45–55	40–60	45–65	40–60	45–65
CaO_{free}	max. 15	max. 7	max. 7	max. 7	trace	trace	trace	trace	trace
SiO_2	10–15	10–18	10–18	10–18	13–18	20–30	15–30	20–40	20–30
Al_2O_3	0.5–2.0	3–8	3–8	3–8	6–10	5–15	2–8	1–10	2–8
MgO	1–2	7–13	7–13	7–13	10–18	2–6	1–5	2–6	1–5
MnO	3–7	4–12	4–12	4–12	max. 2	1–6	max. 1	1–6	max. 1
P_2O_5	1–3	max. 1	max. 1	max. 1	trace	max. 0.5	trace	max. 0.5	trace
CaF_2	max. 0.5	max. 0.5	max. 0.5	max. 0.5	1–10	max. 2	2–4	max. 2	3–4
Cr_2O_3	trace	trace	trace	trace	trace	1–5	max. 1	1–4	max. 1
CaO/SiO_2	3.5–4.5	2.5–3.5	2.5–3.5	2.5–3.5	3–4	2–3	2.5–3.5	1.5–2.5	2–3
Approx. amount discarded (kg/t crude steel)	130–150	140–160	80–150	80–150	20–30	100–130	20–30	100–130	20–30

[a]Percentage content by weight.

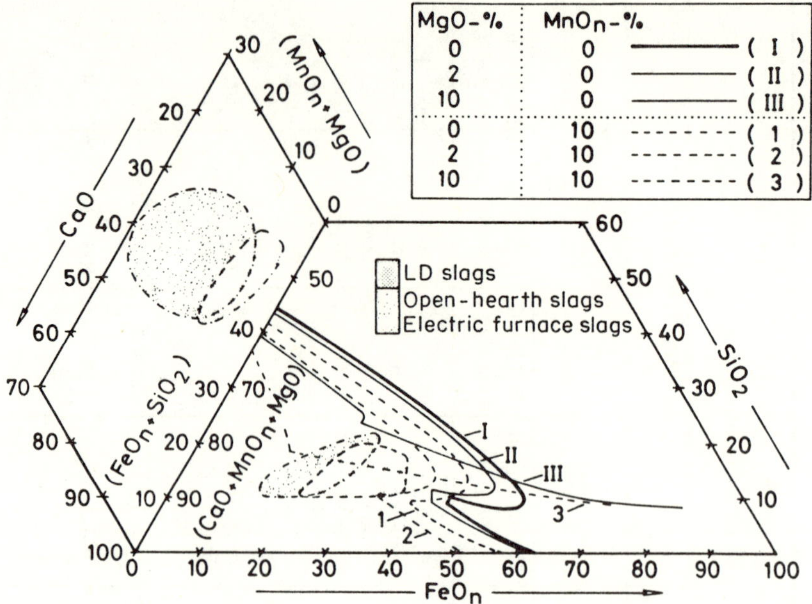

Figure 1. Composition of typical lime–silica steelworks slags in the Fe–CaO–FeO_n–SiO_2–MnO_n–MgO system at 1600° C (% wt)

partly water-soluble, they cannot be dumped just anywhere. According to waste-disposal laws[10] they may only be dumped at an authorized site.

3.8.3 Dust

Dust collected by an arc furnace dust-extractor varies widely in composition. Dust with a high zinc and/or lead content can be processed by leaching or roasting to recover the metal.[11–13] Dust that contains little zinc or lead can be re-charged into an arc furnace as a solid oxygen-carrier and in order to make use of the iron and lime it contains. Dust that cannot be recycled is often pelletized with quicklime added to reduce abrasion and improve resistance to crushing in transportation and dumping.[14] More details about dust from other electric furnace processes can be found in section 13.2 in Chapter 13.

3.8.4 Steelworks Refractory Waste

This comprises all refractory rubble from furnaces, ladles, post-treatment systems, and casting pits.[1] Expensive refractory building-materials, such as magnesite, chrome–magnesite, and bauxite bricks are refurbished and re-used for less-important applications. Fireclay rubble from casting plates and ladles can be separated and then used as a slag-liquefier. Where this is not possible, it can under certain conditions be used as ground-levelling material. The high cost of disposal in an authorized dump is very rarely necessary or economically defensible.

3.9 LITERATURE REFERENCES

3. Energy, Raw Materials, and Consumables

1. Möllenkamp, F, W., Die Zukunft des Elektrostahls. *Klepzig-Fachber.* **76** (1968), 581–6.

3.1 Electric Power

1. *Große Anfrage zur Energiepolitik* Bundestagsdrucksache 8/570 dated 8. 6. 1977.
2. Klätte, G., Nicht ohne Kernenergie. *RWE-Verbund* 1977, No. 2.
3. *Strom für die Bundesrepublik.* Vereinigung Deutscher Elektrizitätswerke e. V. Frankfurt, 1976.
4. VDWE-Schriftenreihe, *Grundzüge für die Beurteilung von Netzrückwirkungen.* Vereinigung Deutscher Elektrizitätswerke e. V. Frankfurt/M. 1976.
5. VDE 0160/T. 2/10/75 Bestimmung für die Ausrüstung von Starkstromanlagen mit elektronischen Betriebsmitteln. Einrichtungen mit Betriebsmitteln der Leistungselektronik in Starkstromanlagen (BLE).

3.2 Industrial Oxygen, Nitrogen, and Argon, and Fuel for Additional Energy Supplies

1. Hollemann, A. F., and E. Wiberg, *Lehrbuch der anorganischen Chemie*, Berlin, 1958, p. 63.
2. D'Ans, J., and E. Lax, *Taschenbuch für Chemiker und Physiker.* Bd. 1. Berlin/Heidelberg/New York, 1967.
3. Baumgärtner, K., Sauerstoffbereitstellung für Großverbraucher in Chemie und Stahlindustrie. *Chemie-Ing.-Techn.* **46** (1974), 887–93.
4. Baumgärtner, K., Oxygen supply of iron and steel works (German). *Stahl u. Eisen* **89** (1969), 67–77.
5. Hollemann, A. F., and E. Wiberg, *Lehrbuch der anorganischen Chemie.* Berlin, 1958, p. 33.
6. as[20]
7. Herning, F., and E. Wolowski, Kompressibilitätszahl und Realgasfaktor von technischen Gasen. Gas u. Wasserfach. *Gas/Erdgas.* **105** (1964), 450–6.
8. DIN – Taschenbuch 20. *Mineral- und Brennstoff-Normen.* Berlin/Köln/Frankfurt, 1972.
9. Foerst, W., *Ullmanns Encyklopädie der technischen Chemie.* 3rd edition, Vol. 15, Munich/Berlin/Vienna, 1969.
10. Foerst, W., *Ullmanns Encyklopädie der technischen Chemie.* 3rd edition, Vol. 6, Munich/Berlin/Vienna, 1969.
11. Foerst, W., *Ullmanns Encyklopädie der technischen Chemie.* 3rd edition, Vol. 10, Munich/Berlin/Vienna, 1969.
12. Foerst, W., *Ullmanns Encyklopädie der technischen Chemie.* 3rd edition, Vol. 17, Munich/Berlin/Vienna, 1969.

13. Recknagel-Sprenger, *Taschenbuch für Heizungs- und Klimatechnik*. E. Sprenger. München/Wien, E. Springer. Munich/Vienna, 1977.
14. Shell AG, West Germany, verbal information.
15. Turbo fuel A. ESSO, Extract from ESSO specification.
16. Hansen, W., *Ölfeuerung. Brennstofftechnische Einrichtung und Anwendung*. 2nd edition Berlin/Heidelberg/New York, 1970, p. 7.
17. *Stahleisen-Kalender 1977*. VDEh, the Association of German Iron and Steel Engineers. Düsseldorf, 1976.
18. Recknagel, H., and E. Sprenger, *Taschenbuch für Heizungs-, Lüftungs- und Klimatechnik*. See chapter entitled: Verbrennung, Vergasung und Entgasung: Zundtemperaturan von Brennstoffen in Luft. Munich/Vienna, 1970. p. 166.
19. Ivernel, A., and M. Just: Oxygène et combustion applications aux brûleurs industriels. *Rev. Gen. Therm.* **4** (1965), 527–51: **5** (1966) 251–9.
20. Kipker, R., Transportspeicherung von technischen Gasen. In *Fortschritte bei der Anwendung technischer Gase in der Metallurgie*. Essen, 1977. (Vortragsveröffentlichungen. Haus der Technik. H. 360.) pp. 9–13.
21. Zentrale Versorgung von Betrieben mit technischen Gasen für Schneiden, Schweißen und verwandte Arbeitsverfahren. Gemeinschaftsarbeit der Arbeitsgruppe 2. 'Technische Gase' und Arbeitsgruppe 3. 'Gasschweißgeräte' im Technischen Ausschuß des Deutschen Verbandes für Schweißtechnik e. V. 2nd edition Düsseldorf, 1973.

Queneau, Paul E., Oxygen technology and conservation. *Metallurg. Trans.* **8B** (1977), No. 3, 357–69.

3.3 The Charge Materials: Scrap, Pig Iron and Sponge Iron

1. Rupp, H.-J., Entwicklungstendenzen in der internationalen Schrottwirtschaft. Cologne, 1964. (Diss. Univ. Cologne.)
2. Vom Schrott zum Stahl. Bedeutung des Rohstoffs Schrott für die Eisen- und Stahlerzeugung. Bundesverband der Deutschen Schrottwirschaft and others Düsseldorf 1977.
3. Martin, K. H., Die Bedeutung des Schrottes für die Roheisen- und Stahlerzeugung. *Stahl u. Eisen* **77** (1957), 259–64.
4. *Die Deutsche Schrottwirtschaft*. Published in honour of the 30th anniversary of the Bestehens des Bundesverbandes der Deutschen Schrottwirtschaft e. V.(B.D.S.) Düsseldorf, 1976.
5. Amelung, E., and H. W. Kreutzer, Entwicklung der Schrottversorgung und ihre Bedeutung für die Stahlindustrie. *Stahl u. Eisen* **98** (1978), 715–17.
6. Oelsen, O., Entwicklungstendenzen von Stahlerzeugung und Schrottwirtschaft. *Schrottbetrieb* **23** (1972), Nov., 4–14.
7. Kootz, Th., K. Lenzmann and A. Altgeld, Gegenüberstellung der Stahlgewinnungsverfahren aus technischer und wirtschaftlicher Sicht mit Betrachtungen über die weitere Zukunft. *Thyssenforsch.* **5** (1973), No. 3, 59–77.
8. Kootz, Th., Eisen und Stahl — gestern, heute und morgen. *Thyssenforsch.* **5** (1973), 28–34.
9. Felcht, K., and H. Külke, Qualitative und quantitative Anforderungen an Stahlschrott unter besonderer Berücksichtigung der SM- und Elektrostahlerzeugung. *Neue Hütte* **21** (1976), 451–5.
10. *Bedeutung geringer Beimengungen bei der Stahlerzeugung und-verarbeitung*. Part I: Erzeugung. Freiberg. Forsch.- H. Reihe B. No. 192. 1977.

11. Amelung, E., and H. Schutt, Kupfer- und Zinngehalte im Schrott und ihre Auswirkungen auf den Stahl. *Stahl u. Eisen* **93** (1973), 740–1.
12. Neuhaus, H., H. J. Langhammer and W. Münstermann, Contribution to the question of melting loss in steel production processes (German). *Stahl u. Eisen* **84** (1964), 1041–52.
13. Amelung, E., Aufbereitung von Altschrott. *Stahl u. Eisen* **92** (1972), 419–20.
14. Weber, R., Recycling und Automobilbau — Gesichtspunkte für zukünftige Entwicklungen. *Automobil-Ind.* **4** (1975), 95–102.
15. Durrer, R., and G. Heintze, Hot metal in the electric arc furnace. *Iron Steel Eng.* **36** (1959), No. 10, 35–42.
16. *Stahleisen-Kalender 1976*. Düsseldorf, 1975. p. 102.
17. Ottmar, H., H. Schenck and W. Dahl, Technical and economic effects of the substitution of scrap by sponge iron in the arc furnace (German). *Stahl u. Eisen* **97** (1977), 731–41.
18. *Direktreduktion von Eisenerz. Eine bibliographische Studie*. 4th edition published by order of the EEC Commission. Düsseldorf, 1976.
19. Kalla, U., and R. Steffen, Direct reduction: progress and plans. *Iron & Steel Internat.* **50** (1977), 307, 309–11, 313–14, 317, 319–20.
20. Kalla, U., and R. Steffen, On the present situation of direct reduction processes (German). *Stahl u. Eisen* **98** (1978), 1211–12.
21. Klingelhöfer, H.-J., and G. Rottmann, State and development of Midrex direct reduction process (German). *Stahl u. Eisen* **98** (1978), 1213–18.
22. Pantke, H.-D., and G. H. Lange, Progress in direct reduction by the Purofer process (German). *Stahl u. Eisen* **98** (1978), 1219–23.
23. Quintero, R. G. Present status of development and application of the HYL process (German). *Stahl u. Eisen* **98** (1978), 1224–32.
24. Reuter, G., W. Schnabel and H. Serbent, Production of sponge iron with the aid of coal by application of the SL/RN process (German). *Stahl u. Eisen* **98** (1978), 1232–7.
25. Janssen, W., G. Meyer, K.-H. Ulrich and K.-H. Vopel, The Krupp–Codir process for the production of sponge iron (German). *Stahl u. Eisen* **98** (1978), 1237–42.
26. Cloran, L., and K.-H. Ulrich, Armco direct reduction process (German). *Stahl u. Eisen* **98** (1978), 1242–3.
27. Meininghaus, F., Use of sponge iron in an integrated iron and steel works (German). *Stahl u. Eisen* **96** (1976), 1245–51.

Scrap

Schauwinhold, Dieter, Entwicklung des Schrottaufkommens und der Stahlgewinnungsverfahren. Rückschau bis 1950 und Vorausschau bis 1990. Clausthal 1978. 62 pages, 68 illustrations, 3 tables in Appendix. 8°. = Dr.-Ing.-Diss. Techn. Univ. Clausthal.

Dial-an-expert: the new dynamics of big time scrap brokering. *Metal Producing* **17** (1979), No. 7, 54–7.

Thornton, D. S., Requirements of scrap for steelmaking and the increased use of scrap processing. *Iron & Steel Intern.* **50** (1977), No. 5, 321–3, 325, 327.

Petrone, Rocco A., Recovery of ferrous scrap from municipal solid waste. In *11th Annual Conference, Rome, Italy, 10–12 Oct. 1977, Report of Proceedings*, International Iron & Steel Institute, Brussels, 1978, pp. 136–47.

Scrap management for electric furnaces. *Brit. Steelmaker* **45** (1979), No. 1, 6–8, 10.

Burlingame, R. D., Trends in scrap quality for the 1980's. *Iron & Steelmaker* **6** (1979), No. 2, 14–18.
Dowsing, R. J., Scrap management in a volatile world market. *Metals & Mater.* 1979, No. 4, 33–8.
Iron and steel scrap: its significance and influence on further developments in the iron and steel industries, United Nations, Economic and Social Council. — Geneva: UN 1978, III, 223 pages, illustrated; 4°
Avila, Felipe, and Hector Lopez Ramos, Latest developments in the HYL process. In *Symposium on Raw Materials, Feed & Energy Sources for the Iron and Steel Industry*, held in Philippines, 9–13 Oct. 1978. Singapore 1979, pp. 463, 465, 467–81, 483.
Sanderson, K. G., Recovery and preparation of scrap for alloy steelmaking. *Ironmaking and Steelmaking* **6** (1979), No. 6, 305–7.
Wilkinson, H. N., and J. Peace, New scrap beneficiation and handling techniques. *Ironmaking and Steelmaking* **6** (1979), No. 6, 301–4.
Mandel, Allen J., Scrap is truly an uncontrolled, free-market commodity. *Iron & Steelmaker* **5** (1978), No. 1, 22–8.
Brown, James W., US scrap supplies seen sufficient for EF growth despite export controversy. *Metal Producing* **16** (1978), No. 9, 51–3.

Sponge Iron

Jaco, Jr, Charles M., The future of direct reduction. *Metal Producing* **16** (1978), No. 9, 44–7.
Keaton, David L., and Jeffrey B. Bradley, Direct reduction. Direct use of lignite as an energy source for steelmaking. Paper given at 2nd ILAF Direct Reduction Convention, Macùto, Venezuela, July 1977, 16 pages, 10 tables (AKS E46.11), See also *Metal Bull. Monthly* No. 93, 1978, 53, 55.
Reuter, G., W. Schnabel, and H. Serbent, Direct use of lignite as an energy source for steelmaking. *South East Asia Iron Steel Inst. Quart.* **8** (1979), No. 1, 46–52.
Kaneko, Dentaro, and Kiichi Narita, Evaluation of raw materials for gaseous direct reduction processes. *South East Asia Iron Steel Inst. Quart.* **8** (1979), No. 1, 19–30.
Sponge iron for export markets. *Metal Bull. Monthly* No. 93, 1978, 45, 47–9.
Morrison, A. L., J. K. Wright and K. McG. Bowling, Microstructure of metallized iron ore pellets reduced by coal char in a rotary kiln simulator. *Ironmaking & Steelmaking* **5** (1978), No. 1, 39–44.
Goodman, Roger J., World direct reduction capacity to 1980. *Skilings Min. Rev.* **68** (1979), No. 11, 1, 12–13, 16.
Ulrich, Klaus Herbert, Transportation of sponge iron — a technical and economic challenge. *Metallurg. Plant & Technol.* **1** (1978), No. 2, 2, 4, 6, 8, 10, 12.
NSC Direct Reduction Process. Report of the workshop on small-scale iron and steelmaking including sponge iron production. Organized by ESCAP and UNIDO, Bangkok, 16–20 May 1978. *Stahl u. Eisen* **98** (1978), No. 23, 1250–1 (AKS E46.10).
The Hockin process. *Metal Bull. Monthly* No. 93, 1978, 55, 57; *Stahl u. Eisen* **98** (1978), No. 23, 1251.
Pantke, H. D., The Purofer Process. *South East Asia Iron Steel Inst. Quart.* **7** (1978), No. 2, 6–17.
Chatterjee, Amit, B. D. Pandey, C. C. Ojha, M. Tiwary, M. N. Poddar and P. K. Chakravarty, The development of the TISCO technology for sponge iron production in rotary kilns. *South East Asia Iron Steel Inst. Quart.* **7** (1978), No. 4, 64–70.

Sciarretta, Claudio, The Klinglor metor direct reduction process. *South East Asia Iron Steel Inst. Quart.* **7** (1978), No. 1, 6–14.
Dixon, James, M., Direct reduction catches fire. *Iron Age Metalworking Intern.* **18** (1979), No. 2, 44MP3–44MP4, 44MP6–44MP8.
Muraki, Junjiro, Nobunao Nishida, Naoki Otsuki, Yukiaki Hara and Kunio Hachisuka, Nippon Steel direct reduction process. *Nippon Steel techn. Rep.* 1978, No. 12, 114–27.
Albert jr., Anthony A., Application Of Accar Technology At Sudbury Metals Company and Niagara Metals Limited. *Canad. mettalurg. Quart.* **18** (1979), No. 1, 97–104.
Wilson, K., The SL/RN Process at the Griffith Mine. *Canad. metallurg. Quart.* **18** (1979), No. 1, 105–9.
Faucher, A., A. H. Marquis and T. E. Dancy, Review of the Sidbec–DOSCO direct reduction operations. *Iron & Steelmaker* **6** (1979), No. 5, 35–40.
Marquis, A. H., A. Claveau and A. Faucher, La fabrication de fer préréduit pour utilisation au four à arc électrique. *Canad. min. mettalurg. Bull.* **72** (1979), No. 806, 120–8.
Claveau, A., A. Faucher and M. A. Lafrenière, Performance review of Sielbec–DOSCO's reduction plant after four years of operation. *Canad. metallurg. Quart.* **18** (1979), No. 1, 85–95.
HYLSA unveils new-generation HYL direct reduction plant. *Metal Producing* **15** (1977), No. 9, 34–6.
Labee, Charles J., HYLSA's Puebla 2 — a fantastic success. *Iron Steel Eng.* **54** (1977), No. 10, 25–9.
Hamilton, John G., and Laurence Amaya, HYLSA's Puebla II direct reduction plant. *Iron Steel Eng.* **55** (1978), No. 5, 25–30.
Braga, Claudio H. M., An Appraisal of Usiba's HYL Plant Performance. *Canad. metallurg. Quart.* **18** (1979), No. 2, 239–43.
Price, J. Federico, and Patrick W. Mackay, Evaluation of Raw Materials For HYL Direct Reduction Operations. *Canad. mettalurg. Quart.* **18** (1979) No. 1, 117–24.
Direct reduction of iron ore: a bibliographical survey, Commission of the European Communities. — London: Metals Society, VIII, 633 pages, illustrated, 8° 1979.
Direct reduction and its application. *Steel Times* **208** (1980), No. 1, 64–68; No. 2, 125–28.

3.4 Alloys

1. Legierungstechnik. *Stahl u. Eisen* (a review) to be published shortly.
2. Schuh, R., and G. Spiegel, *Formate und Analysen von Desoxid at ionsaluminium.* Radex Rdsch. 1977, p. 139–47.
3. Sames, C.-W., *Die Zukunft der Metalle.* Frankfurt/M 1971. (Suhrkamp-Wissen. 17.)
4. *Metallurgie der Ferrolegierungen.* 2nd edition R. Darrer, G. Volkert and K.-D. Frank. Berlin/Heidelberg/New York. 1972.
5. Leitner, F., and E. Plöckinger, *Die Edelstahlerzeugung.* 2nd edition by E. Plöckinger and H. Straube. Vienna/New York, 1965.
6. Houdremont, E., *Handbuch der Sonderstahlkunde.* 2nd edition, 2 volumes. Berlin/Göttingen/Heidelberg and Düsseldorf, 1956.
7. Rapatz, F., *Die Edelstähle.* 5th edition. Berlin/Göttingen/Heidelberg, 1962.
8. Omelin-Durrer, *Metallurgie des Eisens* 4th edition, Vol. 7, H. Trenkler and W. Krieger. Berlin/Heidelberg/New York, to be published.

9. Figures taken from Deutsche Normen; leaflets obtainable from Beuth-Vertrieb GmbH, Burggrafenstr. 4–10, 1000 Berlin 30 and Kamckestr. 2–8 5000 Köln 1.
10. Figures from information provided by GFE, Gesellschaft für Elektrometallurgie mbH Düsseldorf, 1977.
11. Figures from information provided by Climax Molybdenum GmbH Düsseldorf, 1977.
12. Figures from information provided by INCO, International Nickel Deutschland GmbH Düsseldorf, 1977.
13. Figures from information provided by Süddeutschen Kalkstickstoff-Werke AG, Trostberg/Obb., 1977.

Dowding, Michael F., The world of metals. *Metals & Mater.* 1978, No. 7, 27–36.
Ferro-Alloys. *Metal. Bull. Monthl.* No. 92, 1978, 37, 39, 41, 43, 45, 47, 49–50.
Ferro Alloys Survey 1979. A world survey of ferro alloy markets and directory of producers. [5th edition] Ed. by John Parry. Compiled by Ruby Packard. London: Metal Bulletin 1979. 170 pages, many illustrations and tables.
Netschert, Bruce C., The future availability of raw materials. *J. Metals* **30** (1978), No. 8, 11–15.
Irving, Robert R., Converting coal to oil: a big materials challenge. *Iron Age 222* (1979), No. 5, 39–42.
Walters, C. N., Tailored ferro-nickels for the alloy steelmaker. *Steel Times* **206** (1978) No. 8, 685–8, 690.
Kersaint, Charles de, Les ferronickels (French). *Ind. min.* **60** (1978), No. 2, 79–84.
South African metals and minerals. *Metal Bull. Monthly* 1978, Suppl. April, 67, 69–71, 73, 75, 77–83, 85–87, 89–93, 95, 97–101, 103, 105, 107–9, 111–15, 117–19, 121–3, 125–8.

3.5 Electrodes And Resistance Rods

1. Lefrank, P., Herstellung von Kunstkohle und Elektrographit unter besonderer Berücksichtigung der Entwicklung von Graphitmaterialien für Stranggießanlagen. Paper given before the Fachausschuß: 'Kontinuierliches Gießen' in Nürnberg on 8 May 1973, Deutsche Gesellschaft für Metallkunde e. V., 6370 Oberursel.
2. Mantell, Ch. L., *Carbon and Graphite Handbook*, New York/London/Sydney/Toronto, 1968, p. 9.
3. Pirani, M., *Elektrothermie*. 2nd edition 1960. Berlin/Göttingen/Heidelberg, 1960.
4. Ernst, H., and D. Zöllner: Belastbarkeit von Graphitelektroden. *Elektrowärme*, Issue B, **35** (1977), 231–2; Ernst, H., and D. Zöllner; Arbeiten der deutschen Elektrotechnischen Kommission (DKE) zur Vereinheitlichung der Kennzeichnung von Graphitelektroden für Lichtbogenöfen. *Stahl u. Eisen* **97** (1977), 998.
5. Dönges, E., and O. Vohler, Elektrographit und Kunstkohle. In *Chemische Technologie*. Vol. 1. Anorganische Technologie I. 3rd edition K. Winnacker and L. Küchler. Munich 1970. pp. 465–512.
6. Zöllner, D., Elektroden und Elektrodenvorgänge in elektrothermischen Prozessen, Tagung der Gesellschaft Deutscher Chemiker, Fachgruppe Angewandte Elektrochemie, Erlangen, 7./8. Oct. 1976.
7. Riley, M. W., The new world of ... carbon and graphite. *Mater. Des. Engng.* **56** (1962), Sept., 113–28.

8. Kunstkohle- und Graphitgesellschaft mbH Nürnberg. Company publication entitled 'Elektrographit und Kunstkohle'.
9. SIGRI Elektrographit GmbH, Meitingen. Company publication entitled 'Graphitelektroden für Elektrostahl-Öfen'.
10. Schroth, P., Monitoring and improvement of electrode performance. *Ironmaking & Steelmaking* **3** (1976), 321–32.
11. Eigenschaften von Elektrographit und deren Auswirkung auf den Einsatz von Elektroden in Lichtbogenöfen. Part 1. Semmler, J., Die Temperaturabhängigkeit verschiedener Eigenschaften von Elektrographit. *Stahl u. Eisen* **87** (1967), 364–8. — Part 2. Nedopil, E., and H. Storzer, Measurement of the temperature distribution in the tract of electrodes used in the electric arc furnace (German). *Stahl u. Eisen* **87** (1967), 368–73.
12. Kegel, K., J. Granz and H. Kirchhoff, Maßnahmen zur Stabilisierung von Lichtbögen in Drehstromlichtbogenöfen zum Stahlschmelzen zur Vermeidung von Netzrückwirkungen. *Elektrowärme* **26** (1968), 356–67, 453–65.
13. Etterich, O., 70 years steelmaking in the arc furnace (German). *Stahl u. Eisen* **95** (1975), 1028–31.
14. Harms, F., and K. Fröbodt, Installation of hollow electrodes in large electric arc furnaces (German). *Stahl u. Eisen* **85** (1965), 72–7.
15. Kegel, K., W. Lauterbach, B. Reichelt and D. Zöllner, Der Einfluß der Dotierung von Graphitelektroden auf das Verhalten von Lichtbögen in Lichtbogenöfen zum Stahlschmelzen. In *VII. Internationaler Elektrowärmekongreß* 1972, Warsaw, Report No. 103.
16. Lowry, J. D., and M. P. Schlienger, The anatomy of a non-consumable arc system. In *Proceedings of the 3rd International Symposium on Electroslag and other Special Melting Technology*, Pittsburgh, June 1971, p. III.
17. Holzgruber, W., and M. P. Schlienger, New developments in nonconsumable electrode melting. In *Proceedings of the 5th International Conference on Vacuum Metallurgy and Electroslag Remelting Processes*. Munich, 11–15 Oct. 1976, pp. 107–9.
18. Zöllner, D., F. Schieber, W. Lippert, F. Rittman and Chr. Zöllner, Neueste Entwicklung auf dem Gebiet der Elektroden für Lichtbogenöfen. *Elektrowärme*, Issue B. **38** (1980), 246–8.

Electrode connection *à la* IHI: another new frill in EF steelmaking. *Metal Producing* **16** (1978), No. 10, 53.

Koga, Minoru, Yoshio Yanagi, and Akinori Nakamura, Development of the electrode automatic connection device for an arc furnace. *Engng. Rev.* **11** (1978), No. 4, 49–56 (AKS E46.14).

Electrode insertion the easy way. *Iron Age Metalworking Intern.* **17** (1978), No. 11, 16MP15.

Kurzinski, Ed., Electric furnace steelmaking with argon — hollow electrode practice. *Iron Steel Eng.* **55** (1978), No. 12, 62–4.

3.6 Refractory Materials

1. Konopicky, K., *Feuerfeste Baustoffe*, Düsseldorf, 1957.
2. Harders, F., and S. Kienow, *Feuerfestkunde*, Berlin.
3. Feldhus, H. G., *Feuerfeste Stoffe in der Stahl- und Eisengießerei*, **Berlin**, 1970.
4. Chesters, J. H., *Refractories Production and Properties*, London, 1973 (ISI publ. No. 154).

5. Chesters, J. H., *Refractories for Iron- and Steelmaking*, London, 1974. (Metals Society Book No. 155).
6. Agst, J., *Die feuerfesten Baustoffe*. 4th edition. Kapellen/Moers 1969. (Study folder No. 11)
7. Stahl-Eisen-Werkstoffblatt 910–64. Eigenschaften feuerfester Steine. Aug. 1964.
8. Stahl-Eisen-Werkstoffblatt 915–1974. Schamottesteine für Hochöfen und Winderhitzer. June 1974.
9. VDI-Richtlinie 3128. Sheet 4. Physikalische Stoffeigenschaften nichtmetallischer Werkstoffe (Ofenbaustoffe). Jan. 1977.
10. Majdič, A., and G. Routschka, Feuerfeste Baustoffe für den Elektrolichtbogenofen zum Stahlschmelzen im Spiegel der Literatur. *Keram. Z.* **25** (1973), 529–36, 602–10.
11. Gelsdorf, G., and F. Schellberg, Moderne Baustoffe für hohe Temperaturen: Massen, Feuermörtel und Feuerkitte. Ziegelind. **28** (1975), 436–46.
12. Arbeiten und Empfehlungen der Vereinigung europäischer Hersteller von feuerfesten Erzeugnissen (PRE); Feuerfeste Erzeugnisse 1972. PRE, 8023 Zürich, Löwenstr. 31 (New edition in print).
13. *DIN Standards and Draft Standards*. Beuth-Verlag GmbH. Cologne — republished every year. (Standards for refractory building materials are under Groups 500, 501, and 502.)
14. Stahl-Eisen-Werkstoffblatt 916–73. Kennzeichnung ungeformter feuerfester und feuerbeständiger Erzeugnisse durch einen Kennziffernschlüssel. Oct. 1973.
15. Einheitsformate für feuerfeste Steine, Schamottesteine für den Feuerungsbau, 1971. Obtainable from: Verband der Deutschen Feuerfest-Industrie, An der Elisabethkirche 27, 5300 Bonn.
16. Vereinheitlichung der Steinformate für Lichtbogenofendeckel, Vereinheitlichung feuerfester Baustoffe für Gießpfannen. Berichte des Stahlwerksausschusses des VDEh. Obtainable from Verlag Stahleisen mbH.)
17. Richtlinien für die Beladung von Paletten mit feuerfesten Steinen. Verband der Deutschen Feuerfest-Industrie, An der Elisabethkirche 27, 5300 Bonn.
18. 23. PRE-Empfehlung — 1976. Maßtoleranzen für geformte dichte und wärmedämmende feuerfeste Erzeugnisse.
19. 31. PRE-Empfehlung — 1977. Empfehlung zur Prüfung der technischen Vorschriften für die Lieferung feuerfester Erzeugnisse. 'Arbeiten und Empfehlungen der PRE'. PRE, 8023 Zürich, Löwenstr. 31.
20. DIN 51061. Part 2. Prüfung keramischer Roh- und Werkstoffe; Probenahme; Keramische Rohstoffe und feuerfeste ungeformte Erzeugnisse. Draft Sept. 1976. Part 3 Probenahme; Feuerfeste Steine. July 1973.

Hicks, James C., Refractory raw materials — status of supply. *Amer. ceram. Soc. Bull.* **56** (1977), No. 9, 774–6.

Hubble, D. H., and R. L. Wessel, Design and selection of refractories used in continuous casting. *Ironmaking & Steelmaking* **4** (1977), No. 5, 285–91.

Houseman, D. H., Refractories for iron and steelmaking. *Steel Times* **205** (1977), No. 11 (Intern.-No.), 178–9, 182, 184, 186.

Chaklader, A. C. D., Model refractories. *Trans. & J. Brit. ceram. Soc.* **76** (1977), No. 6, 138–41.

Marion, Robert H., and J. Keith Johnstone, A parametric study of the diametral compression test for ceramics. *Amer. ceram. Soc. Bull.* **56** (1977), No. 11, 998–1002.

Knoche, Ch., M. Koltermann and G. Stolte, 28 Refractories for vacuum steelmaking. *Trans. & J. Brit. ceram. Soc.* **76** (1977), No. 6, 141–8.

Cross, M., B. A. Lewis, J. K. Forster and G. A. Carter, Strain generation during the firing of deusa silica shapes. *Refractories J.* **53** (1978), No. 1, 10–18, 20–22.
Johnson, W. C., Grain boundary segregation in ceramics. *Metallurg. Trans.* **8A** (1977), No. 9, 1413–22.
Esnoult, M., and J. Mercier, Nouveautés dans le domaine des réfractaires basiques électrofondus pour fours éléctriques. *Circ. Inform. techn.* **35** (1978), No. 1, 39–49.
Lecrivain, L., Développement de réfracteurès pours fours éléctriques d'aciérie. *Circ. Inform. techn.* **35** (1978), No. 1, 51–65.
Binns, D. B., The use of special ceramics in handling molten iron and steel. *Trans. & J. Brit. ceram. Soc.* **77** (1978), No. 1, 1–13.
Refractories — current trends and development. *Steel Times* **206** (1978), No. 5, 402, 404, 407–8, 411–12, 415–16, 418–20, 421–22, 424.
Houseman, D. H., Refractories and the newer steelmaking processes — 28 continuous steelmaking processes. *Steel Times* **206** (1978), No. 5, 457–62.
Greaves, E. I., Stress–strain properties of high alumina and basic refractories and their use in arc furnace roofs. *Refractories J.* **53** (1978), No. 3, 13–20, 22.
Baker, Bruce H., and Richard L. Shultz, Aluminosilicate refractories for electric arc furnace roofs and steel ladles refractories and the newer steelmaking process — 29. *Amer. ceram. Soc. Bull.* **57** (1978), No. 7, 667–9, 672–3.
Houseman, D. H., The future — processes, problems and prospects. *Steel Times* **206** (1978), No. 10, 903, 906–11.
Leonard, L. A., Dolomite and silica — survival or revival? *Refractories J.* **53** (1978), No. 5, 12–16, 18–19, 22–4, 26–7.
Baum, Rudolf, and Heinz Zörcher, Dolomite bricks in steel-works ladles. *Metallurg. Plant & Technol.* **1** (1978), No. 4, 38–40.
Leonard, L. A., and G. M. Workman, Modern techniques reduce steel plant refractories consumption. I Ironmaking. II Metal handling steel-making and mill furnaces. *Metallurg. & Mater. Technol.* **10** (1978), No. 11, 581–7; No. 12, 642–4, 646–7.
Hodson, P. T. A., and G. C. Padgett, Current trends in the use of refractories. *Ind. Heat.* **46** (1979), No. 2, 48–50, 52.
Hardy, C. W., Mythology, technology and science. *Refractories J.* 1979, No. 2, 10–12, 14–16.
Gilchrist, J. D., Development in the use of refractory materials. *Metallurgia*, Redhill, **46** (1979) No. 6, 391–4, 396–8, 400–1.
Spencer, D. R. F., Refractories in the eighties. *Refractories J.* **54** (1979), No. 3, 12–17, 19–20, 22–4, 26–8.
Collins, J. Dallas, Refractory use in AOD vessels. *Iron & Steelmaker* **6** (1979), No. 7, 11.

3.7 Additives, Slag-formers, Carburizing Agents, and Casting Aids

1. Schiele, E., and L. W. Berens: Kalk. Düsseldorf, 1972.
2. Berens, L. W., H. J. Hoppe and H. I. Masuhr, Beitrag zur Verbesserung der Kalkauflösung bei der Stahlherstellung. Paper given at the Paritätischen Kalkausschuß 6 June 1968 in Dornap.
3. Obst, K.-H., J. Stradtmann and W. Münchberg, Untersuchungen zur Frage des Nachbrennens von Kalk im Aufblaskonverter. *Tonind.-Ztg. keram. Rdsch.* **93** (1969), 328–31.

4. Bardenheuer, F., H. vom Ende and P.-G. Oberhäuser, Lime solution, slag formation and life of dolomite lining when blowing low-phosphorus pig iron in oxygen converters (German). *Stahl u. Eisen* **88** (1968), 1285–90.
5. Davydova, L. N., Die Güte von Baustählen nach Raffination mit flüssigen synthetischen Schlacken in der Pfanne. *Stal in Deutsch* **3** (1963), 417–22.
6. Zelinskij, V. F., and D. G. Žukov, Selbstkostensenkung bei synthetischer Schlacke. *Stal in Deutsch* **8** (1968), 803–5.
7. Ullrich, W., W. Meichsner, K. Nürnberg, and W. Rudack, Effect of addition of bauxite on the metallurgical reactions in the basic open-hearth process (German). *Stahl u. Eisen* **89** (1969), 918–27.
8. Klages, R., Gleichgewichte zwischen flüssigem Eisen und gesättigten Schlacken des Systems CaO-FeO_n-MnO-SiO_2 unter Berücksichtigung der Schwefelverteilung. Clausthal 1968. (Dr.-Ing.-Diss. Techn. Hochsch. Clausthal.)
9. Piepenbrock, R., Untersuchungen zur Auflösung von Kalk in synthetischen Schlacken unter besonderer Berücksichtigung von Zusätzen zum Kalk. Unpublished diploma thesis, Techn. Hochsch. Clausthal, 1967.
10. Takeda, M., On the oxygen converter. *Tetsu to Hagané* (1962), 1085–8.
11. Wakabayashi, K., and M. Takeda, O. Tsubakihara and K. Sasaki, Verwendung von Kalk mit Zusätzen von Fe_2O_3 im Sauerstoffaufblas-Konverter. *Tetsu to Hagané* **51** (1965), 737–40.
12. Tamamoto, S., K. Iwase and K. Yoshida, Der Einfluß der Zugabe von Walzzunder beim Sauerstoffaufblas-Verfahren. *Tetsu to Hagané* **51** (1965), 740–3.
13. Leitner, F., and E. Plöckinger, *Die Edelstahlerzeugung*. 2nd edition by E. Plöckinger and H. Straube. Vienna/New York 1965. p. 674–82.
14. Jauch, R., A. Choudhury, H. Löwenkamp and F. Regnitter, Paper on Herstellung großer Schmiedeblökke nach dem Elektro-Schlacke-Umschmelzverfahren. In *Internationaler Eisenhüttenentechnischer Kongreß*, Düsseldorf, 1974, [27.-30. May]. Vol. 3. [Düsseldorf] 1974. Report 4.2.2.1. 13 pages.

Mills, Norman T., and Billiyar N. Bhat Development of continuous casting mold powders. *Iron & Steelmaker* **5** (1978), No. 10, 18–24.
Takeuchi, Hidemaro, Hisashi Mori, Toshiaki Nishida, Takashi Yanai and Katsumi Mukunashi, Entwicklung kohlenstofffreier Gießpulver für das Stranggießen von Stählen (Japanese). *Tetsu to Hagané* **64** (1978), No. 10, 1548–57.

3.8 Discard Materials

1. Haucke, M., Disposal of waste products in iron and steel works (German). *Stahl u. Eisen* **90** (1970), 348–54.
2. Kraß, K., and W. Fix, LD-Schlacke — ein neuer Mineralstoff im Straßenbau. *Straße u. Autobahn* **28** (1977), 326–34.
3. Blunk, G., W. Fix and K. Kraß, Möglichkeiten der Verwertung von Stahlwerksschlacken. Paper given at the annual meeting of the Eisenhütte Südwest on 13/14 May 1976 in Saarbrücken.
4. Gutt, W., and P. J. Nixon, Steelmaking slag as a skid resistant roadstone *Chem & Ind*. 1972, 503–4.
5. Emery, J. J., Verwendung von Stahlwerksschlacken beim Bau von Landstraßen. *Silicates Ind*. **42** (1977), 209–18.
6. Niesel, K., and P. Thormann, Die Stabilitätsbereiche der Modificationen des Dicalciumsilicats. *Tonind.-Ztg. keram. Rdsch,* **91** (1967), 362–9.

7. Schwiete, H. E., and others, Existenzbereiche und Stabilisierung von Hochtemperaturmodifikationen des Dicalciumsilicats. *Zement Kalk Gips* **21** (1968), 359–66.
8. Gorges, H., H. Graf, H. Lutz, P. G. Oberhäuser and H. Mülders, Experiences with the AOD process for the manufacture of stainless steels (German). *Stahl. u. Eisen* **96** (1976), 1251–8.
9. Bauer, H., K. Behrens and M. Walter, Vacuum refining of high-chromium steels in the converter (German). *Stahl u. Eisen* **97** (1977), 938–44.
10. Gesetz über die Beseitigung von Abfällen. Bundesgesetzbl. 1977, Part 1, p. 42–51.
11. Meyer, G., K.-H. Vopel and W. Janssen, Investigations of the utilization of dust and sludge from the waste-gas cleaning systems of blast furnaces and BOP steelmaking plants in the rotary kiln (German). *Stahl. u. Eisen* **96** (1976), 1228–33.
12. Maczek, H., H. Rellermeyer, G. Kossek and H. Serbent, Tests for processing metallurgical wastes using the rotary-kiln volatilizing method in an industrial plant (German). *Stahl. u. Eisen* **96** (1976), 1233–8.
13. Sugasawa, K., Y. Yamada, S. Watanabe, K. Kato, K. Matsuda, Y. Sato and T. Kawabata, Direct reduction of metallurgical dust (German). *Stahl u. Eisen* **96** (1976), 1239–45.
14. Amelung, E., K. H. Steinbacher, J. Otto and A. Schneider, On the dust collection of scrap-practice open-hearth furnaces (German). *Stahl u. Eisen* **97** (1977), 674–80.

Electric Furnace Steel Production
Edited by E. Plöckinger and O. Etterich

4 Basic Principles of Electric Furnaces

4.1 BASIC PRINCIPLES OF ARC FURNACES

KLAUS TIMM, HAMBURG

This section deals with the electrical theory of arc furnaces and particularly the reactance and operational behaviour of the high-current circuit. The problem arising when connecting arc furnaces to mains supplies will also be indicated.

4.1.1 High-current Circuit

The high-current circuit of an arc furnace using the Héroult system has three phases star-connected to the three-phase mains supply. As shown in Figure 1, each of the three phases 1, 2, and 3 has a supply (A) from the transformer, a flexible furnace cable (B), a tubular conductor on the electrode arm (C), and an electrode (D). The three arcs (E) burn at the foot of the electrodes against the melt (F) which thus forms the floating star-point of the three-phase system. The geometry of the high-current conductors determines the three reactances; these have to be kept low, and the lengths and distances between the high-current conductors must therefore be short. The reactances will be symmetrical if the distances between the three transformer-to-furnace supply conductors are as equal as possible over their whole length, right down to the melt.

4.1.1.1 Electric arcs

An arc is a gas discharge between two electrodes (anode and cathode) which are supplied by a voltage source. In an arc furnace the arc burns under atmospheric pressure between the end of the graphite electrode and the steel melt. The arc is sustained by an alternating current; this means that it must strike anew every half-cycle, and the melt and electrodes alternate as anode and cathode. The ionization necessary to enable the arc to burn is created in the arc furnace by thermionic emission from the cathode spot at about 4000 K.[1]

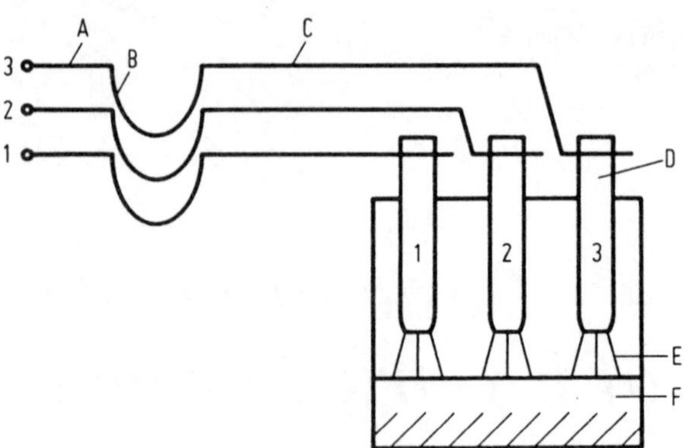

Figure 1. High-current circuit for arc furnace

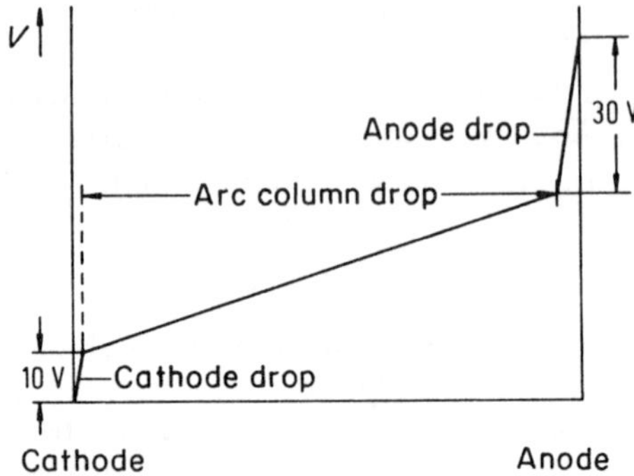

Figure 2. Voltage-distribution in an arc

The arc voltage, between anode and cathode, comprises the cathode and anode volts drops and the voltage drop in the gas column (Figure 2). The anode drop (about 30 V) and the cathode drop (around 10 V) are essentially physical phenomena which occur in small areas of high field-strength in the anode and cathode. A voltage of at least 40 V and a current of 4 kA are required to maintain ionization and produce the heat of the arc.

Calculation of the electrical properties of a.c. arcs is difficult;[2] the dynamic behaviour of the arc in the furnace depends on several variables which are not properly understood. No satisfactory model has so far been proposed. The arc properties most important for the operation of an arc furnace will now be described.

The electrical behaviour of an arc is described by its voltage and current, but it is difficult to measure arc voltage (see section 4.1.2). Figures 3 to 5 show examples of traces recorded at an 80-tonne, 30-MVA arc furnace. The 'a' diagrams on the left show arc voltage and current as functions of time, $v_a(t)$ and

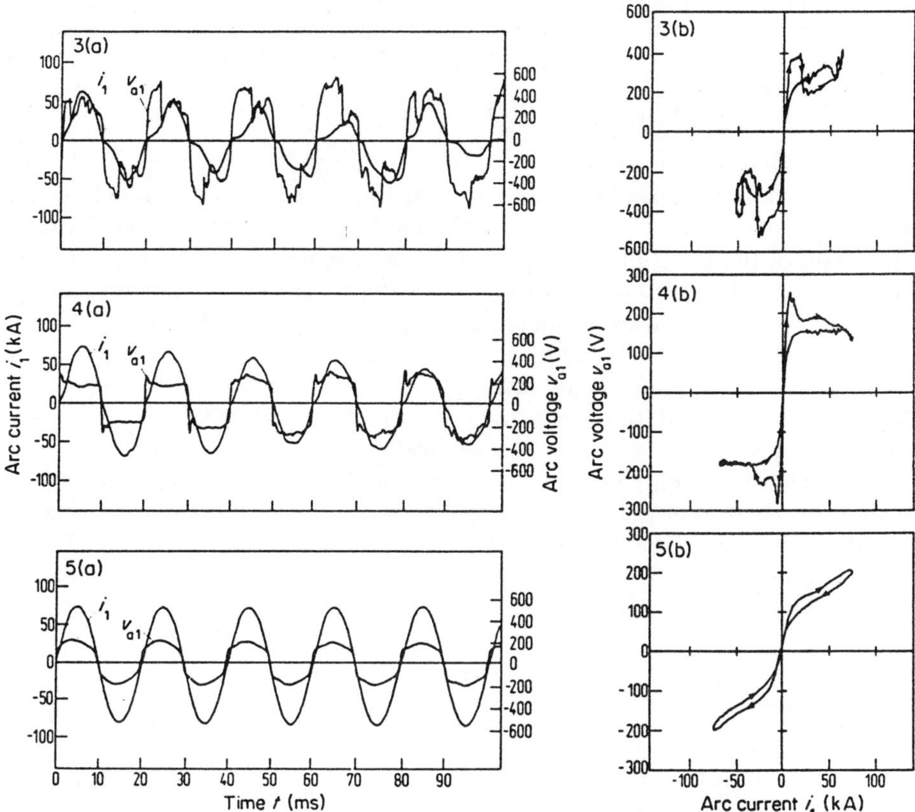

Figures 3–5. Trace records of arc voltage v_{al} and current i_l (diagrams (a) and arc characteristic $v_{al} = f(i_l)$ (diagrams (b)). Figure 3. Measured about 9 min after start of melting the first basket; Figure 4. Measured about 7 min after start of melting the second basket; Figure 5. Measured about 40 min after start of melting the second basket

$i(t)$, respectively; the 'b' diagrams at the right show arc voltage v_a (i) as a function of current. A typical arc is unstable and non-linear, and its current and voltage are in-phase as they pass through zero. This instability and non-linearity are greatest when melting down cold scrap. The delayed and erratic process of striking the arc and the resulting gaps in the current are conspicuous in Figure 3. As melting down progresses, the striking becomes more stable, but the current can still contain low-frequency fluctuations (Figure 4). The temperature and heat of an arc are high with a liquid steel bath, and the thermal conduction is low, so that the arc-characteristic begins to approach the linear behaviour of an ordinary resistance (Figure 5).

The arc voltage waveform is almost rectangular; its amplitude depends on the current and — after overcoming the anode and cathode voltage drops — grows linearly in proportion to the arc length l with a slope of about 1 V/mm:[3,4]

$$v_a = 40\ \text{V} + 1\ \text{V/mm} \cdot l \qquad (1)$$

This rectangular arc voltage waveform is not entirely adopted by the supply current since inductance in the circuit gives rise to inductive reactance which increases with frequency and therefore resists harmonics. In addition, the free star-point of the three-phase system also works against harmonics. Although the current waveform in Figures 3–5 is almost sinusoidal, the noticeable harmonic content[5] causes an increase in reactive power and with it a higher reactance (see section 4.1.1.3).

The magnetic fields of the adjacent phases exert horizontal electrodynamic forces on the arc, with the result that the arc, which has virtually no inertia, is deflected sideways away from the vertical, towards the furnace wall. For symmetrical sinusoidal currents the geometric locus of the power phasor is a circle in a horizontal plane. It is traversed twice in every cycle of the mains frequency in a direction determined by the phase sequence (Figure 6)[6]. It has been confirmed by experiments that the arc can be deflected into circular motion without hindrance from inertia.[7]

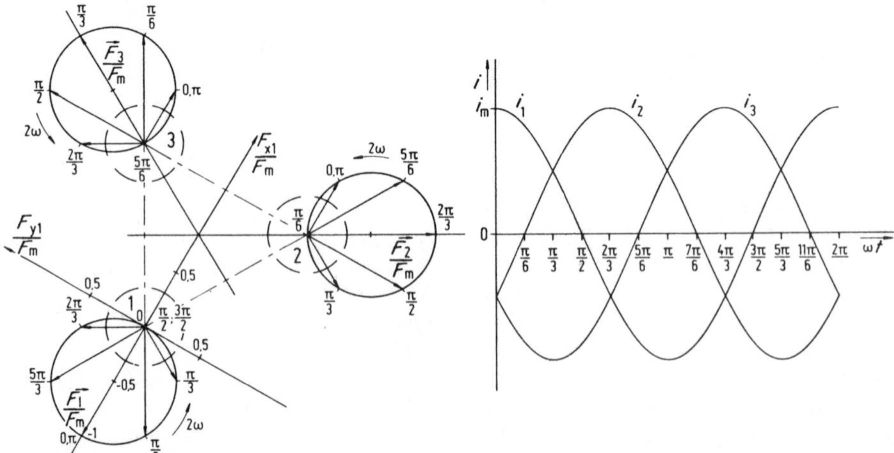

Figure 6. Development in time of electrodynamic forces on the arcs and electrodes for sinusoidal currents and phase-sequence 1, 2, 3[6]

The way the arc fluctuates is especially important during melting down. Large movements of the arc roots on the electrodes cause considerable arc-voltage fluctuations and hence current-changes as well. This causes unacceptable changes in voltage and current in mains supplies which have insufficient power and short-circuit capacity.[5] (See section 4.1.4.2.)

Hydrodynamic forces can act upon the arc cathode even in the absence of an external magnetic field. 'Pinch effect' — necking of the arc on the cathode — produces a recoil force P on the cathode in the direction of the arc axis.[8] If we write r_c for the mean radius of the arc column and r_{ca} for the radius of the cathode hot spot, then we have

$$P = 10^{-7} i^2 \ln r_c/r_{ca} \text{ (N)} \tag{2}$$

The arc displacement increases with the extent of the necking and the square of the current. A 10-kA arc with a necking factor of $r_c/r_{ca} = 4$ experiences a force of 15 N. The force grows to 1500 N at 100 KA with the same necking factor.[8]

In this way high-current arcs produce a desirable stirring effect in the melt (Figure 6).

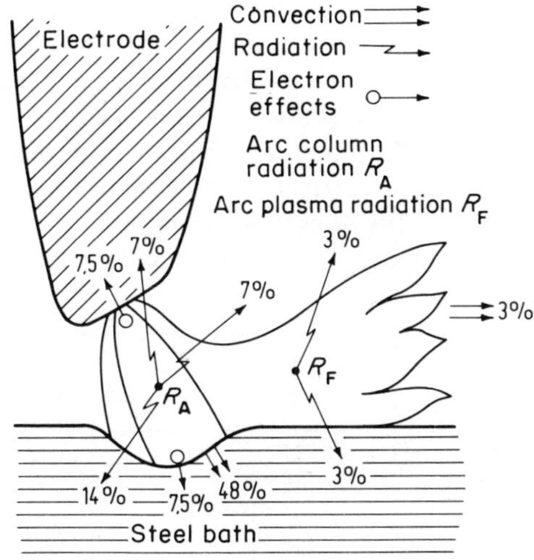

Figure 7. Energy-balance for a 7-kA arc at 143 V a.c.[9]

The arc energy balance picture shows that only part of the arc power is transferred to the furnace charge. Figure 7 illustrates an experiment which shows that electrothermal energy-conversion occurs in the arc column and arc plasma.[9] 72.5% of the electric power of the arc is transferred into the melt; 14.5% goes into the electrode; and 13% to the furnace wall. The proportions depend on the arc length and the slag thickness.

Attention has to be paid to refractory lining-wear caused by the arc. The wall is attacked mechanically by the slag and metal splashes produced by the arc thrust and thermally by the arc radiation. Schwabe defined a refractory index R_E

$$R_E = V_A \, P_A \tag{3}$$

as the product of the arc voltage V_A and the arc power P_A,[10–12] and this has been widely adopted.

ASEA in Sweden have recently carried out basic research into the use of d.c. arcs in electric furnaces for steel.

4.1.1.2 Electrodes (see section 3.5.5)

Graphite is used as electrode material because its low work-function and high melting-point. The electrodes are actually the weakest part of the system at maximum furnace power.[13] Apart from special cases, the largest electrodes available at present have a diameter of 600 mm; these can deal with currents up to 85 kA.[12] The following problems arise if we increase the electrode current-carrying capacity by increasing electrode cross-section:

The alternating magnetic field of the electrode current induces voltages in the conductor which produce eddy currents. In the centre of the conductor these eddy currents oppose the main current; but near the surface, the eddy currents increase the total electrode current. In this way the current is concentrated close to the conductor surface (skin effect,[14–16] Figure 8). Moreover, the cylinder–symmetrical current density distribution becomes distorted in the direction of the next following phase by mutual electrode field effects (proximity effect, Figure 9). Both skin and proximity effects lead to an

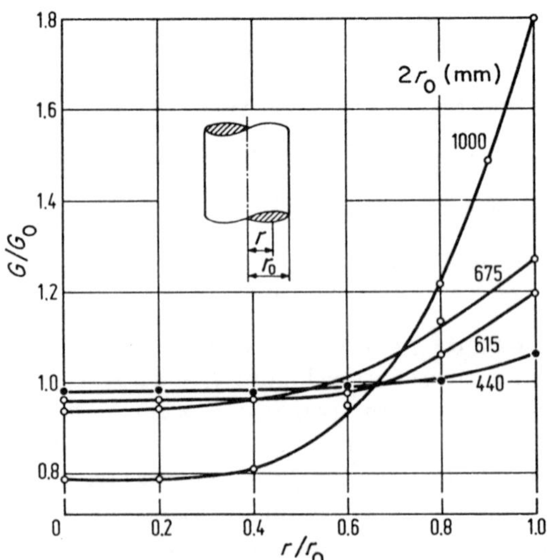

Figure 8. Effective value of the current-density of an electrode (without proximity effect) as a function of radius; parameter: electrode-diameter[16]

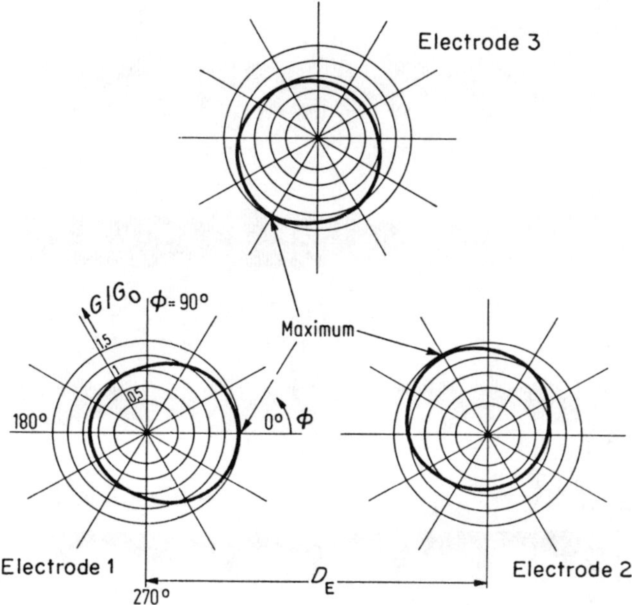

Figure 9. Effective value of cylinder-shell current-density in the electrodes of a three-phase electrode arrangement expressed as a function of phase-angle ϕ; pitch circle diameter $D_p/2r_o = 2.7$; $2r_o = 615$ mm; phase-sequence 1, 2, 3[16]

increase in resistance, which grows as the electrode diameter becomes larger. Therefore, enlarging the cross-section creates zones in the centre of the electrode which contribute little to current-conduction. On the other hand, the specific power-loss density is also concentrated in a particular region of the cross-section due to proximity effect. It is quite possible that the resulting asymmetrical temperature-rise will cause increased burning of the electrode sides and give rise to more crack-formation due to thermal stress.[16]

Current density is more evenly distributed across the cross-section of tubular electrodes; however, although this is one way of achieving a stable arc,[3] tubular electrodes are not universally used.

We must also bear in mind that the electrodynamic forces in the horizontal plane, which deflect the arc towards the furnace wall, also act on the electrodes themselves (Figure 6).[6]

The forces increase in proportion to conductor-length and the square of furnace current. Above all, in the event of a short circuit, there are considerable bending stresses on the electrode mountings which can lead to fracture at the screw-threads. In a high-current arc furnace it is known that a right-hand field can unscrew an electrode attached by a right-hand thread.

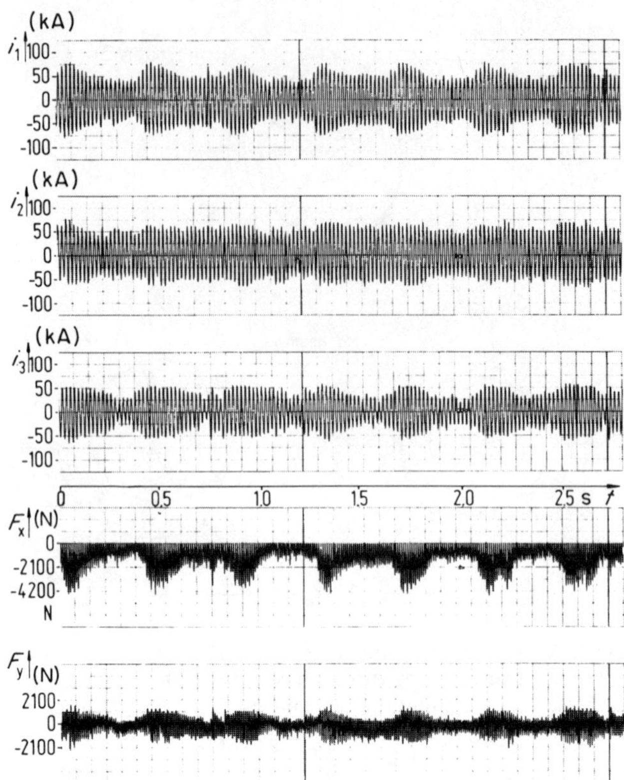

Figure 10. Trace records of currents i_1, i_2, i_3 and forces F_x and F_y on electrode 1 of an arc furnace.[6] Direction of force as in Figure 6

These electrodynamic forces acquire low-frequency components with low-frequency changes in furnace current (Figure 10). These components excite the mechanical system of the electrode and its support-arm into harmonic mechanical vibrations, as shown in the test record depicted in Figure 11.[17] The vibrations cause bending stresses and loosen screwed connections.

4.1.1.3 Single-phase equivalent circuit diagram for an arc furnace

The electrical behaviour of an arc furnace is heavily dependent on the geometrical relationship between the high-current conductors and the resulting reactances. If we want to derive the equivalent circuit diagram for a three-phase furnace, we must first consider the simplified example of a single-phase arc furnace, with a bottom electrode, as shown in Figure 12(a). This case is identical to the symmetrical three-phase unit considered later.

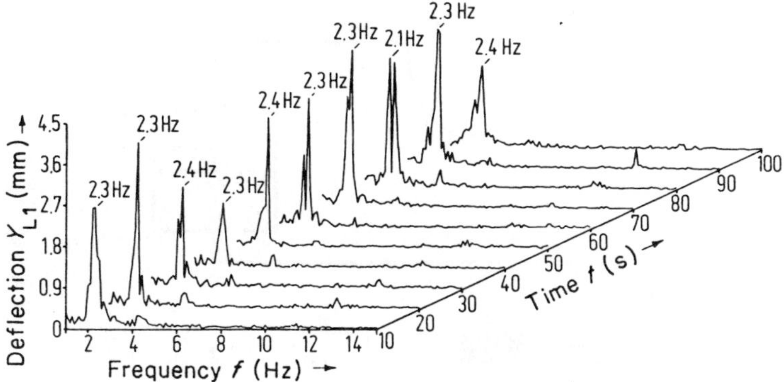

Figure 11. Amplitude–frequency–time diagram of the deflection Y_{L1} (f, t) of electrode 1 in an arc furnace, measured at right-angles to the carrier-arm axis, at the height of the furnace roof, 3 min after the start of melting down the first basket; frequency band 1–15 Hz; resolution $\Delta f = 0.1$ Hz; steps of 10 s per spectrum

The transformer voltage v_{ph} comprises the arc voltage v_A, the voltage-drop v_R over the loss resistance R_R of the circuit, and the induced voltage v_L due to the self-inductance L in the whole current loop 1-M (Figure 12(b)):

$$\begin{aligned} v_{ph} &= v_R + v_L + v_A \\ &= i R_R + L\mathrm{d}i/\mathrm{d}t + v_A \end{aligned} \tag{4}$$

When changing over to the usual linearized equivalent circuit diagram (Figure 12(c)) we must pay attention to the following points. Because the arc voltage is non-linear, besides the 50-Hz fundamental oscillation I_1, the current also contains harmonics I_2, I_3, I_4, ... from the second to the 7th order, where the third harmonic is most prominent.[5] If the transformer voltage is sinusoidal, only the fundamental current oscillation can contribute to the active power P.[14,18] The reactive power Q is obtained from the apparent power S:

$$Q = \sqrt{(S^2 - P^2)} = \sqrt{(Q_1^2 + D^2)} \tag{5}$$

Reactive power Q comprises the fundamental frequency-reactive power Q_1 and the harmonic reactive power D. The reactance X is calculated from the reactive power Q and the square of the effective current:

$$X = Q/I^2 = Q/(I_1^2 + I_2^2 + I_3^2 + \ldots) \tag{6}$$

This apparent operational reactance is not constant but depends on the harmonic content of the current. If the arc is short-circuited ($v_a = 0$), the short-circuit current is purely sinusoidal, and the short-circuit reactance is minimal because $D = 0$. The harmonic content of the current and the operational reactance both increase with rising arc voltage.[19–21] Increased

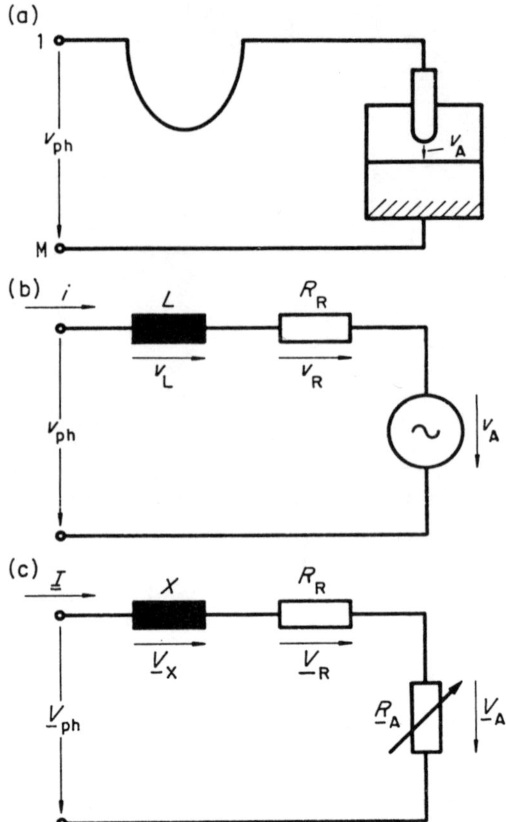

Figure 12. Single-phase arc furnace with bottom electrode. Circuit layout (a), equivalent circuit (b), and simplified linearized equivalent circuit (c)

operational reactance can be observed mainly at the start of scrap-melting when it reduces the power applied to the melt. In another report[22] this power-reduction during the melt-down is explained by statistical variations in linear electrical quantities.

Current harmonics also increase current-displacement effects in the electrodes, so that the loss resistance R_R increases by several per cent.[21]

The equivalent circuit in Figure 12(b) can be simplifed by replacing the non-linear arc by a linear arc resistance R_A. The operational reactance must then be assumed to be 10–15% higher than the short-circuit reactance.[23] The electrical quantities can be described best by using a linearized equivalent circuit (Figure 12(c)) with the help of a complex calculation. Using the loop rule we obtain:

$$\underline{V}_{ph} = \underline{V}_R + \underline{V}_A + \underline{V}_X = \underline{I}\,(R_R + R_A + jX) = \underline{IZ} \qquad (7)$$

where $\underline{Z}$ is the impedance of the current circuit.

The value of the current is:

$$I = V_{ph}/\sqrt{[(R_R + R_A)^2 + X^2]} \qquad (8)$$

This current lags behind the phase voltage by the phase angle ϕ:

$$\phi = -\text{arc tan } X/(R_R + R_A) \qquad (9)$$

4.1.1.4 Three-phase equivalent-circuit diagram for an arc furnace

The equivalent-circuit diagram has proved to be invaluable for describing three-phase arc furnaces.[21,24,25] The reactance of the equivalent circuit is determined by the magnetic coupling of the three loops of the high-current conducters from the transformer. This introduction is based on the idea that any transformer voltage measurement on an arc furnace using a measuring circuit will be inaccurate because of the voltages induced by the strong magnetic fields. This means that the voltages between transformer terminals 1, 2, and 3 and the furnace vessel earth (o) cannot be measured accurately, but the furnace conductor voltages at the transformer can be read without error, because the loop used there is small (see section 4.1.2).

The equivalent circuit of a three-phase arc furnace consists of three phases with inductance, loss resistance, and arc-voltage source connected in series and meeting at a free star-point (o) in the furnace bath (Figure 13(a)).

The effective inductances

$$L_1 = M_{12,13}, \quad L_2 = M_{23,21}, \quad L_3 = M_{31,32} \qquad (10)$$

are the mutual inductances between any two high-current loops. The concept of mutual inductances is illustrated in Figure 14; it arises from the magnetic flux linkage between two different circuits. The voltage induced in loop 1–2 comprises two components which are produced by the fictive circuit currents i_{13} and i_{23}. Thus the mutual inductances act like concentrated circuit elements L_1 and L_2 in phases 1 and 2 (Figure 14(c)).

The equivalent circuit of the three-phase furnace can be linearized — as in the single-phase example — by introducing the operational reactances X_1, X_2, and X_3 (Figure 13(b)). With the phase impedances:

$$\underline{Z}_1 = R_{R1} + R_{A1} + jX_1 \qquad (11)$$

$$\underline{Z}_2 = R_{R2} + R_{A2} + jX_2$$

$$\underline{Z}_3 = R_{R3} + R_{A3} + jX_3$$

we get the following equation system for the furnace conductor voltages:

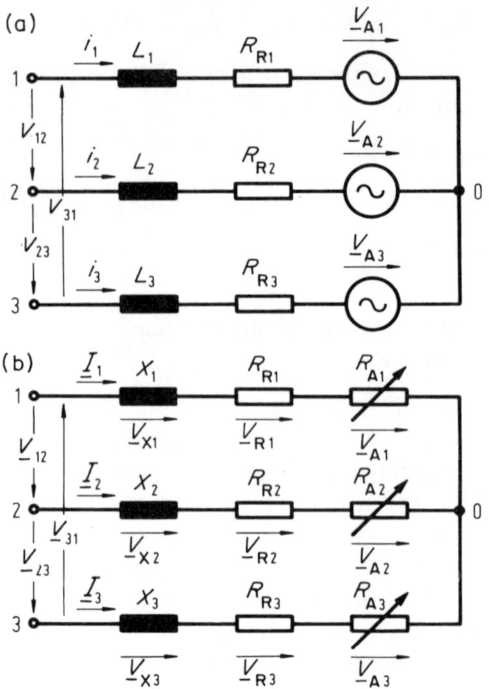

Figure 13. Equivalent circuit (a) and simplified linearized equivalent circuit (b) of a three-phase arc furnace

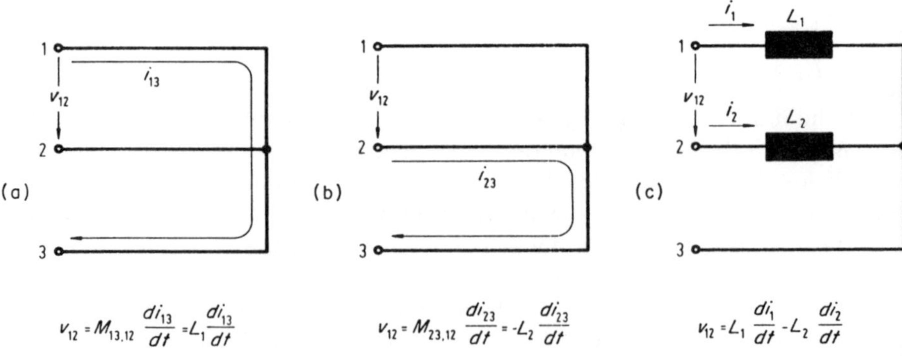

Figure 14. Induced voltages in loop 1–2, caused by (a) circuit currents i_{12} and (b) i_{23} and by (c) superimposition of (a) and (b)

$$\underline{V}_{12} = \underline{I}_1\underline{Z}_1 - \underline{I}_2\underline{Z}_2 \tag{12}$$

$$\underline{V}_{23} = \underline{I}_2\underline{Z}_2 - \underline{I}_3\underline{Z}_3$$

$$\underline{V}_{31} = \underline{I}_3\underline{Z}_3 - \underline{I}_1\underline{Z}_1$$

The furnace phase voltages are:

$$\underline{V}_{10} = \underline{I}_1\underline{Z}_1 \qquad (13)$$
$$\underline{V}_{20} = \underline{I}_2\underline{Z}_2$$
$$\underline{V}_{30} = \underline{I}_3\underline{Z}_3$$

These cannot be measured simply, without error (see section 4.1.2).

Electrical symmetry is especially important in the operation of an arc furnace. If the resistances, reactances, and conductor voltages are balanced, then the furnace currents will be equal, too. The zero potential of the melt is in the centre of the voltage phasor triangle (Figure 15) when the phase voltages are symmetrical: $V_{ph} = V_L/\sqrt{3}$.

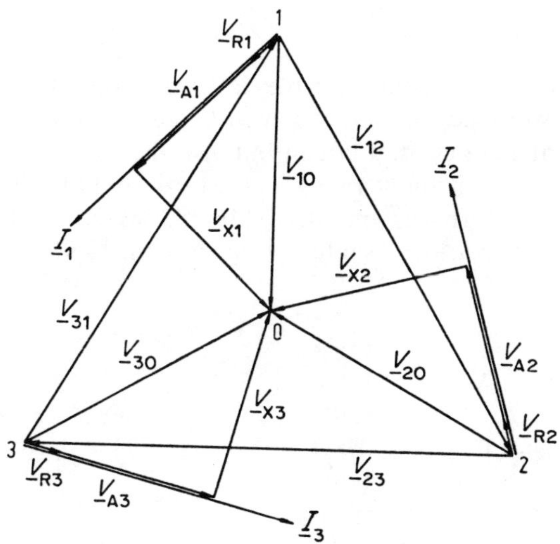

Figure 15. Symmetrical phasor diagram

The furnace current values in the three phases are obtained from:

$$I = \frac{V_L}{\sqrt{(3Z)}} = \frac{V_L}{\sqrt{3}\sqrt{[(R_R + R_A)^2 + X^2]}} \qquad (14)$$

The total apparent power

$$S = \sqrt{(3)} V_L\, I \qquad (15)$$

is divided into reactive power

$$Q = \sqrt{(3)} V_L\, I \sin\phi = 3\, I^2\, X \qquad (16)$$

and active power

$$P = \sqrt{(3)} V_L \, I \cos \phi = P_R + P_A \tag{17}$$

The active power comprises individual power losses

$$P_R = 3 \, I^2 \, R_R \tag{18}$$

and the arc power losses

$$P_A = 3 \, I^2 \, R_A \tag{19}$$

The power factor is

$$\cos \phi = P/S = P/(\sqrt{(3)} V_L \, I_L) \tag{20}$$

4.1.1.5 Optimum reactance

Operational reactance has a decisive influence on efficient arc-furnace operation. In power calculations we have to consider not only the reactance components in the high-current circuit but also the reactance components of the transformers and — if applicable — of the furnace choke. The reactances of ultra-high-power arc furnaces amount to about 3 mΩ at 50 Hz. For example, the reactance of the various conductor sections in Figure 1 is made up as follows:

Transformer	12%
Conductor section A	8%
Flexible cable B	29%
Tubular conductor C	35%
Electrode D	16%

Practical measurement of arc-furnace reactances and loss resistances is carried out under short-circuit conditions, using sinusoidal currents, either on the low- or the high-voltage side of the furnace transformer.[24,26] In the second case the measured amounts are increased by the transformer impedance. Under short-circuit conditions the currents are steady and sinusoidal, and the measured short-circuit reactance is minimal. If we want mainly to know the reactance differences between the phases, then we can determine the individual reactances indirectly in three separate short-circuit tests.[24] However, although the three-phase short-circuit test approximates more closely to the normal working of a furnace,[20] it can only give average impedances for each phase.

The reactances in all the phases must be symmetrical for balanced arc-furnace operation, which means that the high-current conductors must be arranged in a defined configuration. On many of the older arc furnaces the

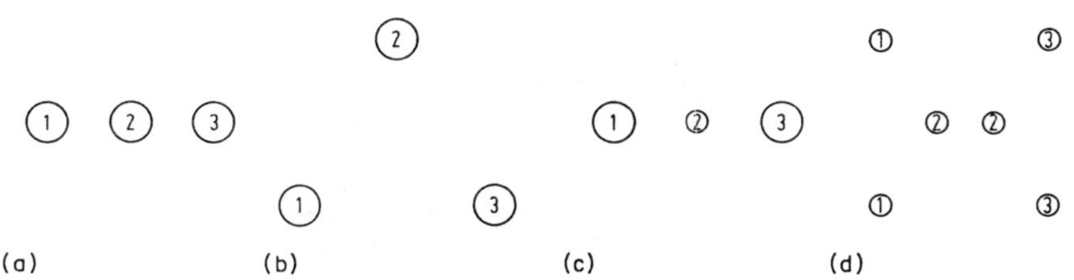

Figure 16. Different arrangements of high-current conductors: (a) co-planar; (b) triangular; (c) modified co-planar; (d) modified co-planar with several conductors (butterfly)

high-current leads to the electrodes are mounted in a straight line for constructional reasons. Such a co-planar conductor arrangement (Figure 16(a)) produces an asymmetrical reactance distribution, with equal reactances in the outer phases and a smaller reactance in the middle phase:[20,24,27,28]

$$X_1 \approx X_3 > X_2 \tag{21}$$

In co-planar systems, the reactance balance factor

$$B_X = \frac{X_2}{(X_1 + X_3)/2} \quad \text{amounts to } 0.68\text{–}0.85 \tag{22}$$

Asymmetrical reactances may cause asymmetrical arc powers and radiation factors, leading to greater attack on the refractory lining near the 'wild phase'. When a change is made to accommodate much higher furnace power or output levels, the layout of the entire high-current circuit must be changed in order to achieve symmetrical reactances. Symmetrical reactances are obtained by triangulating the conductors of the entire high-current circuit (Figure 16(b)).[29] This produces reactance balance factors B_X of 0.94–1.00.

The reactances of co-planar high-current conductors can also be rendered symmetrical by using high-current conductor tubes with different cross-sections. Where there are several conductors the ones for the outer phases can be separated by calculated spaces, while the centre-phase conductors can be set close together[25,30] (Figure 16(c), (d)).

It should also be borne in mind that, in normal operation, the carrier arms may not form an ideal triangle, because the electrodes will be of different lengths; this can cause reactance differences.

4.1.1.6 Balancing

The wild phase is still a problem in operation today, even with modern arc furnaces. One reason is that accurate measurement of individual arc power and radiation factor is very involved (see section 4.1.2). Another reason is

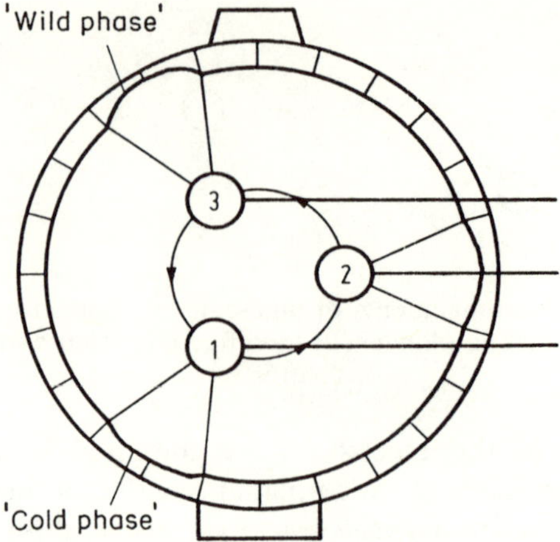

Figure 17. Asymmetrical refractory lining-wear at normal phase-sequence: 1, 2, 3

that the pattern of heat damage in an arc furnace is also governed by non-electric factors, for example by the location of the dust-extraction opening.

We shall now discuss asymmetry in operating an arc furnace with co-planar conductors and unequal reactances. If, with balanced mains supply voltages, we equalize the arc resistances, then we get asymmetrical phase currents. For the normal phase sequence, 1-2-3, I_1 is minimum (cold phase), but I_2 and I_3 reach maximum (wild phases) according to the ratio of resistance to reactance (Figure 17). The currents in the outer phases 1 and 2 exchange values when the phase sequence is reversed.[24] Only one of the three quantities, current, active power, and radiation factor, can be balanced over the arc lengths; the other two will then be asymmetrical. If the arc currents are balanced, the result is asymmetrical wear of the refractory lining,[27] because the radiation factors unavoidably become asymmetrical. If the radiation factors are equal, then the current in only one phase can be fully utilized.[31]

The aim of balancing should be to achieve symmetrical electric values in all three arcs. For arc furnaces equipped with co-planar conductors, this can be achieved either by balancing the reactances or by using asymmetrical transformer voltages.

Reactance differences can also be compensated by passive balancing. This involves rebuilding the conductor installation in triangulated or modified co-planar configuration (see Figure 16). The reactance of the centre phase can also be raised to the value of the outer phases by using a wraparound core with an air gap;[32,33] however, it is not known whether this has been done successfully.

The purpose of active balancing is to equalize differences in reactive voltage components by using asymmetrical supply voltages and thus to achieve symmetrical arc values. However, this makes it more difficult to control the operation, because voltage unbalance introduces an extra degree of freedom, the level of which changes with the working point. Electrical symmetry can only then be achieved if voltage symmetry is calculated on the basis of known reactance values[34,35] or if the working-point setting is monitored with accurate measurements.

Electrical balancing of arc furnaces with symmetrical reactances is simple. Equalization of furnace current means that all other electrical values are equal, too. Because of the way the arc furnace is constructed, the thermal picture does not correspond to the power distribution; therefore, even if the power is balanced, one-sided wear of the furnace lining can still often be observed near one of the electrodes.

Heavy wear of the refractory lining near the wild phase is caused by excessive arc length. Efforts are often made to shorten the arc in the phase in question by adjusting the electrode control so that the current increases. However, it should be noted that changing the arc length in one phase causes only minute changes in current, because the star-point is free. If the arc is shortened too much then another phase can become 'wild'.

4.1.1.7 Circle diagram

A circle diagram describes the dependence of the electrical properties of an arc furnace on arc resistance. It is assumed here that the three phases are balanced, so that a single-phase equivalent circuit may be used (Figure 12(c)).

If the arc resistance R_A is changed, then the real part of the (complex) phase impedance changes according to equation (11). The locus is thus parallel to the real axis, because it is the geometric locus of all complex phasors $\underline{Z}$ in relation to the parameter R_A (Figure 18(a)). $\underline{Z}$ is minimum in a short circuit ($R_A = 0$) and maximum under no load ($R_A = \infty$). The semicircular locus of the complex

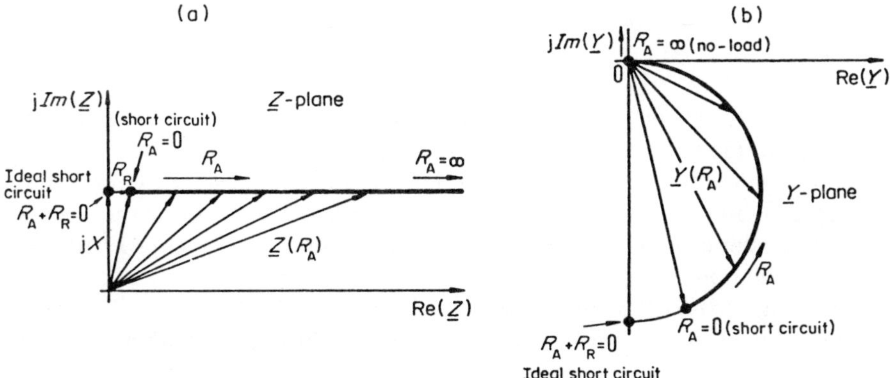

Figure 18. Locus curves of impedance $\underline{Z}$ (a) and admittance $\underline{Y}$ (b) as a function of arc resistance R_A

admittance $\underline{Y} = 1/\underline{Z}$ is then obtained by inverting the $\underline{Z}$ locus curve[14,18] (Figure 18(b)).

According to Ohm's Law

$$\underline{I} = \underline{V}_{ph}/\underline{Z} = \underline{V}_{ph}\underline{Y} \tag{23}$$

At constant-furnace phase voltage the curve of the current locus is similar to the admittance locus curve. Therefore this current locus curve is the well-known circle diagram for arc furnaces (Figure 19(a)), which has been turned 90 degrees agains the $\underline{Y}$ locus curve for the sake of clarity.[11]

If we vary the arc resistance between $R_A = 0$ and $R_A = \infty$, it is clear from the circle diagram that the furnace current I (equation (8)), and its lagging phase position, will change in relation to the phase voltage (equation (10)).

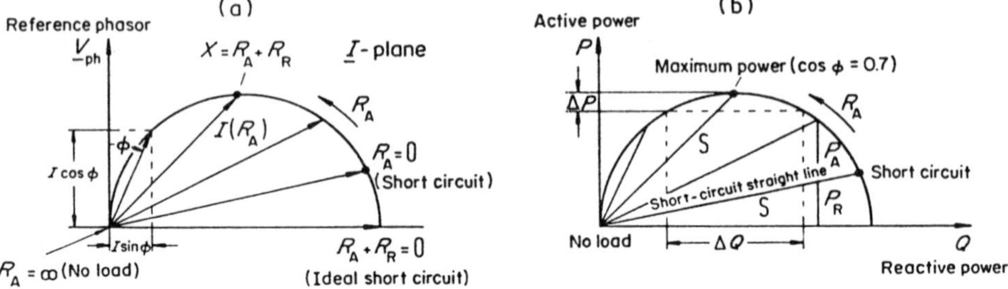

Figure 19. Locus curves for the current (a) and power (b) of an arc furnace

In an ideal short circuit ($R_R + R_A = 0$) a purely reactive current flows ($\phi = 90$ degrees) with a maximum value of $I = V_{ph}/X$, limited only by the reactance. The actual short-circuit current caused by dipping the electrodes into the melt ($R_A = 0$) is slightly less than in the idealized short circuit. The current and the phase-displacement between I and V decrease as the arc becomes longer and R_A increases. At the highest point in the circle the resistance and reactance are equal ($\phi = 45$ degrees) and at no-load ($R_A = \infty$) the current reduces to zero.

The current can be divided into an active $I \cos \phi$ in the direction of the reference voltage and a lagging reactive component $I \sin \phi$ (Figure 19(a)). It is evident from the foregoing that the circle diagram also shows the course of the active phase power on the ordinate; it shows the reactive power on the abscissa, while the distance from zero indicates the apparent power (Figure 19(b)). The course of the power for the whole three-phase system is determined according to equations (15)–(17) by multiplying by 3.

The point at which power input into the arc furnace is maximum is at a phase-displacement angle of $\phi = 45$ degrees ($\cos \phi = 0.707$). This is the preferred operating point for many arc furnaces. At this point the resistance and reactance are equal, as are the active and reactive powers. The short-circuit straight line in Figure 19(b) divides into effective power P_A and power loss P_R. Therefore, at maximum arc power the power factor cos ϕ is more than 0.7.

The maximum reactive power in an idealized short circuit corresponds to the diameter of the circle in Figure 19(b):

$$Q_{max} = 3V_{ph}^2/X = V_L^2/X \tag{24}$$

Therefore the maximum active power is

$$P_{max} = V_L^2/2X \tag{25}$$

The equation indicates the maximum active power available at given conductor voltage and reactance.

Figure 20 illustrates in another way the relationship between the different types of power and the current, which is related to theoretical short-circuit current.[25,29,36] Note that the maximum arc power occurs at a lower current than the overall maximum power. The radiation is greatest with long arcs (cos $\phi = 0.85$). This is why melting is preferably carried out at a low power factor using short arcs.[11,36]

4.1.2 Electrical Measurement Technique on the High-current Side

Measuring electrical values on the high-voltage side of the furnace installation is no problem using the well-known three-phase measurement method.[37] We shall now consider the difficulty of electrical measurements on the high-current side of the arc furnace.

To monitor the electrical conditions in the individual arcs it is essential to measure the current, the phase voltage, and—if possible—the active power in each phase (Figure 21). The currents I_1, I_2, and I_3 are measured by using high-current transformers[38] (cores surrounding the high-current conductors), or with transformers in the furnace-transformer intermediate circuit (Figure 26), or with magnetic voltmeters (Figure 24). The furnace phase-voltages V_{1M}, V_{2M}, and V_{3M} are usually obtained between the secondary transformer leads and the melt (Figure 21). Voltage-separation between the high-current and instrument circuits, and appropriate matching, are achieved using voltage transformers.

In most cases electrode control is based on the impedance principle. The control deviation is derived from measured values of phase currents and voltages, using equivalent voltages converted by the instrument system, and the differences between these signals. The arc length is adjusted through the electrodes until the control deviation is zero, that is, until the specified impedance, and with it a definite energy intake, is achieved.

When measuring voltages on the high-current side, we must remember that the strong alternating magnetic fields can induce voltages in the control-circuit

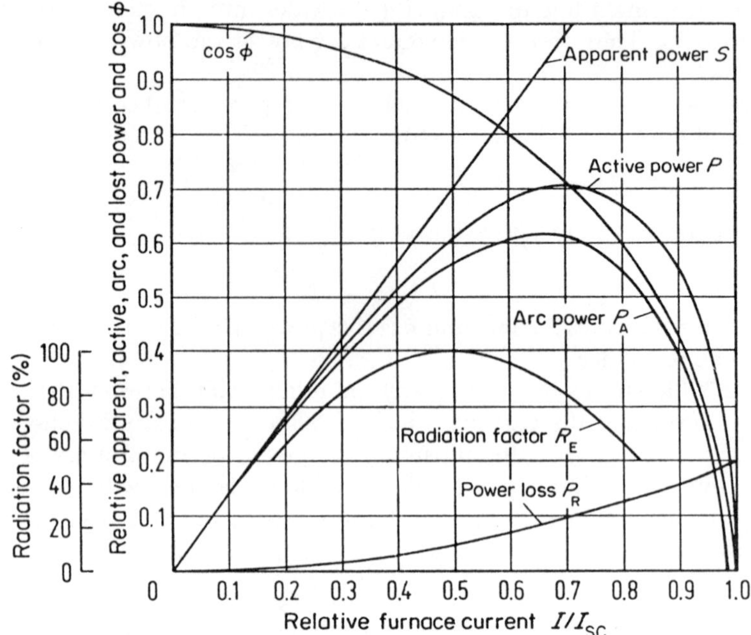

Figure 20. Furnace power diagram

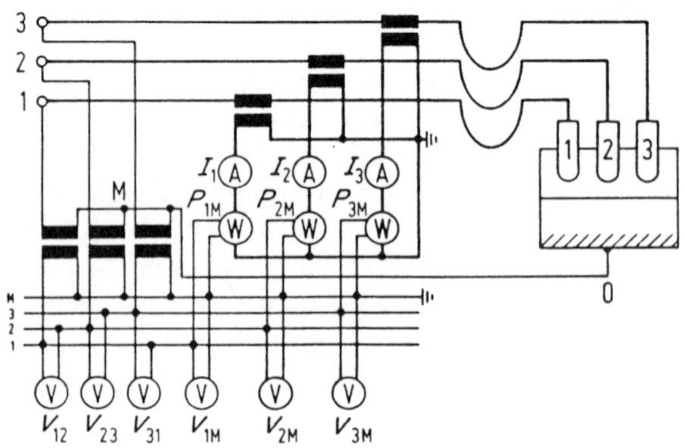

Figure 21. Measuring electrical values on the high-current side of the furnace

loop. Thus, measurements of arc furnace phase-voltages referred to the star-point *M* are inaccurate because a measuring-circuit loop must unavoidably be formed. Therefore, the active powers P_{1M}, P_{2M}, and P_{3M} (Figure 21), determined by the triple-power measuring circuit, are also incorrect. On the other hand, the phase-to-phase voltages V_{12}, V_{23}, and V_{31}, between transformer leads in close proximity, can be measured accurately because there is no measuring-circuit loop.[27,28,39,40]

When the effect of the measuring-circuit loop on the furnace phase-voltages is investigated we find that the interference voltages induced change with every new position of the loop. In an example of measurements,[28] the voltage and current phasor diagram (Figure 22) shows that the star-point potential is different (M_1, M_2, and M_3) for every new measuring-circuit loop. The fact that the measuring loop cannot be avoided prevents fault-free determination of the phase-voltages v_{10}, v_{20}, and v_{30} against the star-point 0 of the bath. In consequence the measured voltages

$$v_{1M} = v_{10} + v_{0M} \tag{26}$$

$$v_{2M} = v_{20} + v_{0M}$$

$$v_{3M} = v_{30} + v_{0M}$$

differ from the phase-voltages by an error v_{0M}, which acts like a star-point displacement and can cause appreciable measuring errors. In the example under consideration, the measured errors in the active phase powers can be as high as 36%. This false active power indication makes it difficult to deal with the wild phase in arc furnaces.

In the arc furnace equivalent circuit derived earlier (Figure 13) the effect of the current-less measuring conductor to the furnace vessel is shown as a voltage source v_{0M}. In this way we obtain the extended arc furnace equivalent circuit in Figure 23.

In reference 40 it has been shown that the faulty voltage-indication depends on the furnace currents and on the mutual inductances $M_{12,3M}$, $M_{23,1M}$, and $M_{31,2M}$ between the high-current loops and the measuring-circuit loops. One way of expressing the error voltage is:

$$v_{0M} = M_{12,3M}\mathrm{d}i_1/\mathrm{d}t - M_{23,1M}\mathrm{d}i_3/\mathrm{d}t \tag{27}$$

Fault-free determination of arc voltages is desirable for the efficient operation of arc furnaces. However, one cannot imagine an induction-free circuit for measuring the arc voltages between the electrodes and the bath. In addition to the arc voltages wanted, the measurable phase-voltages contain induced components as well as active voltage-drops due to resistance in the feeders. Therefore, the undesirable active and reactive voltage components must be compensated[39–42] when determining the arc voltages.

One compensated measuring circuit, which will now be described, is called the Clausthal measurement system.[39] The system is based on equations governing the inductive effects of the closed-current paths in closed measuring loops, where explicitly no erroneous voltage occurs. For example, the voltage between transformer terminal 1 and the measuring lead M is:

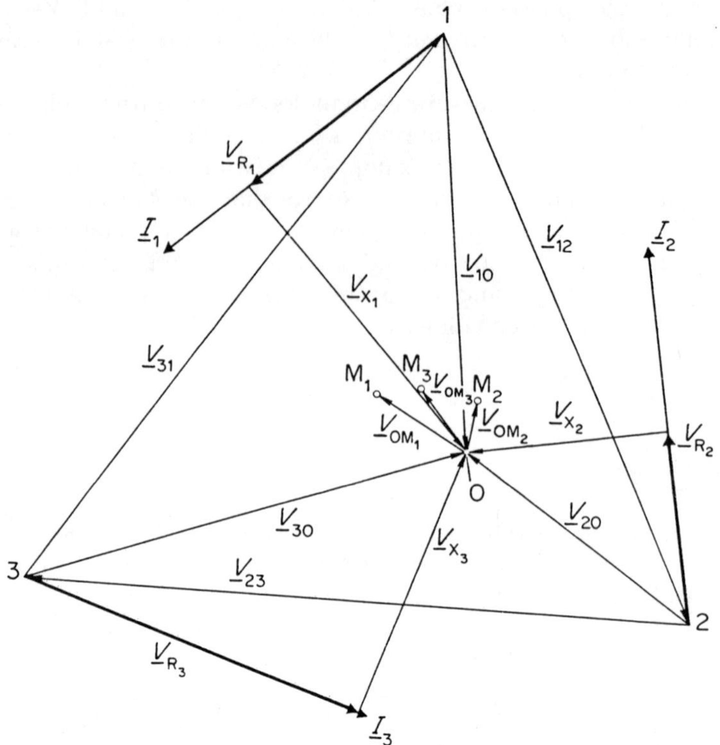

Figure 22. Phasor diagram of the currents and voltages for an arc furnace.[28] 0 is the actual star-point where the load is immeasurable. M_1, M_2, and M_3 are the star-point potentials measured. False indicated voltages are $V_{OM1} = 62$ V, $V_{OM2} = 31$ V, $V_{OM3} = 45$ V

$$v_{1M} = R_{R1}i_1 + v_{A1} + M_{21,1M}\mathrm{d}i_2/\mathrm{d}t + M_{31,1M}\mathrm{d}i_3/\mathrm{d}t \tag{28}$$

The measured voltage v_{1M} includes two induced-voltage components as well as the active voltage-drop in the feeder and the arc voltage v_{A1}. When we solve the above equation for the required arc voltage, we get:

$$v_{A1} = v_{1M} - R_{R1}i_1 + M_{12,1M}\mathrm{d}i_2/\mathrm{d}t + M_{13,1M}\mathrm{d}i_3/\mathrm{d}t \tag{29}$$

Accordingly, the required arc voltages v_{Av} can be determined from the measurable voltages v_{vM}, the currents i_v, and their derivatives in time $\mathrm{d}i_v/\mathrm{d}t$. The feeder resistance R_{Rv} and the mutual inductances M have to be evaluated by balancing as furnace constants.

The analog computer circuit in Figure 24 meets equation (29). The currents

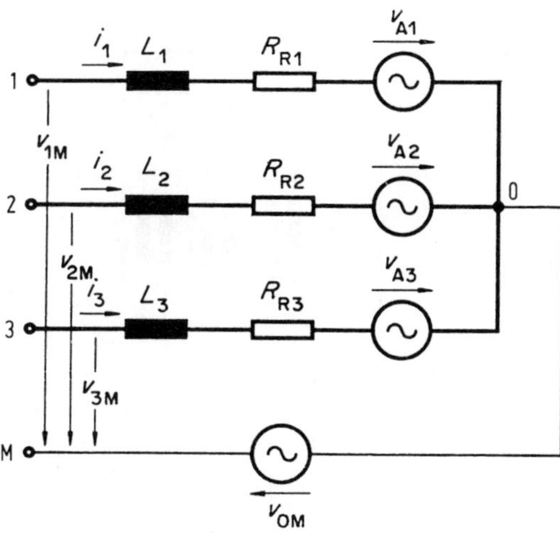

Figure 23. Extended equivalent circuit for a three-phase arc furnace with measuring loop

are determined with magnetic potentiometers (Rogowski coils).[14] These consist of narrow non-magnetic strips wound evenly with insulated wire. When a Rogowski coil encloses the high-current conductor, the voltage induced in the individual turns is proportional to the magnetic potential, which is equal to the magnetomotive force. The induced voltage is determined by the mutual inductance K between the high-current conductor and the Rogowski coil, as well as from the rate of change of current with respect to time $\mathrm{d}i/\mathrm{d}t$. A signal proportional to the current is obtained via an amplifier, which acts as integrator.

In the summing circuit modules in Figure 24, the phase voltages $v_{\nu M}$ are superimposed on the $\mathrm{d}i_\nu/\mathrm{d}t$ and $i_\nu(t)$ signals. For example, the initial voltage in phase 1 is:

$$v_1 = \frac{R_o}{R_M \ddot{v}} \cdot v_{1M} - \frac{R_o K}{R_{11} RC} \cdot i_1 + \frac{R_o K}{R_{12}} \cdot \frac{\mathrm{d}i_2}{\mathrm{d}t} + \frac{R_o K}{R_{13}} \cdot \frac{\mathrm{d}i_3}{\mathrm{d}t} \sim v_{A1} \tag{30}$$

When comparing this with the original equation (29) it is apparent that the initial voltage v_1 is proportional to the actual arc voltage v_{A1}. The level of the furnace voltage is matched to the usual voltage range of operational amplifiers by the proportionality factor $R/(R_M \ddot{v})$. The measuring circuit can be balanced simply under single-phase conditions.

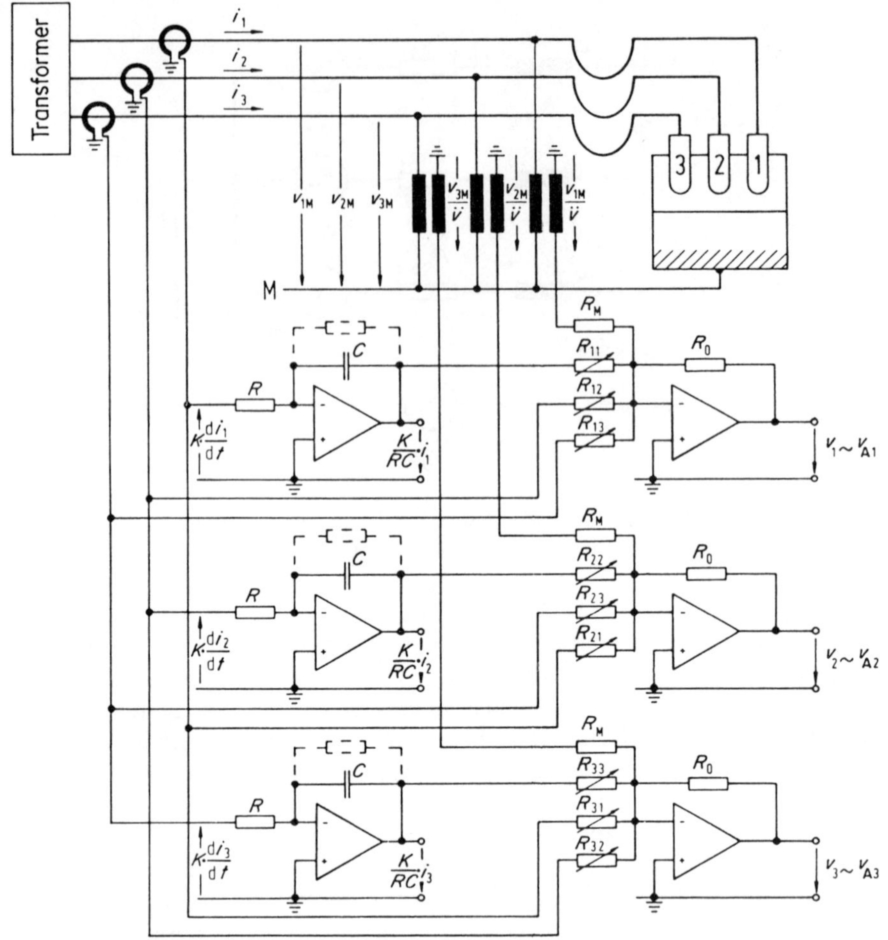

Figure 24. Compensating circuit for arc-voltage measurement

In this way the method described enables instantaneous arc voltages and currents to be determined. The arc voltages and currents plotted in Figures 3–5 were measured by this system. These values can easily be processed further, and the effective values of arc voltages, currents, and powers can also be found by adding extra electronic modules. Thus, all the electric arc values of interest can be determined and used for efficient operation of the furnace.

Another method, which became known as the Röchling–Burbach measuring system, is used in a similar manner for compensating active and reactive voltages.[42] Here the induced voltages are derived for all three furnace currents by using differentiating modules.

4.1.3 Furnace Transformers

The purpose of furnace transformers is to adjust the electrical energy supplied to suit the voltage and current values required by the arc furnace.

4.1.3.1 Transformer principles

We shall not deal here with the basic principles of transformers; these may be found in the relevant literature.[43]

4.1.3.2 Furnace transformer requirements

Arc furnace transformers have to meet the following special requirements:[44–7]

(1) They must be able to take a maximum rated current of up to 120 kA on the secondary side[13] (short-circuit current requirements are higher);
(2) They must be able to provide low voltages of 200 to 800 V on the furnace side from high voltages on the supply side of up to 110 kV[45] and 154 kV[46] and lately up to 220 kV;
(3) Furnace voltage must be adjustable to suit particular demands at a high switching rate (up to 800 per day);
(4) Transformer output can be up to 162 MVA;[13]
(5) They must be robustly built to withstand frequent short circuits;
(6) The inductances in the three phases should be equal (triangulated secondary system);
(7) The three phase-voltages may be capable of separate adjustment to combat the wild phase.

Voltage-adjustments must be performed on the primary side because of the small number of secondary turns (one to three or more) and the high furnace currents. There are two types of transformer according to the type of control used: the directly controlled furnace transformer (two windings) and the indirectly controlled furnace transformer (intermediate circuit transformer). A special variant of the intermediate circuit transformer has the furnace switch in the intermediate circuit.

4.1.3.3 Directly controlled furnace transformers

On the two-winding transformer (Figure 25), voltage control is carried out stepwise by changing the number of primary turns. Increasing the number of

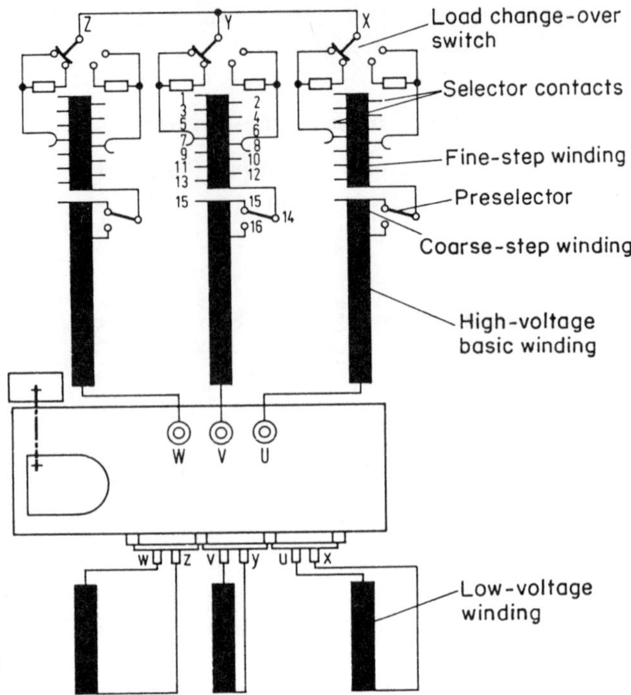

Figure 25. Directly controlled furnace transformer

turns at constant primary voltage reduces the magnetic flux in the iron core and thereby reduces the secondary voltage. The control range can be widened by delta-star or parallel-series connection of the high-voltage winding.

The voltage steps are not equal. When the tap changer-switch is in the position for maximum secondary voltage (least number of turns on the primary side) the sum total of the voltages from the step turns and the basic winding acts on the open end of the step winding. The high total voltage can pose problems for the windings insulation and the tap change-switch. A directly controlled furnace transformer is therefore not suitable for direct connection to high-tension mains above 30 kV.

4.1.3.4 Intermediate circuit transformers

An intermediate circuit transformer (Figure 26) consists of a main transformer and an auxiliary transformer. Stepped voltage control is achieved by means of an intermediate circuit; this is more expensive than direct control.

Its functioning will be explained by a single-phase representation. The circuit (Figure 27) contains a furnace switch in the intermediate circuit, which is

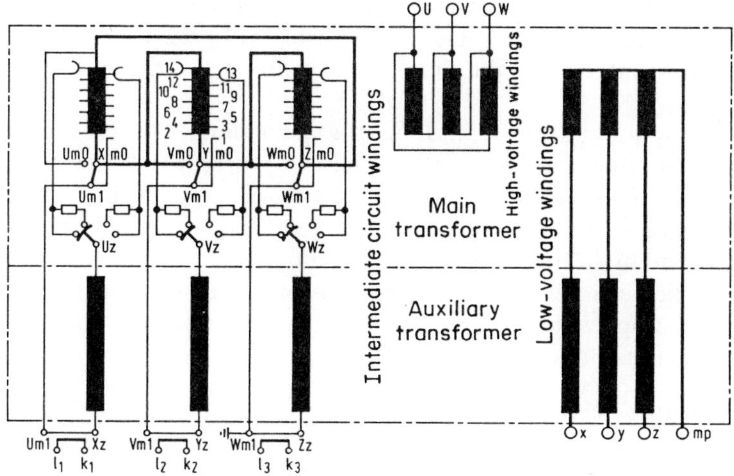

Figure 26. Intermediate circuit transformer

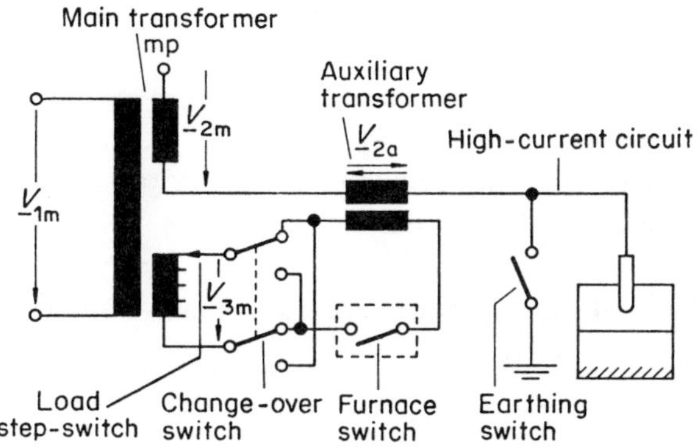

Figure 27. Single-phase circuit diagram of intermediate circuit transformer with furnace switch in intermediate circuit

closed in normal operation. The main transformer generates a secondary basic voltage V_{2m} at a fixed reduction ratio. The excitation voltage V_{3m}, which can be adjusted under load by means of a step-switch, is generated in the tertiary winding and connected to the auxiliary transformer. The secondary windings of the main and the auxiliary transformers are connected in series. According to the position of the change-over switch, the adjustable voltage of the auxiliary transformer V_{2a} is added to or subtracted from the fixed secondary voltage V_{2m}.

It is preferable that the furnace load switch should be in the intermediate circuit, especially when the transformer is connected to high-voltage mains of

110 kV and over,[44–7] because reliable low-voltage switches are available for the intermediate circuit. When the furnace switch is opened, the intermediate circuit is interrupted, and the auxiliary transformer is then working under no-load.

The high main reactance acts on the secondary side of the auxiliary transformer, so that only the small magnetizing current can flow through the furnace circuit. This means that maintenance work can safely be carried out on the arc furnace if the residual voltage caused by the magnetizing current is reduced to zero by short-circuiting and earthing.

The complicated intermediate circuit transformer not only can be worked off high-tension supplies, but also offers still more advantages compared with the two-winding transformer, as follows:

(1) Equal voltage steps.
(2) Free selection of the voltage for the intermediate circuit (and with it optimum operation of the step and load switches), which can operate at relatively low power.
(3) Current measurement using relatively inexpensive transformers in the intermediate circuit (Figure 26), because the current in the intermediate circuit corresponds exactly to furnace current.
(4) Capacitors and filter circuits can be connected into the intermediate circuit to improve power factor.
(5) The high-voltage transformer (110 or 220 kV to 30 kV), which was formerly essential, becomes redundant.

4.1.4 Effects of Arc Furnaces on Mains Supplies

Operation of arc furnaces on the mains causes difficulties mainly when high power is required from a mains supply which has a low short-circuit capacity.

4.1.4.1 Static voltage-drop in the bus bar

Static behaviour will be discussed first. The effect of an arc furnace, with a reactance X, on the mains supply with a no-load voltage v_o and a generally highly inductive impedance X_m (where $X_m \gg R_m$) is considered with the help of Figure 28.[5,48] The current drawn by the arc furnace causes a static volts drop ΔV in the bus bar, the critical common connection, linked to other consumers, where:

$$\Delta V/\sqrt{3} \approx IX_m \sin\phi \tag{31}$$

From equations (16) for the reactive power Q of an arc furnace and for apparent short-circuit power of the mains (S_{sc}):

$$S_{sc} = V_o^2/X_m \tag{32}$$

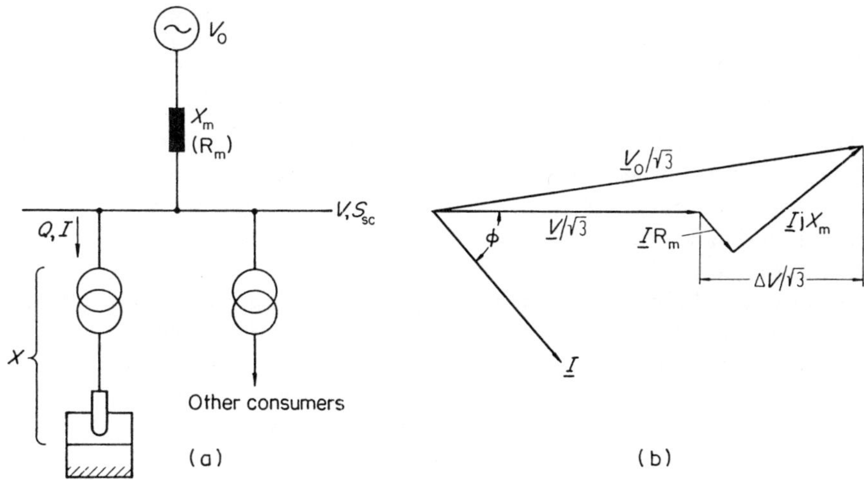

Figure 28. Three-phase mains with arc furnace (a) and its phasor diagram (b)

we obtain the approximate relative voltage-drop in the bus bar:

$$\Delta V/V \approx Q/S_{sc} \tag{33}$$

In particular, the maximum voltage-drop in the bus bar can be determined unambiguously when the electrodes, for an instant, carry no current, and, at the next moment, are short-circuited in all three phases:

$$\Delta V/V = X_m/(X + X_m) \tag{34}$$

The static voltage-drop in the bus bar is the reason that the furnace transformer high voltage is often less than the rated voltage when the arc furnace operates at rated current. The transformer cannot then supply its rated power.

4.1.4.2 Interference by arc furnaces on mains supplies

The current drawn by an arc furnace fluctuates considerably, especially during the melting-down phase. As yet there are no clear explanations for the unsteady behaviour of the arcs. The variation in arc length may be purely incidental. Some of the reasons quoted are the effects of thermal ionization and of sharp projections in the scrap as well as magnetic deflection of the arc. However, load-variations can also be observed at particular frequencies between 1 and 10 Hz. The electrode control system[49] or mechanical resonance of the electrode and its carrier arm[6,17] may also possibly have an effect (Figures 10 and 11).

The circle diagram in Figure 19(b) shows that, around the working point, arc-resistance variations mainly produce fluctuations in mains reactive power ΔQ, compared with which the active power changes ΔP are small. Apart from the static voltage reduction ΔV, there will also be dynamic voltages changes ΔV_d in the bus bar, of magnitude:

$$\Delta V_d/V \approx \Delta Q/S_{sc} \quad (35)$$

The voltage fluctuations also affect other consumers fed from the same bus bar (Figure 28). Large voltage-changes affect voltage-sensitive users. Above all, there are the effects of electric-light flicker on the human eye, which have been studied in a series of papers.[50–3] The eye is most sensitive to flicker-frequencies of about 5 Hz.

Sensitivity to flicker also depends on the type, power, and rated voltage of the lamps and the shape of the voltage-interference curve. There is a difference between the level at which the flicker is perceptible and the threshold at which it becomes intolerable. An international standard for flicker curves, something like that for acoustics, would be desirable; such a standard should take into account some concept of 'tolerability' for mains interference.

4.1.4.3 Interference-free connection of arc furnaces to the mains

Mains-voltage fluctuations caused by arc furnaces escape notice when, as equation (35) shows, the short-circuit power-availability at the mains connection is high enough. The result of an international inquiry into mains reactions to arc furnaces[54] shows that individual countries use different criteria for assessing mains reactions. Experience shows that an arc furnace can operate without causing interference if the short-circuit power at the mains terminal is at least 80 to 100 times higher than the furnace power.[12] As higher short-circuit powers are available only at higher voltages, high power arc furnaces generally have to be operated independently from the rest of the works power supply and require a separate high-tension mains supply of 110 kV or more, unless special systems are installed to maintain the voltage. Figure 29 shows the relationship between the melt-down power and the required mains short circuit capacity.[12] When there are several arc furnaces of the same size, Z in number, a factor $\sqrt[4]{Z}$ has to be taken into account.

4.1.4.4 Measurement and assessment of flicker

As in the case of evaluation of mains interference caused by arc furnaces, individual countries have different methods for assessing voltage fluctuations in the critical bus bar. British,[55] French,[56] Japanese,[57] and German[58] flicker-measuring instruments are defined. Different nations have different ways of treating the crucial question of the relationship between flicker-voltage and nuisance (frequency-dependence) and the assessment of permissable limits of flicker-voltage and interference.

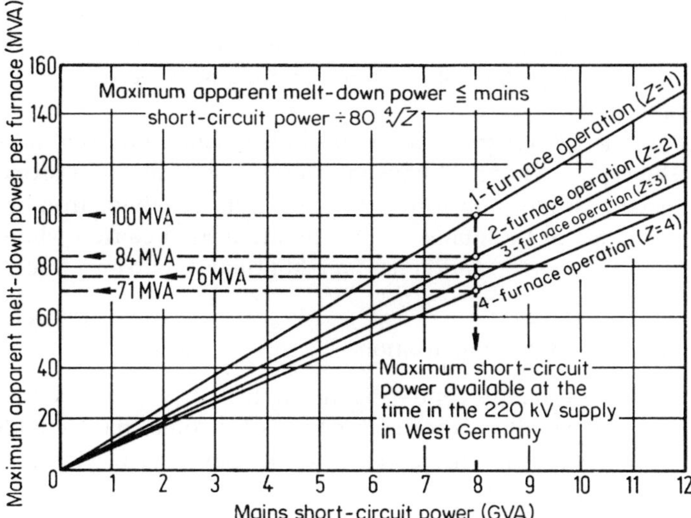

Figure 29. Maximum apparent transformer power in relation to mains short-circuit power for simultaneous operation of up to four arc furnaces

Difficulty in flicker-assessment arises in attempts to measure and analyse a time-dependent, stochastic signal structure using a simple analogue process. These signals may be analysed more efficiently using statistical methods in line with probability theory.[49]

4.1.4.5 Methods of reducing flicker

Repercussions on mains supplies can be moderated by altering conditions in the mains. According to equation (34), we can increase the furnace reactance X by installing a choke; however, this leads to undesirable power-reduction. Otherwise, we have to reduce the mains reactance X_m by reinforcing the mains supply. But variations in reactive power demand ΔQ can also be smoothed — according to equation (35) — by dynamic reactive current compensation (see section 4.1.4.6). Apart from expensive mains-reinforcement methods, there are other practical possibilities for reducing flicker which will now be briefly described.

Mains interference mainly occurs at the beginning of the melt-down phase when the electrodes bore into the scrap. One simple operational way of minimizing flicker is to reduce melting power during the critical interference period and to match it with the mains short-circuit power.[59]

Flicker can also be reduced to some extent by adjusting the working point of the circle diagram. A stabilizing effect is achieved if we work with a short, high-current arc and a low power factor.[52,60]

The arc can also be steadied by using doped electrodes to reduce electrode work-function, or hollow electrodes, or injecting argon into the arc.[60] However, these attempts at stabilization are not cost-effective[61] and have not become established.

It is known that the tendency of arc furnaces to cause interference depends heavily on the type of charge. Mains interference is reduced in proportion as the charge becomes finer and softer. When pellets of sponge iron are melted down the reactance in the mains is 50% less than that caused when melting scrap.[52] Increasing slag-formation has an additional stabilizing effect.

4.1.4.6 Reactive power compensation

A series of works has recently been published about reactive power compensation. Above all, advances in power electronics have enabled static reactive power compensation systems to be developed. Reactive power compensation must:

(1) Improve power factor,
(2) Reduce voltage variations (flicker),
(3) Produce fewer harmonics, and
(4) Balance asymmetrical loads.

These requirements can be met more or less satisfactorily by various methods. The following descriptions assume that the compensation system is connected to the bus bar.

Fixed capacitor installation[5,62] A fixed capacitor in parallel compensates arc-furnace inductive reactive power and improves the power factor sufficiently to meet the power-supply requirements. Mechanical switching of capacitor stages can be used to match the changing reactive power requirements of the arc furnace. The effect capacitors have of increasing voltage can be used to improve the static voltage-level at the mains terminals. However, a fixed capacitor will not reduce voltage-fluctuations caused by the furnace.

When calculating the size of the capacitor installation, consideration must also be given to possible resonance in the mains (see section 4.1.4.7).

Synchronus phase shifter[5,48,62,63] Reactive power machines have the effect, on the one hand, of increasing short-circuit power in the bus bar; on the other hand, mains-voltage fluctuations can be damped by shock-excitation. The method has proved itself and does not produce harmonics; however, it is expensive to install and maintain.

Saturated choke[5,48,62,64] With the saturated-choke method the total reactive

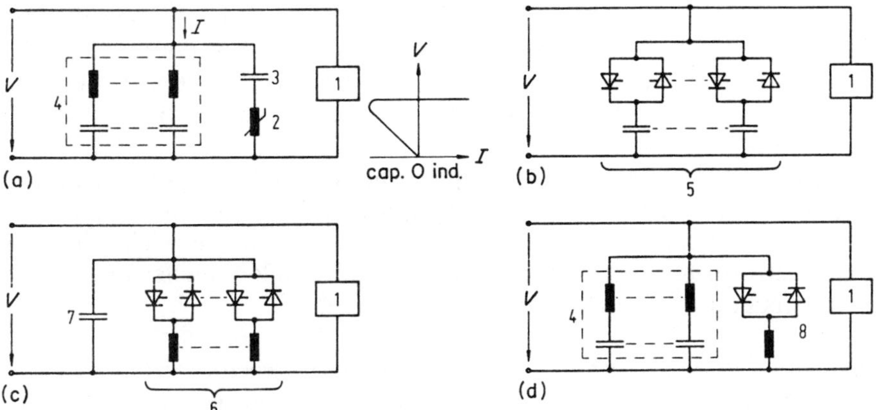

Figure 30. Basic circuits for reactive power compensation in arc furnaces. (1) Using static elements: (a) saturated choke (2), with series capacitor (3), with fixed capacitors and filter circuit (4); (b) thyristor-operated capacitor bank (5); (c) thyristor-operated set of chokes (6), with fixed capacitor (7); (d) thyristor-controlled choke (8) with fixed capacitors and filter circuit (4)

power uptake of the arc furnace and the saturated choke are kept constant (Figure 30(a)). A choke and a capacitor are connected in such a way that the choke characteristic illustrated in Figure 30(a) has a zero rise in the saturation region, and the voltage at the consumer remains constant. The current through the choke varies according to the reactive power to be compensated; the current is least when the arc furnace works all out and is greatest at no-load.

The advantage of this scheme is that the magnetic circuit itself constitutes the sensor, and the reaction time is only one or two cycles. The disadvantages are the low power factor of the whole installation and that strong harmonics are generated. Therefore, a fixed capacitor with filter circuit must be connected in parallel with the system. The choke is sensitive to changes in the mains voltage-level, and must be connected to a regulating transformer.

Thyristor-switched capacitor bank[5,48,62,65,66] This method compensates the reactive power requirement directly by switching on or off at high speed a suitable number of capacitor stages by means of opposed parallel thyristors (Figure 30(b)). Special precautions have to be taken to avoid surges when switching in capacitor stages. This is done by pre-charging the capacitor stages selected partly with the positive and partly with the negative peak values of the alternating voltage; the capacitors are switched in when the current passes naturally through zero. In this way the 'dead' time is reduced to no more than half a cycle (10 ms).

The advantages of the thyristor-switched capacitor bank are that it has low losses, produces no harmonics, and makes per-phase compensation possible.

Thyristor-switched bank of reactors[5,62,66] Input of reactive power from the mains can be kept constant by switching in or out opposed parallel, thyristor-switched chokes in a similar way to thyristor-switched capacitors. Switching inductances using thyristors is less problematical than with capacitors.

The reactive power is compensated to the required value by using a fixed capacitor. The advantages are the short reaction time of 10 ms, the possibility of per-phase compensation, and the absence of harmonics. The drawback compared with the thyristor-switched capacitor bank is higher losses in the chokes under no-load conditions in the arc furnace.

Thyristor-controlled reactor[5,48,62,66–8] A single large reactor can also be used in each phase in place of a bank of switched reactors. Stepless control of reactive power is achieved by varying the thyristor control phase-angle (Figure 30(d)); this is done in the short reaction time of 10 ms. The reactor can be made as a high-leakage transformer with a high short-circuit voltage in the 55–70% range. The reactive power demand is compensated by a capacitor bank. The phase-angle control produces harmonics, so that the capacitors have also to be developed as filters.

4.1.4.7 Voltage harmonics

Because of the non-linear arc resistance, an arc furnace acts as a source of current harmonics of the second to the seventh order, especially during the melt-down period (see section 4.1.1.3). Voltage harmonics are produced in this way through the impedance of the mains supply; the amplitude depends on the value of the harmonic currents supplied and the effective impedances at the harmonic frequencies.[5,53,67]

As the circuit of a furnace installation with a fixed capacitor in Figure 31 shows, the harmonic current I_v of the arc furnace forms a parallel tuned circuit, consisting of capacitor C with reactive power $Q_C = V^2 \omega C$, and mains inductance $L = X_m/\omega$, resulting from the mains short-circuit power S_{sc}. The resonant frequency of the tuned circuit f_o, in relationship to the mains frequency f amounts to

$$f_o/f = \surd(S_{sc}/Q_c) \tag{36}$$

When this tuned circuit resonates at a certain harmonic frequency, its reactance is high and a harmonic voltage arises which is, however, damped by the resistance R of the resistive component of the supply system consumers' equipment. The Q-factor of this tuned circuit is low at times of full load, and no resonant peaks occur. But in slack periods with combinations of low load with high resistance and Q-factor values, harmonic voltages must be expected at levels sufficient to cause appreciable interference.

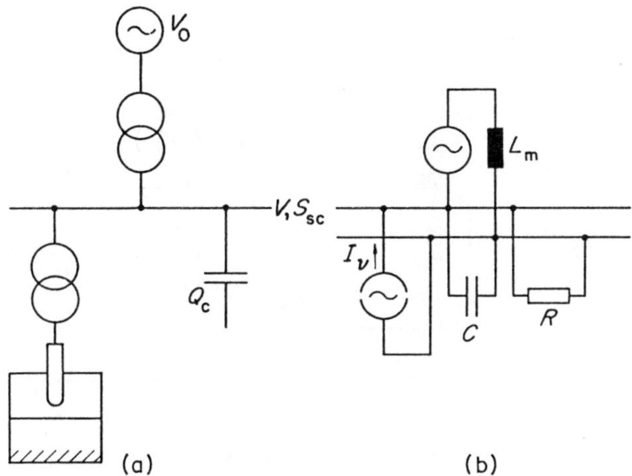

Figure 31. Arc-furnace installation with parallel capacitor (a) and equivalent circuit (b) for evaluation of resonant frequency

Harmonic voltages can be avoided by selecting parallel capacitors in such a way that the resonant frequency lies between harmonic frequencies. Above all, filter circuits should be used to short-circuit critical harmonics with series-resonance in reactive current-compensation systems liable to produce harmonics.[67]

4.2 BASIC PRINCIPLES OF INDUCTION CRUCIBLE FURNACES

Fritz Hegewaldt, Dortmund, and Klaus Timm, Hamburg

4.2.1 Transmission of Inductive Energy

When a piece of conductive material is inserted into a coil carrying an alternating current I_1 (Figure 1(a)), changes in magnetic flux induce voltages in accordance with the induction law. Due to the conductivity of the material, eddy currents I_2 will flow, which heat the material according to Joule's Law. Depending on the extent of the induction effect, heating may occur just on the surface of the material, or throughout the cross-section, perhaps sufficiently to melt the material.

The current I_2 in the material to be heated produces a secondary magnetic field which opposes the primary magnetic field and thus weakens the effect of the induction in the centre of the material. The secondary current will be displaced towards the outside by this superimposition. This current-displacement (skin effect) causes a current-density gradient which decreases from the

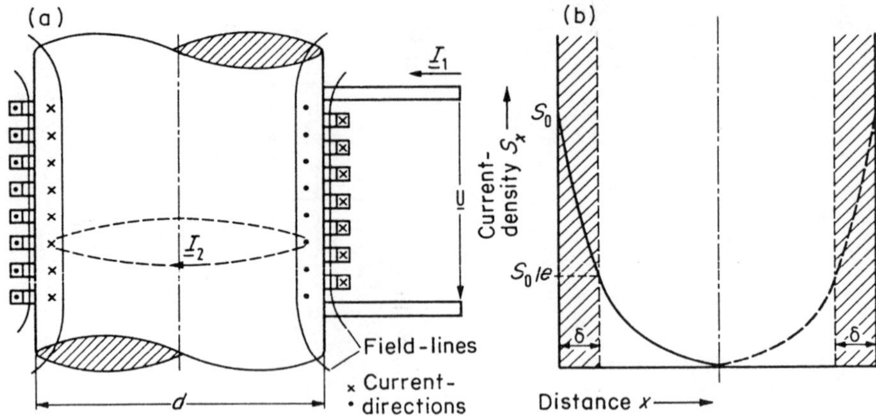

Figure 1. Principle of an induction crucible furnace. (a) Course of the field-lines and generation of eddy currents I_2; (b) course of the current-density

surface towards the centre. If, for simplicity, we assume there is an even electromagnetic wave, we get a current density S_x at a distance x from the surface (Figure 1(b)), where:

$$S_x = S_o \, e^{-x/\delta} \tag{1}$$

δ represents the penetration depth, which is that distance under the surface at which the current density has dropped to the eth fraction of the initial value:[1,2]

$$\delta = \sqrt{[\rho/(\pi\mu f)]} \tag{2}$$

ρ stands for the resistivity of the material, μ for the permeability and f for the frequency. As Figure 2 shows, the penetration depth in metals is very small and decreases with rising frequencies. It should be noted that the material characteristics are temperature-dependent. The relative permeability of ferromagnetic materials changes by several orders close to the Curie point.

According to Figure 1(b), the penetration depth can be interpreted as the equivalent thickness S_o of a conductive layer in which a current of constant initial current density flows. Thus, the material to be heated can be regarded as a hollow cylinder of wall thickness δ, which acts like a short-circuited transformer winding.

The energy absorption is proportional to the square of the induced current. 98.2% of the field energy is displaced to within a depth of 2δ:

$$P_1 = \rho m A H^2/\delta = \sqrt{(\pi\mu\rho f)} m A H^2 \tag{3}$$

Here, H is the magnetic field strength on the surface of the melt and A is the cylinder surface area at the height of the heating coil. The factor m depends on the ratio of the body diameter to penetration depth, and can be seen in

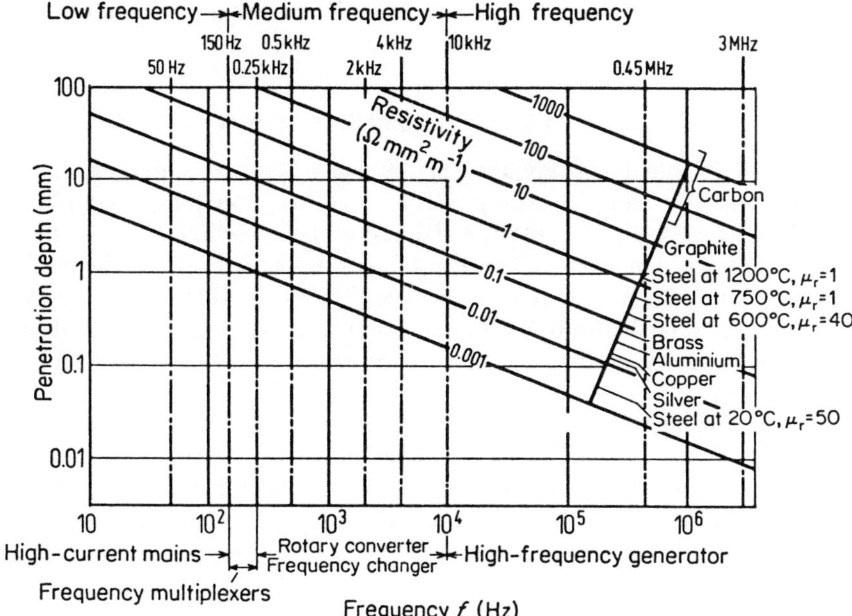

Figure 2. Penetration-depth of the current produced by the induced voltage in relation to frequency for various conductors in order of resistivities

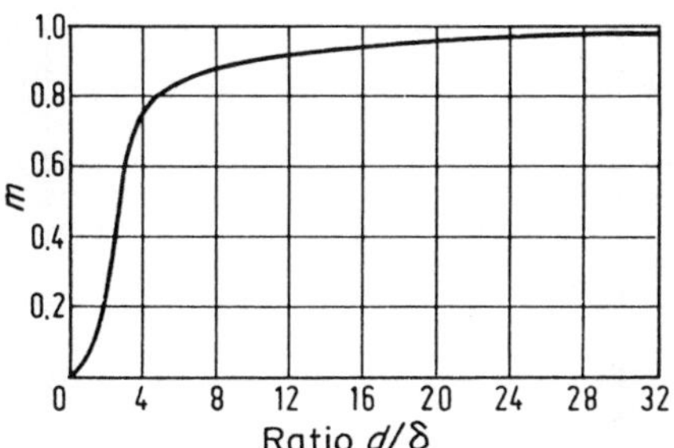

Figure 3. Power-absorption of cylinders for inductive heating after equation (3) in relation to ratio of diameter d to penetration depth δ

Figure 3. For economical operation, d/δ should be 6 or more. It is important that power absorption should be influenced either

(1) For specified frequency by the choice of the dimensions of the charge, or
(2) For specified dimensions by selecting the frequency.

In this case it should be borne in mind that the important parameter of a melt is its size and not the nature of the material charged.

4.2.2 Field Shape and Bath Motion

It is difficult to calculate eddy currents by using Maxwell's equations; it can only be done for simple arrangements such as a cylinder–symmetrical, infinitely long closed coil.[2–5] When the geometry is complicated we have to return to numerical methods.[6–9] Figure 4 illustrates as an example the field of a mains-frequency induction crucible furnace. The magnetic flux created by the coil current goes partly through the charge and partly through the crucible wall. The magnetic return path is formed by the external laminations which also screen the structure.

Current-carrying conductors in an external magnetic field are acted upon by electrodynamic forces, which are perpendicular to the current density phasor $\vec{S}$ and to the induction phasor $\vec{B}$. In an induction crucible furnace, the magnetic field force $\vec{F}$ mainly acts radially into the charge and constricts the melt (Figure 5).

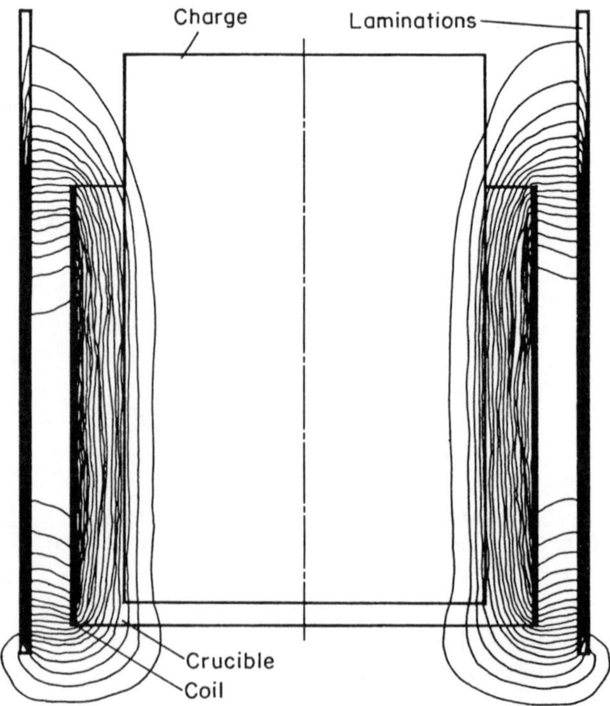

Figure 4. Field of a coil energized by a.c. in a mains-frequency crucible furnace[9]

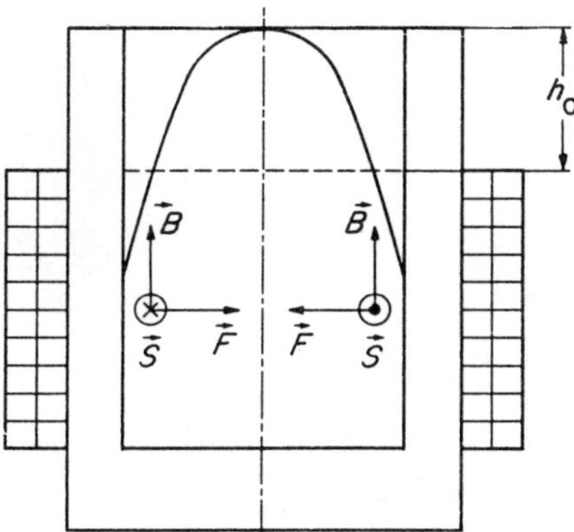

Figure 5. Origin of the static mound h_0 produced by the radial electrodynamic forces $\vec{F}$

The pressure on the melt[1,3,10]

$$p = \frac{\mu\delta P_i}{2\rho A} = \frac{1}{2}\cdot\frac{P_i}{A}\sqrt{\left(\frac{\mu}{\pi\rho f}\right)} \tag{4}$$

grows in proportion to the induced power and drops with $1/\sqrt{f}$. The melt can avoid this force only by rising to the top, so that a static mound is formed as shown in Figure 5. The resulting increase h_o in the height of the bath accordingly produces a hydrostatic counter pressure from the quotient of pressure and the specific weight γ of the melt:

$$h_o = p/\gamma \tag{9}$$

The increase in the height of the bath is anticipated by a freeboard on the crucible above the coil. This also leads to increased energy efficiency, because maximum power intake is possible only when the crucible is completely full.[9]

In an ideal induction crucible furnace stretched out lengthwise, the effective electrodynamic force field is eddy-free. The forces, which are radial all over, can only produce a mound on the melt but cannot cause any flow in the melt. However, with a finite coil length, the magnetic field at the coil end receives an ever-stronger radial component, so that the forces towards the coil ends have additional components, increasing axially and reducing radially (Figure 6(a)). As the sum of the forces in a closed circular stream is no longer zero, pressure differences arise which cause flow in the bath. The speed of the bath motion depends on the strength of the eddies, the path length, and the kinematic viscosity of the melt. If the current in the coil is even, we get circulation of the melt in the form of counter-rotating eddies (Figure 6(b)). This bath motion is characteristic of induction crucible furnaces. It is desirable because it stirs lighter matter into the melt and results in ever-renewing contact between the

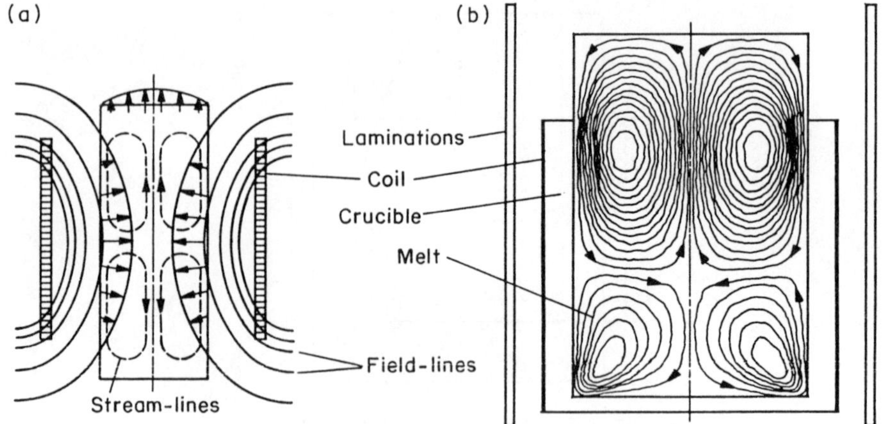

Figure 6. Origin of bath motion. (a) Field in a coil of infinite length and resulting direction of forces $\vec{F}$; (b) flow-lines in melt in single-phase, mains-frequency induction crucible furnace[9]

melt and the slag. However, the bath motion limits the power that can be applied to the furnace. When melting steel or casting iron the bath motion can affect rising carbon monoxide bubbles in such a way that entrained iron is ejected and attacks the crucible wall above the melt.[11]

The mechanical forces in the magnetic field can be increased so much that the melt leaves the crucible walls and — in extreme cases — the charge levitates in the space of the field.[12] However, application of this 'crucibleless' melting is confined to small-sample melts.

4.2.3 Active, Reactive, and Apparent Power

The endeavour to achieve the highest possible melt-down power — and through it, shorter melt-down times — is limited not only by material constants but also, on the one hand, by manageable coil currents and voltages and on the other, by the amount of bath motion that can be tolerated. Figure 7 demonstrates that the power limit of the mains-frequency induction furnace is not determined by electrical limits but by the limit of tolerable bath motion (lower curve). The specific power limit of mains-frequency furnaces for melting iron is about 300 kW/t, but this can be raised by increasing the frequency. The increase in melt height grows linearly with the inductive power and decreases with $1/\sqrt{f}$. Thus, by raising the frequency from f_1 to f_2, we can increase the induced power by a factor of $\sqrt{(f_2/f_1)}$ for the same height of melt (Figure 8).

Inductive reactive power is necessary to build up the magnetic field in the coil. Part of the magnetic flux goes through the charge (Figure 4). Reactive and active power in the charge are equal for large diameters $d/\delta \gg 1$.[8] The crucible wall

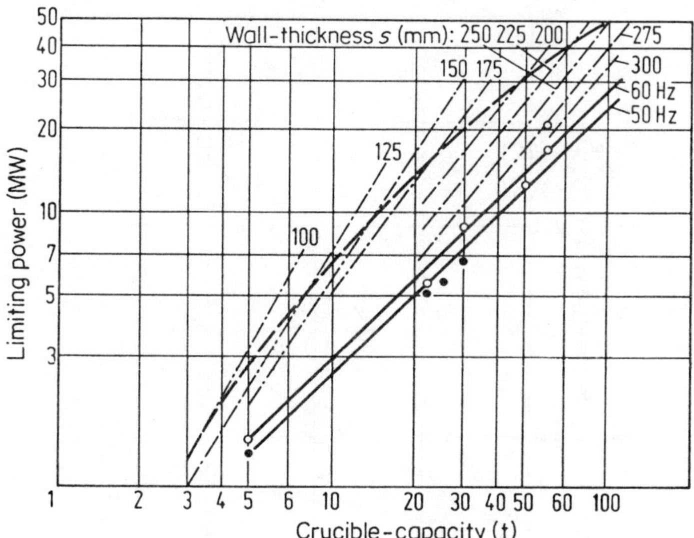

Figure 7. Induction furnace power limit in relation to maxium crucible capacity for grey cast-iron melting.[11] --- Electrically limited values with regard to allowable heating of the laminations (self-cooled) and crucible wall-thickness (coupling-gap). The thickness of the radial laminations is about 10% of the coil-diameter; covering factor (the ratio of lamination-width to lamination-stack division) is 0.5 to 0.65. Lamination quality: V_{10} at 1.1 W/kg. -·- Electrically limited values on the basis of crucible wall-thickness permissible for the size of furnace; —— Empirical limiting curves for bath motion at 50 or 60 Hz; • 50-Hz furnaces; ○ 60-Hz furnaces

should be as thin as possible to minimize the fraction of the flux that leaks through the crucible. For a crucible wall thickness of 2δ the reactive power is already five times more than the active power; for molten iron, $2\delta = 19$ cm (Figure 9).

The electrical limits for an induction crucible furnace are determined by the apparent power

$$S = VI \tag{6}$$

which is the product of the maximum permissible voltage (3000 V) and the current in the furnace coil. The active power is:

$$P = S \cos \phi \tag{7}$$

The greatest absolute power for mains-frequency furnaces can be obtained at limited apparent power, because the power factor $\cos \phi$ diminishes with

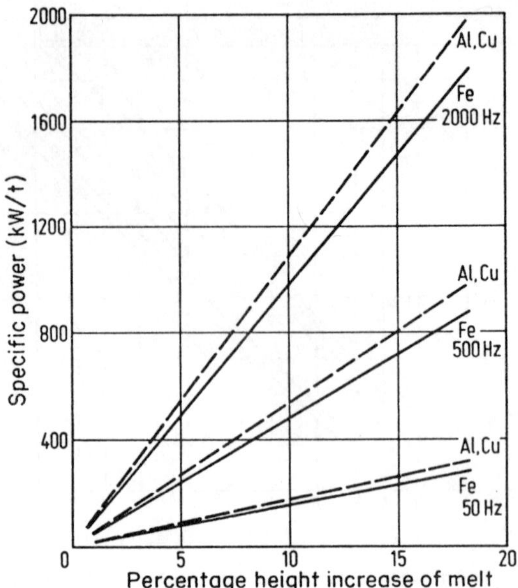

Figure 8. Specific power in relation to percentage increase in melt-height at various frequencies[1]

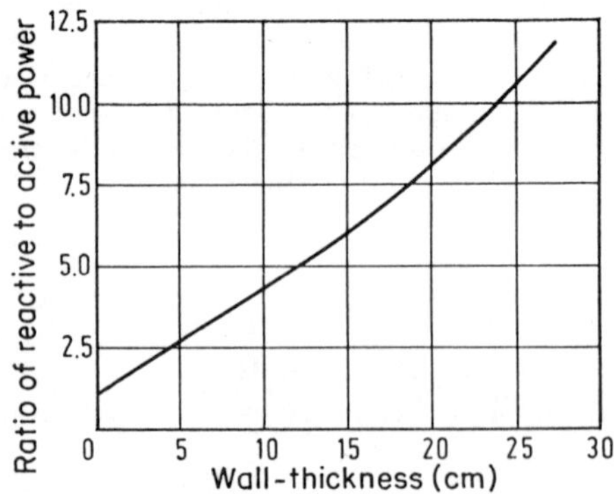

Figure 9. Ratio of reactive power to active power of a 30-t mains-frequency crucible furnace in relation to wall-thickness[9]

increasing frequency. On the other hand, maximum specific power is achieved with medium-frequency furnaces (Figure 10).

The inductive reactive power of an induction furnace is always fully compensated to $\cos \phi = 1$ by connecting a capacitor in parallel to the coil. As shown in Figure 11, the cost of compensation grows with furnace capacity,

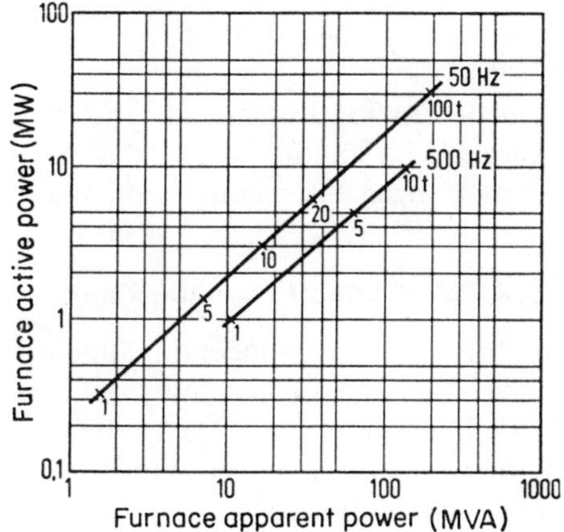

Figure 10. Maximum furnace active power at limited apparent power[13]

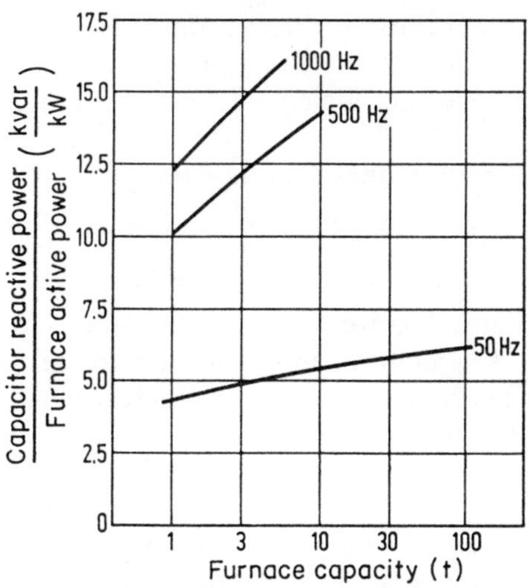

Figure 11. Ratio of capacitor reactive power to furnace active power with complete reactive power-compensation for induction crucible furnaces in relation to furnace capacity at various frequencies[13]

because the crucible wall thickness grows with the size of the furnace and the frequency.

4.2.4 Efficiency

The efficiency of an induction furnace is

$$\eta_e = \frac{P_i}{P_i + P_R} \tag{8}$$

The copper losses P_R can be kept low by rational dimensioning of the coil.[14] The highest efficiencies attainable for melting steel and cast iron are about 80% and are virtually independent of furnace-size and frequency.

4.2.5 Connection to the Mains

In comparison with arc furnaces, the connection of mains-frequency induction crucible furnaces to the mains supply presents no particular problems. In principle, the coil of an induction furnace can also be wound for three phases. Such a moving-field furnace[11] is, however, hardly suitable for melting on account of the high rate of circulation in the bath; on the other hand, it can be used in tandem with other melting systems as a reaction vessel for refining.

A balancing circuit (Figure 12)[1,9] is used to connect a single-phase induction furnace coil to three-phase mains. The furnace coil, compensated with a capacitor in the RS phase to $\cos \phi = 1$, is arranged in the delta circuit, so that only the equivalent resistance R is effective (Figure 13). A capacitive reactance and an equal inductive reactance are connected to the other two phases ST and TR.

If we suppose that for three-phase mains with normal phase sequence,

$$\sqrt{3}R = \omega L = \frac{1}{\omega C}, \tag{9}$$

notwithstanding unbalanced phase currents in the outer supply conductors, the currents $\underline{I}_R$, $\underline{I}_S$, and $\underline{I}_T$ are balanced and in phase with their mains star voltages $\underline{V}_{RM}$, $\underline{V}_{SM}$, and $\underline{V}_{TM}$. There is thus a symmetrical resistance load on the mains. The compensating and balancing elements can be matched in relation to the load for the difficult operation states.

Frequency-changers are used when a high frequency is necessary because of specific power, bath-motion strength, melting time, or a charge of small lumps. There are a number of possible schemes for frequency changing[1] (see Figure 2). Thyristor converters are now of special importance for producing medium frequencies.[15]

4.3 BASIC PRINCIPLES OF REMELTING FURNACES

Klaus Timm, Hamburg

4.3.1 Principles of Vacuum Arc Furnaces

The process in the arc in a vacuum arc furnace is fundamentally different from that in the arc in an electric steel arc furnace. The arc in a vacuum arc furnace

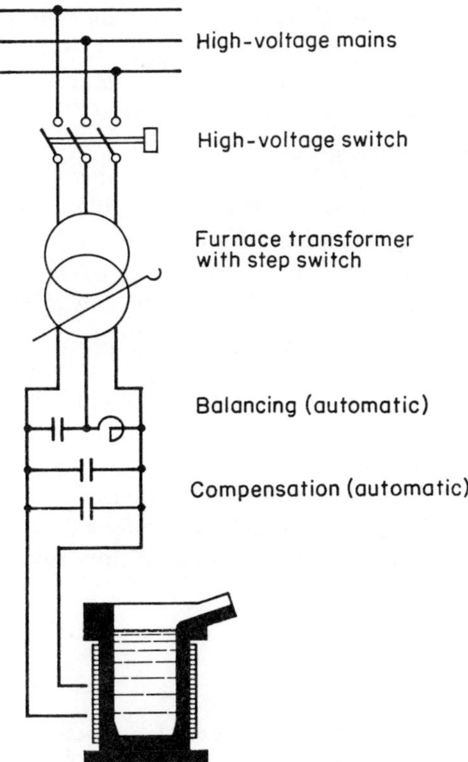

Figure 12. Mains connection of mains-frequency induction crucible furnace[9]

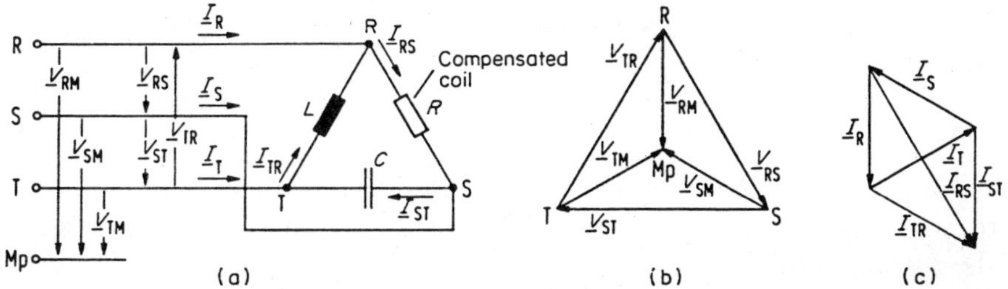

Figure 13. Balancing of induction furnaces. (a) Circuit; (b) voltage-phasor diagram; (c) current-phasor diagram

burns in a gas-free space between two metal electrodes. This arc is essentially an ionized gas column, consisting of molecules of evaporated electrode material. The arc is energized by d.c. because an arc cannot be produced by mains frequency a.c. in a high vacuum at gas pressures below 1 μbar.[1] A current of at least 900 A is required in order to sustain an arc between metal electrodes.

The arc voltage is about 22 V, or 35 V when melting under argon; it is virtually independent of arc current and length. Arcs are possible at lower arc voltages of about 20 V, but they wander a great deal at the root. As the steel is not melted in such cases, one speaks of 'cold arcs'.[2] Transition from cold arcs to a normal 'hot arc' is not clearly understood.

It is necessary for this process that the distance between the melt-down electrode and the melt is kept as small as possible, around 15 to 25 mm for medium-size steel furnaces, and as constant as possible; this distance will be 40 mm for large furnaces. The arc cannot be controlled by the voltage alone because the potential gradient in the arc is very low in a vacuum, and for this reason complex control systems have been evolved.[23]

The arc voltage remains about constant at around 22 V for the usual current of between 1 and 20 kA. The choice of current value determines the melting power and with it the metallurgical result. The optimum current value depends on the crucible diameter and the steel quality.

Originally the high-current supply was controlled by transductors, but thyristors are used today as controllable semiconductor rectifiers. The thyristor system keeps the current constant at a value prescribed by a time-planning device.[2]

4.3.2 Principles of Electro-slag Remelting Furnaces

Electro-slag remelting (ESR) is performed by melting down a metal ingot electrode in a hot slag bath in water-cooled mould. The remelted slag between the surface of the immersed electrode and the molten ingot bath acts like a resistance where the conduction mechanism is ionic.[4] The heat liberated into the molten slag bath covers the thermal requirements of the process. The electrical conditions between the electrodes follow Ohm's Law. The resistance of the slag bath is:

$$R_s = V/I = \rho l/A \qquad (1)$$

The gap l between the electrodes is thus obtained from the voltage V, the current I, the resistivity ρ of the slag melt, and the average cross-section area A of the current path in the slag, and is:

$$l = (V/I)\cdot(A/\rho) \qquad (2)$$

Figure 1 shows how optimum remelting conditions are obtained as current and power are increased with increasing ingot diameter. The voltage-drop decreases although the electrode gap increases, because the current density becomes less. More detailed investigations into the electrical values in the molten slag and their influence on the shape of the electrode tip are described by F. W. Thomas.[5]

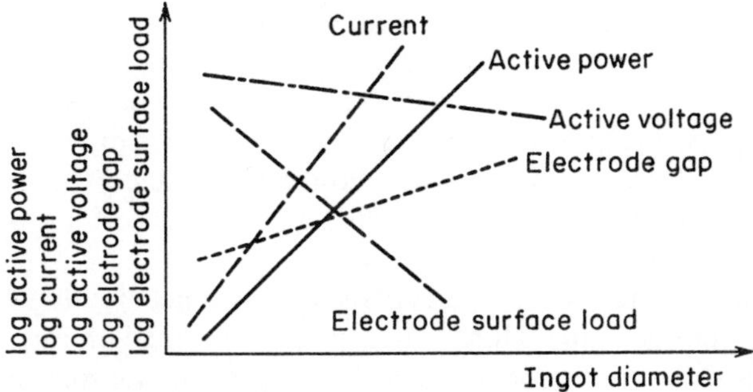

Figure 1. Relationship between electrical parameters and ingot-diameter in the slag bath during electro-slag remelting

When remelting with direct current, undesirable electrochemical effects accompany the thermodynamic reactions between the metal and the slag. For this reason, single-phase a.c. is used for smaller-scale applications and three-phase a.c. for larger installations. Melt-down voltages range between 40 and 140 V and electrode currents from 2 to 25 kA. Large ingots are produced by simultaneously melting down several electrodes.

The equivalent circuit diagram for an ESR furnace[4,6] in Figure 2 shows that we have to take into account the resistance and reactance in the current loop as well as the resistance in the bath. In large ESR installations operated at high

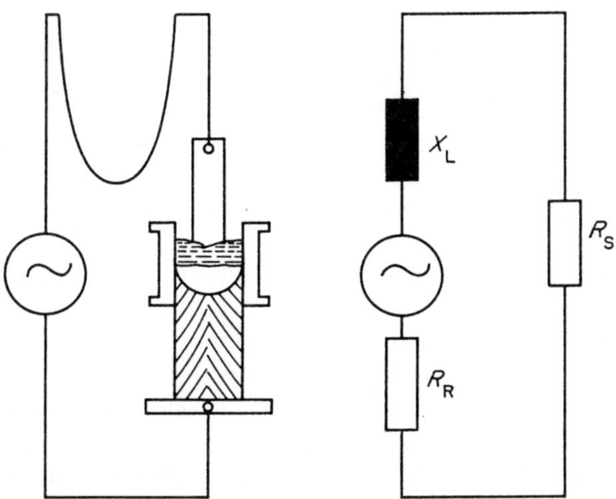

Figure 2. Schematic diagram of an a.c.-operated ESR installation, R_s = slag bath resistance; R_R = resistance of conductors; X_L = inductive reactance of conductors

a.c. current values, the resistance component of the current loop and the inductive reactive power requirement become all too apparent because of current displacement in the steel electrode and ingot. The resistance losses and the apparent power required can be decreased by lowering the a.c. frequency. For this reason, the use of thyristor converters, that produce a.c. at low frequencies from 0 to 10 Hz, has proved to be advantageous[7] for power supplies for large ESR furnaces.

Recently, in another process, electro-slag resistance heating has been used to melt down sponge iron.[8] Here, three-phase electrical energy is supplied to the slag bath in a reduction furnace by means of non-melting Söderberg electrodes made from graphite.

4.3.3 Principles of Plasma Furnaces

The term plasma[6] is used to describe a high-temperature gas state which produces particle-ionization and hence electrical conductivity. Gas-discharge plasma consists of an electrically conductive mixture of inert and ionized molecules and atoms in all states of excitation, which, together with free electrons, emit radiation. Plasma with a low degree of ionization is used for heating purposes at temperatures between 2000 and 20 000 K. Complete ionization of gases only takes place at temperatures of several million K, e.g. in nuclear fusion.

To convert a gas into plasma it is necessary to supply sufficient energy to the gas to cause ionization. One way of supplying this energy is to use an electric arc burning between two electrodes. The gas is blown through the hot arc and thus receives the energy needed for conversion into plasma; this is why one speaks of a plasma arc. Argon is preferred as the gas to be ionized.

Energy can also be supplied by an alternating electric field of sufficient strength between two electrodes (capacitive plasma). A third possibility for producing plasma is to transfer energy to a gas via a high-frequency electromagnetic field inside a cylindrical coil (inductive plasma).

Figure 3 is a schematic illustration of three types of arc plasma generator.[6,9] The arc may burn towards a ring-shaped anode nozzle (Figure 3(a)), or the anode may consist of the work-piece itself (Figure 3(b), in which case the nozzle is subjected to a potential between those on the cathode and the workpiece. If the cathode, nozzle, and workpiece are all related in the electrical system, then the arc supplying the energy to the plasma burner can also be powered by a.c. Here a d.c. supply is connected across the cathode and the nozzle to sustain the arc, while a.c. is connected between the nozzle and the workpiece to supply power to the arc there (Figure 3(c)).

A d.c. plasma burner with transmitting arc and argon as the ionization gas are preferred for plasma melting or remelting of steel. One 10-tonne plasma furnace[10] has three plasma burners supplied with d.c. at up to 6 kA at voltages

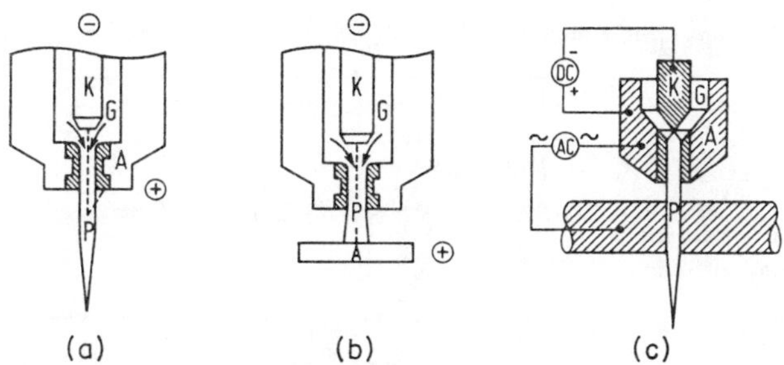

Figure 3. Three plasma generators. K = cathode, A = anode, G = gas, and P = plasma. (a) Non-transmitting arc; (b) transmitting arc; (c) with a.c.

between 200 and 450 V; its maximum burner power is 3 MW. The arc between the plasma burner and the scrap charge measures some 0.6 to 1.2 m in length. The bottom electrode (anode) located in the hearth consists of a water-cooled block of copper. A 30-t plasma furnace under construction at the time of writing is intended for melting steel at an installed power of 24.3 MVA.[11]

4.3.4 Principles of Electron Beam Melting Furnaces

Electron beams are used for a variety of technical applications ranging from production of sub-microscopic structures for microelectronics at beam powers of a few centiwatts up to melting large ingots at several megawatts for refining. The following deliberations are confined to the power range above 50 kW used for melting, vaporizing, and heat treatment.

The energy source for all electron beam systems is the electron gun. This generates the electrons, and then shapes the beam and directs it at the spot where the process takes place. Four main electron beam systems have been developed for electron beam melting furnaces.[12] An example of the basic design of a high-power electron gun is illustrated in Figure 4.[6,13] The beam production system consists of a heated cathode, a control electrode, and an anode. Electrons are emitted at a pressure of about 0.1 μbar, accelerated, and gathered into a beam. The beam then passes through one or more magnetic lenses which have to guide the beam through the electron gun with almost no loss and adjust the power density to suit the task. The electron beam is controlled and directed on to the process spot at the right time by magnetic deflection.

The acceleration voltage and current density must be high in order to project the electron beam over long distances with little loss at the relatively high pressure of a few tenths of a μbar. Acceleration voltage is limited to 45 kV, since at higher voltages the tendency towards breakout from the beam production system increases, and additional safeguards are needed to screen

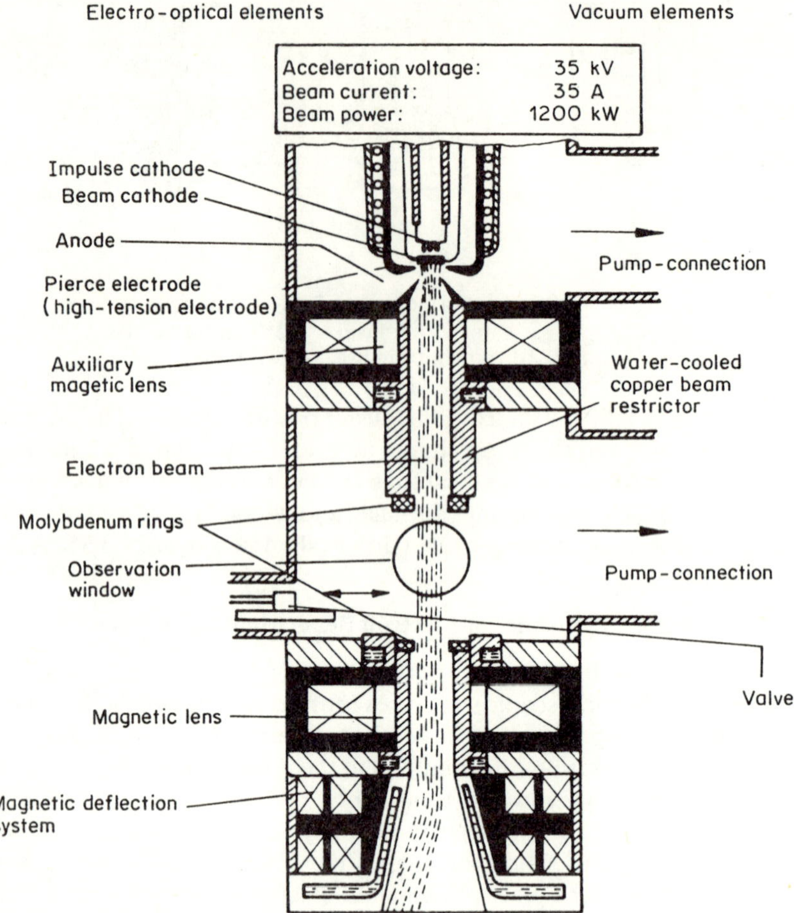

Figure 4. Schematic illustration of a high-power electron gun for use in melting furnaces[13]

against X-rays produced. The highest current densities of 10 A/cm^2 are achieved with tungsten cathodes.

Figure 5 illustrates a vacuum melting furnace with two electron guns. A large proportion of the energy in the moving electrons is converted into heat on the melt-down electrode; this causes degassed molten metal to drip from the melt-down electrode into the melt puddle which forms the ingot. The efficiency of this method of producing heat amounts to some 60–70%, according to the material being melted and the process used.

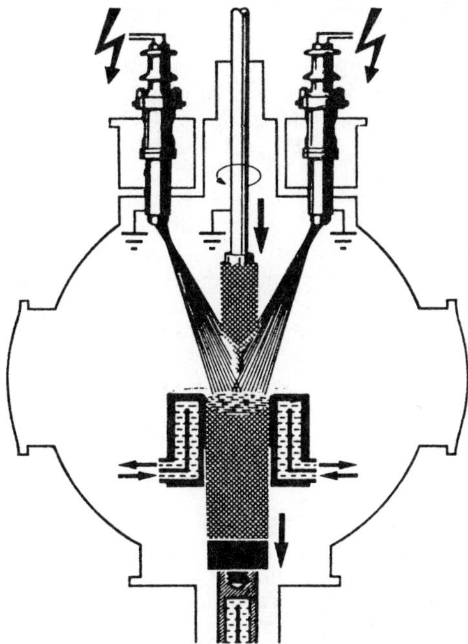

Figure 5. Design for a melting furnace with two electron guns[14]

4.4 LITERATURE REFERENCES

4.1 Basic Principles of Arc Furnaces

1. Muller, H., Physical characteristics of electric furnace arcs (German). *Elektrowärme* **20** (1962), 3–11.
2. Mayr, O., Beiträge zur Theorie des statischen und dynamischen Lichtbogens. *Arch. Elektrotechn.* **37** (1943), 588–608.
3. Schwabe, W. E., Efficiency improvement in electric steel furnace operation. In *4th Internationaler Elektrowärmekongreß*. Stresa, 25–9, March 1959. L'Union Internationale d'Électrothermie, Paris, Stresa 1959. Report No. 220.
4. Bowman, B., G. R. Jordan, and F. Fitzgerald, The physics of high-current arcs. *J. Iron Steel Inst.* **207** (1969), 798–805.
5. Sundberg, Y., Der elektrische Lichtbogenofen als Verbruncher am Versorgungsnetz. *ASEA-Z.* **21** (1976), 99–111.
6. Timm, K., and H. Faber, Electrodynamic stresses acting on arc furnace electrodes (German). *Elektrowärme*, Issue B, 34 (1976), 241–5.
7. Dunski, Ch. V., Research work on the arc furnace at the Electro-Heat Laboratory at the University of Liège (German). *Elektrowärme* **25** (1967), 208–16.
8. Bowman, B., Convective heat transfer and power balance in high current free-burning arcs. *Electrowärme*, Issue B, **30** (1972), 87–93.

9. Jordan, G. R., A. T. Sheridan, R. W. Montgomery, M. Denby, and B. D. Howes, Basic properties of high-intensity electric arcs used in steel making. British Steel Corporation. ECSC Convention No. 6210.93/8/801.
10. Schwabe, W. E., Arc heat transfer and refractory erosion in electric steel furnaces. *Proc. Electr. Furn. Steel Conf.*, Electr. Furn. Steel Comm., Iron Steel Div., Amer. Inst. metallurg. petrol. Eng. **20** (1962), 195–206.
11. Ottmar, H., A. Oerter, and D. Ameling, Der Zusammenhang zwischen den elektrotechnishen und wärmetechnischen Grundlagen bei Hochleistungs-Elektrolichtbogenöfen. *Radex Rdsch.* (1973), 519–27.
12. Ottmar, H., and D. Ameling, Der Hochleistungslichtbogenofen für die Stahlherstellung — Begriffsbestimmungen und Leistungsgrenzen. *Stahl u. Eisen* **94** (1974), 125–32.
13. Ciotti, J. A., and C. G. Robinson, Northwestern's 400-ton arc furnace. *J. Metals* **25** (1973), No. 8, 17–21.
14. Kupfmüller, K., *Einführung in die theoretische Elektrotechnik*. Berlin/Heidelberg/New York, 1973. 10th edition.
15. Dunski, Ch. V., Répartition du courant dans les électrodes d'un four triphasé à arcs en régimes équilibre et deséquilibre. *Elektrowärme* **20** (1962), 503–11, 616–25.
16. Orth, G., Current displacement in graphite electrodes for arc furnaces (German). *Elektrowärme*, Issue B, **34** (1976), 25–30.
17. Timm, K., H. Faber, H.-G. Kunze, and K. Norden, Dynamic behaviour of electrodes and supporting arms in arc furnaces (German). *Stahl u. Eisen* **98** (1978), 695–700.
18. Moeller, F., H. Fricke, H. Frohne, and P. Vaske, *Grundlagen der Elektrotechnik*. 16th edition. Stuttgart, 1976.
19. Bretthauer, K., and A. A. Farschtschi, Magnetic field problems of the power leads of arc furnaces (German). *Elektrowärme*, Issue B, **32** (1974), 33–7.
20. Schwabe, W. E., and C. G. Robinson, Development of large Furnaces from 100- to 400-ton Capacity. In *VIIth UIE-Congress*, Warsaw, 1972, L'Union Internationale d'Électrothermie, Paris. Report No. 105.
21. Bulajic, R. P., Mesure des paramètres électriques d'un four à arc. In *VIIIth UIE-Congress*, Liège, 1976, L'Union Internationale d'Èlectrothermie, Paris. Report No. Ia/7.
22. Bowman, B., Electrical characteristics of arc furnaces allowing for current swings. See Ref. 21. Report No. Ia/10.
23. Das DK-EW-Studien-Komitee 'Statistik-Wirtschaft-Entwicklung' (SWE). *Elektrowärme*, Issue B, **32** (1974), 164–8.
24. Bretthauer, K., and K. Timm, Contribution to the theory of the three-phase arc furnace (German). *Elektrowärme* **28** (1970), 115–20.
25. Sundberg, Y., Der Hochstromkreis von Lichtbogenöfen. *ASEA-Z.* **17** (1972), 77–83.
26. Svensson, E., Das Messen von Impedanzen in Lichtbogenöfen. *ASEA-Z.* **17** (1972), 84–5.
27. Bretthauer, K., and K. Timm, Unsymmetrie-Einflüsse auf den Verschleiß von Lichtbogenöfen und ihre Messung. See Ref. 20. Report No. 406.
28. Bretthauer, K., A. A. Farschtschi, and K. Timm, Investigation into asymmetry and measuring errors with a UHP electric arc furnace (German). *Stahl u. Eisen* **93** (1973), 761–5.
29. Ciotti, J. A., Experience gained in the USA with high-capacity arc furnaces, particularly with their triangulation (German). *Electrowärme* **28** (1970), 555–60.

30. McGee, L., and J. O. Sparrow, Optimum design of arc furnace secondary conductors. See Ref. 20. Report No. 402.
31. Bretthauer, K., and A. A. Farschtschi, Balancing arc furnace variables (German). *Elektrowärme*, Issue B, **34** (1976), 245–51.
32. Austrian Patent 181859 dated 10 May 1955.
33. Düchting, W., Grundlagen für die Symmetrierung von Drehstromlichtbogenöfen. *Elektrowärme* **19** (1961), 285–91.
34. Timm, K., Beitrag zur Symmetrierung von Drehstromöfen. *Elektrotechn. Z.*, Issue A, **94** (1973), 204–8.
35. Beckius, I., Dreiphasige Lichtbogenöfen mit unterschiedlichn Phasenreaktanzen und einzeln einstellbaren Phasenspannungen. *ASEA-Z.* **17** (1972), 27–32.
36. Ottmar, H., Performance capabilities of a 50 ton arc furnace (German). *Stahl u. Eisen* **86** (1966), 201–7.
37. Stöckl, M., and K. H. Winterling, *Elektrische Meßtechnik*. 5th edition. Stuttgart, 1973.
38. Dmochowski, Z., Hochstromwandler für Lichtbogenöfen. See Ref. 21. Report No. Ia/5.
39. Bretthauer, K., A. A. Farschtschi, and K. Timm, Measuring electric quantities of arc furnaces in electric melting shops (German). *Elektrowärme*, Issue B, **33** (1975), 221–5.
40. Bretthauer, K., and K. Timm, Secondary-circuit measurements in three-phase furnaces. *Elektrowärme* **29** (1971), 381–7.
41. Schiffarth, J., Optimization of arc power in electric arc furnaces (German). *Elektrowärme* **20** (1962), 18–26.
42. Eichacker, K., and K. Konrad, Accurate arc control in a 100-tonne furnace (German). *Elektrowärme*, Issue B, **32** (1974), 335–9.
43. Taegen, F., *Einführung in die Theorie der Elektrischen Maschinen*. Braunschweig, 1970.
44. Feyertag, H., Transformatoren für Lichtbogenöfen. *Klepzig Fachber.* **82** (1974), 133–7.
45. Kaempf, P., E. Markworth, and J. Mühlenbeck, 110-kV-Lichtbogenschmelzofen mit Lastschaltung im Zwischenkreis. *Stahl u. Eisen* **94** (1974), 393.
46. Iwansaki, Z., T. Kubota, and T. Anzai, Steel melting arc furnace transformer equipment directly connected to 154 kV power source with tertiary load switching system. *Fuji Electric Rev.* **17** (1971), 16–23.
47. Brehler, R., Ofentransformatoren zum Speisen von Lichtbogenöfen mit Ofenschalter im Zwischenkreis. *Siemens-Z.* **50** (1976), 8–17.
48. Frank, H., and S. Ivner, Thyristor-connected capacitors to compensate the reactance of electric arc furnances (German). *Elektrowärme,* Issue B, **32** (1974), 313–22.
49. Schönberger, W., Analysis of flicker voltages at electric arc furnaces using the correlation technique (German). *Elektrowärme,* Issue B, **34** (1976), 31–5.
50. Simon, K., Das Flackern des Lichts in elektrischen Beleuchtungsanlagen. *Elektrotechn. Z.* **38** (1917), 453–55, 465–8, 475–6.
51. Carjell, U., Ein Beitrag zur Beurteilung des Lichtflimmerns bei Netzspannungsschwankungen. Aachen (1972). (Diss. Techn. Hochsch. Aachen.)
52. Schwabe, W. E., and R. Kasper, Flicker caused by UHP-arc furnaces using scrap and directly reduced materials. In Meeting CNBE-Brazilian National Committee-UIE. Sao Paulo 8–10, Sept. 1976.
53. Grundsätze für die Beurteilung von Netzrückwirkungen. Vereinigung Deutscher Elektrizitätswerke. Frankfurt/M. 1976.
54. Navries, K. B., International poll on the effect of steel melting arc furnaces on the

supply network (German). *Elektrowärme*, Issue B, **32** (1974), 344–53.
55. Dixon, G. F. L., and P. G. Kendall, Supply to arc furnaces; measurement and prediction of supply-voltage fluctuation. *PROC. Instn. electr. Engrs.*, Pt. C, **119** (1972), 456–65.
56. Meynaud, P., Contribution à l'étude expérimentale du flicker provoqué sur les réseaux d'énergie électrique par des fours à arc. *Rev. gen. l'électr.* **76** (1967), 251–62.
57. Zinguzi, T., Progress in electric power supply to arc furnace loads in Japan. In *6. Internationaler Elektrowärmekongreß.* 13–18 May 1968. Brighton. L'Union Internationale d'Èlectrothermie Paris. Report No. 145.
58. Vogel, O., Meßtechnische Grundlagen zur Flickerbewertung. See Ref. 21. Report No. Va/10.
59. Ohmoto, M., A. Kawashiro, and M. Aoshika, Newly developed automatic optimum power control equipment in arc furnaces. *IHI Engng. Rev., Tokyo*, **4** (1971), 46–58.
60. Kegel, K., Design and physics of high-capacity arc furnaces (German). *Elektrowärme* **28** (1970), 551–4.
61. Pautz, J., Influence of argon on the stability and performance of electric arc furnaces (German). *Elektrowärme,* Issue B, **34** (1976), 235/40.
62. Jäger, S., Power factor correction schemes for industry (German). *Elektrowärme,* Issue B, **32** (1974), 326–34.
63. Webs, A., W. Kaufhold, and B. Kulicke, Rückwirkungen von Drehstrom-Lichtbogenöfen in elektrischen Versorgungsnetzen. *Elektr.-Wirtsch.* **71** (1972), 222–8.
64. Clegg, E., A. J. Heath, and D. J. Young, The static compensator for the BSC Anchor Project. In *IEE-Conference: Sources and Effects of Power System Disturbances*, Institution of Electrical Engineers. London 1974.
65. Sundberg, Y., Arc furnace flicker suppression with thyristor-switched capacitors. See Ref. 20. Report No. Va/5.
66. Schröder, D., Spannungsstabilisierung in Drohstromnetzen; ein Überblick über die Grenzen der Dynamik von Verfahrer der Leistungselektronik. In *IFAC-Symposium* Düsseldorf, 1974.
67. Wanner, E., and W. Herbst, Statische Blindleistungskompensation für Lichtbogenöfen. *BBC-Mitt.* **64** (1977),108–18.
68. Kaempf, P., Elektrische Ausrüstung für Lichtbogenschmelzöfen. *Elektr. Ausrüstung* **15** (1974), 19–22.

4.2 Basic Principles of Induction Crucible Furnaces

1. Brokmeier, K.-H., *Induktives Schmelzen*. Essen, 1966.
2. *Elektrowärme, Theorie und Praxis.* Union Internationale d'Èlectrothermie, Paris. Essen, 1974.
3. Esmarch, W., Zur Theorie der kernlosen Induktionsofen. *Wiss. Veröff. aus dem Siemens-Konzern* **10** (1931), 172–96.
4. Simonyi, K., *Theoretische Elektrotechnik*. 5th edition. Berlin, 1973.
5. Hegewaldt, F., Induktives Oberflächenhärten. *BBC-Nachr.* **43** (1961), 434–56.
6. Kolbe, E., and W. Reiss, Eine Methode zur numerischen Bestimmung der Stromdichteverteilung in induktiverwärmten Körpern unterschiedlicher geometrischer Form. *Wiss. Z. Techn. Hochsch. Ilmenau* **9** (1963), 311–7.
7. Reichert, K., A numerical method to calculate induction heating installations (German). *Elektrowärme* **26** (1968), 113–23.

8. Hegewaldt, F., Problems in the design of induction melting furnaces (German). *Elektrowärme* **28** (1970), 197–217.
9. Doetsch, E., and F. Hegewaldt, Use of the crucible-type induction furnace for steelmaking (German). *Elektrowärme,* Issue B, **33** (1975), 248–53.
10. Mühlbauer, A., Electrodynamic forces acting in the bath of induction furnaces (German). *Elektrowärme* **25** (1967), 461–73.
11. Neumann, F., and F. Hegewaldt, Considerations on limiting power values when melting iron and steel in crucible-type induction furnaces (German). *Stahl u. Eissen* **94** (1974), 53–63.
12. Majdic, A., W. Krombach, F. R. Block, and A. Theißen, Electromagnetic suspension melting — a contribution to the crucible-free melting processes (German). *Elektrowärme* **27** (1969), 216–21; **29** (1971), 349–54.
13. Brokmeier, K.-H., unpublished.
14. Hegewaldt, F., Steel melting in coreless induction furnaces (German). *Elektrowärme,* Issue B, **36** (1978), 39–46.
15. Matthes, H. G., Der Thyristorumrichter, ein moderner Mittelfrequenzgenerator für Anwendungen in induktiven Elektrowärmeanlagen. *Techn. Mitt. AEG/ TELEFUNKEN* **67** (1977), 283–92.

4.3 Basic Principles of Remelting Furnaces

1. Krall, F., and G. Ogiermann, Vakuum-Lichtobogenöfen. *Elektrowärme* **19** (1961), 226–31.
2. Bardahl, N., Stromversorgungen für Lichtbogen-Vakuumschmelzanlagen. *Siemens-Z.* **47** (1973), 48–52.
3. Gruber, H., Vakuum-Lichtbogenschmelzen mit Anschmelzelektrode. *Metall* **12** (1958), 901–12.
4. Machner, P., and W. Holzgruber, Das Elektroschlacke-Umschmelzen mit niederfrequentem Wechselstrom. In *VII. UIE-Congress,* Warsaw, 1972. L'Union Internationale d'Èlectrothermie, Paris. Report No. 168.
5. Thomas, F. W., Elektrische Großen im Schlackebad von Elektroschlacke-Umschmelzöfen. Hanover 1975. (Diss. Techn. Univ. Hanover.)
6. *Elektrowärme, Theorie u. Praxis.* L'Union Internationale d'Électrothermie, Paris. Essen, 1974.
7. Thomas, F. W., and W. Schmidt, Einsatz von Direktumrichtern für das Elektro-Schlacke-Umschmelzverfahren. *Siemens-Z.* **47** (1973), 676–80.
8. König, H., and G. Rath, Possibilities of steelmaking with high-sponge-iron charge in the electroslag resistance furnace (German). *Stahl u. Eisen* **97** (1977), 12–17.
9. Schoumaker, H. R. P., Plasma heating ovens. In *VIII. UIE-Congress,* Liège, 1976. L'Union Internationale d'Électrothermie, Paris. Report No. Ib/1.
10. Fiedler, H., Zum Stand der Plasmaschmelztechnik in der DDR. *Neue Hütte* **21** (1976), 69–72.
11. Meyerson, C. I., V. A. Khotin, Ch. Fiedler, and V. Lachner, Plasma steelmaking in ceramic crucible furnaces of up to 30–T capacity. See Ref. 9. Report No. Ib/4.
12. Dietrich, W., H. Stephan, and H. Gerstner, Elektronenstrahlen als technische Wärmequelle. *Elektrotechn. Z.,* Issue B, **15** (1963), 560–4.
13. Schiller, S., G. Jäsch, R. Schroller, H. Förster, and A. v. Ardenne, Erzeugung und Nutzung leistungsstarker Elektronenstrahlen. See Ref. 9. Report No. IIc/3.
14. Basdahl, N., Hochspannungs-Stromversorgung für Elektronenstrahlkanonenöfen. *Siemens-Z.* **46** (1972), 170–4.

Electric Furnace Steel Production
Edited by E. Plöckinger and O. Etterich

5 Electric Arc Furnaces

HEINRICH OTTMAR, BOUS/SAAR

5.1 GENERAL CONDITIONS FOR THE USE OF ELECTRIC STEEL PROCESSES

Industrial steel production started over a hundred years ago with the conversion of liquid pig iron into steel using the air (Bessemer) and open-hearth processes. Later, electric furnace steel processes found favour for making costly, high-quality, and special steels. Today, arc melting furnaces and electric steelworks are used in growing numbers and capacities all over the world for the production of ordinary, high-tonnage steels. This development trend is clearly seen in production statistics, and there are several reasons for the change. Electrical energy can be converted efficiently and with a very high energy concentration in the arc into heat for melting. This opens up the possibility of obtaining high outputs from small furnaces. The steady increase in electrical energy-generation and the expansion of the mains-electricity network make it possible to power and operate increasingly large furnaces. In recent years new technology, operating methods and steelworks concepts have enabled arc furnaces to be used economically for mass-tonnage steel production.

The high rate of growth of electric steel production is demonstrated by comparing absolute production figures. Total world crude-steel production rose by a factor of 2.9 in the years 1952–72. In the Federal Republic of Germany the figure, 2.8, is practically the same. In the same period world electric steel production increased by a factor of 6, while in West Germany it increased by a factor of almost 9. In 1977, the Federal German share of crude steel production was 13%.

For the foreseeable future, the best raw material basis for electric steelmaking in Europe will be scrap; however, this means that the availability of scrap will limit maximum possible electric steel production. At present, scrap is the basis of about 46% of all crude-steel production in West Germany. If basic oxygen and electric processes were the only ones available for making steel, then oxygen blast would account for 78% of the steel made from scrap available in the Federal German economy, and electric processes would account for 22%.

This situation can only be fundamentally changed by new raw materials which are suitable for open-hearth steelmaking processes. The manufacture of direct reduction products instead of pig iron, which has been discussed for many years as a competitive process, could lead to still greater growth in electric steelmaking. However, it is questionable whether ores and fuels available in Europe could permit larger production units for sponge iron to be operated economically. The development of high-capacity furnaces and suitable technologies has greatly favoured the application of electric steelmaking processes for high-tonnage steels. Tap-to-tap times of under 2 h are now easily attainable and the use of jet-burners enables melting times to be shortened considerably more. The increasingly high thermal stressing of furnace linings associated with those techniques can be combated by using improved refractories and, above all, by the introduction of water-cooled elements, especially in the furnace wall. However, the efficiency of each method must be carefully checked on the basis of the costs of energy and refractory materials. Considerable progress has also been made in improving the current carrying-capacity of electrodes, which helps to reduce specific electrode consumption.

In common with the development of high-power technology, the new ladle metallurgy processes and sponge-iron processing have also improved electric steelmaking productivity (see section 5.7 and Chapter 7).

5.2 ARC FURNACE DESIGN AND GUIDELINE PUBLICATIONS

Ulrich Siegers, Bous/Saar

5.2.1 Design Criteria

The design of an electric arc furnace is determined mainly by three factors:

(1) The weight of the charge;
(2) The diameter of the shell;
(3) The electric power supply required.

While electric power consumed is a measure of melting performance, the diameter of the shell is a decisive factor for furnace serviceability, because of its association with lining life. The latest techniques, such as the use of jet-burners, building in water-cooling elements, and employing sponge iron in the charge, all have a profound effect on conventional design criteria and are fast gaining in importance. However, we will first deal only with conventional operation using scrap charges and refractory linings in the furnace. This conventional operation will remain with us for the near future, and it constitutes the basis for new techniques to which we can turn as required.

We shall now deal with the relationship between electrical power supply and process time, and for this we have to take into account the relationship between charge weight and specific electrical power consumption for melt-down. The specific melt-down current depends on the quality of the scrap, which in turn is governed mainly by the iron content, the heat of reaction of the accompanying alloying elements and other metals present, and by the amount of slag required for the given type of scrap (see section 5.2.1.4). If the melt-down programme has been determined, and the time, voltage, current, and transformer apparent power for the melt-down period are known, then we can evaluate the average power factor cos ϕ and the maximum transformer apparent power (Figure 1).

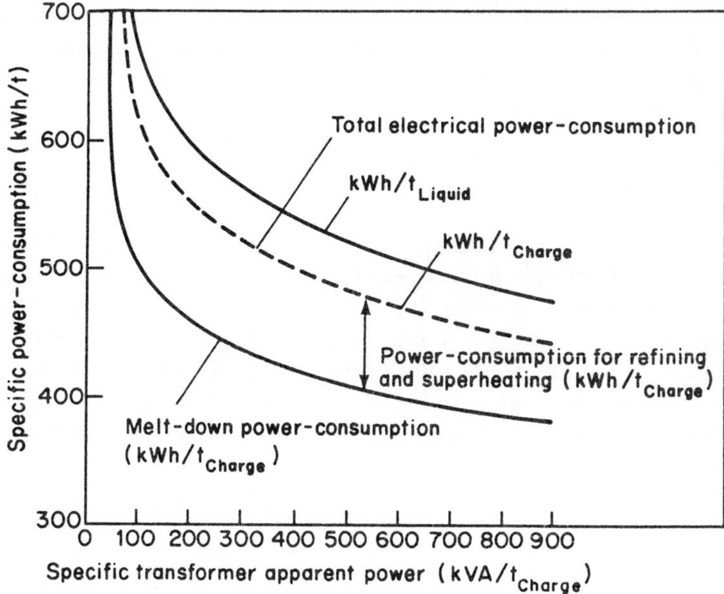

Figure 1. Relationship between specific electrical power-consumption for melt-down and specific transformer apparent power for melt-down, refining, and superheating

The maximum transformer apparent power S is given by:

$$S = \frac{\text{Specific melt-down power consumed} \times \text{charge weight}}{\text{Melt-down time} \times \text{average } \cos\phi \text{ during melt-down}} \qquad (1)$$

If we assume that:
charge weight E = 100 tonnes, specific power consumption = 400 kWh per tonne of charge, required melt-down time = 1h, mean power factor cos ϕ = 0.74 (at maximum apparent power)

then we obtain transformer apparent power S:

$$S = \frac{400 \times 100}{0.74 \times 1} = 54\,054\ \text{kVA}$$

Furnace-availability depends to a great extent on the durability of the refractory lining. This in turn depends on the material quality and the process regime (e.g. the refining method, the amount and composition of the slag and the tapping temperature), but it is also essentially determined by the proximity of the arc to the wall. Optimum shell diameter can be calculated from the arc radiation if calculation is based on the radiation value for a furnace of similar size with good lining durability. Arc radiation depends on secondary voltage and the proximity of the arc to the wall. If we know the electrode pitch circle diameter (d mm) (from the electrode diameter and the constructional design of the electrode holder) we can then calculate the diameter (D mm) of the furnace shell from the following equation:

$$D = 2\left(\sqrt{\left(\frac{c \cdot V^3}{RE_w}\right)} + \frac{d}{2} + W\right) \tag{2}$$

Where c is a constant, RE_w is the refractory wear index, V is secondary voltage (V), d is the electrode pitch circle diameter (mm), and W is wall thickness (mm). According to this equation, the shell diameter does not depend on either charge weight or output weight. Up to now we have considered shell diameter without taking account of any relationship with wall durability. If we base calculation of refractory wear index RE_w on data from a 60 t-furnace with comparable and equally favourable characteristics, after the equation:

$$RE_w = \frac{V^3_{max}}{a^2} \tag{3}$$

(where a is the distance of the arc from the shell wall (mm)), then we can use this refractory wear index in designing larger furnaces of equally good durability.

For example, we can calculate the shell diameter of a 100 t-furnace where $V_{max} = 600$ V, on the basis of a 60 t-furnace with a wall life of 240 melts using about 3 kg of wall brick and 3 kg patching material per tonne produced, allowing only 2 min patching time per melt. If, for the 60 t-furnace $V_{max} = 460$ V and the distance of the arc from the wall $a = 1630$ mm, then

$$RE_w = \frac{V^3_{max}}{a^2} = \frac{460^3}{1630^2} = 37$$

for constant $c = 1^{2,3}$

Maximum secondary voltage $= 600$ V

Electrode pitch circle diameter d = 1300 mm
Wall thickness W = 350 mm
the shell diameter of the 100 t-furnace will be:

$$D = 2[(1 \times 600^3/37)^{0.5} + 1300/2 + 350] = 6832 \text{ mm}$$

For the sake of simplicity, a few factors were ignored in the above example:

(1) Effective radiation does not depend on the rated voltage but on the arc voltage;
(2) Maximum wear does not occur during the maximum voltage stage (which usually occurs during melt-down) but during refining at reduced voltage;
(3) Power factor cos ϕ, which greatly influences radiation, is also reduced; in equation (2) it is concealed in constant c;
(4) The intensity and duration of the radiation emitted by the arc depend a great deal on the metallurgical process (slag depth, boiling, etc.). On account of their long oxygen-blowing times, very large furnaces cannot usually cope with elevated carbon levels;[4] there is thus no shielding afforded by frothing slag or boiling steel, and the arc burns on a flat bath all the time.

If we assume that these metallurgical conditions apply to all furnaces, and if we base the calculation on the refractory wear index RE_w for the Bous furnace at the time of heat transfer, then, using appropriate simplifications, we can derive the relationship between furnace shell diameter and secondary voltage shown in Figure 2.

This example is based on the idea, common until now, that shell diameter and durability both depend on the maximum secondary voltage used. This theory is especially plausible in scrap melting, if detectors installed in the hot zones of the wall signal the appearance of radiation generated by the arc during the maximum voltage stage, and react by causing the voltage to be switched to lower levels. This could be identical to the 'rated secondary voltage for heat transfer' illustrated in Figure 2. It is assumed that the radiation that becomes effective can only arise because of a mixture of variables active during the furnace cycle. If we insert the results for the 60-t Bous furnace into this diagram, and if we assume that the mode of operation is the same as that, for example, for the two 360-t American furnaces,[5] then we can read off working voltages of about 540 V and 640 V for the furnaces. At these settings we should be able to expect a life similar to that for the 60-t furnace. In fact, the lining life is considerably shorter in these furnaces because of the effects of slag and decarburization; this highlights the importance of these parameters. Also, choosing arc voltage as the reference value, as was done by Bowman,[6] does not improve certainty (Figure 3).

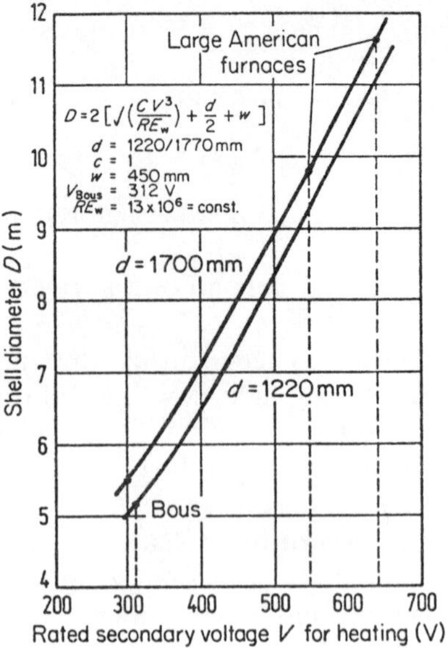

Figure 2. Shell diameter in relation to refractory wear index RE_W and related secondary voltage for heat transfer

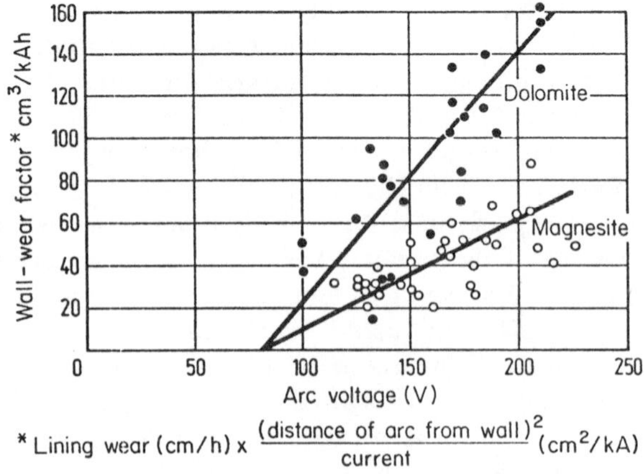

Figure 3. Wall-wear factor in relation to arc voltage (after Bowman)

If we develop the above argument further, then we can predict the resultant wall lives of a particular furnace by varying the arc voltage in equation (2).

Figure 4 shows the dependence on secondary voltage that could apply for the Bous furnace. Here a relationship with the known values (life H = 236 melts

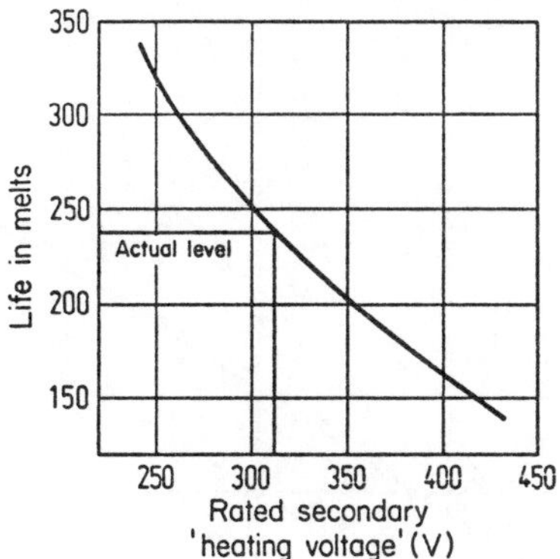

Figure 4. Wall-life for various values of 'heating voltage' for the Bonus furnace

for $V = 312$ V and corresponding $RE_w = 13 \times 10^6$) is assumed. The wall lives read from this diagram correspond approximately with actual figures. However, at voltages below 226 V and cos ϕ of 0.6, radiation is so low that there is no wear due to arc radiation in the Bous furnace.

At this point the curve ought to change into a straight line parallel to the Y axis. For furnaces of different orders of size the relationship must be determined from the appropriate data.

The wall height determines the furnace volume and has a quite marked effect on wear and on the costs of the refractory wall and roof materials and of the electrodes.[1] The furnace volume and with it the density of scrap indicate the number of baskets of scrap required and the power-off time for charging.

The effects of the various parameters for different wall heights are known in the main. Thus roof wear decreases as wall height increases because of the increasing distance from the radiant arcs and the melt. On the other hand, both wall material and electrode consumption increase at the same time; here the greater refractory volume in the one case and increased operating length in the other are important.

With scrap density other factors are important. The more dense the scrap, the worse are the arcing conditions; however, if the scrap is too light, too many back-charges are required.

Data for 89 central and northern European furnaces built between 1950 and 1971 are plotted in Figure 5.[7] The following logarithmic equation

$$K_H = 164.64 \times \ln K_D - 843.24 \tag{4}$$

was derived with a 70% confidence level by regression analysis. It agrees well in the shell diameter region of about 3.6 to 6.5 m with another equation, also derived by linear regression analysis, with a confidence level of 61%:

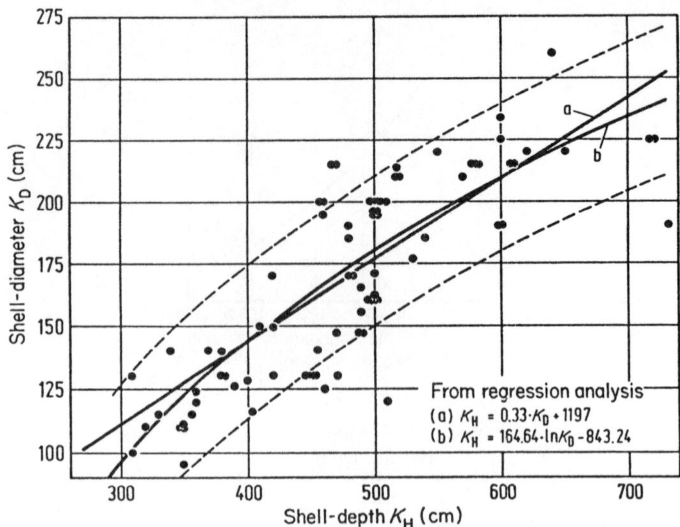

Figure 5. Shell diameter K_D and depth K_H for furnaces of capacity over 10-t built in central and northern Europe from 1950 to 1971

$$K_H = 0.33 \times K_D + 11.97 \tag{5}$$

First estimates of K_H using these equations should be within deviations of no more than $\pm$ 0.3 m.

5.2.1.1 Specification of maximum annual production

In designing all parts of an installation it is important first to determine the maximum specified annual production. Today's technology enables installations to be designed to match specified production requirements very precisely. If the purchaser of a furnace has it in mind to reserve spare capacity for the future and does not inform the constructor of the furnace or installation, then great difficulties and extra costs will certainly arise when he later attempts to expand production.

For existing installations, secondary steelmaking transfers the refining from the furnace to the ladle with considerable saving of time and improved metallurgical possibilities.

5.2.1.2 Specification of batch sizes and numbers of furnaces

The production rate envisaged during the planning stage for a steelworks can be achieved using several small furnaces or fewer large ones. If several small furnaces are used, this choice offers the advantages of better adaptability and higher overall availability in the event that a furnace breaks down. The costs of dividing up the melt and storage of materials are low. Fewer larger furnaces have, on the other hand, the advantage of lower conversion costs, higher ingot production, and better analysis and temperature control. Moreover, the shop-floor area and overall investment level are lower, even though larger cranes and taller buildings, etc. are needed.

Furnaces which feed continuous casting plants have to be specially checked so as to ensure that melting time, post-treatment time, and casting time match the number and size of furnaces and that continuous casting or sequential casting are possible. The ratio of the number of furnaces and casting frequency to the number of continuous casting units determines the melting time necessary for sequential casting.

$$\frac{\text{Number of furnaces} \times \text{casting frequency}}{\text{Number of continuous casting units}} = \text{melting time for sequential casting} \quad (6)$$

This means that for an installation with two furnaces, one continuous casting unit, and a casting frequency of 1 h, the melting time for the casting sequence is 2 h. A steelworks with three furnaces and two continuous casting units and the same casting frequency will, on the other hand, have a melting time for the casting sequence of 1.5 h. As the latter grows, sequential casting becomes more difficult, even if the number of continuous casting units is increased.

5.2.1.3 Time required and time sequence

With knowledge of the specified annual production target and hence the number of furnaces and batch sizes we can determine the tap-to-tap time; however, for this we must also take annual working time into consideration:

$$\text{Tap-to-tap time} = \text{Furnace batch size } (t) \times \text{number of furnaces} \times \text{annual working time } (h) \div \text{specified annual production } (t) \quad (7)$$

In this the annual working time does not include total time on lining maintenance. If the furnace is operated for three shifts 5 days a week, lining maintenance can be carried out at weekends and need not be included in the equation. In completely continuous operation we have to know the lives of the wall and hearth and the times required for lining repairs in order to work out the total time attributable to lining maintenance. Because these times have a marked effect on the productivity of the installation (Figure 6) one must attempt to minimize lining maintenance time as far as possible. It is possible to arrange to carry out lining repairs in furnace working time by using split or replaceable shells. This enables a wall to be replaced in 3–6 h instead of 8–24 h under conventional arrangements. If wall-replacement time is cut too short, care must be taken that sufficient time is allowed for necessary servicing and mechanical maintenance. Experience shows that 4–8 h must be allowed where

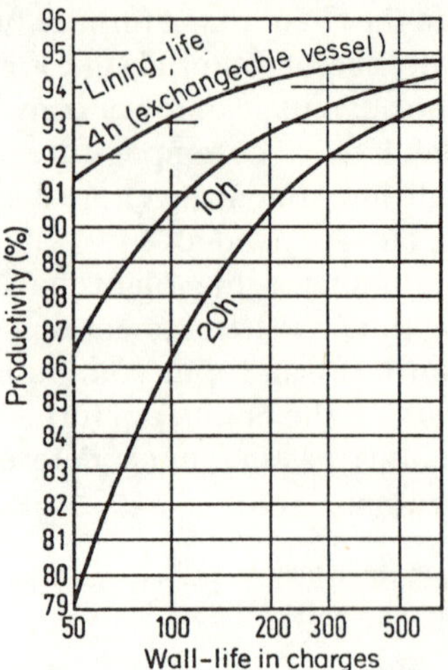

Figure 6. Plant-utilization on productivity related to wall-life and lining-life (with 5% diversion allowance)

wall-life is, say, 3 weeks. In older installations one must allow as much as one shift per week down-time. Longer wall-lives involve correspondingly longer maintenance times.

Special attention must be paid to power-off times. These are particularly governed by works organization, personnel training standards and the quality of scrap and the electrodes. With a well-organized operation, furnace power-off time can be planned as 30 min. For instance if three baskets are charged per melt, then power-off time can be apportioned (in minutes) as follows:

Patching and charging the first basket	10
Charging the second and third baskets	8
Electrode servicing	3
Time lost electrically	4
Tapping time	5
Total idle time per melt	30

Difficulties with materials and insufficiently trained personnel lead to extended power-off times. In particular, heavy patching and extended charging times incurred with bulky scrap and other unfavourable circumstances can increase the total power-off time to over 60 min. In well-organized plants the power-off time due to both electrical and mechanical stoppages is in the region of 1–3% of total melting time.

We will merely mention here that several furnaces operating together can affect one another and can impose heavy peak demands on the mains. When a new works is planned these effects should be taken into account in optimizing material flow and adequate consideration to the electrical design. The conditions to be met by the consumer, laid down in electricity supply contracts, must also be examined in the planning stage.

5.2.1.4 Electrical planning

If the furnace power-off time and the times for lining repairs and interruptions are known, the electrical design can then be started. The total 'power-on' time can be divided into times for melting down, oxidizing, refining, and superheating. When determining melting time and the times for refining and superheating using equations (1) and (2), we took account of the reduction of the specific power consumption as process time is shortened. Figure 1 shows the theoretical curves on which operating values are based.[8] For example, a power consumption of 400 kWh per tonne of charge is expected if available melt-down power is 600 kVA per tonne of charge. This corresponds to the values for central European steelworks under normal scrap conditions. If scrap with a lower heat of reaction and/or higher non-metallic content is used, this figure can rise to 600 kWh per tonne of charge and more.

The average power factor $\cos \phi$ during melt-down, used in equation (1) to calculate melting time, is adjusted to a lower value because radiation attacks the wall when specific transformer apparent power is increased. Figure 7 shows the mean $\cos \phi$ adopted here during melt-down.[8] It also shows the resulting mean specific active power for melting plotted against specific transformer power.

Refining and superheating times also are evaluated by using the mean power-utilization during refining and superheating plotted in Figure 7. The mean power-utilization in this part of the process is defined as the ratio of actual adjusted active power to maximum transformer apparent power. In this particular part of the process we must especially remember the radiation attack on the lining, and therefore we often can no longer work at maximum apparent power. To adjust for these difficulties, the mean active power is calculated using the assumed mean power-utilization.

The maximum melting power and maximum power factor $\cos \phi$ possible during melting govern the specification of the furnace transformer secondary voltages. In this study we assume that the maximum $\cos \phi$ possible decreases as specific transformer power increases, as shown by the straight line in Figure 7. From this we obtain the curve for maximum possible active melting power, which will be used later to determine the characteristic value C_2—maximum

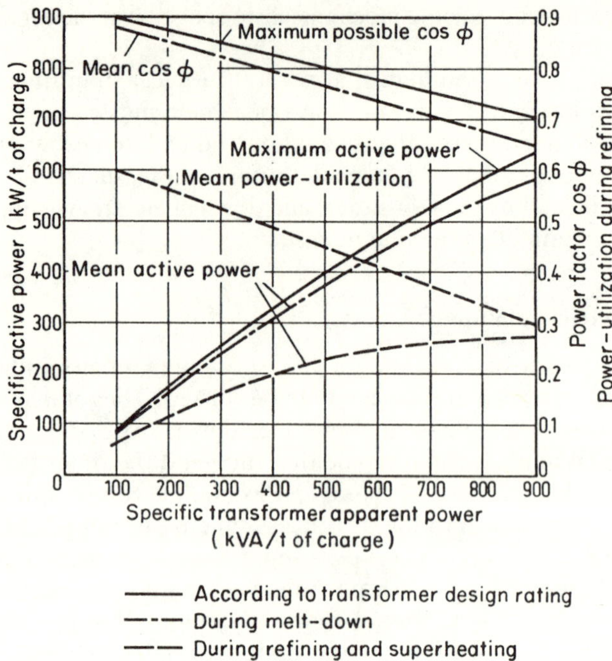

Figure 7. Variation of specific active power and power factor with increasing specific transformer apparent power

active power utilization. The curves assumed here apply for the usual type of furnace transformer design. In a different design approach for high specific power transformers the maximum secondary voltage range is selected so that a high maximum cos ϕ is attainable.[2,9] For sponge iron charges the requirements are different and will be covered in section 5.7.

A nomogram derived by a working group of the German Committee for Electro Heat[10] summarizes all the important design data. The presentation in Figure 8 is based on the following assumptions:

(1) Average melting time for all furnaces is 1 h. In order to achieve this melting time, a mean specific transformer power was selected of 720 kVA per tonne tapped (taking into account a power consumption of 450 kWh per tonne of charge and a cos ϕ of 0.707). Tap-to-tap times of 1.4, 2, and 3 h are illustrated to account for different metallurgical processes.

(2) Secondary voltage and current limits have been chosen so that a maximum rated apparent power of S = 100 MVA is not exceeded. For modern arc furnace operation transformers must be designed so they can sustain rated apparent power for the entire power-on period.

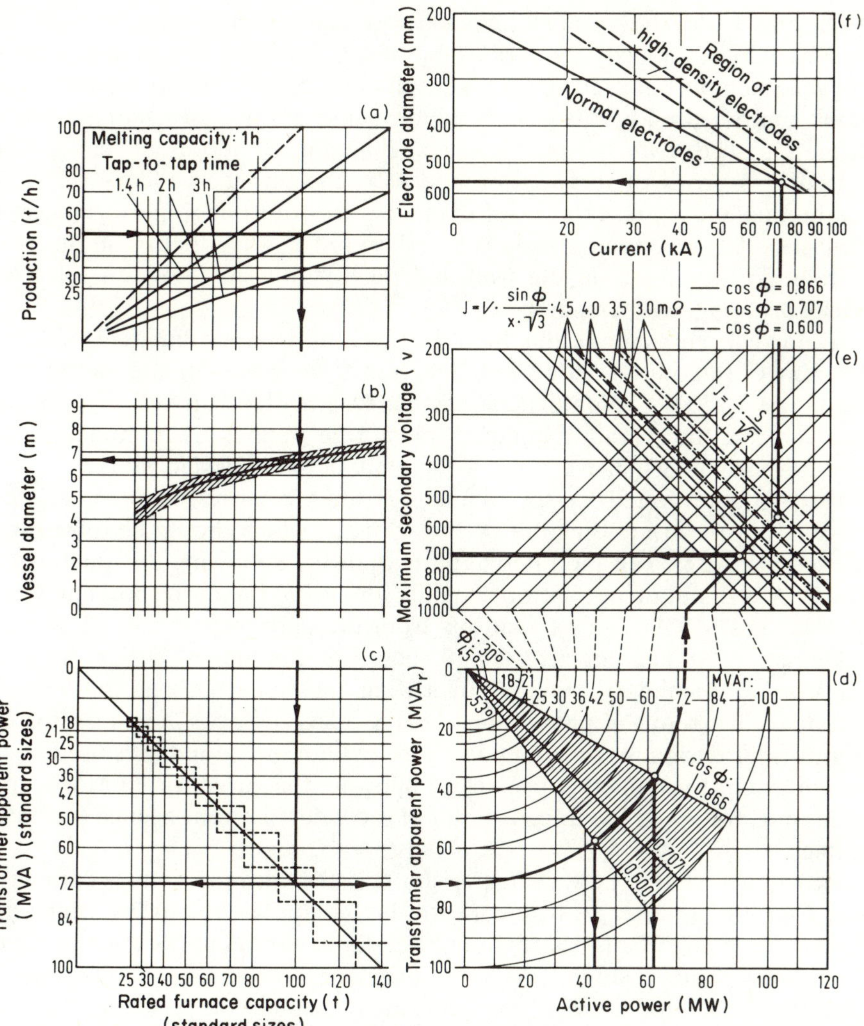

Figure 8. Nomogram for planning principal characteristics of modern arc furnaces (net melting time 1 h; specific transformer apparent power 760 kVA per tonne of charge[11])

(3) Paragraphs (1) and (2) above result in a maximum rated furnace capacity of 140 tonnes.

(4) Progression of nominal furnace capacity should be in steps of about $\sqrt[4]{2}$.

The nomogram depicted in Figure 8 consists of six part-diagrams A–F. The content of the nomogram and its use will be described by means of following actual example.

Starting from the production capacity required per furnace (t/h), the necessary rated capacity for various tap-to-tap times can be found in diagram A. The example selected of 50 t/h results in a rated capacity of 100-t for a tap-to-tap time of 2 h. Diagram B (below A) shows the corresponding vessel diameter. The two broken lines in diagram B represent earlier and present-day series of furnaces produced by the furnace-builders, while the continuous line represents vessel diameters for future high-performance arc furnaces. This recommendation is based on present-day theory on the effect of radiation intensity on the wall and the relationship between refractory brickwork, slag, and temperature.

In our chosen example the 100 t furnace has a vessel diameter of 6.7 m; diagram C (below B) shows that the transformer name-plate rating for it should be 72 MVA. Nominal ratings increase in steps of $\sqrt[4]{2}$. The broken lines indicate the spread of nominal transformer ratings and nominal furnace capacities which, when averaged, form the basis of the design value.

Having determined vessel diameter, rated capacity, and transformer power, we can turn to diagram D (lower right), which shows the relationship between active power and rated apparent power. Following the quadrant for 72 MVA to its intercept with the radius for cos ϕ, the corresponding values of MW and MVAr can be read off from the appropriate axes. The phase angle ϕ of 30 degrees, cos ϕ = 0.866, sets the upper design limit for transformer secondary voltage. In setting the upper design for limit for secondary current, a phase angle of 53 degrees (cos ϕ = 0.6) for the transformer design is recommended here (within the boundaries used by the design group for an actual project). Following the curve for 72 MVA up to diagram E brings one to a log/log plot of secondary current against secondary voltage.

Reactance is the key characteristic for the current–voltage design of transformers. In the present context this comprises the transformer plus furnace reactance from the high-voltage input terminals to the tips of the electrodes. This includes an allowance for working reactance due to non-sinusoidal operation. This working reactance is greater than the reactance measured sinusoidally in a short circuit test.[12]

Specification of the transformer maximum secondary voltage is based on the family of inclined linear characteristics derived from the following equation:

$$I = \frac{V \sin \phi}{X\sqrt{3}} \tag{8}$$

For these parameters a range of reactance values between 3.0 and 4.5 mΩ and values of cos ϕ of 0.6, 0.707 and 0.866 were selected. The secondary voltage and current values can be read off from the abscissa and ordinate respectively, corresponding to the reactance and cos ϕ values calculated for the projected design.

In this example the maximum secondary voltage is selected on the basis of $\cos \phi = 0.866$ and a total reactance of 3.5 mΩ. This reactance should include an additional component (usually 10–15%) to simulate the average non-sinusoidal conditions during melting which improve with time. Selection of maximum secondary current is from the family of declining linear characteristics derived from the equation

$$I = \frac{S}{V\sqrt{3}} \tag{9}$$

For the above example we adopted $\cos \phi = 0.6$ and a total reactance of 3.0 mΩ. The lowest possible working reactance (corresponding approximately to short-circuit reactance) is used for this working point.

Finally, diagram F at top right shows the relationship between maximum secondary current and nominal electrode diameter. The solid line represents the secondary current curve for standard electrodes of various diameters. The dash-dotted line is for existing high-density graphite electrodes. The broken line shows the possible outcome of current developments. The diagram caters for electrodes of up to 600-mm diameter, although there are furnaces in operation using larger electrodes with diameters of 650 or 700 mm. However, these are special cases and will remain so for some time, so we did not extrapolate the lines to include these diameters. In our example, standard 600 mm electrodes are rated up to a secondary current of 80 kA.

5.2.1.5 Influence of new technologies

New techniques include continuous charging, e.g. with sponge iron, accelerated melting with oxy-fuel burners, the partial replacement of wall and roof refractories by water-cooled sections, and by secondary steelmaking. Some techniques are insufficiently proved to enable one to make definite recommendations about the effects of these technologies on the physical and electrical design of new arc furnaces. Economical implementation of the first three processes mentioned is decidedly influenced by the costs of energy and raw materials, which differ greatly between the raw materials exporting countries and the present-day industrialized countries.

Continuous charging of sponge iron is almost always effected through a fifth hole in the furnace roof. Widespread trials in which the flow was divided between three apertures were not successful. Disadvantages in construction outweighed the hoped-for metallurgical and refractory life-improvements. The water-cooled roof, recently developed in a West German steelworks, could overcome some of these constructional difficulties. However, the variant with three charging apertures could become superfluous if the sponge-iron analysis is matched correctly to the furnace dimensions and mode of operation. The diameter of the feed-hole and that of the feed-chute above the roof can be calculated, from the maximum sponge iron lump size and maximum charging rate, by using well-known equations from conveyer technology.

In various steelworks the furnace vessel was enlarged after a few years using sponge-iron charges.[13] Wall-life had become too short because of increased arc radiation due to prolonged periods of flat-bath conditions. As is shown in the following equation derived by Schwabe and Ottmar,

$$RE_w = \frac{c P_A V_A}{a^2} \tag{10}$$

the radiation load RE_w on the wall varies as the reciprocal of the square of the distance a of the arc from the wall, where P_A and V_A are arc power and voltage, respectively, and c is a constant.

Wear on the refractory lining can be reduced at constant electrical values P_A and V_A not only by increasing the value of a but also by decreasing the factor c. It is generally known that the arc can be shielded to a greater or lesser extent by varying metallurgical process methods.[14] For sponge-iron charges a furnace may be operated as shown in the upper part of Figure 9, depending on the metallization at different high-carbon throughput rates and sponge iron charge rates (t/h). In this diagram the lowest limit of carbon throughput rate shown is about 10 kg/m^2 of carbon per hour, while the uppermost limit shown is 45 kg/m^2 per hour. Here, the bath diameter was taken to be 4.58 m with a shell

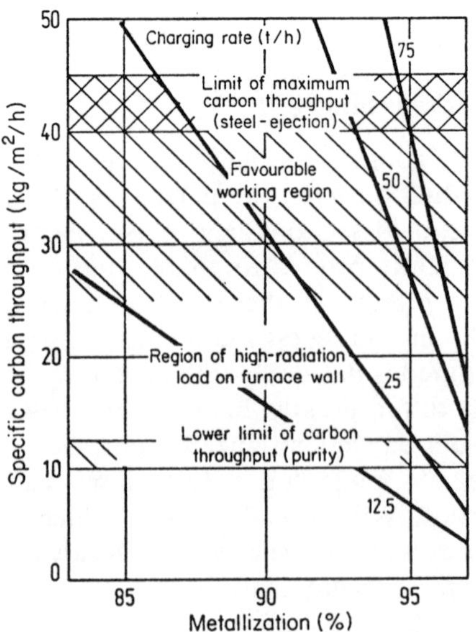

Figure 9. Performance limits for sponge-iron charge of various degrees of metallization

diameter of 5.18 m and 300-mm wall thickness. If carbon throughput is increased, steel may be ejected. If the boiling action is too weak, steel purity may not be sufficient.

The maximum possible charging rate (t/h) is easily calculated from the available power (kW) divided by the specific heat of solution of the sponge iron (kWh/t) of sponge iron. Figure 9 also shows that metallization η_{met} is the decisive factor in planning shell diameter to optimize charging conditions with sponge-iron charges.[15] Figure 10 shows the mathematical interdependence of carbon throughput, shell diameter, and metallization when charging rate is kept constant at 50 t/h.

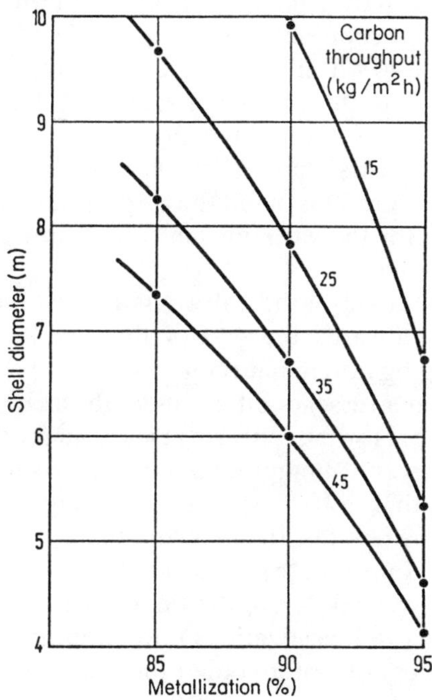

Figure 10. Shell-diameters for different metallization levels (charging rate constant at 50 t/h)

For example, a 70-t, 35-MVA furnace can melt sponge iron continuously, fed at this rate, if it has a specific heat consumption of 560 kWh/t. The power factor cos ϕ is taken in this case to be 0.8, indicating an active power of 28 MW. This corresponds to a type of sponge iron of which η_{met} = 92.2% with 10% slagging. For steel with a final carbon content of 0.1% and sponge iron containing 1.63% carbon, carbon throughput will automatically be 35 kg/m^2 h if the bath surface diameter is 5.3 m. This means that in this case a shell diameter of about 5.90 m is necessary. However, sponge iron reduced to only

85% metallization needs a much larger vessel of over 8.20 m for the same conditions.

If such large vessels are to be avoided for this kind of sponge iron, then the charging rate must be reduced. For the same carbon throughput as in the previous example, halving the charging rate to 25 t/h results in a vessel diameter reduced to exactly 6 m (Figure 11).

Although reducing the charging rate cuts the power supply required by half, it also causes a drastic drop in productivity. If we go right up to the empirical limit of maximum carbon throughput possible (about 45 kg/m^2 h) then we can still operate a 6-m shell using sponge iron of only 85% metallization at a charging rate of 32.2 t/h. If it is assumed that more iron (say, 30%) will be vitrified into slag, and that less oxygen and carbon are converted into CO, then charging rate can be raised further to about 40 t/h; however, iron losses then simultaneously rise from 1.2 to 3.75%. Thus the operating region at a carbon throughput of between 25 and 45 kg/m^2 h, termed favourable in Figure 9, allows sufficient freedom for selection of a suitable shell diameter. The 35 kg/m^2 h carbon throughput on which Figure 11 is based thus represents an average value which allows sufficient room for the variations in sponge iron analyses arising from present production methods.

The melt diameter increases as the wall wears away; therefore, specific carbon throughput decreases towards the end of the wall-life, if the sponge-iron composition and charging rate remain constant.

It is also possible that a vessel of different depth can be used for sponge-iron charges instead of scrap. However, bearing processing costs in mind, the effects of wear on the refractory roof-lining and electrode-consumption tend to oppose variations in vessel depth. For 100% sponge-iron charges and the crater process, it would theoretically be possible to use a very shallow vessel that allows room only for boiling steel. However, the wall and roof linings in such a 'frying pan' would be excessively stressed. Up to the time of writing no conclusive results have come from extensive investigations into optimizing this dimension. The recently reported success of water-cooled furnace roofs[16] tends to confirm a trend towards using shallower vessels for sponge-iron charges in future.

There is no need to increase the rate of furnace-gas extraction for sponge-iron charges compared with scrap. Experience shows that oily swarf, which must always be expected in scrap, causes much more gas than any refining.

As was mentioned before, the sponge-iron composition and charging rate are both interdependent with the electrical power supplied. Therefore these must be known before deciding on transformer power required for arc furnaces intended for sponge iron.

Up to now, furnace transformers were designed by many constructors to be operated for a limited time at 20% overload. This must be remembered, because charges containing a high proportion of sponge-iron require long periods at maximum power. In the previous section we mentioned a power factor

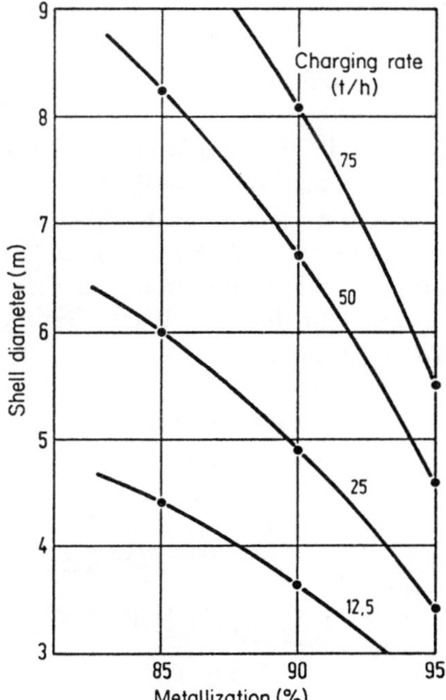

Figure 11. Shell diameters for different metallization levels (specific carbon throughput constant at 35 kg/m^{2}h)

of cos ϕ of 0.8; this goes against the values of 0.7 or less advocated as the optimum by some operators.[17,18] The disadvantages implicit with a long arc (cos ϕ = 0.8) can, however, be compensated by appropriate adjustment of the sponge-iron composition and carbon throughput. In this way we can take advantage of lower electrode currents which lead to reduced electrode wear. A transformer operated in this way does not require excess copper in its windings. The disadvantage is that changing the operating point from 0.707 to 0.8 power factor reduces active power by 4%. However, this is more than compensated by the advantages and can be taken into consideration in planning for active power by increasing the appropriate secondary voltage by about 2%. In this case the current is also increased by only 2% compared with 18% for operation at a cos ϕ of 0.707.

An increase in current of 18% would involve, among other things, selecting a larger electrode diameter; this is usually impossible with existing furnaces, or may only be possible with high capital outlay.

Process-control for sponge-iron charges is particularly simple. If the charge is sufficiently homogenous, the process can be controlled by coupling charging rate to measured active power, without using a computer. The interdependence of these two variables governing the specific heat to melt sponge iron has been mentioned above. For example, the melt-temperature can be raised or lowered during melt-down by making small adjustments to this definite heat input which is specific to the sponge iron.

From the beginning of the 1970s refractory lining materials were partly replaced in Japan by water-cooling; now more than fifty modified furnaces are in service there. The use of this new technique has spread quickly in the last two years. By the end of 1978 over 50% of West German furnaces of over 35-t capacity had been equipped with water-cooled elements.[19] The Japanese system has been developed further in West Germany, with German modifications. The systems in operation today are illustrated in Figures 12–16.[16]

The makers of the cooling elements deal with fears about safety by drawing their customers' attention to the following list of precautions:

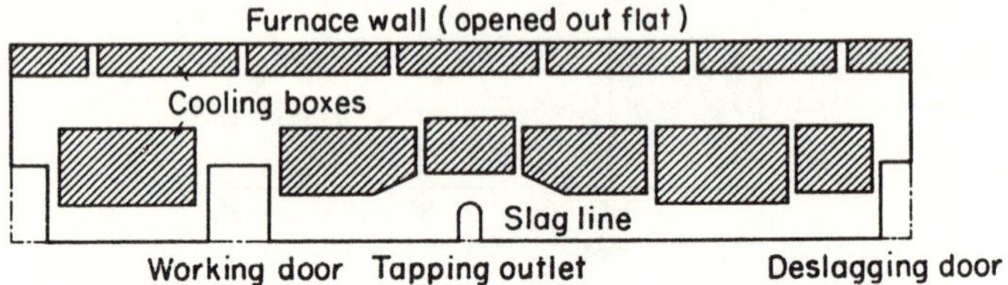

Figure 12. Schematic illustration of the cooling-box installation in a 70-t furnace using the DAIDO cooling system. Shell diameter 5800 mm; transformer power 50 MVA; water-cooled surface area $20m^2$, equivalent to half the total lined wall-surface

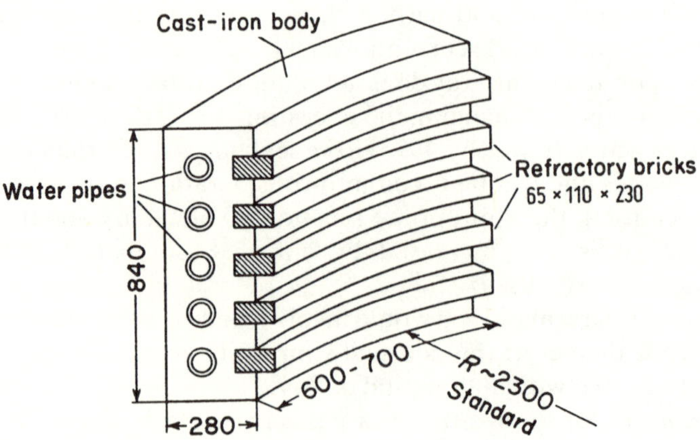

Figure 13. Schematic illustration of a water-cooled element after the IHI Permablock System

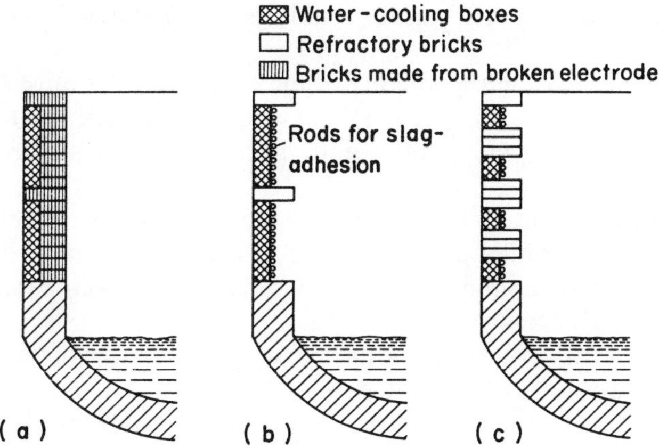

Figure 14. Various arrangements of water-cooling boxes in Japanese arc furnaces

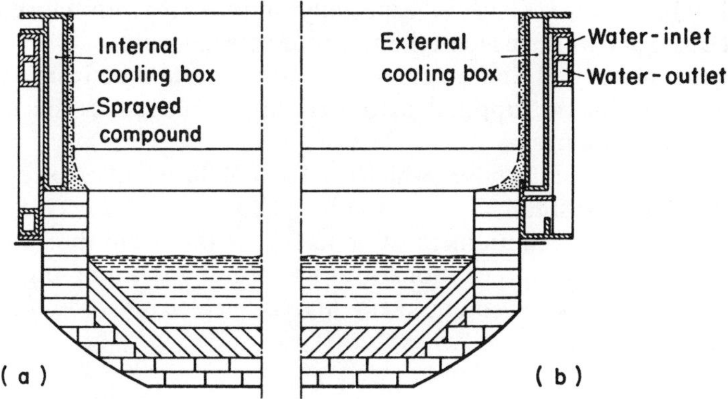

Figure 15. Large-area water cooling with internally mounted (a) and external (b) cooling boxes (Korf–Fuchs system)

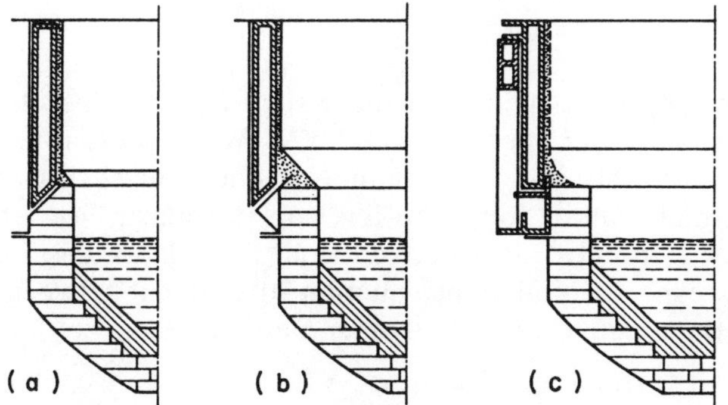

Figure 16. Different ways of arranging external cooling boxes so as to avoid ingress of water in event of leakage

(1) The lowest cooling boxes must be not less than 400 mm above the melt surface.
(2) The cooling boxes in the vicinity of the tapping outlet must be arranged so that liquid steel cannot reach them when the furnace is tilted.
(3) Use of a large tapping outlet and short tilting times increases safety considerably.
(4) There must be a safety hole above the tapping outlet to indicate the level of liquid steel while the furnace is tilted. Thus liquid steel can be prevented from coming into contact with the cooling boxes during tapping.
(5) The walls of the cooling boxes must be of sufficient thickness so that small 'sparkling' flashover arcs do not cause leaks.
(6) The cooling boxes must be arranged in such a way that they cannot be damaged by oxygen lances, especially when the lances are introduced through holes in the furnace wall.
(7) The furnace must be equipped with means to measure cooling water flow and temperature. This is especially important with closed cooling systems, such as are used in the Hamburg steelworks.
(8) It is recommended that an emergency water system be installed, so that water can still be supplied to the cooling elements in the event of a water-pump failure.
(9) There must be safety valves which can blow in the event of excess pressure in the cooling system.
(10) Similar safety precautions must be taken for the event that water enters the furnace; such precautions are also usual for water ingress to the furnace via other cooling elements like electrode cooling rings or water-cooled lances.

After safety, two factors are of economic importance:

(1) Water consumption, and
(2) Electricity consumption.

Water-consumption depends on the system and the operator, and varies from 80 to 450 l/m^2 per minute.[16,20–4] With present-day techniques consumption of 150 l/m^2 per minute seems universally possible. An empirical formula can be found[1] to link cooled wall-surface area M_{cool} (m^2) with capacity T (tonnes); it is derived from the relationship between vessel diameter and capacity (Guideline equation 20) and that between ideal furnace volume and the weight of scrap and other material:

$$M_{cool} = 0.7\pi(0.637\ T^{0.64} - 0.5T^{0.36}) \quad (11)$$

It is assumed here that the weight of scrap and other material is 0.8 tonnes/m^3;

charging is effected three baskets at a time, and 70% of the wall area is cooled. Therefore specific area cooled decreases from 0.36 m^2/t for a 10-t furnace to 0.15 m^2/t for a 300-t furnace (Table 1).

Table 1. Specific cooling area for different furnace capacities after Guideline equation (20)

Capacity T(t)	Cooling area $M_{cool}(m^2)$	Specific cooling area (m^2/t)
10	3.6	0.36
50	12.6	0.25
100	20.9	0.21
200	34.2	0.17
300	45.3	0.15

The cost degression curves in Figure 17 are based on a cooling-water cost[25] of 0.30 DM/m^3, melt-down time of 2.5 h, and water-consumption of 150 l/m^2 per minute.

The same literature quotes the additional electricity-consumption, when 70% of the furnace wall is water-cooled, to be 10 to 25 kWh/t. Here we should bear in mind that all the furnaces mentioned work with relatively short refining times of about 30–40 min (Figure 18). Furnaces that use longer refining periods consume more electricity (Table 2). The wall-damage was equated for comparable cases for the melting time.

The water-cooling boxes themselves cost around 5000 DM/m^2. Table 3 shows the sum for three different refining times of the costs of water, additional

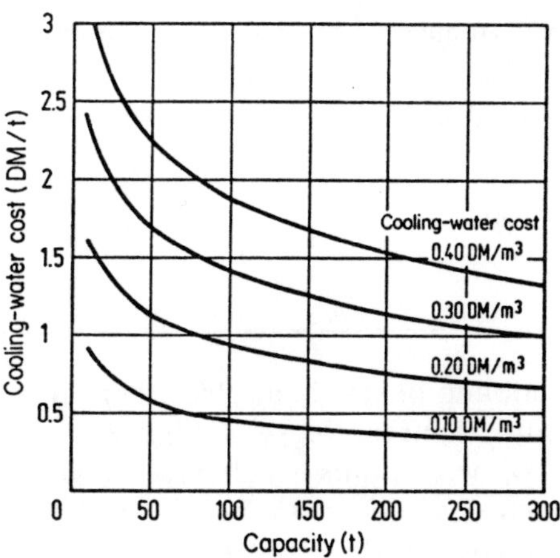

Figure 17. Cost of cooling water for various furnace capacities

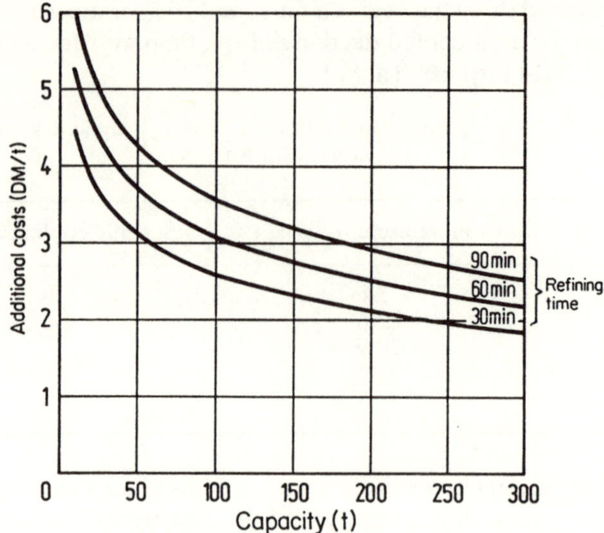

Figure 18. Additional costs for water-cooling boxes in relation to refining time, taking into account the costs of electricity and water and the cost and life of water-cooling boxes

Table 2. Additional electricity consumption for various refining times and furnace capacities with water-cooling

Furnace capacity T (tonnes)	Additional consumption (kWh/t)				
	Refining time (min)				
	30	40	50	60	90
10	20.9	27.9	34.8	41.8	62.7
50	14.5	19.3	24.2	29.0	43.5
100	12.2	16.3	20.3	24.4	36.6
200	9.9	13.2	16.5	19.8	29.7
300	8.7	11.6	14.5	17.4	26.1

electricity, and the cooling boxes themselves; here it is assumed that water costs 0.30 DM/m^3 and electricity costs 4 Pfg/kWh. These composite costs correspond roughly to the amounts which have to be saved through reduced consumption of refractories in order to break even. In this we have not taken into consideration the costs of the installation and of laying on a large water supply; neither have we considered any possible rise in productivity or reduction in roof-wear.[26]

Table 3. Costs for water-cooling boxes priced at 5000 DM/m^2 with a life of 1500 melts

T	t/life	DM/life	DM/t
10	15 000	18 000	1.20
50	75 000	63 000	0.84
100	150 000	104 500	0.70
200	300 000	171 000	0.57
300	450 000	226 500	0.50

In addition, development of the water-cooled roof first produced positive results in the Federal Republic of Germany in 1978.[25] Examples of designs are shown in Figure 19.

5.2.2 Guideline Publications

Guideline publications on the sizes of arc furnaces first appeared in Germany in 1942[27] and were aimed at limiting the number of types of furnace. These guidelines were reprinted in 1950[28] and in 1961.[29] They were succeeded in 1963 by the DTC Guidelines proposed by Harms,[30] in which shell diameters (D), transformer apparent power (T), and capacity (c) were specified. Today all furnace-makers have their own guidelines. Table 4 shows their evolution for a 60-t furnace. At first sight it is surprising that the trend is towards smaller shell diameters, despite the increase in supply power levels available. However, wall-life has remained the same or has even improved. The reason for this lies not only in the development of improved linings but more especially in the new technique of operating at a low power factor cos ϕ.[31,32] Thus arc length and radiation intensity are adjusted according to the state of the melt (see section 5.5.2.1).

The proposed guidelines of 1942, 1961, and 1963 are summarized in Figure 20 together with the present-day guidelines used by various furnace-builders.[1] A log–log plot was chosen because of the exponential relationship between shell diameter D and furnace capacity T:

$$D = kT^a \tag{12}$$

If the furnace-builders were to keep strictly to this equation the log–log plot would show straight lines. Factor k causes parallel displacement of the straight lines in the log–log plot, while an increase in exponent a causes the lines to become steeper. The following equation is valid for the 1963 DTC guidelines:

$$D = 1.8T^{0.280} \tag{13}$$

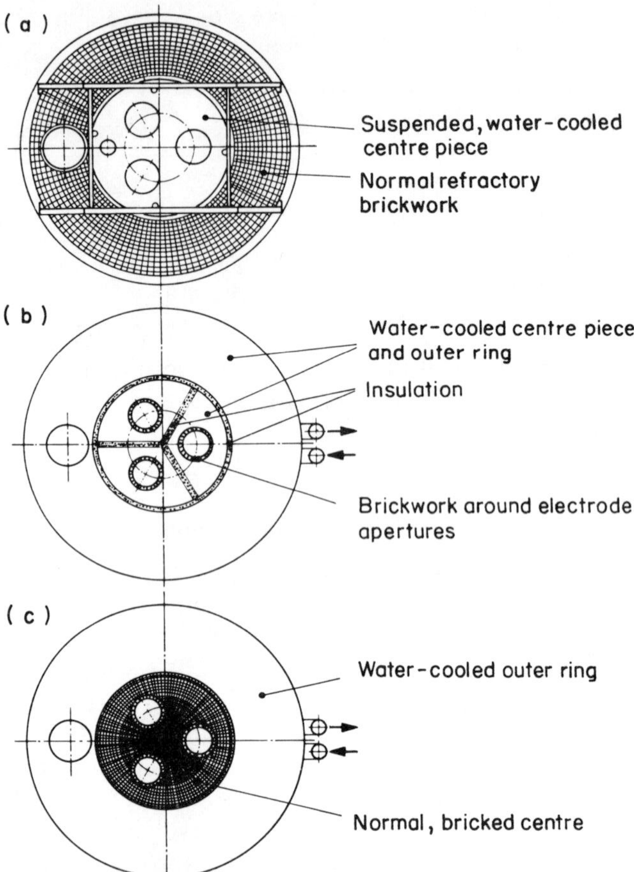

Figure 19. Different roof-cooling systems. (a) Conventional bricked roof with suspended, water-cooled centre-piece; (b) completely water-cooled roof with three water-cooled, insulated sectors in the centre; (c) roof with water-cooled outer ring and bricked centre

Table 4. Evolution of basic design values for a 60-t furnace

Year	Shell diameter (m)	Electrode pitch circle diameter (m)	Distance from pitch circle to wall (for 450-mm thick wall) (m)	Supply power (MVA)
1942–61	6.0	1.70	1.70	15
1963	5.8	1.35–1.55	1.68–1.78	20
1976	5.0–5.6	1.2 –1.4	1.45–1.65	30–40

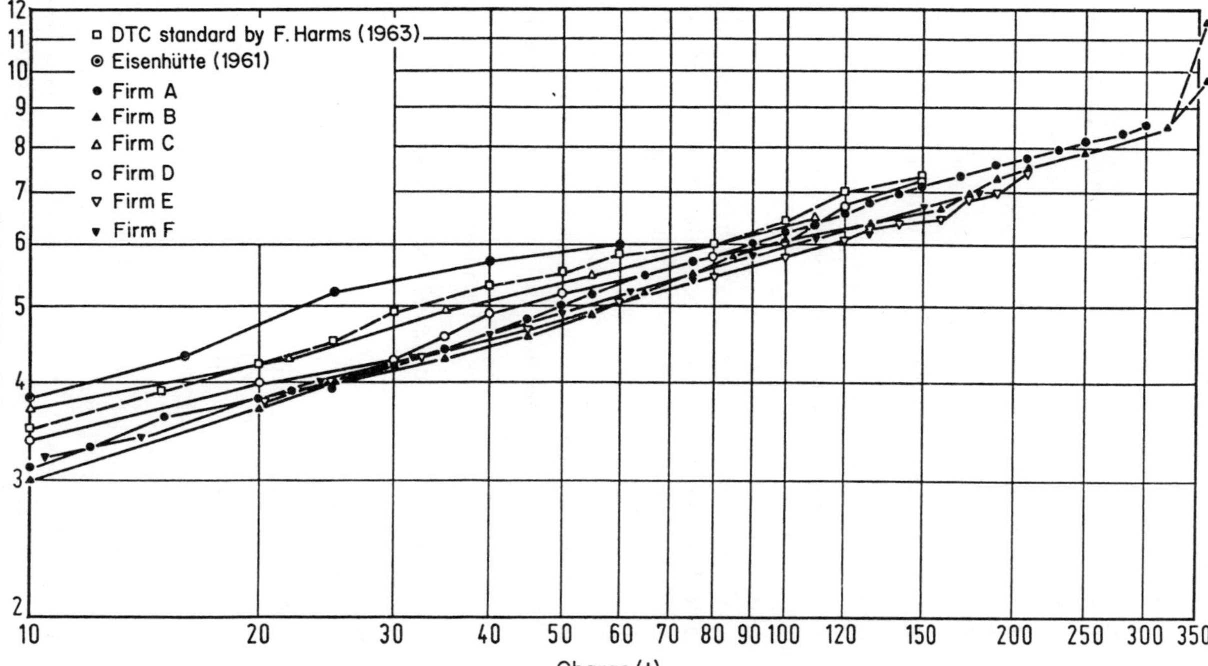

Figure 20. Summary of shell-diameter guidelines published as standards and by various builders

On the other hand, shell diameters today fit the following equation:

$$D = kT^{0.288} \quad \text{where } k = 1.5\text{–}1.65 \tag{14}$$

The curve for the 1961 guidelines cannot be expressed with this kind of function. If we assume that the curve follows this equation between 10 and 25 t, and that it continues to do so after a break at 25 t, we then obtain the following equations for up to 60 t:

$$\text{10–25 t:} \quad D = 1.74T^{0.34} \tag{15a}$$

$$\text{25–60 t:} \quad D = 3.11T^{0.16} \tag{15b}$$

The table mentioned in the literature does not cater for furnace capacities above 60 t.

Only the series of furnaces made by firm F follows the above equations exactly, although all the others appear to be based on them. The jumps in all the series cannot be explained, but stricter adherence to the mathematical equations would appear to make sense.

When a series of furnaces is planned, the mathematical relationship to be followed should be determined first, followed by the question of the size of the steps in the series. For example, furnace-builder A proposes 27 different furnaces ranging from 10 to 300 t capacity. The jumps in size are rather arbitrary here. The capacity difference between 35- and 45-t models is about 29%, while between 50- and 55-t it is only about 10%. From 75- to 90-t the jump is again 15 to or about 20%, while from 130- to 140-t it is only about 10 t or 7%.

There are two sound, alternative ways of planning a series of furnaces:

(1) A constant percentage size difference between types;
(2) A progressively reducing percentage difference between the size of different types.

Several series which follow these two principles are listed in Table 5. All the series involve furnaces in the capacity range from 10 to 300 t. Series A to D show constant percentage differences, following a suggestion by H. Ottmar.[32] Series A shows that the differences between the smaller capacities are too large; however, on the other hand, the number of models seems acceptable. Factor f in Table 5 must be increased in order to obtain larger jumps among the smaller capacities; this reduces the number of types but makes the difference between the larger capacities still greater.

A similar result is seen if factor f is reduced linearly as capacities increase (series E). If we want to meet all requirements we can change factor f hyperbolically, e.g. (series G):

$$f = \frac{2}{3.33 + n} + 1 \tag{16}$$

Table 5. Capacities in tonnes in different furnace series

	f=const.				f=f(linear)	f=f(hyperbolic)	
No.	A	B	C	D	E	F	G
	f=1.1854	f=1.2215	f=1.2750	f=1.36235	f=1.5−0.0208 n_i*	$f = \frac{1.72}{3.58 + n_i} + 1.02$*	$f = \frac{2}{3.33 + n_i} + 1$*
1	10	10	10	10	10	10	10
2	11.9	12.2	12.8	13.6	15.0	15.4	16.0
3	14.1	14.9	16.3	18.6	22.2	22.1	23.4
4	16.7	18.2	20.7	25.2	32.4	30.0	32.2
5	19.7	22.3	26.4	34.4	46.5	39.1	42.3
6	23.4	27.2	33.7	46.9	65.9	49.6	53.9
7	27.7	33.2	43.0	63.9	92.0	61.5	66.8
8	32.9	40.6	54.8	87.1	126.5	74.7	81.2
9	39.0	49.6	69.8	118.7	171.4	89.5	96.9
10	46.2	60.5	89.0	161.7	228.5	105.8	114.0
11	54.8	73.9	113.5	220.2	300.0	123.7	132.5
12	64.9	90.3	144.7	300.0		143.3	152.4
13	77.0	110.3	184.6			164.6	173.6
14	91.2	134.8	235.3			187.8	196.3
15	108.1	164.3	300.0			212.8	220.3
16	128.2	201.1				239.8	245.7
17	151.9	245.6				268.8	272.5
18	180.1	300.0				300.0	300.7
19	213.1						
20	253.1						
21	300.0						
	21	18	15	12	11	18	18

* n_i = 0 starting at 10 t

where f is the factor
n is the number of the step in the series starting with n = 0 at 10 t.

This proposition follows the three requirements:

(1) Sensible step-sizes at all capacities;
(2) Limitation of number of types to a sensible number;
(3) Logical relationship

If we assume that the physical and electrical relationships between the refractory wear index (described by Schwabe[31]) and the wall-lining life are correct, then it should be possible to calculate the characteristics of a series of furnaces of equal life.

A similar proposal was made by Ottmar and Schmeiduch in 1970.[33] If we follow this suggestion further[1] we can determine the connection between life, shell diameter, and capacity (see Figure 21). The empirical life equation is:

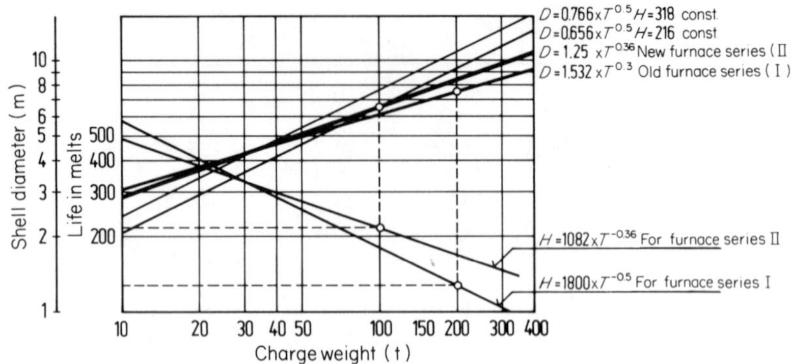

Figure 21. Relationship between shell diameter and life for various guidelines, depending on charge weight

$$H = \frac{T^{-0.5}}{c} \tag{17}$$

(where H is life, T is capacity, and c is a constant) which relates to the furnace series in which

$$D = 1.532T^{0.3} \tag{18}$$

(where D is diameter).

This is supported by American data for magnesite linings. On inversion, equation (17) becomes:

$$D = cT^{0.5} \tag{19}$$

If, for example, we take c to be 0.766, then this line intersects with that for the known series (18) $D = 1.532T^{0.3}$ at $T = 32$ t.

A furnace of this capacity has a wall-life, evaluated from the empirical equation, of:

$$\begin{aligned} H &= zT^{-0.5} \\ &= 1800 \times 32^{-0.5} \\ &= 318.2 \text{ melting cycles} \end{aligned}$$

If we want to build a 100-t furnace with the same life, then we need a shell diameter of

$$D = 0.766 \times 100^{0.5} = 7.66\ \mathrm{m}$$

It is just as easy to evaluate the shell diameter for a given capacity and life. It should be noted that shell diameter here is based on the distance of the arc from the wall, and that any change in the ratio of the electrode pitch circle diameter to the shell diameter must be taken into account. A shell diameter of 7.66 m for a 100-t furnace appears large compared with the dimensions of about 6.1–6.4 m which have so far been common. The hearth radius and melt depth expected for a 100 t charge in such a vessel show that this size is larger than is sensible. The logical step to take for a higher capacity is to choose a reduced, defined life and diameter. With regard to productivity (see Figure 6), we should have a relationship as shown in Figure 22. From this we can see that productivity is reduced by about 0.8% for the example of the 100 t furnace. This new series (II) corresponds to the equation:

$$D = 1.25T^{0.36} \tag{20}$$

If we now plot this in Figure 21, we get a life of 216 melting cycles for the 100-t furnace with a diameter of 6.56 m. The furnace series proposed in Table 6 can be derived from the series equation (20) and the step sizes between individual capacities (16). The additional transformer apparent power can be calculated, as in Figure 8, from the specific value of 760 kVA/t.

We cannot exactly say today how much the new technologies, mentioned in section 5.2.1.5, will change the guidelines. The operating methods on trial for these technologies are still, in the main, under development, and it seems that the optimum operating method is not yet clear. In any case, the possibilities mentioned have a good chance of leading to further improve-

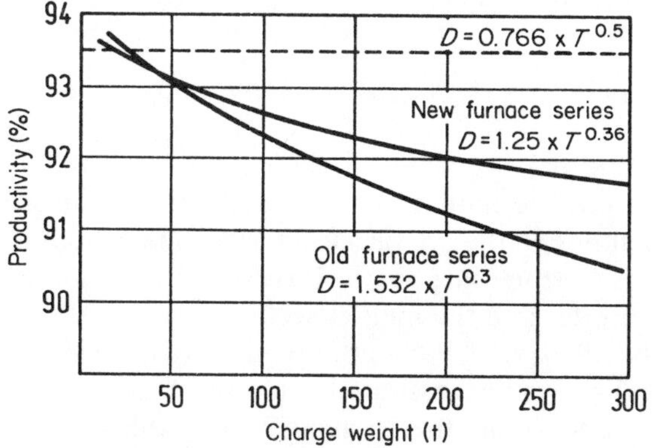

Figure 22. Productivity in relation to durability for a furnace-life of 10 h and various furnace series

Table 6. Furnace guidelines (from equations (16) and (20) and Figure 8)

Capacity T (t)	Shell diameter (D) (m)	Transformer apparent power (MVA)	Specific transformer apparent power kVA per tonne of charge
10.0	2.86	7.6	760
16.0	3.39	12.2	760
23.4	3.89	17.8	760
32.2	4.36	24.5	760
42.3	4.81	32.1	760
53.9	5.25	41.0	760
66.8	5.67	50.8	760
81.2	6.09	61.7	760
96.9	6.49	73.6	760
114.0	6.88	86.6	760
132.5	7.26	100.7	760
152.4	7.63	113.0[a]	742
173.6	8.00	113.0[a]	651
196.3	8.36	113.0[a]	576
220.3	8.72	113.0[a]	513
245.7	9.07	113.0[a]	460
272.5	9.41	113.0[a]	415
300.7	9.75	113.0[a]	376

[a] Apparent power limit after Ottmar and Ameling[8] at 113 MVA.

ments in the economics and extent of application of the arc furnace process in steelmaking.

5.3 MECHANICAL DESIGN AND CONSTRUCTION OF FURNACES

Hans Essmann, Schermbeck, Horst Laufer, and *Ludger Zangs, Essen*

In spite of the rapid development of electric steelmaking, and its effects on furnace construction, the modern electric arc furnace still works on the old Héroult system, although the arc furnace has become a melting machine. The necessary heat is fed to the charge in the form of thermal radiation from the arcs between the electrodes and the melt as well as from resistance heating of the melt itself. In the 1950s and 1960s, the possibility of using high specific and absolute electrical powers necessitated constructional changes to accommodate the altered conditions; these changes are today incorporated by all furnace-builders in their standard models.[1–4]

With relatively high furnace outputs, the question of refractory wear has assumed a greater importance than before. In order to minimize refractory wear and to keep it as even as possible, the secondary conductors are usually laid out in a trefoil configuration (triangulated) and the electrode pitch circle is made as small as possible. The latter can only be achieved by making the centre electrode arm shorter than the others. Moreover, austenitic material must be extensively used, with or without water-cooling in many places, in order to deal with inductive effects. The problem of the heating up of constructional elements, especially of the electrode arms and masts and the roof-support beams, needs special attention with high-performance furnaces.

In order to keep refractory wear as low as possible, it is necessary to keep arcs as short as possible at high power; this is made easier by using the shortest possible triangulated secondary conductors. This leads to optimized low impedances, so that short arcs can be used. In West Germany, high-current conductors to the transformer are now predominantly of triangulated design, and low-tension connections are also triangulated.

Since the end of the 1970s water-cooled wall elements have become thoroughly established, particularly for high-performance furnaces (see section 5.3.2.1); this has enabled more flexible operation with regard to arc length. There is therefore no longer such great concern about lining wear, since refractory materials are still used only just above the actual slagging area.

A series of production aids has been devised and further developed to optimize operation of high-power furnaces. The induction stirring coil deserves mention in this respect; in the early days it was used principally for making special high-grade steels. This stirring coil can be used only when the bottom of the furnace vessel is made from austenitic material.

Automatic oxygen-blowing is coming more and more into widespread use with the larger high-performance furnaces. This is accomplished by means of lances, mounted either on the furnace or close beside it, which advance automatically. These lances can either be introduced down through the roof or laterally through the vessel wall. Most lances used are self-consuming. Water-cooled lances have not yet become established for this purpose, since the bath becomes too agitated and, besides, the problem of safety has not yet been completely solved.

Powdered lime and carbon are often fed pneumatically into the furnace, if this seems expedient for refining. More recently, deslagging machines have been introduced; these simplify and speed up the removal of slag.

Various machines have been developed for nippling graphite electrodes on the furnace — also for large furnaces. These can become established if they are reliable in service and require as little attention as possible. For small and medium-sized furnaces, nippling is at present mostly performed on the furnace. On larger furnaces it is more and more becoming the practice to remove the electrodes and effect nippling on a special stand beside the furnace.

It is now possible to build additional oil– or natural gas–oxygen burners into the furnace if it is necessary to shorten the melting time and if power supplies are not adequate. In this way electrical energy is replaced by primary energy, and at the same time scrap in the 'dead zones' is melted down without delay.

Present developments enable us to expect large furnaces of up to 400-t capacity to be built in larger numbers in the foreseeable future. Such furnaces may need between 120- and 150-MVA power supplies. We can also imagine even higher powers, if the local electricity supply and regulations will allow this.

5.3.1 General Construction and Components

Figure 1 is a schematic illustration of one example of the construction of an arc furnace installation; the movements in operation are shown in Figure 2. Figures 3–5 show examples of construction offered by various furnace-builders.

One important component is the furnace vessel containing the melt. The vessel is usually mounted on a structural frame which can be tilted by means of rockers and roller tracks for tapping and deslagging. The vessel is covered by a roof. Three electrodes connected to the three-phase electricity supply are introduced through the roof. The roof and electrodes are carried by a portal which can be swung aside for charging.

5.3.1.1 Vessel and roof

The vessel and roof, which are lined with refractory bricks, have the function of containing the melt for processing and shielding it against heat losses to the surroundings. The vessel has a rounded bottom (hearth) and a cylindrical wall with openings for the tapping spout and for a working or slagging door for deslagging and making additions. Openings may also be provided for blowing-in lime, carbon, or oxygen and for fume-extraction where necessary.

The bottom is made from thick steel plates welded together with appropriate stiffeners. Above the melt there is a cage structure which carries loose mounted plates or most recently water-cooled elements made up from welded boxes and pipes or castings with integral pipework (Figure 6).

The latter elements partly replace refractory materials for the furnace wall. Important parts, such as the vessel upper flange which carries the roof, and the opening for the working door, are water-cooled. The vessel is fixed in a rocker frame and thus can be tilted. Vessels may be exchangeable or fixed; exchangeable vessels may be made in one or more parts. While a non-exchangeable vessel has to have its hearth and wall re-lined *in situ*, an exchangeable vessel is removed in one piece for re-lining and is replaced by a second vessel (Figure 7). Since the hearth and wall wear at different rates,

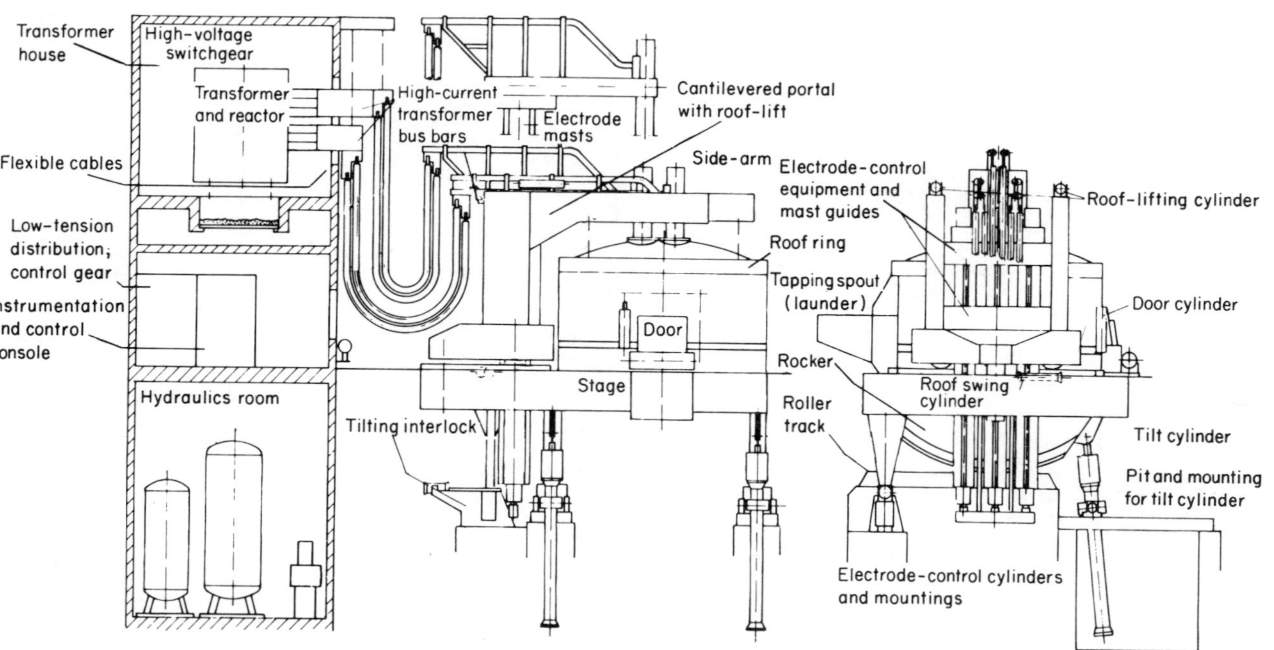

Figure 1. General arrangement of an arc-furnace installation

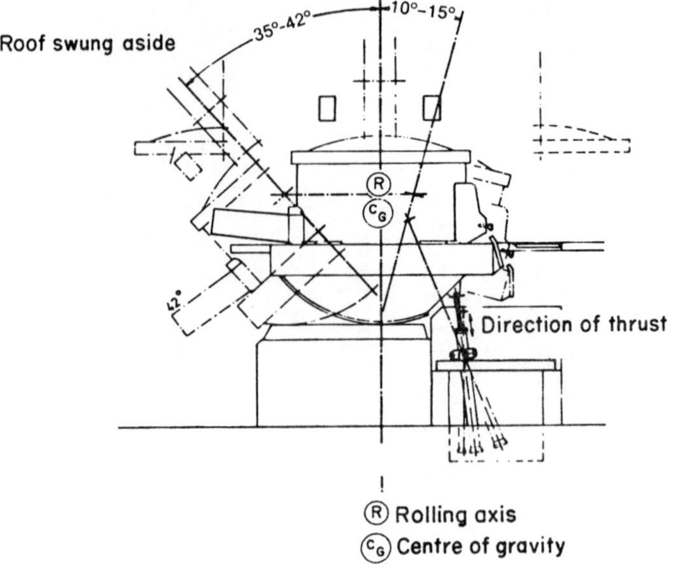

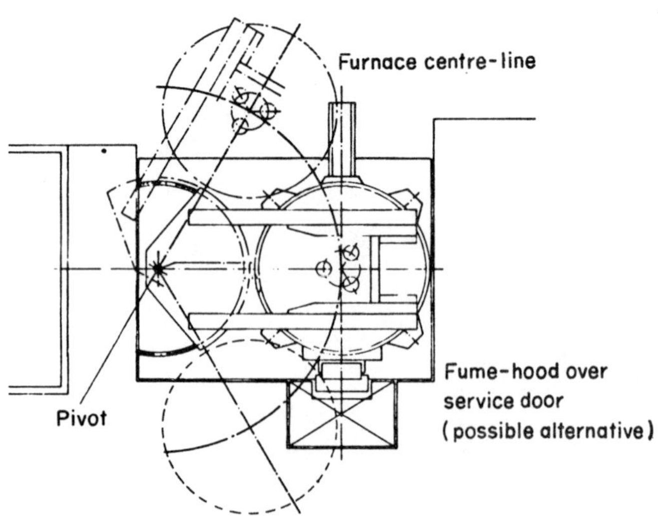

Figure 2. Arc-furnace movements in operation

construction is sometimes modular, so that only the cylindrical upper part of the vessel need be exchanged.

The vessel is enclosed on top by means of its roof. The roof is almost always fitted with a water-cooled bezel ring which mates with the vessel upper flange

Figure 3. 200-t arc furnace with 7.9-m diameter vessel, 100 MVA melting power, and exchangeable vessel. (Works photo by MAN division, GHH Sterkrade)

Figure 4. 85-t arc furnace with 5.8-m diameter vessel, 48 MVA melting power, shown with cantilever-portal and roof-lift mounted upon tilting-platform system. (Works photo by MAN division, GHH Sterkrade)

Figure 5. 45-t arc furnace with 4.8-m diameter vessel, 30 MVA melting power; furnace vessel has detachable upper part. (Works photo by Krupp)

Figure 6. 100-t arc furnace with 6.8-m diameter vessel, 75 MVA/60 MW melting power. Vessel mounted in cage structure with suspended water-cooling elements and detachable upper vessel part. Equipped with three natural gas/oxygen burners. (Works photo Krupp)

Figure 7. 6.4-m diameter exchangeable vessel for a 100-t arc furnace. (Works photo by Mannesmann Demag)

and is fastened to it (Figure 8). The roof ring usually carries a self-supporting refractory dome, which, like the part of the vessel above the melt, may be replaced by water-cooled elements. The roof is provided with ports for the electrodes and usually also has a fourth hole for direct fume-extraction. More openings may be provided for feeding in materials in small lump sizes. The openings may be lined by cooling rings.

The roof is suspended from rods on the portal and is swung aside for charging. There used to be other configurations in which either the vessel was drawn out or the portal complete with roof was moved forward. The entire roof is replaced when re-lining is carried out.

5.3.1.2 Tilting device

The purpose of the tilting device is to tilt the furnace accurately for deslagging or tapping. This is nowadays done by means of rockers and roller tracks. One design is shown schematically in Figure 2. The rockers are rigidly fastened to the frame carrying the vessel and portal. The roller track is embedded in the foundation and incorporates either a row of holes or a toothed rack along the side for straight, non-slip operation. The rockers are usually shaped like segments of a circle. During tilting, the tapping spout has to move first in such a way that the ladle can move to meet it. The rocker shape may be derived from a higher-order curve in order to simplify this procedure.

The maximum tilt angle is 35–42 degrees for tapping and 10–15 degrees for deslagging. Most furnaces are constructed in such a way that the vessel centre of gravity causes a rearwards tilting moment in the deslagging direction. The

Figure 8. 100-t arc furnace with 6.4-m diameter vessel, 66 MVA melting power. (Works photo by Mannesmann Demag)

forces necessary for tilting are usually transmitted hydraulically by means of a jack flexibly mounted on the foundation.

In the normal (zero tilt) position the vessel is held by an interlock mechanism.

5.3.1.3 Portal with roof-lift and swing system

The portal carries the roof, the electrode arms, and the electrodes themselves, together with their connectors. The portal is nearly always a welded box structure.

There are several possible ways of arranging the portal with its roof-lift and swing systems; these are shown in Figure 9. In case (a) the roof-lift and swing device stands on a special plinth. However, it can also be arranged, as in (b), together with the furnace vessel on a common platform. In both cases (a) and (b) the portal is lifted and swung away complete with the suspended roof. However, in many cases the roof is lifted separately by means of bars in the portal ((c) and (d)).

The portal is swung about a kingpin (c). For very large furnace diameters and heavy roofs it is possible to make the portal into a full arch (d). It should be

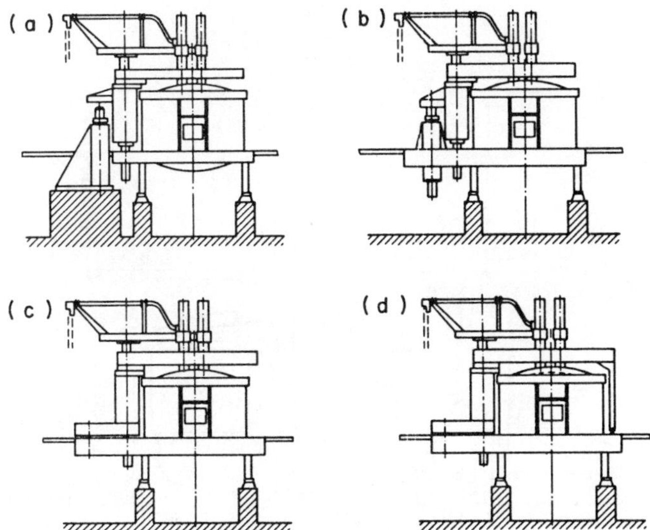

Figure 9. Various arc-furnace configurations. (a)–(c) Cantilever portals for shell diameters of up to 9 m and different ways of mounting the hydraulic lifting-ram; (d) portal configuration for shell diameter of more than 9 m

mentioned that in cases (b), (c), and (d) the tilting device must be locked in the zero position to counter the tilting moment during swivelling.

5.3.1.4 Electrode mast, arm, and bus tubes

The function of this structure is to carry and move the conductors necesary for carrying electric power to the furnace. Figures 10 and 11 show examples of the most important parts. The electrodes are attached to the carrier arm by means of contact pads and clamps. Current is fed to the electrodes through the contact pads; contact area and pressure are of decisive importance for this. The carrier arm and clamps are water-cooled. The electrodes must be moved up and down for starting and stopping and for control during operation. For this purpose the electrode arm is attached to a guide-pillar which is supported in guide-rollers by the portal. As shown in the drawing, the electrodes can be moved hydraulically or mechanically. Details about control can be found in section 5.4.

Current is fed through high-current water-cooled flexible cables from the low-voltage side of the transformer (Figure 1) to the electrode arm and then via high-current water-cooled bus tubes to the electrodes. In order to achieve the lowest possible reactances, the conductors must, among other things, be as short as possible.

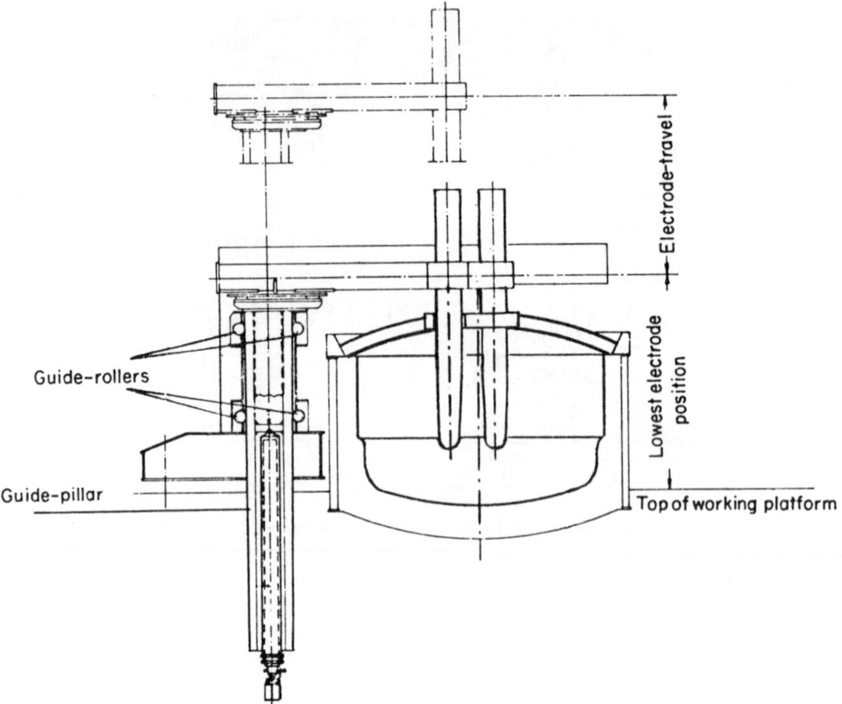

Figure 10. Example of an electrode-lifting device

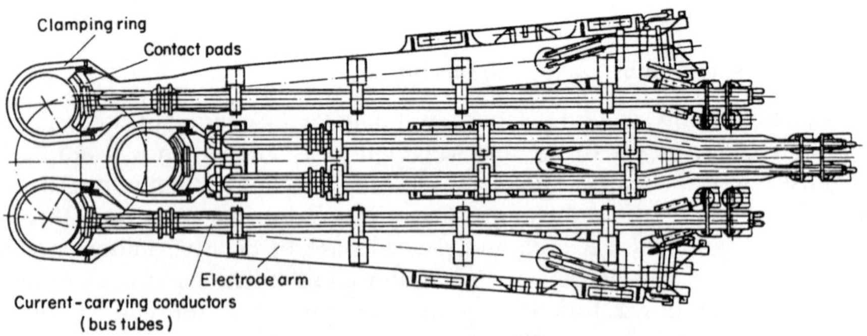

Figure 11. Example of electrode-carrier arms with clamps

The high-current tubes and cables are usually arranged in the cross-section as an equilateral triangle (triangulated) in order to balance the mutual induction effects between the phases.

5.3.2 Special Types of Construction

The rapid development of electric steelmaking processes and its effect on the design and construction of arc furnaces has resulted in designs which a few years ago were special becoming regarded as standard equipment today. This is true, for instance, of triangulation of the whole high-current conductor system so as to balance reactances as far as possible. In this way, reactance asymmetry can be reduced to below 5%. Other designs for the high-current conductor system are not as important as triangulation and are limited to exceptional cases. These include the 'knapsack circuit' in which bifilar high-current conductors are used as right up to the electrode clamps at which the secondary delta is completed. Another method is to arrange the high-current conductor paths for the individual phases in geometrically different configurations. Further examples of this rapid development of special application into standard equipment are the modular and exchangeable furnace vessels already mentioned in section 5.3.1. However, the following forms of construction and additional equipment may still be considered today to be special equipment.

5.3.2.1 Water-cooled elements in place of refractory linings

A few years ago in Japan development began of the use of water-cooling for highly exposed areas of the furnace wall. This involves using cooling boxes made from welded steel plate after the Daido system and cast-iron bodies incorporating refractory bricks cast in place after the IHI Permablock system (section 5.2.1.5). Meanwhile, after thorough testing, several West German steelworks successfully introduced various systems for cooling about 75% of the ring wall area above the slag zone (Benteler-GHH, DEMAG Rohrwand, Korf–Fuchs). Following the successful experiences with water-cooled wall elements, partly water-cooled roofs are currently on trial; these have a centre piece of refractory material but are otherwise water-cooled all over.

The results so far have been good, and we can expect that water-cooled elements will become standard features of furnace walls and roofs — under specified conditions — in the foreseeable future. However, increased cooling-water costs appear to make this more economical for high-power, fast-melting furnaces. There may well be a tendency to use water-cooled elements side-by-side with conventional refractory linings using much-improved materials. Water-cooling would then be used for high- and maximum-performance designs, while refractories would be used more for normal-performance designs.

Further improvement in economics can be expected under specified conditions if the process can be combined reliably with energy recovery.

5.3.2.2 Bottom-pouring arc furnaces

At the Thyssen Edelstahlwerke AG works in Witten, conventionally operated 50-t arc furnaces have about 70% water-cooling of the wall above the sill line. The results there have been good, and this led to thoughts of extending the use of water-cooling in the wall as far as is technically possible. The outcome was that in 1979 a tilting furnace was converted to near-static configuration, with a bottom-pouring system for tapping and tubular water-cooling elements over the entire wall area down to a level 200 mm above the sill line. This furnace can still be tilted by up to 12 degrees for deslagging through the working door. Figure 12 shows the furnace lining with the cooling system before and after conversion.

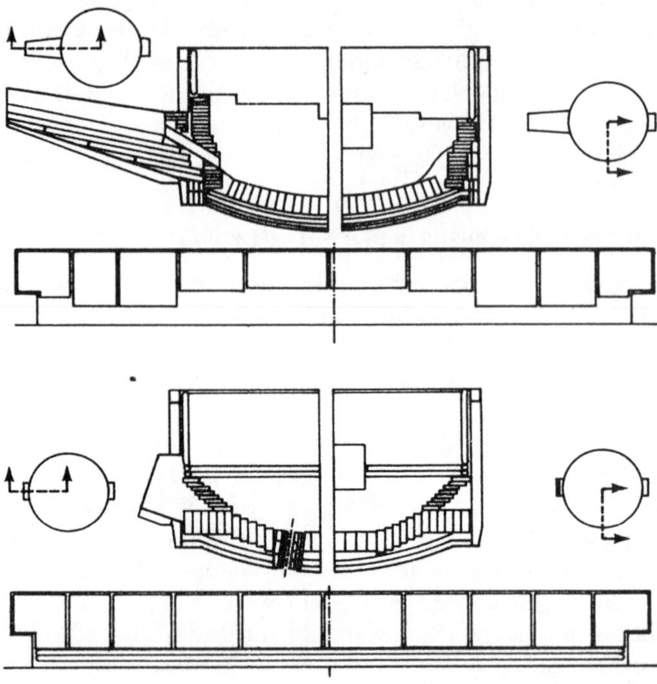

Figure 12. Lining of the electric arc furnace, in each case shown in two sections, and the arrangement of the water-cooled tube wall elements (furnace shell top edge to sill line); top: tilting furnace, bottom: bottom-pouring furnace

Figure 13 is a schematic representation of the bottom-pouring system. Tapping is initiated by swinging the closure flap out of the way. As a rule the bridge of sintered filler material breaks through by itself, and the melt flows out through the tap hole. For high-chrome melts, in about 50% of all cases runout is initiated by a short burn-off using oxygen. Runout lasts 50–90 s, depending on the diameter of the tap hole.[4b,4c]

In 1981, the Thyssen Edelstahlwerke AG works in Witten was putting a new 110-t arc furnace into service; this has a shell diameter of 6.8 m, uses a transformer apparent power of 75 MVA, and is configured as a static,

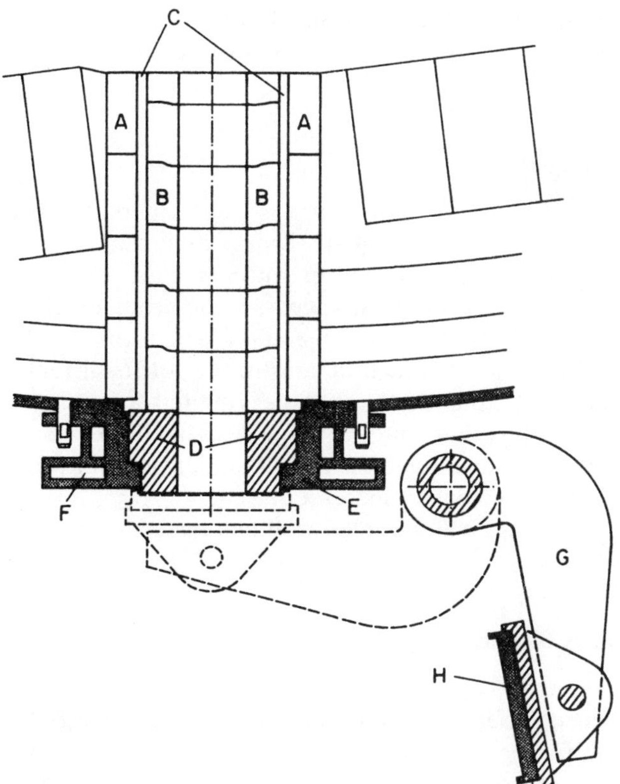

Figure 13. Schematic diagram of bottom-pouring system. A MgO-settling stone; B MgO-wearing tube; C magnesite sealing compound; D graphite terminal brick; E end-ring; F water-cooling; G closure flap; H graphite plate

bottom-pouring furnace. It will take over production from the Martin steelworks. The furnace is built by MM-DEMAG working together with Thyssen Edelstahl AG to further development of bottom-pouring technology.

Using all the possibilities afforded with new furnaces constructed exclusively for bottom-pouring has economic and technical advantages for the installation. The structural weight is less, and reactances can be reduced by using shorter cables. Advantages can also be expected in collecting exhaust gases during melting and tapping, since enclosing a static furnace is less problematical than for a tilting furnace. Since tilting is much reduced with a bottom-pouring furnace, the electrodes are subjected to less bending stress and thus should break more rarely. The short, enclosed runout channel enables tapping temperature to be cut, thus affording an immediate saving in energy. Experience in operating the new furnace will provide more information.

5.3.2.3 Jet-burners

Experience in the use of oxy-fuel burners (jet-burners) has been building up over the last few years in several steelworks in Japan as well as in a few in Europe. Special natural gas–oxygen burners were developed in a West German steelworks; this technique bypasses the problems that can arise with fuels containing sulphur. In high-powered electric furnaces, the advantage of using jet-burners is not so much in substitution of fossil fuel for part of the electrical energy but rather that the charge is melted down more evenly and in a shorter time.[5] On the other hand, with furnaces having less powerful electrical energy supplies the advantage is more that jet-burners provide additional energy and that productivity is increased.

Generally speaking, water-cooled wall elements are needed when jet-burners are used.

5.3.2.4 Rotary furnace vessels

As high-power technology has developed, so the importance of rotary furnace vessels has receded because rotation requires the electrodes to be withdrawn and, with fast furnaces, the time lost cannot be regained. A rotary furnace vessel can only be considered for exceptional cases when large quantities of extremely light or extremely heavy scrap or sculls have to be melted down.

5.3.2.5 Furnaces tilted about the spout

Sometimes small, exact amounts of molten material have to be tapped from a relatively large furnace; additionally, there are cases where the ladle must not be moved during tapping because of the production process. In such cases a furnace is required which can be tilted about its spout, so that the launder does not move horizontally or vertically during tapping. Here the furnace vessel sits on a specially constructed rocker or swing and pivots about an axis across the lip of the spout; it is lifted by two hydraulic jacks at the other end. Such furnaces are used for remelting and for storing ferro-alloys in liquid form.

5.3.2.6 Special furnaces for melting sponge iron

At present, sponge iron is melted in normal arc furnaces in the highest possible proportion in relation to the total charge. There are now justified doubts about whether the standard arc furnace is the best device for melting 100% sponge iron economically.[6] Thus attempts are being made in various places to develop an ideal furnace for sponge iron.

5.3.2.7 Furnaces with six electrodes

With their present current limits, the largest possible electrodes of 600 mm and also 700 mm diameter can handle powers of up to around 120–150 MVA. Furnace capacity can thus be increased up to 350 to 450 tonnes, as has been demonstrated with a few furnaces of this sort of capacity already built in the USA. However, for the foreseeable future, these furnace sizes and powers must be regarded as exceptional, required for casting purposes and sometimes for special batch sizes. The size of these furnaces is already so great that the use of six electrodes seems desirable. However, it would be better to abandon the present furnace concept for such cases and to seek more suitable solutions.

5.3.3 Design of System Components

The steady increase in electrical power density and the reduction of melting cycle times has increased the demands on all furnace components. With specific furnace power levels of up to 650 to 700 kW/t, melting cycles of under 2 h and ladle-refining following immediately, the arc furnace has become a fast-melting machine, inexpert handling of which can result in extra operational disruption and unnecessary down-time.

Therefore, some of the most important objectives in arc-furnace design are robustness, reliability, and ease of maintenance. Besides mechancial and thermal loads, we must also pay ever more attention to the inductive effects of the powerful magnetic fields, the vibration caused by high alternating currents,[7] and the problems of masses which move as fast as possible. In the following sections we will deal with the different requirements for the various parts of the furnaces.

5.3.3.1 Furnace steelwork (rocker frame, shell, roof-ring)

The rocker frame and rocker are mainly subjected to normal mechanical stresses. In addition, the cross-girders are attacked by thermal radiation from liquid steel and slag during tapping. Badly exposed parts are usually protected by heat shields.

As well as being subject to mechanical stresses, the furnace vessel suffers from severe thermal loading which has to be taken into account in design by attention to shape, choice of materials (resistance to thermal fatigue), and welding technique. Expertly installed water-cooling improves vessel-rigidity. One good way of improving vessel-rigidity is to use loose-plate cage construction. This is also an advantage for the installation of water-cooled wall elements and jet-burners.

When deciding on the geometrical shape of the lower part of the furnace it is necessary to optimize the shell diameter, bath depth, bottom radius, tilt angle, and the position of the spout; these are all interdependent for a given capacity. The smallest possible bottom radius is desirable for constructional reasons, but this implies a relatively deep bath. On the other hand, a relatively shallow bath is desirable for metallurgical reasons, since with a deep bath the differences in temperature and analysis throughout the melt will be more marked, and there is a greater risk of delaying the boil.

The rocker frame, rockers, and vessel are generally designed so that the furnace exhibits a tilting moment towards the slagging side because of the position of the centre of gravity.

The inclination of the spout is important, and that is governed by its mounting point. Next, the spout should be as short as possible for metallurgical reasons. This requirement can always be met by moving the ladle by means of the ladle car. However, if the ladle is suspended from a crane during tapping, a longer spout is necessary so that the crane cables cannot touch the bus tubes on the electrode arms when the furnace is tilted.

The roof-ring suffers particularly heavy thermal loads. The choice of steel to withstand thermal fatigue, the use of intensive water-cooling, and attention to welding technique are all of great importance.

5.3.3.2 Roof-lift and swing mechanism

With the usual sort of swing roof design, the most important load on the roof-lift and swing mechanism is the moment of force needed to raise the roof or the entire upper structure complete with roof, according to furnace type. Mechanical loads are dependent on basic furnace design; however, the roof carrier arms are also exposed to great heat and require water-cooling as well as additional heat shields near to the electrodes.

5.3.3.3 Electrode masts, arms, and connectors

In designing the electrode masts and arms (mast or pillar and carrier arms) we have to consider not only simple mechancial stresses but also heat from radiation and particularly from induction. In addition, the resonant frequency of the dynamic system comprising the electrodes, carrier arm, and mast has to be taken into account.

Above all, it is of decisive importance to be fully conversant with the problems of induction heating and vibration in large, high-powered furnaces. Induction heating is dealt with by correct structural design and partly by using non-magnetic steel. Above all, it can be controlled by water-cooling the electrode clamps and electrode arms as well as the flange connecting the electrode arm to the electrode mast and, sometimes, the upper part of the electrode mast, too. The system consisting of the electrodes and their carrier arm and mast is subject to mechanical resonance caused by the forces generated by the electric current; its frequency is mostly in the flicker region of between 2 and 12 Hz.[7] Resonance produces risks of fatigue fractures and negative feedback in the electrode control system. These dangers can be countered through careful design of the electrode mast roller-guides and the electrode carrier-arm. However, the effects of all these measures are limited because the main dimensions cannot be changed at will, and because the most important parts of this system — the electrodes — must not be affected structurally.

The electrical conductors between the electrode clamps and the transformer secondary terminals are usually water-cooled copper bus tubes and flexible copper cables.[8] These are best for minimizing both weight and reactance.

5.3.3.4 Furnace and electrode drive systems

The quality of an electrode control installation depends not only on the electrical control unit but, to a great extent, also on the electrode drive and transmission elements. The drive and transmission must therefore meet the requirements of adequate acceleration, shortest possible response times, and appropriate damping. These requirements may be met by hydraulic or electromechanical control systems.

Within the limitations imposed by the sizes of the masses to be moved, all furnace movements must be as quick and smooth as possible. This applies not only to the beginning and end limits of movement but also to intermediate stops as are common in tilting the furnace and swinging the roof. The requirements are best met by hydraulic systems, and that is why these are so widely used today.

All furnace movements involve unlocking and relocking the power-operated interlocks between the various components. For safety and in order to prevent damage, these links must function as precisely as possible under extreme and varied conditions, and all movements must always be exactly repeated. These links and their mountings must be given particular attention in design. In addition, all interdependent movements must be electrically interlocked.

5.3.4 Guidelines for Design

If we study the general considerations for arc furnace design, and particularly the criteria and recommendations in section 5.2, then we can distill the main points and plot them graphically and in tabular form — as in Table 1. Figure 14 shows the relationship between vessel diameter and required capacity,

Table 1. Guideline values for arc furnace design

Capacity (t)	Vessel diameter (mm)	Electrode diameter (mm)	Transformer power (MVA)	Maximum vessel hight above sill (mm)
10–20	3200–3700	300–350	6–12	1600
20–30	3700–4500	400–450	10–20	1800
30–40	4300–4800	400–450	15–25	2100
40–50	4600–5200	400–500	18–30	2300
50–60	4900–5500	450–500	20–36	2300
60–70	5200–5800	450–550	25–40	2500
70–80	5500–6200	500–600	35–50	2500
80–90	5800–6400	500–600	40–60	2750
90–110	6000–6800	550–600	50–70	2750
110–130	6400–7200	600	60–80	2750
140–160	6700–7600	600	80–110	2750
190–210	7500–8400	600	80–120	2750
230–270	8000–9000	600–(700)	80–120	2750
280–320	8500–9600	600–(700)	100–120 (150)	2750
330–370	8800–10 000	600–(700)	100–120 (150)	2750
380–420	9500–11 000	600–(700)	100–120 (150)	2750

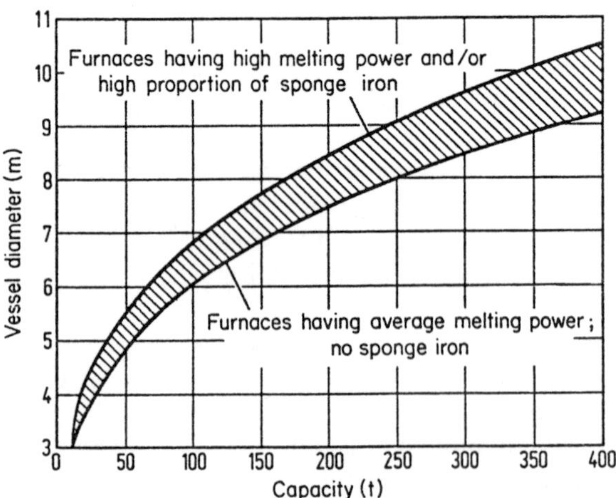

Figure 14. Dependence of vesel diameter on capacity and melting power

depending also on the charge and metallurgical objectives. According to the requirements, any furnace capacity between the limits illustrated can be regarded as 'reasonable'. High proportions of sponge iron in the charge (and here metallization has to be considered; see section 5.7) as well as high melting powers have the effect of enlarging the required vessel diameter.

Table 1 shows the vessel diameter, supply power, electrode diameter, and the upper limit of vessel height above the sill for various classes of vessel capacity.

It should be explained here that the figures for transformer power and electrode diameter in Table 1 are based on the power supply used all over Europe, which has a frequency of 50 Hz. High transformer power values and electrodes of greater diameter should be used for larger furnace diameters, provided that the electricity supply can stand it and it can be justified on metallurgical grounds. In any case, the quality of the electrodes available must always be considered.

The maximum possible height of the vessel above the sill is determined by the volume of scrap to be charged, and the stability of the electrode column when the furnace is tilted; it is valid for the largest diameter electrodes mentioned. Electrodes of 700 mm diameter are seldom used at present; data about them can be obtained from the manufacturers. Transformer powers of 150 MVA are shown in brackets in Table 1, because the power supply has to be examined in each individual case to see whether this power is obtainable in practice.

5.4 ARC FURNACE ELECTRICAL EQUIPMENT

Josef Schiffarth, Bocholt-Barlo

5.4.1 Purpose and Design

This section deals with the practical design of arc furnace electrical equipment with due regard for the basic electrical principles of arc furnaces, the problems of connection to the mains supply (see Chapter 4), and the aim of providing a reliable melting plant with the best power-consumption and the highest output possible.

The electrical equipment comprises (Figure 1)

Furnace switchgear
Static and dynamic reactive power compensation
Furnace transformer, with supplementary reactor if required
Electrode control system
Instrumentation for furnace monitoring and control
Automatic furnace control to optimize the charging process
Furnace auxiliaries

5.4.2 Furnace High-voltage Switchgear

Arc furnaces must always be connected to high-voltage mains because of the high short-circuit power required. Small and medium-sized furnace installa-

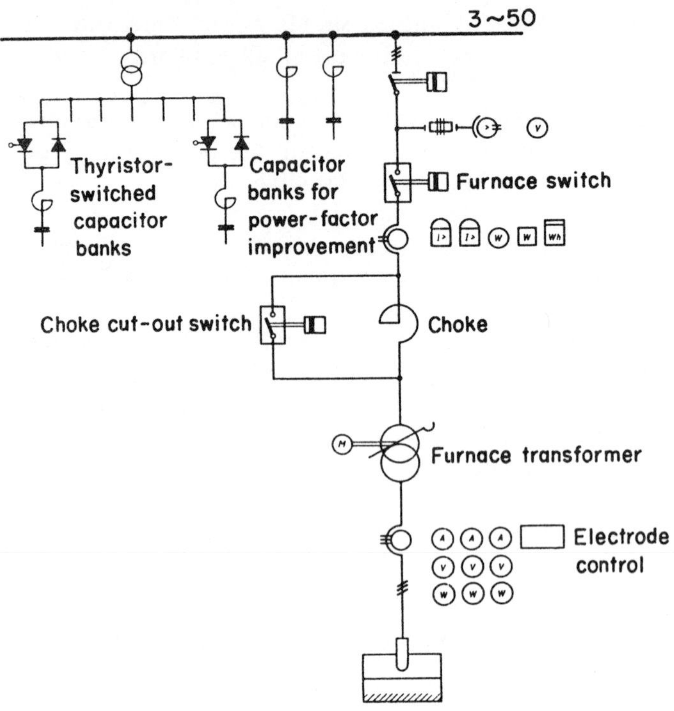

Figure 1. Schematic circuit diagram by Brown Bovery Co., Baden

tions use supply voltages of 6 to 30 kV, while large furnaces with capacities of up to 300 t and supply power of up to 100 MVA may be connected direct to mains supplies of up to 220 kV or may be connected to mains at even higher voltages via an intermediate transformer (e.g. SIDOR). When the furnace transformer is connected direct to high-voltage mains, the furnace transformer and switchgear have to be designed for this voltage. Difficulties can arise if the furnace transformer and high-voltage switchgear are installed close beside the furnace. If the furnace switchgear is of the high-speed type, operated by compressed air, and is situated outside the steelworks, then environmental noise may be unacceptable because of the high frequency of operation; noise-reduction measures will then be necessary.

For indirect connection the furnace transformer is connected to the mains via its own step-down transformer and associated switchgear. This offers the advantage and possibility of installing the furnace circuit-breaker in the intermediate medium-voltage circuit between the two transformers.[1–3]

The capital cost of the installation and buildings and the costs of losses in the transformers and connectors have all to be considered before deciding whether to connect the furnace transformer directly or indirectly to the mains.

5.4.2.1 Furnace switch

Furnace circuit-breakers perform 50 to 80 operations a day, according to the charging procedure. This extremely high operating frequency thus gives rise to stringent specifications for this type of switch. Above all else, these concern the switch-actuation system and kinematics; but the contacts and arc extinction system are also highly stressed. To assess the suitability of such a switch, we must also examine its suppression of transient over-voltages when switching small inductive currents, as well as the maintenance costs, and operating noise.

Most furnace switches in service use either air pressure for actuation and extinction or have vacuum-switching tubes. The air-blast switch, with cast resin insulators, is built around a compressed-air cylinder, on to which the various components are welded. The valves are also attached direct to the cylinder.[4]

Figure 2 shows the cross-section of a single-stage switch used mostly at lower powers. When switching off, the tubular contact piece first lifts a short

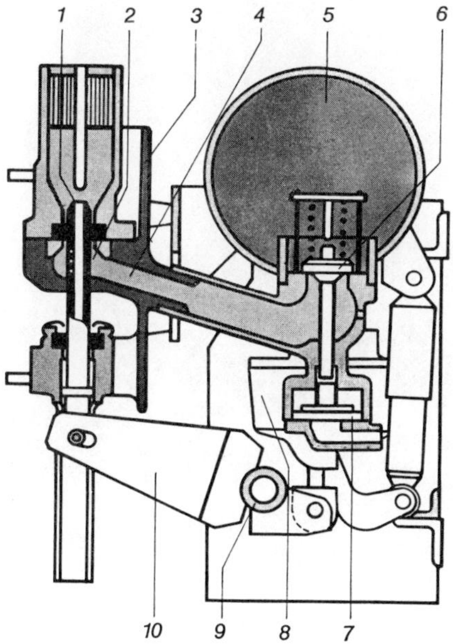

Figure 2. Cross-section of a single-stage switch for an air-blast circuit-breaker system (diagram by AEG).[4] 1 Tubular contact piece; 2 arc-extinction chamber; 3 cast resin insulator; 4 blast tube; 5 compressed-air vessel; 6 blast valve; 7 blast-valve actuator; 8 air blast actuator; 9 switch shaft; 10 insulated lever

distance for arc extinction; it then moves further, beyond the visible extent of the insulation. The tubular contact piece is moved by an insulated lever, actuated by a piston driven by compressed air. After opening the blast valve, the arc-extinction air flows through the blast tube into the arc-extinction chamber and then mainly along the arc through a silencer which also acts as a cooler for the hot, ionized air. Part of the arc-extinction air flows through the tubular contact-piece, down into the open air.

The retro-blast is particularly intense because the air flows across the arc and cools its root. Figure 3 illustrates a cast resin 1000-MVA circuit-breaker.

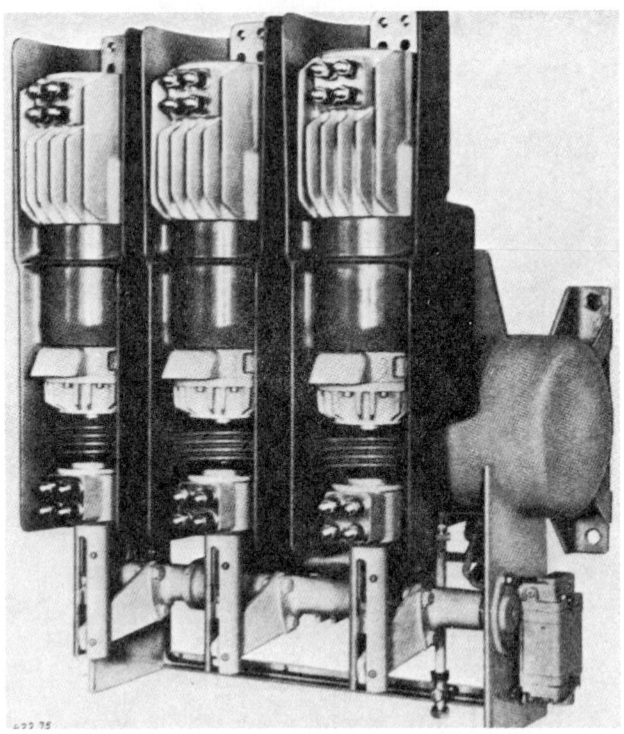

Figure 3. Type 10 cast resin circuit-breaker, 2500 MVA, 1000 A, in the 'off' position. (Works photo by AEG)[4]

Because of the high frequency of operation, the furnace switch contacts, which operate under pressure, are made from very hard copper alloys, while the mechanism incorporates special shock-absorbers. This enables a contact life of 30 000 to 50 000 operations to be achieved; contact-life is even longer for lower load and supply currents. However, this is only possible if compressed air quality remains consistent for each operation.

Steps have to be taken to reduce voltages with highly inductive components, which can arise when furnace power is switched off.[5,6] Damping resistors are used with medium-voltage switches for this purpose; in addition, voltage-surge arrestors protect the rest of the electrical system from excessive voltages when switching off. The type of surge arrestor has to be particularly carefully chosen for furnace work.

Vacuum circuit-breakers are being used for switching furnace systems in America, and also more recently in Europe. These circuit-breakers, which are guaranteed for 40 000 switching operations, are actuated electromagnetically via a system of levers. The disadvantage of this type of circuit-breaker is that it cannot switch mains short-circuits; therefore an extra power switch, which protects for these short-circuits, must also be used.

In addition, when a vacuum circuit-breaker is used, care must be taken to see that the furnace transformer is protected against transient over voltages by means of an RC circuit.

5.4.2.2 Reactive power compensation

Reactive power compensation reduces the negative effects of fluctuating wattless power input to the furnace in the following ways. Improved furnace power factor raises mains voltage because of reduced inductive reactive power. In this way, if power factor is constant, furnace power increases as the square of mains voltage. If voltage rises by 10%, power input to the arc goes up by around 20%. Moreover, a poor power factor signifies a higher load on the supply network and generators for a specified active power, and this must inevitably lead to higher electricity costs. Substantial improvement in electricity costs can be achieved relatively cheaply and simply by installing a capacitor bank for static power factor correction.

The main purpose of dynamic reactive power compensation is to reduce voltage fluctuation (flicker); this enables furnaces to be connected to mains supplies whose short-circuit power ratings are appreciably lower than the value otherwise required, which would be 80 to 100 times the furnace power. Dynamic reactive power compensation reduces fluctuations in the range of 3–7 Hz generated by an arc furnace.

Harmonics distort the mains voltage and can cause severe interference in electronic equipment such as control systems, unless capacitor banks are made up from separate filter circuits with appropriate chokes. Reactive power compensation also balances out asymmetrical furnace loads, for instance when a furnace works on two phases only for a while during the melt-down period. In such cases other consumer units, such as three-phase motors and power units, could be severely affected. Asymmetry at the star-point should not exceed 0.5% averaged over a period.

Dynamic compensation systems are made in such a way that enables capacitive or inductive power to be fed to any phase to compensate and balance

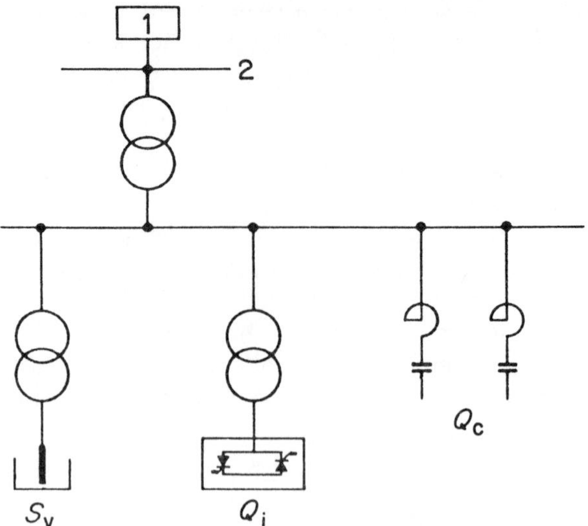

Figure 4. Compensation circuit working on the principle of controlled reactors with power-factor correction. (Diagram by Brown Bovery Co.) 1 mains supply; 2 point of common coupling; S_v consumer apparent power; Q_c captive reactive power; Q_i controlled inductive reactive power

out reactive power fluctuations, balanced or asymmetrical, that arise in furnace operation. Thyristor-switched capacitors, connected in parallel (Figure 4), involve the lowest capital cost. With appropriate control, this method not only provides reactive power compensation for the three phases but also takes care of balancing and deals with limitations in switch-on time. Water-cooled thyristors must use completely de-ionized water. Heating prevents the cooling system freezing during shut-down in cold weather.[6] Any failure of a pair of thyristors is indicated optically. Every series-circuit branch should incorporate at least one spare pair of thyristors in reserve so that defective thyristors can be replaced later during routine servicing. Part of the capacitor bank is connected permanently at the transformer terminals. When the furnace is not operating this fixed part of the bank must not cause a voltage increase of more than 5%, since, according to VDE Specification 0532, the applied voltage may only exceed the rated voltage of the furnace transformer by 5%.[7–14]

5.4.2.3 Furnace transformers

Furnace-transformer requirements were described in section 4.1.3 together with the possible circuits and direct and indirect control of secondary voltage. The following section deals with the practical construction of transformers.

The capacity and desired output of a furnace determine the power, the secondary voltage, and, with it, the relevant current on the secondary side of the furnace transformer. With transformer-rated power, we must bear in mind that, according to VDE specifications, transformers may be subjected to 20% overload for a maximum of 2 h, provided that sufficient time is allowed for cooling down after overloading.[15] It is better to design a transformer for a defined load cycle which can be repeated as often as required and which also incorporates a definite break-period. It is recommended that rated power is defined to be the (highest) power required for melting down; if this is so, there is no time-limit for melt-down power.

Furnace transformers may be constructed as three-phase transformers or as three single-phase transformers. With the latter configuration, it is possible to continue to operate a furnace after failure of one transformer by using two transformers in the required circuit; however, the high-current conductor layout is more complex since the three single-phase transformers must be interconnected outside the transformer housing. Such high-current conductors must be arranged so as to enable the high-current circuit to be reconnected simply after one transformer has failed so that the furnace can continue to operate on two single-phase transformers.

Frequent overloads during the melt-down period place heavy demands on the construction of furnace transformers, and especially on their short-circuit capacity. The core of a furnace transformer is made from grain-oriented cold-rolled electrical steel sheet of high magnetic quality and particularly low specific losses. Five-limb transformers are required where the transformer height is low or when differential LT voltage adjustment is used to enable the furnace to be balanced if the high-current conductors are asymmetrical. The low-tension winding comprises many disc-coils connected in parallel. With the shell construction, the winding consists either of parallel connected rings or of a solid sheet copper cylinder.[17] Since the maximum radius of this cylinder is determined by the eddy-current losses, in extreme cases the current density permitted for thermal reasons determines the axial measurement, and with it the height of the transformer.

All secondary windings have a common feature in that they must be arranged outside the high-tension windings. This is necessary because of the space required for the interconnecting bars for disc coils or rings connected in parallel, whereas, with copper cylinders, extending over the entire length of the windings, the ends can only be brought direct to the front, to the terminal blocks in the wall of the transformer housing. Thus the high-tension winding is mounted directly on the core, and so more space is required. Above all, it is essential that transformers to be connected direct to the high-tension mains-supply must have a wide insulating gap and/or an earthed screening winding between the core and the high-tension winding.

It is essential to adopt special methods to fix the low-tension winding securely so as to prevent any risk that axial forces produced by the frequent current-surges might cause the windings to lift or 'hammer'. The necessary mechanical security under dynamic load is achieved by prestressing the winding.

Apart from short-circuit withstandability, local temperature-rises at all high-current junctions and in adjacent solid components are important. Therefore, on assembly, transformers are first subjected to safety tests at almost maximum rated current for about 10 min with no oil outside and inside the tank, in the course of measurements to establish short-circuit voltage. Hot spots at particularly critical locations are detected by using thermal sensors and are suitably dealt with. In particular, the high-current secondary terminals in large transformers are located on the side of the housing and are triangulated and water-cooled (Figure 5). With primary-circuit furnace transformer tap changing, the high-tension winding must be designed for the range of voltage expected at the free end of the winding. Large furnace transformers connected

Figure 5. 100-MVA furnace transformer. HT voltage 115 kV; LT voltage 830 to 271 V; maximum current on low-voltage side 79.9 kA. (Photo by Trafo-Union.)

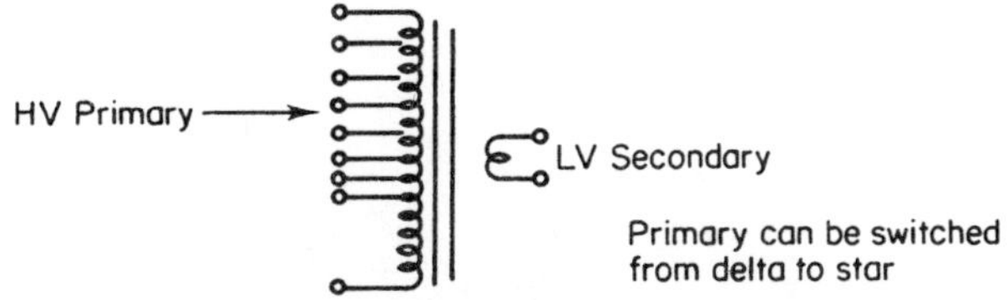

direct to maximum voltage mains are therefore provided with a voltage tap changer in the intermediate circuit. Overvoltage arresters must be provided at the furnace transformer input terminals.

Secondary voltage is today changed almost exclusively by means of a load-tap step switch. No-load voltage switching, particularly used with small furnaces, inevitably entails lost time and hence production when the furnace is switched off. Furnace load-tap switches are either fitted with reinforced contacts and strengthened mechanical actuation, or they may incorporate other means of withstanding high switching frequencies. These features enable load-tap switches to effect 50 000 to 100 000 operations before refurbishing.

Distributing high-current conductor bushes over the whole side of the transformer tank causes increased heating of the tank wall, especially with large furnaces, and it may be necessary to incorporate aluminium panels into the wall. For lower currents, normal high-current bushes are used to carry the secondary winding through the top of the tank; however, this design only allows the bushes to be set in a straight line, so that the high-current conductors to the arcs will have unequal inductances. For low-power transformers, heat losses can be dissipated by convection cooling using radiators; however, this system often leads to problems with indoor installations because of the large amounts of air required. Thus, high-power furnace transformers are usually equipped with oil–water heat exchangers (Figure 6). These operate at low oil temperatures in order to minimize local overheating of the high-current windings caused by overloads in service, thus prolonging furnace transformer life. Pumps without stuffing boxes transfer the hot oil from the furnace transformer to the heat-exchanger, where the oil gives up its heat to the water. The water-outlet temperature should not be allowed to exceed 40°C if possible, since above this temperature substances dissolved in the water begin to precipitate out and obstruct the pipes and attack them chemically.[18–27]

5.4.2.4 Reactors

In smaller furnace-installations, supplementary reactors are connected in series with the furnace transformer primary. Because of their extra inductance, these chokes create favourable arcing conditions, especially during the melt-down period. After a liquid bath has formed in the melt-down period, the reactor can be by-passed using the on-load switch provided. The inductance of

Figure 6. Furnace transformer. HT voltage 35 kV; LT voltage 451 to 195 V; maximum secondary current 58.5 kA. (Photo by Brown Bovery Co, Baden.)

the reactor is almost independent of current and constant up to three-phase electrode short-circuit if the air gap is correct, so that the potential drop produced by the reactor inductance is proportional to the current. Normally, the reactors have tappings so that the reactance can be matched to prevailing conditions. This matching can be carried out during the commissioning period for a new furnace. The reactor by-pass switch should be of the same design as the furnace switch so that the by-pass switch can be used in place of the furnace switch should the latter break down. A reactor is not necessary for medium and large-sized furnace installations above 7.5 MVA power.

5.4.3 Electrode Position Control and Running Adjustments

The electrode control system keeps the electrodes in each phase at the right distance from the scrap or the melt, so that the power of the arc is optimized for production and the requisite metallurgical processes as well as for economical wall-life. Electrode control requirements for the melt-down period are quite different from those for the ensuing refining periods.

The control systems of the three electrodes of a three-phase arc furnace are coupled through the common power supply system. Control of one electrode must not interfere with the other electrodes and must not cause any system-instability. This is achieved by treating electrode control in arc-melting furnaces as differential impedance control. For this, the electrode currents and voltages must be measured as close to the arcs and the melt as possible. The measured current and electrode voltage enable the electrode control system to derive the control signal from an impedance variation:

$$\Delta Z = \Delta V - k\Delta I$$

Factor k accounts for interlinking of transformer voltage and specified impedance. The electrode control system must fulfil the following requirements.

(1) Selected power must be kept constant during the melt-down period.
(2) Different up and down electrode speeds.
(3) Maximum electrode up-speed as soon as current rises when scrap is contacted in order to avoid tripping the furnace-breaker.
(4) Current interruptions should be avoided by using a high-quality control system with good dynamic response. The control system must behave differently when melting down scrap than when the melt is liquid.
(5) The operating points, that is, the impedance values per phase, must be selected automatically for each secondary voltage step of the furnace transformer during operation.
(6) It must be possible to intervene in the event of abnormal operating conditions, e.g. the presence of non-conducting material in the scrap.
(7) Electrodes must rise up on switch-off or power failure.
(8) Manual control of electrodes.
(9) Protection of the electrode arms against excessive vibration.
(10) Servicing and setting up should be simple to do.

The electrodes are moved by electromechanical or electrohydraulic actuators. An electromechanical actuator here consists of a winch, whose motor or eddy-current clutch is energized by the control unit. With electrohydraulic actuation, the valve-releasing pressurized fluid to the electrode lift ram is operated by a Ferraris controller (Figure 7) or a solenoid controller (Figure 8) to give the desired electrode movement. These systems achieve electrode speeds of up to 150 mm/min.

Electronic control offers ways of meeting the requirements for low-inertia systems having fast response time, with minimum delay and optimum electrode speed in relation to control demands.

Figure 9 illustrates an electromechancial control system using electronic components.[28–32] Here a unidirectional three-phase motor drives an eddy-current clutch which varies the drive torque according to the excitation. The electrode mast is supported by a slung cable and steadied by rollers in a guide. The electrode control system varies the excitation of the eddy-current clutch which drives a reduction gearbox and winch to raise and lower the electrode.

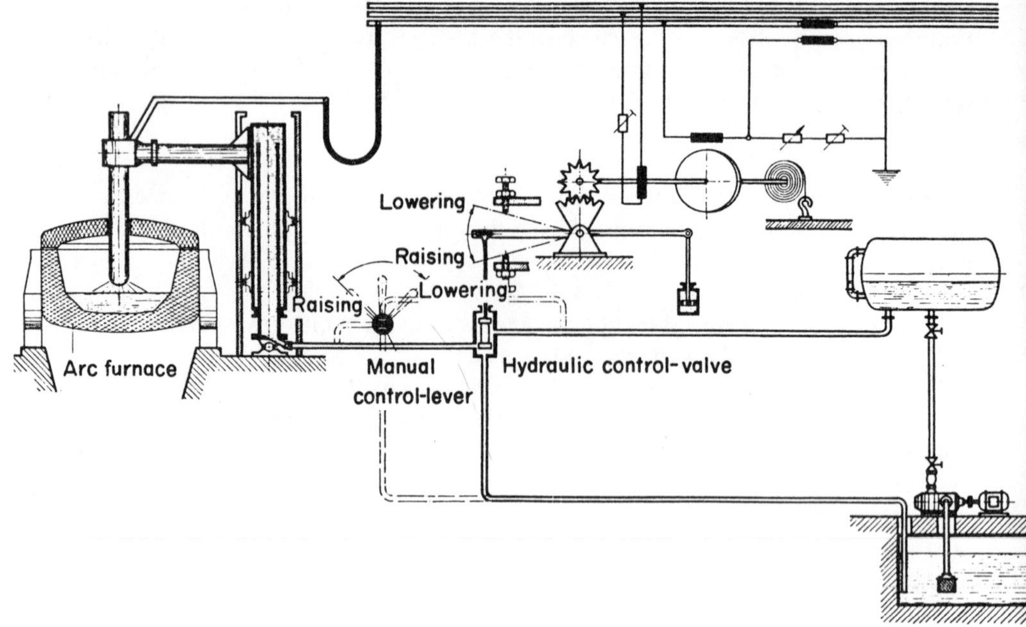

Figure 7. Electrohydraulic electrode system with Ferraris controller

The strength of excitation governs the speed of the electrode. When the electrode is at rest the eddy-current clutch is still energized by holding current. If the excitation drops below holding torque the electrode column sinks under its own weight; if the excitation is greater, then the electrode is raised.

The electrode speed and mode of operation can be separately set for each function, manual or automatic, according to production requirements. A tachogenerator monitors electrode-movement and produces a voltage directly proportional to electrode-speed. If the electrode drops faster than the maximum-speed setting, then a brake is actuated which stops the drive and the electrode. The brake is multi-surface, spring-operated and is held off electromagnetically during normal operation. In manual operation the brake acts as a stop, i.e. it operates when the eddy-current clutch excitation almost reaches the holding level; the brake stops the electrode at once in event of interruption of the auxiliary supply.

The motor-speed controller in the electronic system works on the proportional, integrating, differentiating (PID) principle. This ensures that rotational speed exactly corresponds to the value specified. The rotational speed controller provides the set value for the eddy-current clutch excitation. The current controller measures this excitation current and compares it with the specified value; this system acts as an extra stand-by control circuit.

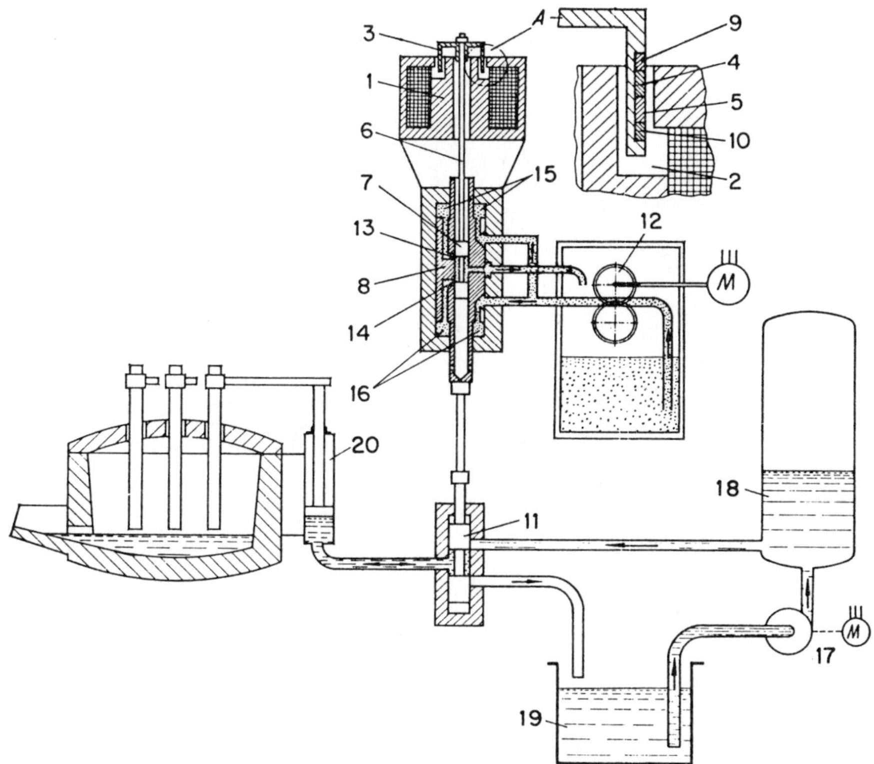

Figure 8. Electrohydraulic electrode control system with moving coil. 1 Solenoid; 2 air-gap; 3 moving coil; 4/5 control windings; 6 push rod; 7 small piston; 8 slave piston; 9/10 spring coils; 11 hydraulic spool valve; 12 oil pump; 13/14 oil-pressure inlets; 15/16 pressure chambers; 17 hydraulic pump; 18 hydraulic accumulator; 19 hydraulic reservoir; 20 actuating cylinder; A Enlarged detail of moving coil and solenoid

The response time of the eddy-current clutch to current changes is very much reduced by a thyristor output stage in the controller. The controller dead-time cannot be shortened any further without exceeding the stress limits for the electrode arms.

The moving coils 4/5 in Figure 8 are energized by rectified electrode current and voltage signals in opposition. The moving coil operates in the field of a solenoid and is governed by the control signal; it operates the main control valve via an oil servo in order to move the electrodes up and down as required. When the furnace is switched off, additional springs raise the electrodes to the highest position.

Figure 7 shows a control system using the Ferraris principle. The energizing winding is connected across two phases of the furnace transformer secondary.

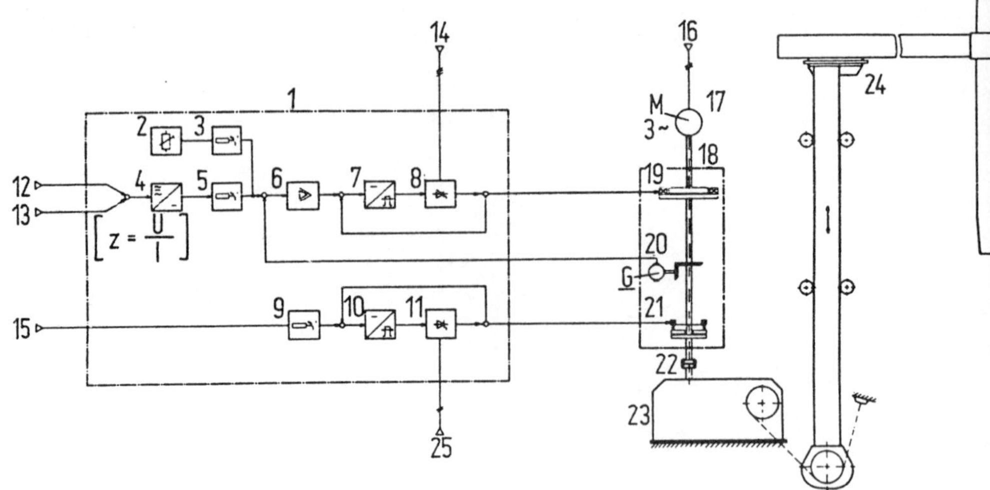

Figure 9. Schematic circuit diagram of an electromechanical electrode control system with electronic control unit[28]. 1 Electronic control unit; 2 set value adjustment—manual operation; 3 control relay—manual operation; 4 difference signal generator; 5 control relay—automatic operation; 6 motor-speed controller; 7 current controller; 8 thyristor output stage for clutch control; 9 brake control relay; 10 brake current controller; 11 thyristor output stage for brake; 12 arc voltage set value; 13 actual arc current; 14 supply for the clutch thyristor output stage; 15 brake-control signal; 16 electrode motor supply voltage; 17 electrode winch motor; 18 control of transmission of winch power; 19 electromagnetic clutch coil; 20 tacho-generator; 21 multi-plate spring-operated brake; 22 mechanical clutch; 23 high-power gearbox; 24 electrode mast, arm, and electrode; 25 supply voltage for the brake tyristor output stage

The electrode voltage and the electrode current signals from the other phase are connected in opposition across the Ferraris controller operating coil which is offset at 90 degrees. A control demand causes the hydraulic control valve to raise or lower the electrode. When the furnace is switched off or if power fails, a return spring operates the control valve to demand electrode lift to the highest position. Springs inside the control unit vary sensitivity as required, according to control demand. Vibration is damped by built-in oil dashpots.

All the control systems described can achieve response times of a few milliseconds with minimum delay and very high sensitivity. When assessing the quality of an electrode contro system, the entire system comprising the control unit and mechanical actuation for the electrodes must be considered. The quality of a control system can be ascertained and monitored by means of the following practical test. An a.c. potentiometer attached to the electrode arm signals arm-movement to an oscillogram. This oscillogram indicates the time taken for an electrode to come to rest after receiving a control demand at maximum lifting speed from rest — without oscillation. The oscillogram also allows the path travelled by the electrode between control demand and coming to rest to be determined. This path must be as short as possible, especially for furnaces in which the melt must not pick up carbon from electrode-immersion. This test is suitable for checking out new furnaces on acceptance as well as for monitoring the control unit and the moving and stopping of the electrodes in normal service.

Arc-furnace control can be improved at maximum arc power, by operating the control on electrode current and the electrode voltage signal which has been corrected to correspond to the true arc voltage for the phase. Additional circuits are required for evaluating the arc voltage (see section 4.1.2 and literature references 14, 37, 39 and 42).

The control variable, the ratio of electrode voltage to electrode current, has a particular value corresponding to optimum furnace working current for the secondary-voltage level in use. In modern furnaces this optimum working-point is adjusted automatically at each secondary voltage level.

In section 4.1.1.7 the maximum active power P_{max} for given line voltage V_L and reactance X is stated to be:

$$P_{max} = \frac{V_L^2}{2X}$$

Under the above assumptions the optimum working-point at the highest arc power possible occurs at a phase-displacement angle ϕ which is half the phase angle for short-circuit. However, it is also affected by the stresses on the furnace lining, the extent to which the molten steel is shielded by scrap, the depth of the slag layer, and the danger of electrode breakage after charging scrap. Figure 10 is a circle diagram for a 150-t furance of 78 MVA rated power with a secondary-voltage range of 640 V to 240 V; it shows the possible limits of variation of the working-point.

The refractory wear index RE, defined by W. E. Schwabe to be the product of arc power and voltage, should be included in these considerations. The phase-displacement angle at short-circuit has been determined to be $\phi_{SC} \simeq 81$ degrees, i.e. cos $\phi_{SC} \simeq 0.15$. The maximum possible melting current is about 78 kA. A typical scrap process uses the following working points:

(A) The furnace is started at the highest voltage-tapping. Power factor is about 0.75, and current is about 10% less than rated current. The arc should be very long and should burn a broad channel in the scrap to protect the electrodes against breakage when scrap slides into the melt. In particular, this setting protects the hearth from damage at maximum power until an adequate molten pool forms.

(B) After 5–10 min the rated current is set at the highest voltage-tapping, and the relevant specified value is set on the control unit. The furnace melts

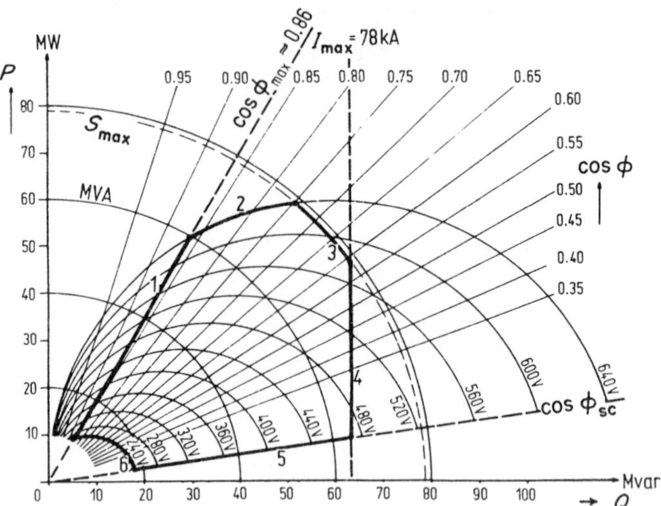

Figure 10. Limits of variation of working point.[28] 1 cos ϕ max ≈ 0.86; 2 maximum furnace transformer secondary voltage; 3 maximum apparent power S; 4 maximum electrode current; 5 arc short circuit cos ϕ_{sc}; 6 minimum furnace transformer secondary voltage

down the first and subsequent baskets at this working-point. The radiation factor here is relatively high.

(C) 20–15 min before the end of the melting-down period for the last basket, the current is increased to the maximum possible rated value which lowers the radiation factor. At the same time the power factor at this new setting drops to about 0.6.

(D) After the end of the melting-down period and sampling, the temperature must be raised. At a voltage setting of 520 V, for instance, active power drops by about 30% and cos $\phi = 0.6$. The radiation factor drops sharply as well. The arc is so short at this stage that the greatest possible proportion of the power is transferred direct to the melt from the arc, and the wall is spared as far as possible.

(E) If metallurgical refining is required, the operation can proceed at a lower voltage tapping between 240 and 440 V and at cos $\phi = 0.5$–0.6; the radiation factor is then very low. The melt can also be kept at a constant temperature using these settings.

This mode of furnace-operation starts automatically when the secondary voltages are set, and the system lends itself to programmed control of the furnace switch.

Similar working points are selected for sponge-iron processing, only here the furnace operates at higher constant currents.[33–41]

Where water-cooled wall elements are used, the radiation factor need not be considered as it has to be for purely refractory walls. Operation can therefore proceed at better, higher-power factors, especially in phases (B) and (C), in order to achieve higher active powers which lead to somewhat shorter charging times.

Electrode-fracture due to non-conducting material in the scrap contacting the electrodes can be prevented with hydraulic electrode-control systems by monitoring the pressure in the electrode-operating cylinder. When the electrode contacts a non-conducting object, the downward movement of the electrode causes a pressure drop in the cylinder. This can be used to signal the control system to raise the electrode. Similar control arrangements are also possible with electromechanical electrode control systems.

As electricity costs in West Germany are divided into demand and unit costs, it becomes ever more attractive to avoid uncontrollable power-peaks and to make better use of power-consumption at rates and prices agreed with the electricity-supply undertaking. The metering period over which peak power demands are costed is 15 min for most heavy consumers; therefore the arc furnace should be switched on and off very frequently by an automatic demand limiter if it is to be treated as a heavy but controlled consumer and costed favourably. The furnace circuit-breaker cannot be operated so frequently for this purpose; however, the electrode control system can be used to cut power to the peak power-consumption rates agreed with the electricity supply authority.[42]

5.4.4 Stirring Coils

A stirring coil, which is mounted under the bottom of the vessel, creates a changing magnetic field which moves the molten steel near the surface between the furnace door and the spout and causes a return flow past the furnace walls towards the door. The stirring coil consists of an iron core and a two-pole winding, similar to that used in the stator of a rotating electrical machine. The windings consist of water-cooled, hollow copper conductors, embedded in position. The stirring coil is protected from the hot vessel by a layer of refractory materials.[43–6]

Stirring the melt with the coil speeds up metallurgical processing and simplifies deslagging. However, transfer of metallurgical refining to the ladle makes stirring coils uneconomical in many cases.

5.4.5 Instrumentation and Control Panels

The most important instruments for arc-furnace operation are clearly arranged at a switchboard or control desk, which should be positioned so that furnace-operators can see the instruments at the same time as watching over the furnace.

The instruments display the following arc furnace values:

(1) Arc current in each phase, which is measured either via current transformers on the secondary conductors, in the furnace transformer intermediate circuit, or with magnetic volt meters.
(2) All secondary voltages, which are taken at the transformer terminals. When taking measurements, it must be remembered that strong magnetic fields around the high-current conductors affect the readings. If necessary, arc voltage determined by means of compensation circuits can be displayed (see section 4.1.2, 'Clausthal measurement'). Active power in each phase is derived from arc current and phase voltage. The instruments should be provided with suitable damping for clearer readings.
(3) A line voltage on the high-tension side before the furnace switch.
(4) A line voltage on the low-tension side of the furnace transformer.
(5) Total furnace active power on the high-tension side (on indicator and recorder). Because of marked fluctuations during the melt-down period, it is recommended, here too, that a resettable integrator be used for monitoring the furnace.
(6) Electrical power consumed, measured on the high-tension side of the furnace transformer. The meter cumulatively displays the total amount of electricity used, while the amount of electricity consumed per charge is displayed on a second indicator.
(7) An indicator to give advance warning of furnace load-shedding required to limit peak electricity-demand for the whole works. Furnace-operators can respond to this warning by carrying out necessary metallurgical work or deslagging while furnace power is cut.

The control console also incorporates the most important furnace-control switches. The furnace circuit-breaker control switch can only function if all the necessary interlocks on the furnace and in the electrical system have first been operated. The operator sets the secondary voltage required from the furnace transformer on a preselector switch. The power-level selector then responds to this demand automatically. All the services such as water-cooling pumps, hydraulics, and oil pumps are switched on and off by means of push-buttons on the control console. Dust-extraction system controls and instruments are on another part of the control console. All the vital functions for the dust-extraction system are interlocked with those for the furnace. A programmed control unit adjusts the current in the individual furnace phases to suit each secondary-voltage stage, so that the optimum furnace working-point is always achieved automatically (Figure 11). This unit controls the various secondary-voltage stages and power-on times throughout the entire charging process in relation to a dominant value, such as electricity consumed per tonne of scrap charged.[47–9]

The general layout and space used by the transformer building and arc furnace services are illustrated in section 5.3 (Figure 1) Multi-storey construction

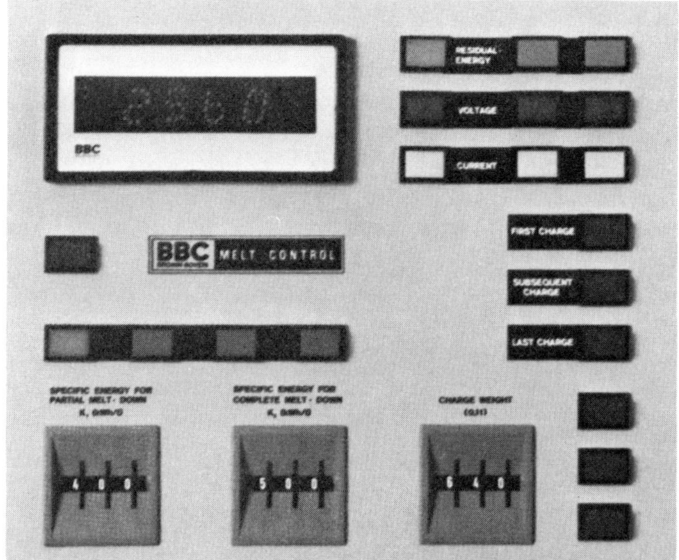

Figure 11. Furnace control panel with programmed control unit. (Photo by Brown Bovery Co.)

reduces the floor-space required in the melting shop appreciably. For the protection of furnace-operators the control console or control desk should, if possible, be positioned in a sound-proofed room inside the transformer building, within view of the furnace.

5.5 FURNACE-LININGS AND DURABILITY

Walter Klein, Staufenberg, and Günter Schmeiduch, Schwalbach

5.5.1 Fundamental Considerations on Furnace-linings

Walter Klein, Staufenberg

The last few years have seen not only a considerable increase in arc furnace energy-input per unit time but also variation in the type of fuel used and the mode of its application. This led to increased exposure of the furnace refractory lining to wear, with reduced durability and an increase in lining cost per tonne of steel. This development, which greatly affected the economics of arc-furnace processes, could only be countered by using better refractory materials, better construction and process techniques, and finally by the use of water-cooled elements. If we want to find the best lining for an arc furnace with a specified steel-production programme, we have to consider a number of criteria:

(1) The insulation and thermal storage capacity of refractory brickwork must match its wear-resistance. High thermal insulation guarantees low energy losses from the furnace vessel. However, brickwork with good thermal insulation properties also results in a high average brick temperature and deep penetration of foreign substances into the brick, lowering the refractoriness of the brick and decreasing its wear-resistance.
(2) For maximum availability of the furnace, i.e. for the shortest repair or re-lining times, it is advantageous for individual parts, such as sections of a wall or even the entire furnace vessel, to be replaceable.
(3) The types of refractory for the wall and/or the roof must be compatible, and their choice is governed by the mode of furnace operation: continuous or intermittent, single or multi-basket charges, or continuous feeding of the charge materials.
(4) For many years now, basic refractory linings have proved satisfactory for arc-furnace steel-production. Acid linings are nowadays used only for producing steel castings and grey iron castings (see Chapter 11). Also for the roof linings, high alumina or basic refractories have increasingly replaced ceramically bonded silica materials.

5.5.1.1 Hearth-linings

The bottom lining of a basic arc furnace consists of a backing lining which supports the actual 'hearth', usually laid on top of an insulation layer. Backing linings, consisting of two or three flat courses, with a soldier course of fired magnesite bricks on top, are well proven (Figure 1).

Depending on its intended purpose, the working hearth is lined either with a refractory material or with basic bricks. The following are used:

Crespi–Hart hearths of special granulated dolomite.
Special granulated magnesite with sinter additions.
Tempered refractory bricks containing carbon.
Fired refractory bricks with or without coal tar.

Crespi granulated dolomite is applied in layers without any additions and de-aerated by using free-falling manual compaction tools. Recently, special vibrating apparatus has also been introduced, but in any event grain-size segregation must be avoided. The completed hearth is covered with thin metal sheets to avoid damage by the first charge, melting of which must proceed slowly, so that a rigid bowl can be formed by sintering.

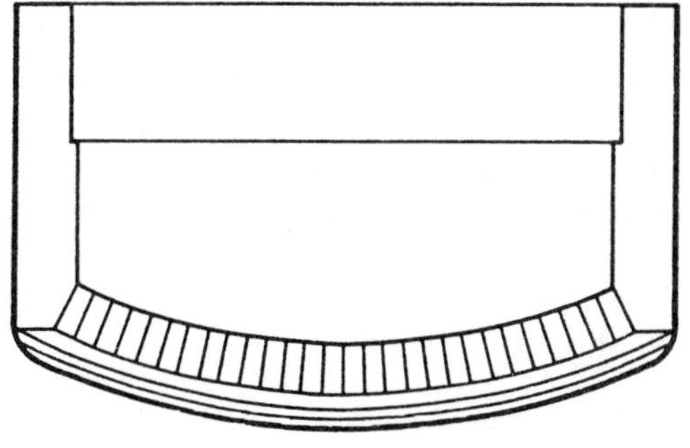

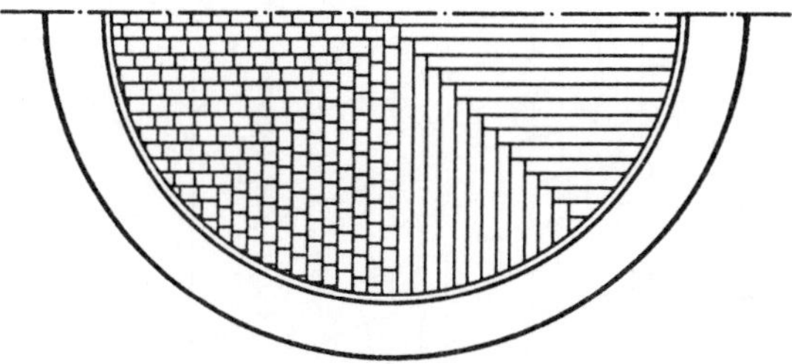

Figure 1. Schematic diagram of the lining of an electric furnace hearth with a hot face of cross-bonded bricks

Initial sintering of magnesite hearths is more difficult and is usually undertaken by the supplier.

Tempered bricks containing carbon pass through a state of minimum strength at lower temperatures. 'Weeping' or 'blistering' often seen depends on the thermal history of the bricks.

For safety reasons, heavy-duty arc furnaces are frequently lined with ceramically bonded bricks. In order to ensure a tight lining it is necessary to see that the bricks conform to dimensional tolerances and that tar-impregnated bricks have a smooth, clean surface. Only under these conditions can a satisfactory lining with tight joints be achieved. It is considered good practice to fill the joints with grout.

The following figures are a good guide to lining thicknesses (mm) for common types of hearth:

Working linings	400–350
Backing linings	250–200
Total hearth thickness	650–550

5.5.1.2 Wall-linings

The wall-lining is built up from a rigid horizontal base — the top of the hearth-backing lining. The furnace wall itself can be built up from layers of individual bricks or from larger composite blocks. The former method is usually more labour intensive. The best method, on time-saving grounds and also for repairs, is to use ready-made components (vibrated or rammed blocks) or prefabricated blocks called 'prefabs' (cemented or clamped assemblies of individual bricks), as shown in Figure 2.

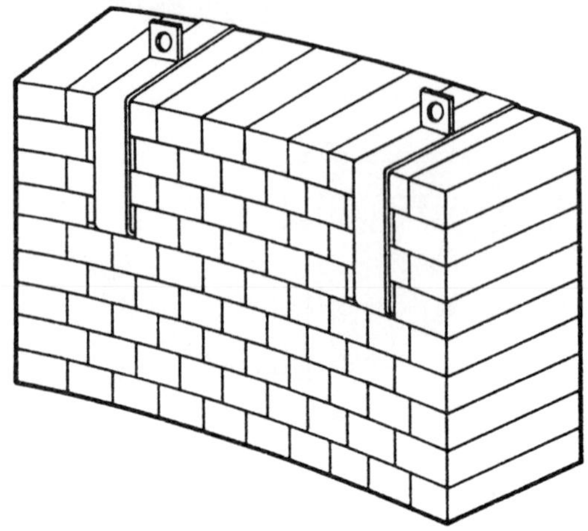

Figure 2. Cemented prefabricated wall block

These wall-elements can be inserted block by block or as a ring (consisting of a number of blocks) in one sweep (Figure 3). With very high walls it is advisable to build the wall from two staggered rows of blocks. The gap remaining between the furnace shell and the lining is completely filled with a fine granulated refractory material with a good thermal conductivity.

These large blocks can be made of different types of refractory bricks according to the wear-characteristic of the furnace, in the same way as manually laid brick linings are made from individual bricks. Because of their good properties, tar-bonded or tempered, fired or high-fired bricks are well proven for the lower parts of the furnace wall. Chemically bonded bricks (also with carbon) are used, either alone or in combination with other brick types, for upper wall-linings above the slag-line. Any steel reinforcement must not touch the molten steel because of the danger of breakout; the change in metal level during furnace-tilting must not be forgotten.

There is no cheap universal lining brick because the load on an arc furnace refractory lining varies locally. The best, most economical lining must be found experimentally for each type of production programme. A change in

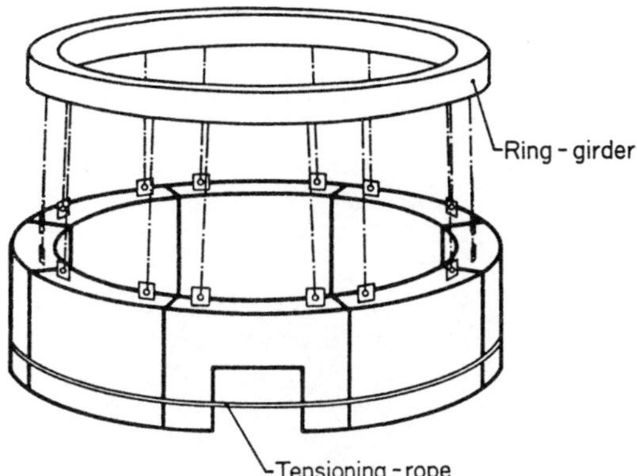

Figure 3. Assembly scheme for a complete circular wall of individual bricks

wall-thickness can often provide a more economical solution. If the furnace wall is steel-reinforced, and/or if the stability of individual bricks or whole blocks of bricks is improved, then parts of the wall subject to excessive wear ('windows') can be replaced during the life of the whole lining, or the wall may be allowed to wear down to a very small thickness.

Not so long ago, on renewal of the lining, the wall was cooled with fans, so that the residual lining could be wrecked 8–24 h after shutdown. Today the residual lining can be partly or totally removed by telescopic grabs immediately after switch-off. Exchangeable furnace vessels are re-lined at any convenient time, irrespective of melting operations. Usuall wall-thicknesses are (mm):

Upper wall region: 200–300
Lower wall region: 350–450

Furnace doors generally have water-cooled frames; the doors themselves may be lined with cheap, standard-shape refractory bricks or rammed with refractory compound.

Like the hearth, the tapping launder has a working lining above a backing refractory brick lining. The working lining can consist of individual bricks, monolithic prefabricated sections called runner blocks, or can be of refractory ramming. Development started originally with fireclay and progressed to high-alumina, magnesite, or dolomite materials, depending on the composition of the molten steel and slag, tapping-time, and number of taps per day. Carbon additions to the working lining reduce the wetting action of the steel and pouring slag, thus avoiding undesirable penetration.

5.5.1.3 Roof-lining

The old type of roof-lining, bricked in rows with domed arch bricks (double tapered bricks with rectangular bottom cross-section, usually in a sectional

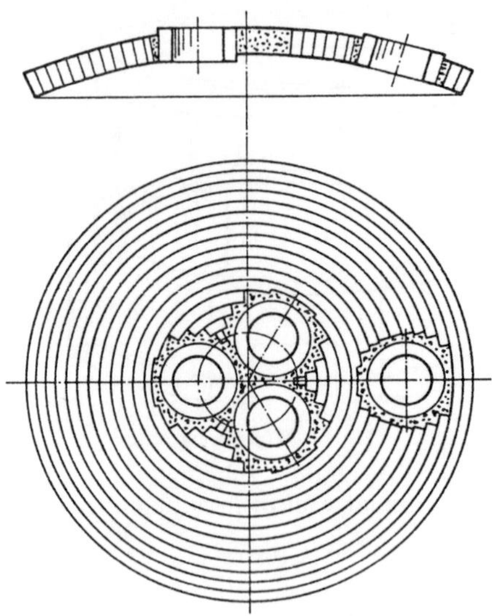

Figure 4. Electric-furnace roof with ringed courses and rammed centre

outer ring), was long ago superseded by a water-cooled skewback ring with rings of domed side-arch bricks of varying taper in conjunction with domed-arch bricks (Figure 4). These domed side-arch bricks (tapered in three directions) are mixed with rectangular-section bricks in definite proportions for each ring.

Under the auspices of the Verein Deutscher Eisenhüttenleute (the Association of German Iron and Steel Engineers), the steel and refractory industries laid down standard sizes for bricks for arc-furnace roofs (new 1977 edition of the steelworks committee's report No. 414a); these standards have proved very helpful to both parties. The spherical radii were chosen in such a way that a rise of 11–13% can be used for all current roof-sizes, and present experience has shown that this is suitable.

The openings for the electrodes, gas-extraction, and charging are also lined with standard-shaped bricks, the backs of which are corrugated to facilitate keying up with the ramming. The remaining gaps are packed with a plastic refractory compound. It is possible in this way to build a roof quickly and satisfactorily without having to use complicated and expensive brick shapes for the centre; in addition, the stress distribution is optimized. The roof thickness is in the region of 200–300 mm.

Flat roofs with suspended refractory bricks are under development for special applications. However, they require a different superstructure (Figure 5). Just as with furnace walls, water-cooled elements make the roof much more durable, since only the centre is lined with refractory materials.

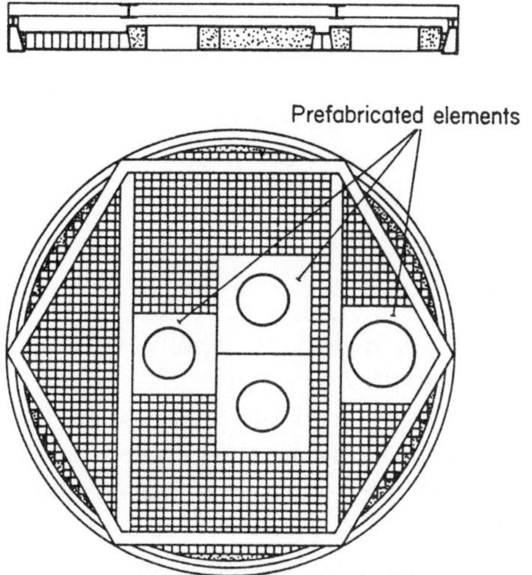

Figure 5. Design for a flat roof

Classic silica bricks, which have excellent thermal shock resistance above 700°C and an upper temperature-limit of about 1670°C, are widely used for furnace roofs; however, a large number of other refractory materials are also used.

High-alumina bricks (mostly containing bauxite) are used in intermittently operated and heavy-duty furnaces. Relatively expensive corundum bricks (e.g. high-fired bricks) are now also used, but no conclusive reports are as yet available. When acid refractory materials are used in the furnace roof, care must be taken to protect the basic-lined furnace walls from attack by dripping acid materials, e.g. by the use of basic drip edges.

Fully basic roofs, made from dolomite-based materials, are now widely established, partly because they are cheaper. However, they must be protected from water and damp; long shut-down periods lead to failure caused by hydration.

Roofs made from MgO/Cr_2O_3 compositions work well in continuous operation in which temperature-fluctuations are kept as small as possible. However, when this type of construction is used, account must be taken of increased brick-weight and volume-changes. Zircon-bearing bricks are not used for roofs because they are too expensive. Mixtures of different bricks are often used in rings or sections and not for economic reasons alone.

Expansion-allowances for roof linings are derived from various criteria: it depends mainly on the roof clearance as well as on the degree of transformation in the bricks, on the tendency of fired products to expand in volume, on the coefficients of thermal expansion, and on the way the furnace is operated. The expansion allowance needed may be between zero and 1.3 times the value of the volume-expansion; it is calculated from a value which depends on the temperature in the centre of the brick.

We should point out in this connection that there is no need to insert millboard at the back of the lining to allow for expansion, either of the walls or of the roof. If this is still done behind wall-linings, then the reason is mostly to obtain quicker parting for wrecking.

5.5.1.4 Water-cooling boxes

In very high-power arc furnaces, especially with additional fuel–oxygen burners to shorten the melt-down period, it is not possible to obtain a satisfactory lining-life with the available refractories. About ten years ago the first attempts were made to overcome the problem by the use of water-cooled wall-elements. Today welded or cast-iron wall-cooling elements or cooling blocks, faced with a protective refractory layer, are used mostly in high-power furnaces. The various designs and stringent safety precautions are treated in depth in section 5.2.1.5. The development of water-cooled furnace roofs is also mentioned there.

5.5.1.5 Programmed spraying

This is a method of spraying refractory protective coatings on to the lining or parts of the lining. Such protective coatings should resist wear for at least one melting cycle.

5.5.2 Wear Problems

Günter Schmeiduch, Schwalbach

The costs of producing steel in arc furnaces are determined to a considerable extent by the cost of the lining and of running repairs (refractories and wages); they also influence the productivity and thus the total costs of continuously operated furnaces.

Operational efficiency is reduced by interruptions, equipment servicing, and, above all, by time taken in maintaining the linings. If we assess the time unavoidably lost in delays at 4%, and for special work on the system — e.g. replacements, major repairs, and hearth-linings as 1% of the operating time, then operational efficiency, is still also affected by the cost of wall-lining maintenance.

Figure 6 in section 5.2 shows the percentage productivity of an arc furnace in relationship to the wall-life and the time spent on the lining. Here, the down-time is quoted as 5%, while the lining times have two different values, i.e. five and ten times the tap-to-tap time. The range, 80–94%, of the productivity values in this diagram shows how much productivity depends on the wall-life and on the time spent on the lining. Like melting-times, these figures merit particular attention.

5.5.2.1 Furnace-wall deterioration

The load on the refractory lining, and thus lining-life, are affected by a number of factors, but there is no agreement about their relative significance. In the following paragraphs we will name only the most important factors and briefly explain their effects:

(1) Physical deterioration through radiation and erosion;
(2) Chemical deterioration;
(3) Thermal shock;
(4) Mechanical stressing.

The connection between physical deterioration through radiation and erosion and the thermal and electrical conditions will now be considered using the refractory wear index *RE* defined by W. E. Schwabe.[1] The radiation emitted by the arc is directly proportional to the power P_a and the length of the arc. The arc voltage V_a is a measure of the arc length; thus

$$RE = P_a \times V_a$$

RE acts as a measure of the deterioration of the refractory lining. This measure involves the following simplifications:

(1) The arc length is not absolutely proportional to the arc voltage. The arc voltage would have to be reduced by the anode and cathode drops which occur with gas discharges.
(2) The arc does not strike vertically downwards but is slightly inclined, so that maximum radiation is directed at the wall slightly above the horizontal, and the radiation towards the furnace-centre is correspondingly directed downwards.

Notwithstanding these simplifications, *RE* (as defined) is a physically sensible quantity.

We can show by mathematical conversion of P_a and V_a that, to a first-order approximation:

$$RE = cV^3$$

According to the geometrical circumstances, we have, for the thermal load on the wall-lining:

$$RE_w = cV^3/a^2$$

where a is the distance between arc and wall. This dependence of the refractory index on the third power of the voltage is responsible for the reduced life of large furnaces with high transformer powers, when a^2 does not change with the corresponding V^3. We can also see from this relationship that the selected vessel diameter will have a decisive effect on the life of the wall. The thermal load on the wall can be calculated from the relationship between vessel diameter and the charge weight T.[2]

$$RE_w = c \times T^{0.9}$$

This shows that the radiation effect increases with increasing charge weight T. However, comparison of the actual reduction in refractory lining-life for furnaces of increasing charge weights shows that the life — obtained from practical results — grows with $T^{0.5}$, which is slower than the theoretically established value for the thermal load, which is proportional to $T^{0.9}$.

The relationship between the electrode pitch circle diameter and the vessel diameter also has a decisive influence on the wall-wear in an arc furnace. For example, if we change the ratio in a furnace from 4.0:1 to 3.5:1 then the radiation factor will increase by 11%. We must take care that the ratio of vessel diameter (taking into account the lining) to the pitch circle diameter must be selected in such a way that the furnace wall will have optimum life for maximum charges.

From 1970, automation in electric steelmaking provided the possibility of computer control of the electrical operation in such a way that furnace-wear can be controlled. In fully balanced furnaces, arc power and arc length should be changed in a controlled manner during the melting process, so that the diminishing protection provided by melting scrap is compensated. In Figure 1, current, voltage, arc power, arc length, and the change in wear coefficient are plotted schematically in relation to the specific applied electric power for a one-basket charge.[3] Shortly after power-on, the power and length of the arc are set at maximum and kept there until the scrap around the wall begins to melt. From then on, the arc length is progressively reduced while the current is changed in steps, either separately or together with the voltage.

With this method of operation the refractory wear index is steadily decreased and matched to the progressive melt-down of the scrap. The radiation factors, which decrease very quickly with power factor cos ϕ and are a measure of lining attack, are shown in Figure 2 for various furnace voltages.

In the planning and building stages of a new arc furnace, we must — as a matter of principle — aim at optimum economic operation with regard to its lining life, and consider automating energy-input and using a computer to control voltages and currents.

According to B. Bowman,[4] it is not arc-radiation but what is called 'arc-flare', blown at the wall by electromagnetic forces, which constitutes the greatest thermal load on furnace wall-linings.

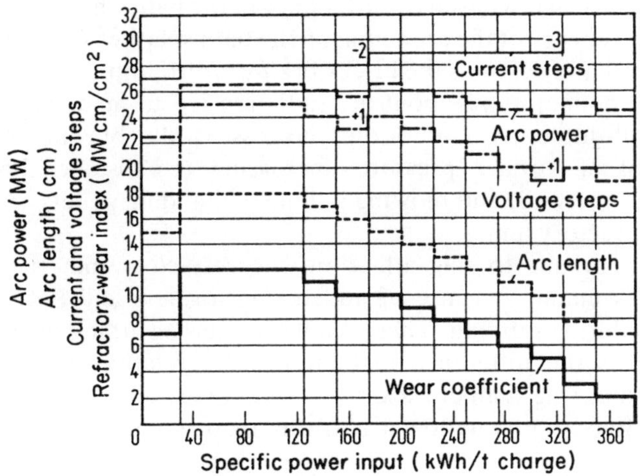

Figure 1. Wear coefficient in relation to arc power and arc length (1 basket charge)

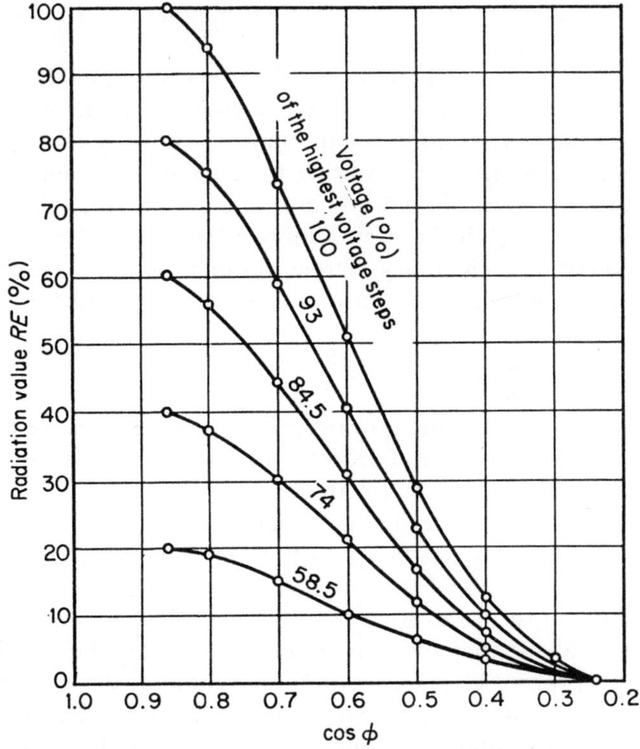

Figure 2. Relationship between radiation value and cos ϕ and voltage[2]

Heavy erosion can be caused by turbulence of the bath when oxygen is blown through it. The energy of the arc acting on the bath splashes steel and slag on to the furnace wall and destroys the brickwork by infiltration.

Theoretical deliberations about arc radiation led to many suggestions for improving wall-durability. Contouring the vessel walls, blowing in protective gases, control of furnace pressure, increasing the distance between the electrodes and the wall, and blowing in lime and adding pellets are all said to protect the refractory lining.

Chemical wear due to slag attack on the refractory material in electric furnaces arises mainly because of added slag formers and FeO, SiO_2, and MnO formed during refining. These oxides in the slag attack the furnace wall all the more as the difference in their basicity relative to the brickwork intensifies and as slag temperature rises. Moreover, the physical state of the slag has an effect, e.g. whether it is frothy oxidizing slag or mirror-smooth refining slag.

Trials of the effect of alloying elements on slag-formation with refractory materials in arc furnaces have shown that chrome alone exerts a decisive influences on the mechanism of slag-formation. Chromium oxide, together with lime, forms a protective layer with a high melting-point, which delays slag attack with dolomite and magnesite.

In addition to the reaction between the liquid slag and the refractory, there is also a chemical attack on the refractory lining via the gas phase. The composition of the furnace atmosphere changes several times during a melting cycle. Continual change from oxidizing to reducing atmospheres and back, together with infiltration of oxidizing dust and temperature changes, in time causes embrittlement of the brick structure and spalling.

The addition of supplementary charges during a melting cycle often leads to thermal fatigue. The high thermal stresses arising in the refractory material can lead to flaking of the bricks. This constitutes a not inconsiderable part of the lining deterioration and is especially marked in vessels which are not in continuous production.

The maximum surface temperatures measured on refractory linings in arc furnaces lie between 1400° and 1850°C. Temperature peaks of 2000°C and more occur in the walls near the electrodes.

Mechanical damage to the lining is mainly caused by charging of scrap. Pieces of scrap which are too heavy or too long can damage the lining quite severely.

Figure 3 shows a record of wear per melting cycle.[5] Wear is greater at the beginning of a wall life than towards its end. The wear melting cycle decreases as furnace life progresses because of the increase in distance between the electrodes and the wall. The way heat is conducted to the exterior may also be a contributory cause; conduction increases as the bricks become thinner. Increased distance from the heat source and improved heat-conduction produce a lower wall-temperature and with it a smaller load on the bricks. In

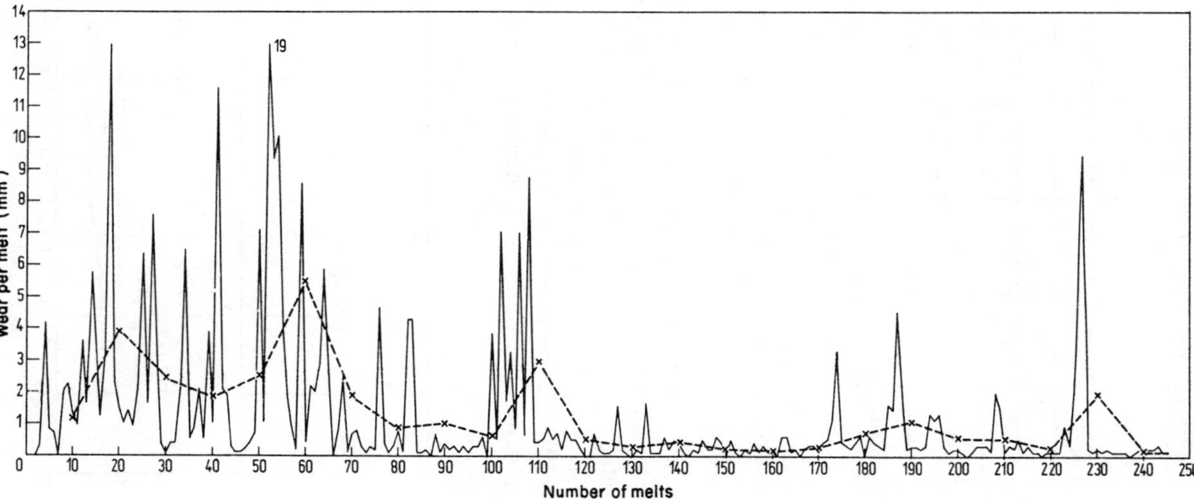

Figure 3. Wall-wear (mm) per charge in relation to furnace-life[5]

addition, the extremely rapid heating up of an arc furnace (practically without preheating) causes cracks in the brickwork, which can lead to spalling during the first few melts. The bricks only attain some stability as furnace life progresses.

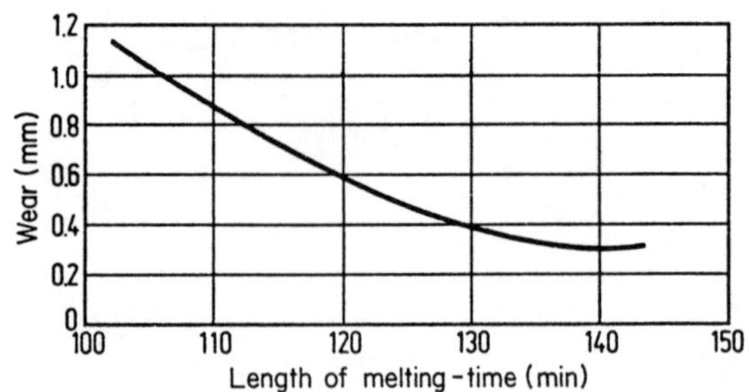

Figure 4. Average wear of wall (mm) during the melt-down period in relation to the melt-down time (60-t furnace[6])

Investigations into the effect of process parameters showed (Figure 4)[6,7] that lining wear depends on the length of the melt-down period. During oxidation, wear is greatly affected by melt-down carbon content (Figure 5)[5]. The total wear can be reduced to low values, as experienced during melt-down, by a melt-down carbon content above about 0.45%. The reason for this is that with a vigorous, frothing bath-boil, frothing slag shields the furnace wall from arc radiation.

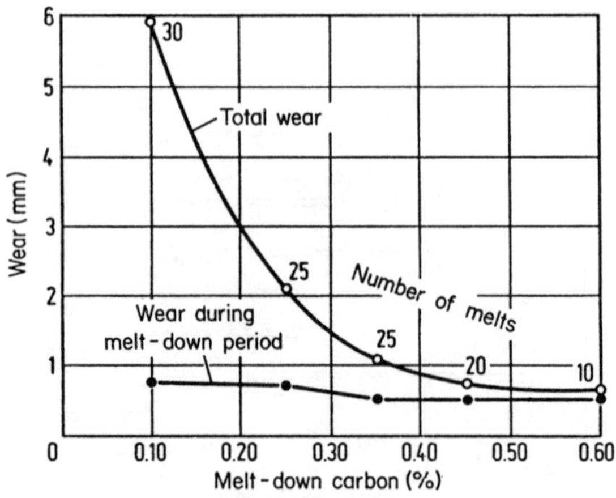

Figure 5. Average wall-wear (mm) during oxidation in relation to melt-down carbon content (60-t furnace[5])

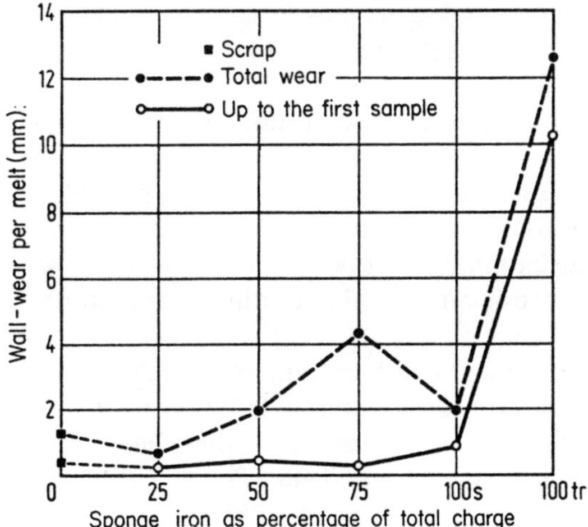

Figure 6. Wall-wear in relation to percentage of sponge iron in the entire charge

Figure 6 shows the deterioration of the wall when melting sponge iron continuously.[8] With 25% sponge iron in a charge, the rate of wear is similar to that for a 100% scrap charge, because, in spite of the higher temperatures for longer times, the furnace wall is shielded by frothing slag.

When the sponge-iron content of the charge is increased to 50–75% the effect of lengthy exposure to high temperatures on the wall also increases deterioration. When melting 100% sponge iron in a liquid bath, a protective coating of slag quickly builds up, but when melting sponge iron without liquid present, the arc radiation hits the wall directly, without restraint. Unlike in other trials, most wear does not occur after the first sample is taken, but during melt-down before the first sample.

When contemplating the refractory wear index, it is at once clear that the parts of the furnace wall opposite the electrodes are most severely exposed. Hot spots develop at slag level; these can be of different intensities because of electrical and thermal asymmetry in the furnace. But the thermal load on the wall at slag level lies between the hot spots; it is distinctly less towards the upper rim of the vessel. Best performances in the hot-spot positions were achieved with:

Fusion-cast magnesite–chrome bricks;
Fused corundum bricks;
High-fired, direct-bonded magnesite–chrome bricks;
Low-iron, tar-impregnated magnesite bricks.

In heavy-duty furnaces the lining above the slag zone is frequently made of:

High-fired directly bonded magnesite–chrome bricks;
Chemically bonded, sheet metal clad magnesite–chrome bricks.

The most important properties of a suitable refractory are:

(1) Adequate refractoriness and the highest possible resistance at high temperatures to the slags expected (infiltration);
(2) High resistance to thermal shock;
(3) High abrasion-resistance at all operating temperatures.

High-power arc furnaces with extremely high thermal wall-loadings can no longer be satisfied by the materials now available. Water-cooled wall elements have proved themselves here; they have been increasingly used since 1970[9] (see sections 5.2.1.5 and 5.5.1.4).

Methods of measuring wear. Wear-measurement is defined here as the brick thickness lost with time or time-related events such as melt-cycle or furnace life. The following well-proven methods are used for wear-measurement:

(1) Residual thickness measurement. This is done on days when there is no production.
(2) Change in slag-analysis. The loss of refractory material is calculated from the magnesia content of the slag.
(3) Heat flow sensors:[10] Sensors are inserted in holes drilled in the lining. Thermocouples measure temperature relative to the furnace-wall surface. This relative temperature lies between the actual wall-surface temperature (in the furnace) and the ambient temperature. When the temperature measured exceeds a threshold value as the wall-lining wears away, the sensor is pulled back, say about 10 mm, and indicates wall thickness remaining. The temperature measured is then compared with a critical value, and if it is above that the computer commands a voltage reduction.
(4) Ultrasonic measurement:[11] A quartz glass rod, which wears away with the wall, is inserted with a special cement. The time taken for ultrasonic sound to travel from a sound source is measured, and the actual distance in millimetres is calculated from the speed of sound. The accuracy is about ± 1 mm. This method allows wear to be determined during operation, and the process itself can be controlled by a computer.
(5) Radioactive method:[12] For this method refractory bricks, each with a piece of radioactive wire embedded lengthways, are inserted in particular regions of the furnace wall. Molten wire gets into the steel melt, together with particles of worn lining, without dissolving in the slag. The amount of wire consumed, and with it the lining wear per melting cycle, is determined from the radioactivity of the steel melt. The wear can be

measured at any time during the process or several times during a melt-cycle.

Wear can also be determined by infra-red[13] and laser methods.[14]

5.5.2.2 Hearth-wear and maintenance

Refractories used for hearth-lining must fulfil the following conditions:

(1) High refractoriness;
(2) High resistance to chemical and mechanical influences.

These properties are met by magnesite and dolomite. The choice of either one of these materials in particular works is governed by local technical and economic conditions. The material can be applied to the hearth by ramming or in the form of bricks, which have been compressed under high pressure.

In operation the hearth can be damaged:

(1) Mechanically,
(2) Chemically,
(3) By the arc,
(4) By water ingress.

Mechanical damage can be avoided if heavy scrap is not laid directly on the bottom of the charge basket, which is lined with a protective layer of swarf or chips; this prevents the heavy scrap from falling directly on to the hearth lining. Skulls and bundled scrap must not be charged with a basket but laid on the hearth by the furnace crane.

Refractory material wear through chemical action has been shown in tests to result from loss of high temperature strength and changes in composition through penetration. Alumina penetrates deepest into the bricks and is responsible for spalling. Lime on the hearth and sufficient basicity of the slag offer a certain amount of protection. Hearth-damage by the arc can be avoided by ensuring that the electrodes are not lowered past limit marks, so that the gap between the electrode-tips and the hearth cannot become too small. If the gap is too small, deep holes can be burnt in the lining before an adequate liquid depth has accumulated.

Water-ingress on a dolomite hearth causes damage which is not easily rectified and which leads eventually to destruction. The burnt dolomite tends to crumble (in the same way as burnt lime) due to hydration from moisture-ingress. Patch-repair of such damage is not recommended.

Hearth maintenance

Minor hearth-damage found after tapping can be patched by applying dolomite or coal tar–dolomite mistures. But holes burned by electrodes advancing too far must be carefully repaired. Any steel left in the holes must be blown out with oxygen or removed by scraping; then the holes are filled with patching compound (e.g. pitch and dolomite). The patches are covered with sheet steel to ensure complete sintering and to avoid further damage. The patching compound sinters hard when the first charge melts down.

5.5.2.3 Roof-wear

Silica-linings are attacked by slag mainly when basic dust settles on the roof-face. Highly reflective slag (refractory slag) can lead to melting by radiant heat alone.

Basic roof-bricks are damaged less by slagging than by flaking,[15] particularly magnesite bricks. In service, magnesite bricks expand by 1–2.5%, causing the roof to buckle. Attempts to deal with this expansion have not been successful so far. Areas subject to heavy wear are known for particular furnace types, and repairs are decided on economic grounds. Most patching repairs are carried out on the centre of the roof.

5.6 TECHNOLOGY AND METALLURGY IN SCRAP CHARGING

Rüdiger Heinke, Siegen, and Rudolf Plessing, Kapfenberg

5.6.1 Basic Furnace-linings

Rüdiger Heinke, Siegen

As electric arc furnaces and melting technology developed, basic furnace-linings were introduced generally for melting special, high-grade, and mass-tonnage steels. Melting in acid-lined furnaces is usually confined to small units (under 10 t) with special production programmes, e.g. for steel and grey-iron castings. At first sight, acid-linings appear attractively cheap; however, this advantage soon disappears because of shorter life, metallurgical drawbacks (especially in dephosphorizing and desulphurizing), and the long time needed to repair the refractories between charges.

The possibility of operating arc furnaces, no longer just as refining units for high-grade steels but as melting machines with only essential refining, has given rise to the following requirements:

(1) Short melt-down times.
(2) Bringing forward metallurgical reactions and effecting them during melting as far as possible, leaving the remaining metallurgical work to supplementary units.

(3) Reduction of auxiliary time, e.g. charging, patching, and electrode-change and assembly.

Short tap-to-tap times not only increase production but also reduce specific energy-costs and electrode-consumption due to lateral burning — to mention just two important factors. The three requirements mentioned above are valid generally, irrespective of whether the arc furnace is used for making high-tonnage or high-grade steels.

The process variants for high-tonnage steels are first distinguished by the choice of charging materials. In the *scrap process* the metallic charge comprises recycled and newly bought scrap. The piece-size and density of the scrap and the way it is loaded into the baskets can appreciably affect the result of the melt.

The sponge-iron process permits not only 100% pre-reduced material to be used, but also large proportions of scrap. Charging can be done continuously or in baskets. The charge proportions[1] (see section 5.7) are determined by the furnace data and by the degree of metallization, slag components, and the carbon content of the sponge iron.

Scrap and pig iron are used in several variants as charge materials for electric furnaces. The pig iron can be added either in liquid or solid form.[2–4] Alloy steels and especially high-alloy steels are melted in different process variants.

Composition melting This variant is always advantageous or necessary if the specified analysis severely limits those elements which arrive in the furnace in the charge and which either cannot be removed at all or can only be eliminated with difficulty by metallurgical means during melting. For example, in the case of phosphorus the problem always arises if the specified analysis also requires a high chromium or manganese content, since here dephosphorizing is not possible because of the high affinity of these elements for oxygen.

The concept of composition melting originates from the original practice of first producing a soft, low-phosphorus oxidizing slag, then adding the alloy elements for the specified steel. Special dephosphorizing can be avoided if low-phosphorus scrap, e.g. packeted car-body steel, is available. This method can be adopted for all steels, but the higher the costs of longer melt-down times and of more expensive clean scrap and alloying elements have to be accepted.

Remelting The increasing raw-material shortage is causing alloy prices to rise faster than alloyed scrap prices. Efforts are therefore being made — as far as scrap supplies allow — to obtain specified alloy compositions mainly by remelting alloy-steel scrap. The same is also true for elements such as chromium and vanadium, which are easy to oxidize although they require costly reducers. This process gives a good yield, especially for higher carbon contents, but it does not preclude oxidation to lower carbon contents by using gaseous oxygen. Quality need not be a problem, even with 100% remelting of scrap, because today steel-purity can be achieved by ladle-treatment on its own.

Duplex process This process is currently in recession because of the increasing trend to carry out metallurgical work in the ladle. Fundamentally, the process consists of selecting the most suitable combination of units for a given melting job, e.g. the converter for oxidizing and dephosphorizing (because of the high oxidization rates possible and the favourable slag effects due to intense mixing of slag and steel under oxidizing conditions), followed by refining in the electric furnace under a reducing slag. However, arrangement of the various furnace units always takes time and causes delays. Moreover, the duplex process requires two or three different melting processes to be available. Nevertheless, if the duplex concept is taken further, so that, for instance, the ladle in the vacuum vessel acts as an independent melting unit in vacuum–oxidation-decarburization (VOD) or the converter does so in argon–oxygen-decarburization (AOD), then the duplex process becomes very important for high-alloy steels, especially stainless and acid- and heat-resisting alloys.

5.6.1.1 Charging techniques

As a rule, baskets are used to place scrap in the furnace. Depending on the nature of the scrap, several baskets may be required for one charge. There are two types of basket: the lamella basket and the clam-shell basket. In the lamella basket, shutter-segments open suddenly and the charge falls into the furnace in about the same layered order as it was packed in the basket. It is an advantage here that this basket can be sunk deep into the furnace, so that the drop is short and the hearth is not damaged by chunks of scrap. The furnace space is filled efficiently because the scrap does not form a conical heap. The main drawback is that the basket does not shut tightly enough to convey particulate material (e.g. carbon, lime or granules) to the furnace without spillage. Additionally, this type of basket cannot be used to level out scrap protruding from the melt. Because the basket opens suddenly, with a backward jerk, there is a possibility that scrap will jam in the basket and lead to damage of water-cooled furnace-elements when the basket is raised again. There are a number of reliable shutter-mechanism designs as well as one with a hemp rope which is burnt through above the furnace.

The clam-shell basket has recently gained popularity. The advantages are that it can be closed tightly; it can be used to level out protruding scrap; it is compact and needs little repair; it can be stood on level ground, and it is easy to handle because there is no manual closure at the bottom as there is on the lamella basket. It seldom hits furnace-walls and is well suited for use where there are water-cooled elements. The opening-height for this type of basket is greater because of the space required by the shutter-shells; this can be a disadvantage, since it is not easy to load heavy scrap directly on the bottom of the basket without the support of ligher scrap. After opening, the original layered packing of the basket does not persist as well as with a lamella basket, and a small conical heap of scrap can form. The disadvantage can be reduced by using an underlay of sheet steel.

In most cases the baskets are loaded according to a definite scheme, e.g. heavy scrap is usually placed on the bottom of the basket to avoid damage to the electrodes during melting caused by heavy pieces of scrap sliding against them. Light scrap in the upper part of the basket allows the electrodes to penetrate quickly into the scrap after the arc is struck, and thus the walls and roof are shielded from arc-radiation. In any case, the first basket must be loaded in such a way that, once the electrodes have melted through, there will be an adequate liquid heel to protect the hearth from arc-erosion. In some steelworks it is the practice to leave a liquid heel of metal in the furnace after tapping; so the problem recedes into the background. Sponge iron can also be charged using baskets.

When a lamella basket is used, additions such as carbon and lime are usually charged from special containers hung on the basket; with shell baskets the shutter closes tightly enough to allow these materials to be charged in the basket. Such materials can also be fed to the melt via pneumatic systems; these have the advantage of optimum timing. Finally, a charging machine may be used.

As part of the endeavour for smooth, uninterrupted production, it is also possible to feed electric furnaces continuously with selected scrap such as shredded scrap or stamping waste. However, continuous feeding of pre-reduced materials has first to be tested and introduced formally (see section 5.7).

Another combined charging method can be used for non-alloy steels; it consists of feeding in scrap together with liquid or granulated pig iron. Liquid addition considerably accelerates the melting-down process, but it is only practicable when carbon content can be adjusted using the normal oxidation process; otherwise, the reduced melt-down time has to be compensated by a lengthy oxidation process which is not good for the lining. The rate of continuous feed of granulated pig iron in scrap melting must be determined by the final carbon content required. With the usual high oxygen supply-rate, this method provides a continuous feed of carbon and silicon for combustion and augments the heat supply for melting without increasing iron-oxidation.

The duplex process, using other melting units such as converters in conjunction with electric furnaces, is now hardly used in this way.

5.6.1.2 Melting techniques

This section deals mainly with the melting techniques for pure scrap. The particular melting technique for sponge iron will be treated in section 5.7. The selection of melting technique for scrap depends, above all, on the following factors:

(1) Transformer power and design;
(2) Wall-lining;
(3) Quality programme;
(4) The types of scrap available.

The given transformer apparent power, secondary voltages, and reactances determine the secondary currents.[5,6] Where furnace-walls are lined only with refractory material, these quantities determine the mode of operation for melting and superheating[7] (see section 5.5.2.1). Bearing this in mind, we try to limit wear of the walls due to arc radiation as far as possible by careful furnace operation. Ottmar and his colleagues have shown[8] how to operate two different furnace designs with the help of circle diagrams (Figure 1). It is clear from this that the highest power factors, formerly most commonly used for furnaces with purely refractory wall-linings, are chosen only for that short period during which the walls are still shielded by scrap. Thus, power should be matched to actual furnace conditions, and the furnace operated as far as possible in a way that enables the melt to progress without damage to the lining due to arc effects. When making full use of apparent power, this is only possible if the entire system, comprising the transformer, high-current conductors, and electrodes, is designed for high secondary currents at low reactances. This theory results in the following plan for melting scrap:

(1) One or more starting voltage steps with relatively short arcs, so that arc-radiation on the walls and roof is reduced immediately after the arc is struck on the scrap.
(2) After the electrodes penetrate into the scrap, apply maximum active power so that the electrodes produce the largest possible craters while boring into the scrap. The furnace-lining is completely shielded by the scrap during the whole of this phase.
(3) After the electrodes produce a small liquid-metal heel on the hearth, the scrap is melted basically from underneath. The empty spaces around the electrodes grow larger but are filled in again by the collapsing scrap. The secondary voltages remain high.
(4) The time arrives when the scrap closest to the wall is melted down. The wall is attacked by radiation in hot spots, and the secondary voltage is reduced while simultaneously increasing secondary current in such a way as to keep apparent power as level as possible so that wall-wear is minimized.
(5) It takes a relatively long time, without the use of melting aids such as oxygen, until the molten cavities, which are shaped like a clover-leaf, widen enough so that they can either be filled with fresh charge or the

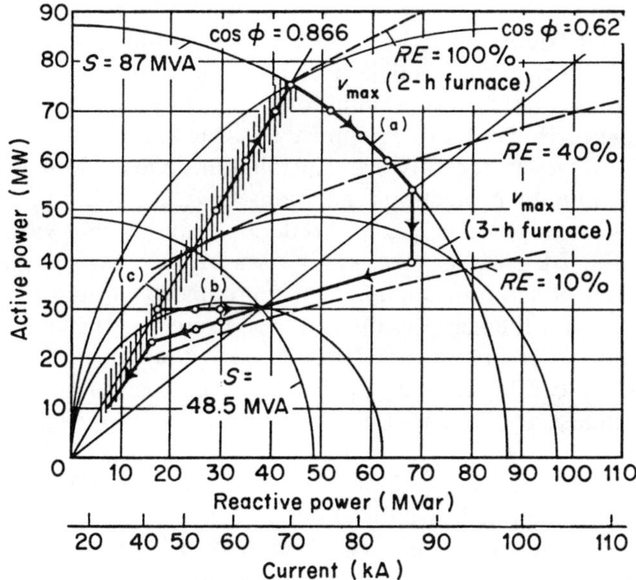

Figure 1. Different modes of furnace-operation in circle diagram for power. (a) Operating range for 2-h furnace; (b) operating range for 3-h furnace; (c) working region common until now

melt-down process ends after the last basket. Therefore it is advisable to lower the voltage while the wall is exposed to attack.

Until the beginning of the 1970s, this melting plan was universally applicable, and transformer design followed the requirements arising from planning and rebuilding electric furnaces. Then, in Japan, the development of water-cooled wall-elements, together with jet-burners, enabled furnaces to be operated at a higher average power factor cos ϕ, with less concern for refractory wear index *RE*. Furnaces with water-cooled wall-elements can be operated for melting scrap at maximum voltage throughout almost the entire melting cycle and still attain refractory wall-lives of over a hundred melts. At the same time, operation at high voltages keeps electrode current relatively low. Today, current density still determines the current limits for large furnaces. The old notion that optimum transformer operation and long furnace wall-life are incompatible no longer applies for scrap processing.

Any description of melting techniques for unalloyed and low-alloy scrap would be incomplete without mentioning additional heat sources, mainly from pre-heated scrap, oxygen as a melting aid, and fuel/oxygen burners.

Scrap preheating was first developed in Scandinavia and Austria, in the first instance to remove all snow and ice before charging the scrap into the furnace and thus to avoid explosions in the furnace vessel. As the economic advantages of this process were recognized, it was developed in time until scrap was substantially preheated to several hundred degrees Centigrade so as to reduce melting time in the furnace or to optimize electrical power utilization[9–12] (see section 10.7). However, recently many installations had to close down because of strict environmental protection regulations.

For scrap of small piece-size, e.g. shredded scrap, another variant of scrap pre-heating, the BBC/Brusa process[13] passes furnace exhaust gases over a continuous feed of scrap or sponge iron, perhaps in an additionally heated rotary furnace, and the scrap is then dropped into the furnace through the dust-extraction outlet in the roof. It is claimed that up to 40% savings of electrical energy can be effected in this way (see section 10.7). Utilization of furnace exhaust gas heat was also tried in a system developed in South Africa.[14]

Oxygen is almost generally supplied to the furnace vessel as an additional energy source for melting low-alloy and non-alloy scrap. We refer to a fourth hot spot where oxygen ignites hot scrap and melts it.[15] But first, oxygen blown in near to the electrodes can help the melting process from shortly after the start. In this way, the carbon (from additions of coal or coke and silicon and carbon from the scrap) is burnt first. Scrap-burning only starts properly after a liquid-metal heel forms and the scrap is heated sufficiently under the influence of the oxygen. In some cases the melting process is assisted by the use of consumable oxygen lances which are inserted manually through the furnace main and auxiliary doors; in this way up to about 25 Nm^3* of oxygen are used per tonne of steel.

In 1973 a significant increase in productivity for given transformer output was achieved in Japan through the use of oil–oxygen jet-burners (mostly three in number) fixed in the furnace walls between the hot spots. This arrangement is shown in Figure 2. These burners were introduced with the development of water-cooled wall-elements, which are virtually indispensable in such cases.[16–18] This method is being increasingly used not only with oil and kerosene as the fuel but also with natural gas. In the case of kerosene, fuel is used at a rate of 6–10 l/t together with twice the volume of oxygen. The thermal efficiency is said to be 60–70%, so that electricity can be saved at a rate of up to 70 kWh per tonne.

These burners direct additional energy to help in the melting-down process in the parts of the furnace where arc-radiation is least effective, thus balancing the melt. Therefore, when arc-radiation begins to affect the nearest parts of the wall, the charge in between has already melted down sufficiently to enable more scrap to be inserted. It can be seen that these methods, used together with

* Nm^3 here signifies 'normal' cubic metres — i.e. measured at 0 °C and 780 mm of Hg.

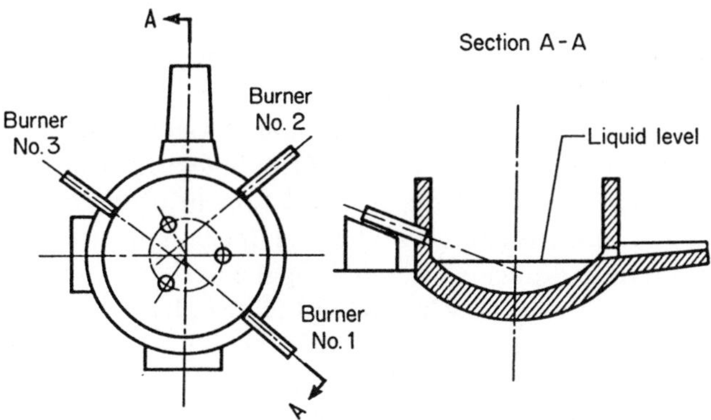

Figure 2. Possible arrangement of three auxiliary fuel–oxygen burners

water-cooled wall-elements, allow the highest voltages to be used throughout the entire melting cycle. Time to melt down scrap distant from the arcs is thus cut. This is why there are second thoughts about the paramount importance of Schwabe's radiation index, and there is a return to melting at very high average power factors. High currents combined with significantly reduced voltage are now only used in periods when the bath is flat.

Melting aids such as oxygen and extra burners are, of course, not suitable for high-alloy chrome–nickel, chromium, and manganese scrap, since the possible saving in time would be offset by uneconomical loss of chromium or manganese. It is best that no oxygen should be used for remelting low-carbon, high-alloy scrap. When large quantities of chromium alloy, high in silicon and carbon, are charged, oxygen-refining can begin even before all the scrap has melted down, since the silicon burns off first, causing a rapid temperature-rise.

5.6.1.3 Metallurgy

In the course of the last few years, electric arc furnaces have been relieved of much metallurgical work. This has been brought about particularly because of inert gas-treatment in the ladle, vacuum installations, and the possibility of desulphurization outside the furnace in conjunction with basic-lined ladles and post-treatments such as AOD (argon–oxygen-decarburization) and VOD (vacuum–oxygen-decarburization) for rust-, acid-, and heat-resistant steels (see Chapter 7).

Non-alloy and medium-alloy steels Definite oxidizing and refining times are a special characteristic of the 'classic' mode of producing non-alloy to medium-alloy steels. Gaseous oxygen or iron ore are added to create favourable dephosphorizing conditions, i.e. lime slag, rich in iron oxides, which is reactive at low temperatures. A high proportion of the high-phosphorus slag is run off right up to the end of the melting process. This follows the actual oxidation to adjust carbon content; its metallurgical significance is that while CO evolves in the steel, non-metallic suspensions are removed together with hydrogen and nitrogen, which diffuse into the rising CO bubbles. Iron ore, mill scale, or iron-ore pellets can be added during oxidation in order to lower the slag viscosity. Where transformer power is limited, the boiling process is very useful for raising temperature quickly after melting if gaseous oxygen is lanced in. In most cases, desulphurization achieved in the gas phase[19,20] under oxidation conditions in melting and refining with basic slag is not sufficient to enable the charge to be tapped after oxidation and adjustment to the required composition. Furthermore, the degree of purity and deoxidation reached at this time will hardly be adequate for the quality specified. Therefore, in classic operations, the iron-oxide-rich slag is removed by deslagging (if it has not already been run off) and a refining slag is added. The oxygen content of the melt must be removed by appropriate deoxidizing means. For harder steels, the first step is to rise carbon content; this is helped by high temperature and a slag-free melt. Manganese, carbon, silicon, and aluminium effect further oxygen reduction in the sense of predeoxidation. In this way, the specified analysis can be achieved with certainty, after tapping, for example. Whilst carbon forms gaseous oxides which rise quickly through the melt, other oxides precipitate out in liquid or solid form and take much longer to ascend through the melt. Further deoxidation takes place via the refining slag. This consists first of lime, silicon, and fluorspar, to which are added reducers such as carbon, ferro-silicon, silicon carbide, calcium carbide, calcium, silicon, and aluminium. No carbon-bearing materials are added for low-carbon steels in order to avoid possible recarburization. Such slags are termed 'white'. On the other hand, hard steels can be refined with a 'grey' slag containing CaC_2, lime, fluorspar, and carbon. Reducers are only added when the lime–silica–fluorospar slag liquefies. Reducers should be in small lumps, as newly burnt as possible, and should be stored before use in airtight containers so as to avoid absorbing moisture and thereby taking a lot of hydrogen into the steel. If possible, actual refining is carried out in a closed furnace, with the dust-extraction system throttled right back or turned off. Since reduction of metallic oxides rising through the melt and also desulphurizing are purely diffusion processes taking an appreciable time, attempts have been made to speed up the exchange processes. This can be done by building in stirring coils, by stirring with rabbles, or by blowing desulphurizers in under the melt surface using neutral carrier-gases.

After alloying, followed by diffusion deoxidation, the melt is finally deoxidized (blocked) using silicon and aluminium, for instance. Only comparatively small amounts of deoxidation products are then formed. In special cases, aluminium is replaced as a reducer by calcium–silicon, titanium, zirconium, or other materials. 'Classic' furnace metallurgy nowadays has little importance for larger production units. (Further details, together with examples, can be found in the second edition of this book.)

In comparison, the cost of metallurgy in fast, modern arc-furnace melting processes is very low. Here, steels of most qualities, with low- to medium-sulphur requirements, are melted in quick succession; the main aims are to conclude metallurgical processing by the end of melting, to lower carbon content below the specified level, and to reach tapping temperature. Special aspects of sponge iron metallurgy will not be discussed here but will be treated in section 5.7. A special fast metallurgical process has been developed for scrap charges with the object of reducing the steel sulphur content in the furnace below 0.025%. Since the introduction of the intensive use of oxygen as a melting aid, reduction of phosphorus to levels below 0.15% in the presence of basic lime slags no longer presents any metallurgical problem. The situation is different for sulphur which is already reduced to a significant extent in the gas phase under oxidizing conditions with a basic slag;[19,20] however, resulting levels are still too high for most specifications. With this in mind, a process was developed in 1960 whereby slag is made from fine lime, and dissolution takes place rapidly over the relatively large surface area of the fine lime in the presence of fluxes such as FeO.[21,22] The effectiveness of metallurgical slags depends on good reactivity and appropriate composition and, to a great extent, on reaction time and the chemical disequilibrium between the slag and the steel. If care is taken that a slag with a particular sulphur-acceptance is added little by little, and each such addition is run off after a brief reaction time, then an effect somewhat like fractional distillation will be achieved. Slags are exchanged in such a way that new lime is blown in at almost the same time as old slag is run off. This partial desulphurization starts about 10 min before the end of initial melting, with the addition of the fine lime, so that some liquid slag is already present when refining starts. Furnaces with short tapping cycles (e.g. under 2 h), into which oxygen is fed intensively during melting via lances and/or extra burners, have very short refining times (about 10 min) after the scrap has melted down; this is because levels of carbon, silicon, and other elements are appreciably reduced during melting down. The oxidizing process does assist the rise in temperature and affords limited protection to the refractory lining; however, with high-power transformers its object is no longer to provide the temperature-rise necessary for tapping, but mainly to guarantee homogeneous melt-composition and temperature and to lower nitrogen and hydrogen contents. Now, with the advent of ladle metallurgy, oxidation has lost its importance for washing out non-metallic suspensions.

One of the characteristic features of rapid electric steel-melting is that only minimal, if any, predeoxidion of the melt with FeMn or SiMn takes place after oxidation, and tapping follows immediately if the temperature is right. Tapping low down principally avoids slag pouring out with the melt. In this way deoxidizers added in the ladle (e.g. ferro-silicon and aluminium) can be very effective. The furnace slag is either kept back in the furnace or it is run off into the ladle after the steel. There it is either hardened to render it less reactive, or it is tipped out before ladle-refining begins.

High-alloy steels If we wish to describe the present-day technique for melting high-alloy steels, we must first divide it into two groups:

(1) Highly alloyed Cr-, CrNi-, and CrNiMo steels with lower carbon contents under about 0.3%.
(2) Highly-alloyed high-speed steels, tool steels, heat- and wear-resisting steels with high carbon contents and very expensive alloying elements.

The process for melting group 1 steels in electric furnaces changed fundamentally with the addition of special converters (AOD = argon–oxygen-decarburization and CLU = Creusot–Loire–Uddeholm process) and vacuum-refining units (VOD = vacuum–oxygen-decarburization). The furnace has thereby been relieved of almost all metallurgical work, and is now again employed for straight melting work. All the processes, whose metallurgy will be described elsewhere, are aimed at slagging the chromium and then recovering as much chromium as possible from the slag. With conventional decarburization down to below 0.1%, using gaseous oxygen in the furnace, the chromium content of the slag could be several per cent.

The post-treatment can handle a chrome-rich melt with almost any carbon and silicon content, and enable the furnace to be operated without refining unless homogenizing or low transformer output are overwhelmingly important. Thus it is only necessary to melt the charge and raise the temperature high enough so that the molten steel may be safely transferred into a ladle for post-treatment.

Vacuum-refining usually takes place in a casting ladle where the space above the melt is small compared with that in a converter; this limits the carbon content that can be processed. If very large quantities of charge chrome containing 2–4% silicon and 5–8% carbon are used, we have preliminarily to refine the melt in the furnace and to reduce the slag according to the carbon content, so that the refining reaction can be carried out under control in the ladle later on. Apart from this, no other metallurgical work need be done in the furnace with the vacuum-refining process. Both processes prove themselves in different ways. The electric furnace–converter sequence relieves the furnace of much work while vacuum refining is particularly advantageous for producing low-carbon steels (under 0.03%) because there is far less chrome in the slag.

The second group of high-alloy steels contains primarily those whose main alloying elements are so expensive that it is particularly rewarding to optimize their recovery. Among these we must especially note molybdenum, tungsten, and vanadium, which not only are added later to the melt, but are also fed in as recycled scrap. In other cases the specified contents are so high that tungsten and vanadium are inserted with the charge as ferro-alloys so they can have more time to dissolve. In particular, ferro-tungsten and ferro-molybdenum dissolve slowly into an iron melt because of their high melting-points and lie for a long time on the hearth at the bottom of the melt because of their high densities. These dissolve best where the carbon content is high enough so that the refining phase is lively or where there is a stirring coil. Vanadium, as ferro-vanadium, dissolves relatively easily in the melt, but only little gets into the slag, and it is thus usually added later into the deoxygenated melt after refining. In all melts with high-value alloying elements, the basic refining slag must be carefully reduced before deslagging. As well as in ferro-alloys, molybdenum and tungsten also come in molybdate, tungstenate, and vanadium salts or their oxides, or they may be recovered from slag. These can also be economical in view of the high values involved.

The final analysis can be adjusted with almost no loss of material, even for tungsten, molybdenum, and vanadium, by sufficient argon-flushing in the ladle. Metal losses can hardly be avoided in arc-furnace melting because of evaporation in the arc plasma. Enveloping the arc with slag can give some protection, for instance during refining; nevertheless, appreciable amounts of manganese, chromium, molybdenum, vanadium, cobalt, and nickel, according to the melt contents, evaporate as oxides and find their way out through the furnace dust-extraction system.

Another way that heavy metals are lost is by absorption into the furnace hearth. It makes sense to avoid this loss and also the risk of inaccurate compositions by arranging the melting programme in a way that permits these metals to migrate gradually back from the hearth into subsequent melts.

Typical post-treatment processes such as intert-gas flushing, desulphurization, and vacuum treatment for chromium and chromium–nickel steels (AOD, VOD, and CLU) will be described in another chapter.

5.6.1.4 Furnace maintenance

Repairs to the refractory lining or replacement of whole furnace-components or parts, such as the hearth, wall, or roof, all take time and profoundly effect arc-furnace productivity. Naturally, the extent and frequency of repairs to the refractory lining depend on the quality of the materials used and on the effects of the particular furnace design and operation. The principles of furnace-lining design and selection of the relevant materials are dealt with in other sections; we shall deal here with maintenance and the replacement of parts.

For fast furnaces, hearth-maintenance costs are low because times spent at high temperature are short, and the hearth does not soften. More serious damage can be caused by using the wrong melting programme or because of unsuitable transformer design. Small holes are patched with dolomite or a magnesite patching compound. Any steel residue present can either be blown out with oxygen or stiffened (with dolomite, for example) and removed before the filler is applied. Furnaces used for high-alloy melting programmes and those with long refining processes frequently suffer more extensive hearth-damage which must be made good. Here, too, the hearth must be free of any residual steel before large quantities of filler, e.g. dolomite, magnesite, or magnesite–chrome ore are tipped, shovelled, or sprayed on. The layers must not be too thick, so as to ensure that they will definitely sinter right through and bake into the hearth. In order to avoid impact-damage, the repaired surface is suitably covered with steel sheets before charging.

The hearth can grow because of frequent patching close to the wall or because of inexpert limestone-charging and residues of highly basic slag. Remedies recommended are application of a lighter charge and the use of an extended refining period at higher temperatures. Flux or ferro-silicon should not be thrown on to the hearth, because these materials can affect the whole hearth adversely and often lead to the formation of a conical crater in the centre.

For furnaces with exchangeable vessels it is possible to replace the entire hearth in a few hours and then to preserve the undamaged areas of the old hearth and to completely refurbish it. The mechanical cost is high, and the furnace-crane capacity and availability have to be high; however, theneed for running patch-repairs is much reduced, and the saving on refractory material is appreciable.

The furnace wall, which is subject to most attack because of slag and arc radiation, requires most patching. It is important that this work is carried out quickly after tapping. With smaller furnaces, filler can be applied with shovels, but spraying machines and centrifugal 'slingers' are widely used for larger furnaces.

Spraying machines have advantages for local wear, since with this method the material can be applied accurately to the damaged area without the furnace roof having to be opened, and radiation losses during patching remain low. Two machines can be used simultaneously on furnaces with both main and auxiliary doors. The regions either side of the doors can be reached with specially shaped spray tubes. It is important that the wettability, bonding, and grain-size of the material to be sprayed are well adapted.

If wall-wear is even throughout the slag zone, then the use of a slinger is preferred. Although it is true that the furnace roof must be opened for this purpose, a great deal of patching material can be applied around the whole circumference at the desired height in the shortest possible time. 'Segment-slingers' enable slinging also to be used on particular sections of the wall. The radiation losses incurred by opening the roof must always be balanced against the advantages of using a slinger.

The cost of patching has been drastically reduced since the introduction of water-cooled wall-elements; this is particularly true when working with sponge iron, where the walls are mostly unprotected. Refractory wear is then concentrated around the slag line and the hot spots. As with hearths, complete walls can be exchanged by lifting the wall off the hearth and replacing it with a pre-lined upper part, with or without water-cooling elements.

Since furnace roofs in service now generally incorporate exhaust-gas extractors, the roof needs no special servicing in operation apart from pneumatic dusting. Depending on the type of furnace, parts of the roof often wear particularly heavily; it can therefore often be economical to carry out interim roof-repairs such as retamping or renewing brickwork at the electrode ports. The use of additional burners has altered the furnace atmosphere and the thermal load on the roof to such a degree that the durability of high-clay refractory bricks is now much reduced. Therefore, basic roof-linings of magnesite–chrome or dolomite bricks, partly of suspended construction, have now come into use — as also have water-cooled roofs.

Furnace maintenance also concerns the repair and replacement of the sill and tapping launder. The furnace door sill is usually laid on a fragment of broken electrode, which also forms the front edge of the sill. Slag or steel do not stick or bake on to the electrode graphite because of the poor wettability of the latter. It is customary to have a top layer of dolomite towards the furnace interior; this forms a barrier during energetic boiling and can easily be removed if necessary, e.g. for deslagging. The formation of deeper tunnels must be avoided; otherwise steel may run out unnoticed beneath the slag. The barrier or dam must be cleaned and built up again after each melting cycle.

A properly lined tapping launder can be used for several melting cycles in succession without patching. It is also advisable to ensure there is a sharp edge on the end of the launder, so steel pours out smoothly during tapping. Run-blocks are a new development. These are monolithic, silicon-carbide-doped launders which are made to the appropriate size; they can be changed quickly. On account of the silicon carbide, these run-blocks have a low wettability, so that steel accretions hardly ever form in the launder. The launder is usually serviced during melt-down, so these exchangeable launders save melting personnel from much exposure to noise.

5.6.2 Acid Furnace-linings

Rudolf Plessing, Kapfenberg

Of all the steel produced in arc furnaces, only a very small proportion is still melted on acid hearths. This proportion becomes smaller every year, because the advantages of 'acid' melting diminish all the time with continuing improvement in melting and casting techniques for 'basic' arc furnace operation.

The advantages claimed for melting on acid hearths are low conversion costs, shorter melting cycles, and less flaking; however, these are only valid in comparison with 'basic' melting using the two-slag method. These advantages amount to very little compared with those of a basic arc furnace working on the single-slag method.

'Acid' melting can be advantageous for certain types of steel because smaller amounts of chromium, vanadium, and tungsten are lost by burning. These elements burn to acidic oxides which cause less burning of acid hearths than with basic linings during refining (if the oxide slag is removed after refining).

The lower viscosity and better castability of acid steel are a particular advantage in the production of steel castings, and this is why acid-melted arc-furnace steel is still mostly used for this purpose today.[1]

Acid electric furnaces have the following disadvantages. Not all types of steel can be economically produced. High-carbon steels which also have a low silicon content and steels with large quantities of alloying elements that oxidize easily are both difficult to make. Low-carbon steels containing more than 2% manganese and soft, high-alloy chrome steels are virtually impossible to make because of heavy silicon-reduction and severe attack on the furnace lining. Furthermore, phosphorus and sulphur cannot be removed to nearly the same extent with acid as with basic processes; therefore quality acid steels must be made from relatively clean scrap. The scrap must contain only a little rust, since otherwise the hearth will be heavily attacked. The mechanical properties of acid steel, especially tensile and notch-strength across the grain, are inferior to those of basic steels because of the higher proportion of elongated silicon slag inclusions present. Hardening across a large section is better, but the high-temperature limits are tighter than for basic steel. These characteristics, which are typical for acid steel, can be improved, however, by appropriate deoxidation.

Works which generate a lot of their own scrap from basic electric furnaces or from the LD (Linz–Donawitz) process can find the acid process more economical, and there may also be qualitative advantages for certain steels for which silicate slags are preferred to alumina slags.

5.6.2.1 Furnace-lining

The lining of acid arc furnaces generally presents fewer problems than for basic-lined furnaces, despite the fact that the acid coating, especially of the hearth, takes part in the reaction (crucible reaction). This may partly be due to the relatively small capacity of acid furnaces, which is rarely more than 10 tonnes.

The *hearth-lining* consists of an insulating layer of fireclay bricks, usually 200–250 mm thick, then one or two flat courses and a curved course of ceramically bonded silica bricks (containing at least 95% SiO_2) laid dry on top. After the wall-lining is completed, the hearth is covered with tamped pure, dry quartz sand (grain-size under 2 mm, with at least 99% SiO_2 content). This layer is 150 to 200 mm thick. Sinter materials (mixtures of small quantities of clay, boric acid, or slag clay) are now hardly ever used.

Before the first charge, the bottom is covered with steel sheets to protect it from damage when scrap is dropped in. The rammed floor is sintered hard during the subsequent melt-down, which must be carried out slowly because of the high thermal expansion of the acid lining. The old practice of preheating the hearth before the first melt ha proved to be unnecessary.

Silica bricks are always used for the *wall-lining*; expansion gaps are provided at regular intervals to take up the 2–2.5% expansion during heating up. A 15–30 mm gap, filled with fine quartz sand, between the brickwork and the furnace shell, serves the same purpose. Acid cement is only used to bind the silica bricks in the vicinity of the doors and the tapping launder. The wall thickness varies from 300 to 450 mm, depending on the furnace size.

The arched *roof* of an acid arc furnace always consists of silica bricks and is no different from acid roofs on basic arc furnaces (see section 5.5.1.3).

5.6.2.2 Melting technique and metallurgy

Generally, acid electric arc furnaces are used only for remelting recycled scrap — mostly from the same works. This can be done with or without a boiling reaction. We will first describe the boiling method used for melting high-grade steels in preference to remelting without boiling.

We start off in such a way that about 0.20% carbon can be refined down. It takes about an hour to reach the desired carbon level, since decarbonization takes longer than in basic arc furnaces. During melt-down, oxidation of the charge and partial oxidation of iron, manganese, and silicon lead to the formation of a light, liquid slag of SiO_2, FeO, and MnO, which is very aggressive to the lining. This attack on the lining only stops when the slag has become thoroughly saturated with about 50% silica.

After the end of the melt-down, the slag generally contains 20–30% FeO, about 20% MnO, and a few per cent of CaO, Al_2O_3, and Cr_2O_3 (for chrome-alloy steels) as well as SiO_2.

Depending on the carbon content of the melt, refining starts at temperatures of 1570–1600°C. During refining, iron ore is added several times, up to 1% of the charge weight, and oxygen is blown in if necessary. The slag becomes more liquid if small amounts of lime are added at the same time; this enhances the refining effect of the slag by intensifying the activity of the iron (II) oxide.

About 20 min before the specified carbon level is reached, quartz sand is added to the slag to accelerate its silica saturation. If at the same time the melt temperature is raised to 90–120°C above the liquidus temperature of the specified steel, and, using the following endothermic crucible reaction:

$$2[C] + (SiO_2) \rightarrow [Si] + 2\{CO\}$$

an attempt is then made to increase the silicon content in the melt up to the lower analysis-level. This results in a considerable reduction in the amounts of heavy metal oxides in the slag and is advantageous for the output of iron and alloys.[2,3]

Alloys such as ferro-molybdenum and ferro-tungsten must be added to the melt at least 20 min before tapping, and ferro-chrome at least 10 min before tapping, because they do not dissolve easily. Small quantities of ferro-vanadium, ferro-manganese, silicon, and deoxidizing aluminium are added in the ladle, together with the required carburizing media.

The 'art' of acid melting consisted of extensive monitoring of the very temperature-dependent bath reaction for the control of the reduction of silicon from the slag and the lining without exceeding the specified silicon content of the steel. The progress of silicon reduction can be assessed and controlled by frequently sampling and observing the unkilled steel as it hardens and watching the change in slag viscosity and colour (the slag becomes more viscous and brighter as the amount of silicon in the melt increases). The bath reaction, i.e. the silicon reduction, is assisted by increasing carbon content and raising the melt temperature. The reduction phase in the furnace can be clearly recognized outside, because white smoke is emitted from the furnace. It is important to tap the melt at the right moment, since delays can easily lead to the permissible silicon level being exceeded.

When tilting the furnace, care must be taken to avoid slag and steel flowing simultaneously into the ladle, since any free oxide could cause an undesirable reaction with the liquid steel. This is achieved in practice by holding the slag back with rabbles or by suitably bricking up the tapping outlet so that, when the furnace is tilted quickly only steel pours into the ladle at first.

The total quantity of alloy burnt off in melting depends, among other things, on what proportion of the alloy input is charged with the scrap and how much is added in ferro-alloy during finishing. If about 40% of the alloy is fed through the scrap, then we get the total alloy losses shown in Table 1. It should be noted that chromium, tungsten, and vanadium are subject to significantly greater melting loss when producing wrought iron directly in *basic* arc furnaces than in refining in acid arc furnaces. But, with a low-phosphorus charge, significantly better alloy output can be achieved in a basic arc furnace through oxygen refining and subsequent slag reduction than can be attained in an acid arc furnace.

Acid-melting with a boiling reaction has the disadvantages of more severe lining attack and higher melting losses; however, acid furnaces can also be used for *remelting without boiling*. This process is used to advantage for heavy, rust-free scrap charges or if the steel to be melted is so highly alloyed that no boiling reaction occurs. The process is very similar to that for medium-frequency acid induction furnaces. If the charge is free of oxides the melting losses are low, except for manganese. As there is no way of removing hydrogen from the melt in this process, the scrap charged must be free from rust, and subsequent additions must be dried.

Table 1.

Alloy	Absolute melting loss	Remarks
Manganese	Up to 30%	For hard tool steels containing up to 0.25% Mn the loss is almost zero. For steels with 0.40–0.70% C and 0.40–0.80% Mn, the loss is about 15%. For soft steels containing up to 2% Mn the loss is about 30%
Chrome	Up to 20%	For steels up to 2.5% Cr the loss is about 15%; for higher Cr content the loss is up to 20%
Tungsten	Up to 12%	For low-alloy, cold-working tool steels containing up to 1.5% W, the loss is 5 to 10%
Vanadium	Up to 15%	For V content below 0.5%

F. Dobrowsky[4] reported on the conversion of a 7.5-t basic arc furnace to acid operation. This was done due to lower melting costs for the works' own clean scrap recycled from the LD process. Lower sulphur and phosphorus levels could also be achieved by using slag containing relatively high levels of CaO and MnO:

10–16% CaO, 65–70% SiO_2, 0.5–1.5% Al_2O_3,
2–6% FeO, 3–8% MnO, and 0.5–1.0% Fe_2O_3

Relatively low levels of nitrogen (about 0.005% average) are caused by using a low nitrogen charge and 'acid' remelting.

The melting process described involved remelting without refining, and the melting time is about one third shorter than when using a basic furnace. The mechanical properties quoted by Dobrowsky are surprisingly good; both longitudinal and transverse values are almost equal to those for 'basic' steels of the same chemical composition.

5.6.2.3 Furnace-maintenance

Wear of the hearth and wall-lining is heavily dependent on the melting programme. Wear is caused mainly by basic oxides from the charge and the refining agents, by SiO_2 dissolving in the acid slag, and by silica reduced from the hearth-lining by the melt.

After each melt the *furnace-wall* must be made good at the slag level by throwing on clean quartz sand.

Hearth-repairs should only be undertaken at longer intervals, after 100 to 150-mm wear has occurred, since experience shows that thick layers of silica sinter harder than thinner ones. A mixture of broken silica bricks, about the size of a fist, with about an equal quantity of clean quartz sand, applied to the hot hearth, has proved itself well in service. Before charging scrap, this layer must be densified by ramming with iron bars.

If the furnace is maintained as described, the hearth will have a life of about 1000 melts (up to 2000 for cast iron), and the wall-lining will last up to 500 melts. With an acid-furnace lining the silica-vaulted roof reaches a much longer life than in a basic furnace; roof-lining life is of the same order as that of the acid-wall lining.

5.7 TECHNOLOGY AND METALLURGY FOR SPONGE-IRON CHARGES

Heinrich Ottmar and Ulrich Siegers, Bous

Sponge-iron processing is a relatively young technology compared with using scrap to charge arc furnaces. However, extensive research and development in operation in steelworks already clearly indicate the peculiarities of the new raw material — sponge iron.

Many parameters affect the grading and application of sponge iron. These all interact in a complex manner, so that a change in one quantity results in changes to the others. The uniform consistency of sponge iron, compared with its ore basis, is an advantage for the production process and the mode of reduction. This means that, compared with scrap processing, the technological and metallurgical processes can be much easier to control by computer.

5.7.1 Properties of the Raw Materials

5.7.1.1 Physical characteristics

Sponge iron, as a raw material, is obtained from the ore by reducing its oxides without departing from the solid state. Thus it is available after reduction in about the same form and lump-size as the ore used, and it may be briquetted afterwards. It is available today in the following forms:

(1) Pellets;
(2) Lump sponge iron, in the same form as crushed ore;
(3) Fine-grain sponge iron, e.g. from iron sands; and
(4) Sponge iron briquettes.

The solid state simplifies storage and is very well suited to continuous or discontinuous transport, even in large quantities. Briquetted sponge iron can be considered as a special case for improving the transportability and utility of fine-grained sponge, while at the same time reducing reoxidation and material-loss rate and improving storability to a great extent.

After successful reduction, the physical heat content of sponge iron is usually lowered by cooling so as to avoid reoxidation. In specially equipped installations, it is possible to use this excess heat to advantage in post-treatment.

5.7.1.2 Chemical composition

Like its physical characteristics, the chemical composition of sponge iron depends most of all on the ore used. Since the solid form of the ore is retained in the reduction process, the gangue form of sponge iron cannot be lost, unlike with pig-iron production. The oxygen content of the ore is considerably reduced, a certain amount of carbon is added (depending on the process), and the sulphur content can also rise accordingly. The percentages of carbon and residual oxygen in the sponge iron, which are determined by the degree of metallization, depend heavily on the reduction process used and can vary a great deal. Both these quantities, as well as the proportion of gangue, which depend on the ore and perhaps on the bonding medium mixed with pellets, determine the possible use to which the sponge iron may be put.

To sum up, we can say that consistent form and chemical composition, depending on the ore and the reduction process, are of great importance in the successful use of sponge iron in steelmaking. The possibility of continuous charging constitutes a real step towards a continuous steel-production process.

5.7.2 Technology and Metallurgy

In processing all-scrap charges, melting down, oxidizing, and refining the steel all still constitute distinct time-periods in the melting process. With sponge iron the picture is completely different. The melting down and oxidizing phases cannot be practically separated, and the refining phase, together with adjustment of the analysis and the temperature of the melt, usually cannot be distinguished as separated in time. The reason for this different technological and metallurgical procedure for sponge iron is the residual oxygen content of the sponge iron and the carbon contained at the same time in the liquid melt during melting down, particularly in continuous processing. The combination of these elements to form carbon monoxide agitates the melt continuously. The oxygen and carbon contents must be correctly adjusted in relation to one another so that conditions are optimized. If both elements are not already present in the correct proportions in the sponge iron, then one or the other must be added to the liquid melt.

5.7.2.1 Metallization and carbon content of sponge iron

Metallization is all-important for the evaluation of sponge iron. It is defined as:

$$\eta_{met} = \frac{\text{metallic iron in the sponge}}{\text{total iron in the sponge}} \times 100\,(\%)$$

It is thus also a measure of the residual oxygen, i.e. the oxygen still combined with iron. This affects the process in two ways:

(1) By dissolving in the slag, where it raises the FeO content;
(2) In decarburization by forming CO — after the reduction of the FeO.

These relationships can easily be recognized in this example of their effect.[1]

Experience shows that slag whose basicity is about 1.5, and which contains about 15% iron, will result in steel whose final carbon content is 0.1%. This 15% iron in the slag corresponds to an oxygen content of 0.45% relative to the iron in the sponge, where the slag represents 10% of the weight of the metallic charge. The oxygen is thus released by the slag. This corresponds to 1.5% unmetallized iron in the slag. When using sponge iron with, say, 95% metallization, the following circumstances obtain: of the 5% non-metallized iron (reckoned to be FeO), 1.5% enter the slag as FeO without forming any gas. The remaining 3.5% cause gas to evolve by combining with carbon in the molten steel. The amount of carbon required for this can be calculated from stoichiometric principles.

Changes in the metallization, the amount of slag, or the iron content of the slag also alter the amount of carbon required. Given a metallization % η_{met}, an amount of slag %Sl (= proportion of slag in relation to amount of liquid), and a fixed amount of iron %(Fe) in the slag, we can use equation (1) to determine the amount of carbon required to reduce the FeO content of the sponge iron, which, in other words, will not be taken up by the slag. This part is shown in square brackets in equation (1):

$$\text{kgC} = \left[100 - \%\,\eta_{met} - \left|\frac{\%\text{Sl}}{100} \times \%(\text{Fe})\right|\right] \times 0.2227 \times 0.75 \times 10 \qquad (1)$$

Explanation of factors:

0.2227 = Conversion of the proportion of FeO, which is to be reduced into oxygen which reacts with carbon
= 16 ÷ (55.85 + 16);

0.75 = Conversion of oxygen with carbon in the ratio C:O
= 12:16;

10 = the conversion factor required if the weight of carbon C is in kilograms.

The amount of carbon determined in this way must be added to the total quantity which combines with steel in order to arrive at the specified initial content of the sponge iron. It should be borne in mind here that, according to

Vacher-Hamilton, this carbon content depends on the oxygen content of the steel, which in turn depends on the iron content %(Fe) of the slag. Sponge iron which matches the composition described in terms of metallization, carbon content, gangue quality, and the combination of the relationships just described can be termed self-sufficient. It can be used for steel-production without any other additions. Any deviation from this required the addition of oxygen, carbon, or slag formers so that the desired steel can be made.

The literature[2,3] describes the following simple relationship between metallization and carbon content in sponge-iron:

equivalent metallization of 95–7% = % metallization + 5 × (%C content)

This equation only corresponds to basic requirements for sponge iron for universal application; it does not include consideration of changes in the amount of slag, nor in the FeO content of the slag, and it does not cater for variations in the carbon content of the steel to be produced.

5.7.2.2 Gangue and amount of slag

The amount and composition of the gangue in sponge iron determine how much slag arises in the steel-production process. In the basic arc furnace, apart from the amount of gangue, the acid constituents are of greatest importance; these could necessitate the addition of basic slag-formers if the gangue alone cannot provide the slag basicity required.

The gangue content of sponge iron known today ranges from about 2% to over 25% of the total. Therefore the amount of slag arising with the required basicity could range from under 5% to about 50% of the weight of the crude steel. Large amounts of slag, in combination with high lime-consumption, high iron-losses, and high energy-consumption, can severely limit the utility of certain types of sponge-iron.

5.7.2.3 Phosphorus, sulphur, and trace elements

As a rule, sponge-iron grades used today contain only small quantities of phosphorus, sulphur, and unwanted trace elements. Therefore the slag refining to produce the required steel can be relatively simple. If necessary, the high sulphur content (up to about 0.03%) of certain grades of sponge iron can easily be reduced to the required level with the aid of modern ladle refining. With high phosphorus levels, a short dephosphorizing phase is added after melting down; this is quickly completed with simple means. Since it usually contains negligible amounts of undesirable elements such as copper and tin, sponge iron can also be used to improve the utility for steelmaking of low-grade scrap containing high levels of such elements.

5.7.2.4 Nitrogen and hydrogen content

The nitrogen contents of melts produced in arc furnaces by the scrap process are distinctly higher than those for oxygen-blown melts. This difficulty can be almost eliminated by using sponge iron. Figure 1 shows an example of the effect of adding a '100% sponge iron' melt to the liquid heel produced by a previous scrap melt.[4] The nitrogen content of the steel remaining in the furnace is lowered from 0.0064% to about 0.0025% through continuous charging and oxidizing. The nitrogen-drop can be attributed to the vigorous decarburizing reaction, i.e. the flushing effect of the carbon monoxide bubbles formed. The melt illustrated in Figure 1 has a mean decarburizing rate of 4.4 kg C/min; this corresponds to 0.44% C/h and a volume of carbon monoxide of about 8.2 Nm^3 ('normal' cubic metres) per minute. In comparison with other furnaces this gives a specific value, related to hearth area, of 16.5 kg of carbon per square metre per hour.

5.7.3 Productivity

In the sponge-iron process, melting and oxidizing take place simultaneously, so that productivity is largely governed by the charging and melting rates possible. The three quantities which determine productivity are the available active power (specific to the transformer), the heat required to melt the sponge iron (specific to the sponge iron), and the possible CO gas throughput (specific to the furnace).

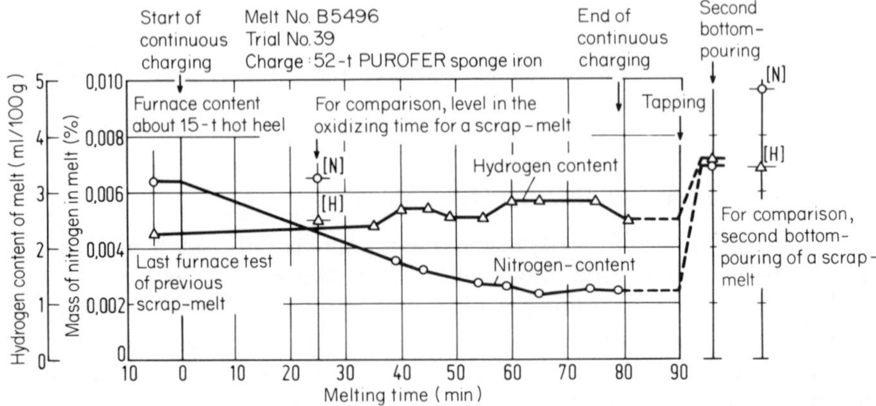

Figure 1. Change of nitrogen and hydrogen levels in a '100% hot heel' sponge-iron melt

5.7.3.1 Available active power and heat-requirement

The maximum productivity, governed by the active power available from the transformer, can easily be determined from the specific heat required per tonne of liquid steel. A particular grade of sponge iron having a high degree of metallization, low gangue content, and a specific heat-consumption of 55 kwh per tonne of liquid steel could be converted into steel at a productivity rate of 1.82 t of liquid steel per hour per MW available active power.

However, sponge iron with a specific heat consumption of, say, 900 kW per tonne of liquid steel[5] can be turned into steel only at a productivity rate of 1.11 t per hour per MW active power (see also section 5.7.4.1).

5.7.3.2 Specific carbon throughput

When assessing productivity the amount of carbon monoxide gas flowing through must not be forgotten. For oxidizing times of one hour or more, a volume of carbon monoxide gas, corresponding to a carbon throughput of 10–15 kg carbon per hour per square metre of melt surface, may be sufficient; this figure is derived from old open-hearth practice in SM furnaces. For modern high-performance arc furnaces, with oxidizing times of 10–30 min, the carbon throughput must be enough to ensure effective flushing (see section 5.2, Figure 9).

Maximum carbon throughput per square meter per hour calls for increased bath-volume and a greater height-difference between the surface of the melt and the sill. It must not be forgotten here that the reaction in which carbon and oxygen combine to form carbon monoxide is endothermic and can cause a greater demand on applied heat than is required for good steel-flushing.

The formation of CO_2 in the steel-slag system is insignificant and can thus be ignored. The greater proportion of CO_2 detected in the gas analysis could be caused by secondary-combustion air infiltrating the furnace and may attack the furnace refractory lining because of the radiation effect of the CO_2.[6]

5.7.3.3 Hot-charging

The possibility of using the latent heat from the freshly reduced sponge iron by charging it hot into the arc furnace is limited by the maximum gas throughput. Thus, sponge iron charged at a temperature of, say, 800°C will save about 150 kWh per tonne.[7,8] But, for a given metallization, it can increase performance only as far as gas throughput will permit. If the sponge iron used is highly metallized, and if the furnace charging rate cannot be increased because of limited installed transformer power, then the energy-saving effect of using hot sponge iron comes fully into its own.

5.7.3.4 Yield

Yield can easily be calculated when sponge iron is used. The slag, formed by the gangue and the added basic slag formers, contains iron oxides, the amounts of which depend on the balance between the slag and the steel melt. Charges containing good grades of sponge iron and steel can be expected to produce yields of 95–8% with final carbon contents of 0.1%.

5.7.4 Processing Costs

Processing costs incurred by the outlay for electrical energy, electrodes, and refractory linings have a substantial effect on the overall cost of steel-production. A few guide-values will now be given.

5.7.4.1 Electricity-consumption

The electrical energy consumed in steel production from sponge iron is distinctly different from that for scrap-processing. The reason for this is that there is no heat of reaction due to oxidation of the impurities and alloying elements such as silicon, manganese, aluminium, and others which are introduced with scrap, as well as partial oxidation of metallic iron.

With the sponge-iron process, it is not only necessary to provide heat for melting and superheating the iron, but also for the unreduced iron oxides and the gangue. In addition, heat is also required for reduction of the surplus, unreduced iron in the sponge which is not taken up in the slag.

Evaluation of many practical results shows that the power-consumption theoretically required for melting sponge iron is given by the following equation:

$$\Sigma \text{ kWh/t} = 396 \text{ kWh/t of liquid steel at } 1600°\text{C} + 13 \text{ kWh\% reduced FeO} + 7.5 \text{ kWh\% slag (basicity about 1.5–2)}.$$

To these calculated electricity-consumption figures we must still add the thermal losses arising from the processing time, the thermal load on the furnace vessel, and the electrical losses.

Altogether, we get a thermal efficiency for the sponge iron process of 85–95%; this is higher than that for melting scrap in an arc furnace, where the best thermal efficiency obtainable is only 80%. These good values for the sponge-iron process may arise from the prolonged carbon-oxygen reaction, peculiar to this process, in which the frothy slag screens the arcs and thus reduces direct heat-losses.[6]

5.7.4.2 Refractory lining

In contrast to the scrap process, continuous charging of sponge iron entails most of the process taking place with a liquid bath; therefore the mean temperature of the furnace is higher. It is clear that chemical infiltration of fine-grained sponge iron, particularly into the roof-lining, raises the wear of the refractory lining. This was the situation during the first few years of sponge-iron processing; however, things have changed appreciably as the result of technical experience since then. On the one hand, the problems of lining-wear have been solved through the use of water-cooled elements in the wall and roof, and on the other, distinct improvements have been made through different charging techniques and matching the charging rate with the sponge-iron analysis. Today we can assert that substitution of sponge iron for scrap need no longer mean higher expenditure on the furnace refractory lining.[6]

5.7.4.3 Electrodes

Processing highly metallized sponge iron at high charging rates requires less electrode-current density than for scrap at comparable power because the energy supply is smoother. Therefore the electrode-tips will burn off more slowly. For a given distance between the roof and the melt, the rate at which the electrode-sides burn away is determined mainly by the composition of the furnace atmosphere, the melting time, and the time the electrodes spend outside the furnace — e.g. for basket-charging and melting down. All these factors make electrode side burning less severe with sponge iron than for scrap-processing. The same is true about electrode fractures, which occur less frequently with sponge iron.

5.7.4.4 Oxygen

The fact that sponge iron contains both elements needed to create motion of the melt, namely oxygen and carbon, could mean that in the most favourable cases no extra oxygen would be needed for processing sponge iron. However, the three most important sponge-iron processes employ predetermined ranges of oxygen and carbon content, and choice within these ranges is further limited by economic constraints; it therefore follows that the resulting grades of sponge iron of different compositions require different processing techniques. This leads to very different claims and results.[2,9]

Thus, sponge iron containing more carbon than the stoichiometric level will need considerably more oxygen than for production of the same grade of steel from scrap. On the other hand, if the sponge iron contains more oxygen than the stoichiometric level, then a carbon-carrier must be added — preferably together with the scrap. The disadvantage of this procedure is that the eventual carbon level is uncertain, and therefore it is more difficult to standardize the melting programme; otherwise sponge-iron processing is distinctly better.

5.7.4.5 Lime

The addition of lime is governed by the amount of silica to be removed, the phosphorus content of the sponge iron, and the sulphur content, except that part which is removed by desulphurization in the ladle.

Most processors of sponge iron aim for a slag basicity of about 1.5 to 2. Table 1 shows the amounts of slag and lime involved; the iron melting losses resulting are also given.

Table 1. Amounts of slag, lime-additions, and iron melting losses in relationship to the iron ore and the slag analysis. (Ignition loss in the ore is taken as 1%. Fe in ore is Fe_2O_3)

Fe-ore	66	67	68	68.5	69
Gangue ore %	4.6	3.2	1.8	1.05	0.33
Gangue ES % (100% acid)	7.0	4.8	2.6	1.53	0.48
Amount of slag % fluid					
for B = 1	18.9	13.0	7.0	4.1	1.3
= 1.5	23.6	16.2	8.8	5.2	1.6
= 2.5	33.1	22.7	12.3	7.2	2.3
Amount of lime %					
for B = 1	7.0	4.8	2.6	1.5	0.48
= 1.5	10.5	7.2	3.9	2.3	0.72
= 2	17.5	12.0	6.5	3.8	1.20
Iron melting loss %					
for B = 1	3.8	2.6	1.4	0.83	0.26
= 1.5	4.7	3.2	1.8	1.03	0.32
= 2	6.6	4.5	2.5	1.45	0.45

5.7.5 Possible Charging Methods

One particular advantage of sponge iron which is in small lumps is that it can be charged continuously into the furnace. This enables the melting and refining periods to be overlapped, carbon and oxygen contents to be correctly adjusted, and a vigorous boil to be achieved. This leads to high productivity and excellent purity.

Many methods of feeding sponge iron into the furnace have been tried, ranging from simple conveyer belts through slingers to charging-cranes. Newly installed steelworks exclusively use conveyers from intermediate silos equipped with weighing systems.

The sponge iron is fed into the furnace through the roof. A system of three holes between the electrode ports and the wall has not proved itself because it weakened the roof.[10] Today the best solution seems to be a fifth hole in the roof. The sponge iron arrives virtually in the centre of the bath and is then drawn immediately into the boiling steel melt.

In Mexican steelworks part of the sponge iron is charged in baskets. The higher capacity of the basket leads to increased productivity.[11] Earlier methods described, in which sponge iron was charged on top of scrap[12] and which involved permanently solid sponge iron in the middle of the bath (iceberg),[13] were not successful.

5.8 LITERATURE REFERENCES

5.2 Arc Furnace Design and Guideline Publications

1. Siegers, U., Die technisch-wirtschaftlich optimale Abstimmung der Bau und Betriebsbedingungen bei der Erzeugung von Massenstahl im Elektrolichtbogenofen auf Schrottbasis. Aachen 1978 (Dr-Ing.-Diss. Techn. Hoch-schule Aachen.)
2. Ottmar, H., A. Oerter, R. Assenmacher and P. Goldstein, The high-capacity electric arc furnace (German). *Stahl u. Eisen* **89** (1969), 466–71.
3. Eßmann, H., Überlegungen zur Auslegung und Planung von Hochleistungs-Lichtbogenöfen unter Berücksichtigung neuerer Betriebserfahrungen. *Nachr. Eisenhüttenind*. GHH Sterkrade 1975, No. 1, 46–56.
4. Robinson, C. G., Verbal communication, Northwestern Steel and Wire, Sterling/Illinois, USA, 1975.
5. Schwabe, W. E., A UHP steelmaking arc furnace with pellet charge and an 11.6 m diameter furnace of low reactance. In *Congrés Union Internationale d'Electrothermie*, Moscow 1977. Report No. 4/01.
6. Bowman, B., Der moderne Hochleistungslichtbogenofen — Betrieb und wichtige Konstruktionsmerkmale. *Klepzig Fachber.* **82** (1974), 128–32.
7. *Anlagendaten von Lichtbogenöfen.* Verein Deutscher Eisenhüttenleute. Düsseldorf 1971. (Fachausschußbericht des Vereins Deutscher Eisenhüttenleute. No. 2006.)
8. Ottmar, H., and D. Ameling, The high-powered electric-arc furnace for steelmaking — definition and limits of productivity (German). *Stahl u. Eisen* **94** (1974), 125–32.
9. Ottmar, H., A. Oerter and D. Ameling, Der Zusammenhang zwischen den elektrotechnischen und wärmetechnischen Grundlagen bei Hochleistungslichtbogenöfen. *Radex-Rdsch*. 1973, 519–27.
10. Vogel, J. M., Das DK-EW-Studien-Komitee 'Statistik-Wirtschaft und Entwicklung'. *Elektrowärme*, Issue B, **32** (1974), 164–8.
11. Etterich, O., 70 years steelmaking in the arc furnace (German). *Stahl u. Eisen* **95** (1975), 1028–31.
12. Schwabe, W. E., and C. G. Robinson, Development of large steel furnaces from 100- to 400-ton capacity. *Elektrowärme*, Issue B, **30** (1972), 321–5.
13. Elsner, E., H. Knapp and H. Voss, The meltdown of direct-reduced iron ore in electric arc furnaces (German). *Stahl u. Eisen* **94** (1974), 1322–30.
14. Ottmar, H., H. Schenck and W. Dahl, Technical and economic effects of the substitution of scrap by sponge iron in the arc furnace (German). *Stahl u. Eisen* **97** (1977), 731–41.
15. Ottmar, H., G. Schmeiduch and U. Siegers, Betriebsparameter von Lichtbogenofenanlagen mit Eisenschwamm-Schrott-Einsatz. *Paper at the 2nd Lichtbogenofentagung Kohászati Gyárépítö Vállalat in Budapest*, Oct. 1978.
16. Ameling, D., R. Assenmacher, E. Elsner and G. Fuchs, Water-cooled sidewall elements in UHP arc furnaces (German). *Stahl u. Eisen* **98** (1978), 429–34.

17. Elsner, E., and H. Knapp, Das Einschmelzen von Eisenschwamm in Lichtbogenöfen. Paper at the Congress of the Association des Ingénieurs Electriciens sortis de l'Institut Electrotechnique Montefiore, Comité Belge de l'Electrothermie et de l'Electrochimie. Report No. 5b, Liège, Nov. 1973.
18. Schwabe, W. E., Electric furnace problems: design and operating requirements for UHP furnaces melting pre-reduced charged materials. In *VIII. Congrès Union Internationale d'Electrothermie*, Liège, 1976, Report No. I.a.8.
19. Anlagendaten von Lichtbogenöfen in der Bundesrepublik Deutschland. Verein Deutscher Eisenhüttenleute. Düsseldorf 1978. (Report from Fachausschuß Verein Deutscher Eisenhüttenleute. No. 2017.)
20. Zangs, L., Technische und wirtschaftliche Aspekte beim Ersatz der feuerfesten Zustellung durch wassergekühlte Elemente in elektrolichtbogenöfen. *Paper at the 2nd Lichtbogenofentagung Kohászati Gyárépítö Vállalat in Budapest*, Oct. 1978.
21. Derp, H., Wassergekühlte Wandelemente der Thyssen Gießerei AG. Unpublished Report from VDEh, the Association of German Iron and Steel Engineers, Unterausschuß Elektrostahlbetrieb 21. April 1977.
22. Grubert, K., Erörterungsbeitrag über Ergebnisse mit wassergekühlten Wandelementen at Borsig GmbH, Berlin. Unpublished paper, see Ref. 21.
23. Pulvermacher, W., Erörterungsbeitrag über Ergebnisse mit wassergekühlten Wandelementen at FKH Bochum. Unpublished paper, see Ref. 21.
24. Harmsen, L., Erörterungsbeitrag über Ergebnisse mit wassergekühlten Wandelementen at Stahlwerk Lingen GmbH & Co. Unpublished paper, see Ref. 21.
25. Scheffler, G., Systematisches Konstruieren und Berechnen von Stahlwerken unter besonderer Berücksichtigung der Investitionen und betrieblichen Verarbeitungskosten. Aachen 1976. (Dr-Ing.-Diss. Techn. Hochschule Aachen.)
26. Assenmacher, R., H. Klein, E. Elsner, D. Ameling and G. Fuchs, Water-cooled roofs for arc furnaces (German). *Stahl u. Eisen* **98** (1978), 1044–7.
27. Typenbeschränkung für Lichtbogenöfen. *Stahl u. Eisen* **62** (1942), 887.
28. Sommer, H., and H. Pollack, *Elektrostahl-Erzeugung*. Düsseldorf 1950 (Stahleisen-Bucher. Vol. 8) pp. 103 ff.
29. Hütte. *Taschenbuch für Eisenhüttenleute*. 5th edition. Akademischen Verein. Berlin and Düsseldorf 1961. p. 654.
30. Harms, F., Use of the arc furnace, with particular reference to the large volume furnace, for steel production (German). *Stahl u. Eisen* **83** (1963), 257–70.
31. Schwabe, W. E., Arc heat transfer and refractory erosion in electric steel furnaces. *Proc. Eletr. Furn. Steel Conf.*, Electr. Furn. Steel Comm., Iron Steel Div., Amer. Inst. min. metallurg. petrol. Eng. Vol. 20. 1962, 195–206.
32. Ottmar, H., Performance capabilities of a 50 ton arc furnace. (German). *Stahl u. Eisen* **86** (1966), 201–7.
33. Ottmar, H., and G. Schmeiduch, Problems of high-capacity arc furnaces — theoretical and practical studies (German). *Elektrowärme* **28** (1970), 179–85.

McManus, George, Electric furnaces push to the fore. *Iron Age Metalworking Intern.* **17** (1978), No. 5, MP19, 21, 25, 27, 31, 33, 37, 39, 41.

Korf develops cooling system for arc furnace sidewalls. *Iron Steel Eng.* **55** (1978), No. 8, 73.

Assenmacher, Rolf, Emil Elsner and Dieter Ameling, Water cooling systems in electric arc furnaces. *Metallurg. Plant & Technol.* **1** (1978), No. 3, 5–6, 8, 10, 12, 14, 16, 20.

Roberts, K. S., An update on the use of water cooled blocks in arc furnace sidewalls. *Proc. Electr. Furn. Conf.* **36** (1978), 198–9.

Nanjo, Toshio, Akinori Nakamura and Takamitsu Yamada, New system saving refractory and electrode consumption in UHP arc furnace operation. *Iron & Steelmaker* **6** (1979), No. 11, 29–34.

Tubes cool electric furnaces. *Iron Age Metalworking Intern. Metal prod. Ed.*, **19** (1980), No. 4, 14MP23.

Markworth, Erich, Possibilities for increasing performance of direct arc furnaces with water circulating vessel linings through operation with higher power factors (German). *Stahl u. Eisen* **100** (1980), No. 10, 532–4 (Stahlwerksaussch, 1041).

5.3 Mechanical Design and Construction of Furnaces

1. Laufer, H., and G. Reimann, Die zunahmende Bedeutung des Elektrolichtbogenofens für die Stahlerzeugung. *Elektrowärme*, Issue B, **31** (1973), 212–7.
2. Markworth, E., Prospects for the use of arc melting furnaces in steel plants (German). *Elektrowärme*, Issue B, **32** (1974), 340–4.
3. Scheffler, G., and G. Dikta, Ultra-modern electric melting shop reduces impact on the environment (German). *Elektrowärme*, Issue B, **33** (1975), 234–42.

4a. Eßmann, H., Limits for design and layout of high-powered large arc furnaces (German). *Stahl u. Eisen* **97** (1977), 577–81.

4b. Bauer, Hannsgeorg, Knut Behrens, Helmut Meyer, and Josef Otto, Entwicklung und Betriebsergebnisse an einem Lichtbogenofen mit Bodenabstich. *Radex Rundsch*, 1980, 187–96 (German).

4c. Bauer Hannsgeorg, Knut Behrens, H. Meyer and J. Otto, Development and operating results connected with electric arc furnace with bottom pouring system. *MPT, Metallurgical Plant and Technology*, Verlag Stahleisen, 1980, 4/80, 26–39.

5. Planung und Inbetriebnahme eines Hochleistungslichtbogenofens mit Prozeßsteuerung für die Edelstahlerzeugung. *Fachber, Hüttenprax. Metallweiterverarb.* 1978, 780–8.
6. Meyer, G., D. Radke, and G. Reimann, Use of sponge iron according to the Krupp sponge iron melting process (German). *Stahl u. Eisen* **97** (1977), 7–12.
7. Timm, K., H. Faber, H.-G. Kunze, and K. Norden, Dynamic behaviour of electrodes and supporting arms in arc furnaces (German). *Stahl u. Eisen* **98** (1978), 695–700.
8. Etter, W., and S. Schulz, Elektroanschluß und Elektrodenregulierung von Lichtbogenöfen — Probleme und Lösungen. *Nachr. Eisenhüttenind.* GHH Sterkrade No. 2, 1977, 34–48.

Zangs, Ludger, Water-cooled linings for direct arc melting furnaces. *Steel Times* **206** (1978), No. 10, 912–16.

Schwabe, W. E., and C. G. Robinson, Characteristics of high-productivity arc furnaces for steel production. *Iron & Steelmaker* **5** (1978), No. 11, 25–8.

Fukumoto, Yukio, Energieeinsparung beim Schmelzen im Lichtbogenofen. *Tetsu to Hagané* **64** (1978), No. 13, 1968–79.

Baum, Rudolf, Arc furnaces for steelmaking. *Stahl u. Eisen* **99** (1979), No. 17, 922–7 (Stahlwerksaussch. 1026).

Assenmacher, R., E. Elsner and A. Ameling, Water cooling system in electric arc furnace. *MPT, Metallurgical Plant and Technology*, 3/1978, p. 5.

Hill, O. K., and C. G. Robinson, Large arc furnaces and the effect of key dimensions on the performance of UHP furnaces. *Iron Steel Eng.* **56** (1979), No. 7, 33–6.

5.4 Arc Furnace Electrical Equipment

1. Brehlar, R., Ofentransformatoren zum Speisen von Lichtbogenöfen mit Ofenschalter im Zwischenkreis. *Siemens-Z.* **50** (1976), 8–17.
2. Kaempf, P., E. Markworth and J. Mühlenbeck, 110 kV-Lichtbogenschmelzofen mit Lastschaltung im Zwischenkreis. *Stahl u. Eisen* **94** (1974), 393.
3. Mizushima, T., Neues Verfahren zum Ein- und Ausschalten des Ofenstromes durch Ofentransformatoren im Zwischenkreis. *5. Congrès Union Internationale D'Electrothermie*. Wiesbaden, 1963, Report No. 126.
4. Huhn, P. J., AEG-Gießharzschalter, ein wirtschaftliches Druckluft-Leistungsschaltersystem für höchste Anforderungen. *Techn. Mitt. AEG-Telefunken* **65** (1975), 35–40.
5. Baltensperger, P., and H. Meyer, Überspannungen beim Abschalten von Hochspannungsmotoren. *Brown-Boveri-Mitt.* **40** (1953), 342–50.
6. Hinterthür, K.-H., and B. Schemann, Bedämpfung von Schaltspannungen mit nichtlinearen Widerständen an Hochspannungs-Leistungsschaltern für 7,2 bis 36 kV. *Techn. Mitt. AEG-Telefunken* **65** (1975), 46–50.
7. Kaempf. P., Elektrische Ausrüstung von Lichtbogen-Schmelzöfen. *Siemens — Elektrische Ausrüstung* **15** (1974), No. 2, 19–22.
8. Webs, A., W. Kaufhold and B. Kulicke, Rückwirkungen von Drehstrom-Lichtbogenöfen in elektrischen Versorgungsnetzen. *Elektr.-Wirtsch.* **71** (1972), 222–8.
9. Frank, H., and K. Pettersson, Steigerung der Elektrostahlproduktion durch Spannungsstabilisierung mit Hilfe thyristorgeschalteter Kondensatoren. *ASEA-Z.* **22** (1977), 30–40.
10. Chit, A., and W. Horn, Neuzeitliche, ruhende Blindleistungskompensation für Industrienetze. *Techn. Mitt. AEG-Telefunken* **66** (1976), 286–90.
11. Wanner, E., and W. Herbst, Statische Blindleistungskompensation für Lichtbogenöfen. Brown, Boveri & Co., AG. publication No. CH-JW 511 790 D.
12. Carjell, U., Ein Beitrag zur Beurteilung des Lichtflimmerns bei Netzspannungsschwankungen. Aachen, 1972. (Diss. Techn. Hochsch. Aachen.)
13. Webs, A., Elektrische Netzverhältnisse bei Zwei-Elektroden-Betrieb eines Drehstromlichtbogenofens mit einem Transformatoraggregat. *Elektrotechn. Z.*, Issue A, **97** (1976), 438–41.
14. Jäger, S., and D. Knuth, Thyristor-controlled reactors to suppress network flicker caused by arc furnaces (German). *Elektrowärme*, Issue B, **30** (1972), 267–74.
15. Weigel, G., Erwärmungsberechnung von Transformatoren bei zeitlich veränderlichter Belastung. *Elektrische Bahnen* (1970), No. 12, 3–12.
16. Nausch, F., Ofentransformatoren. *ELIN-Z.* **16** (1964), 6–16.
17. Grundmark, B., Große Ofentransformatoren. *ASEA-Z.* **18** (1973), 7–12.
18. Jwasaki, Z., T. Kubota and T. Anzai, Steel melting arc furnace transformer equipment directly connected to 154-kV power source with tertiary load switching system. *Fuji Electr. Rev.* **17** (1971), 8–14.
19. Markworth, E., and W. Müller, Connections of furnace transformers with tappings on the high-tension side (German). *Elektrowärme,* Issue B, **32** (1974), 5–11, 137, 340.
20. Bonis, P., and F. Coppadoro, Transformatoren für Lichtbogen-Schmelzöfen. *Brown-Boveri-Mitt.* **60** (1973), 456–67.
21. Bretthauer, K., and K. Timm, Secondary-circuit measurements in three-phase furnaces (German). *Elektrowärme* **29** (1971), 381–7.
22. Mehus, O. H., Induced losses and heating problems in high-current arc furnace systems. National Industri, Norway. 1974.

23. Grundmark, B., Große Ofentransformatoren. *ASEA-Z.* **18** (1973), 7–12.
24. Webs, A., Elektrische Netzverhältnisse bei Zwei-Elektroden-Betrieb eines Drehstromlichtbogenofens mit einem Transformatoraggregat. *Elektrotechn. Z.*, Issue A, **97** (1976), 438–41.
25. Webs, A., Elektrische Netzverhältnisse bei symmetrischem Betrieb eines Drehstromlichtbogenofens mit einem Transformatorenaggregat. *Elektrotechn. Z.*, Issue A, **97** (1976), 96–100.
26. Brekler, R., Ofentransformatoren zum Speisen von Lichtbogenöfen mit Ofenschalter im Zwischenkreis. *Siemens-Z.* **50** (1976), 8–17.
27. Reiplinger, E., Dämmung von Ofentransformatorgeräuschen um mehr als 10 dB (A). Information der Transformator Union 1971.
28. Etter, W., and S. Schulz, Elektroanschluß und Elektrodenregulierung von Lichtbogenöfen — Probleme und Lösungen. *Nachr. Eisenhüttenind.* GHH Sterkrade No. 2, 1977, 34–48.
29. Steinmetz, G., Increased production and cost savings by electronic electrode regulation of arc furnaces (German). *Elektrowärme,* Issue B, **33** (1975), 96–9.
30. Buxbaum, A., A. Chit and H. Utecht, Elektronische Regeleinrichtung für die Elektrodenregelung von Lichtbogenöfen. *Techn. Mitt. AEG-Telefunken* **63** (1973), 232–5.
31. Bretthauer, K., and K. Timm, Secondary-circuit measurements in three-phase furnaces (German). *Elektrowärme* **29** (1971), 381–7.
32. Bretthauer, K., and A. A. Farschtschi, Magnetic field problems of the power leads of arc furnaces (German). *Elektrowärme,* Issue B, **32** (1974), 33–37.
33. Ameling, D. Über den Zusammenhang zwischen den elektrischen Bedingungen und den Strahlungsverhältnissen des Lichtbogenofens bei Einstellung verschiedener Arbeitspunkte im elektrischen Leistungsschaubild. Clausthal 1966. (Dipl.-Arb. Techn. Hochsch. Clausthal.).
34. Ottmar, H., A. Oerter, R. Assenmacher and P. Goldstein, II. The high-capacity electric arc furnace (German). *Stahl u. Eisen* **89** (1969), 466–71.
35. Ottmar, H., and G. Schmeiduch, Problems of high-capacity arc furnaces — theoretical and practical studies (German). *Elektrowärme* **28** (1970), 179–85.
36. Ottmar, H., A. Oerter and D. Ameling, Der Zusammenhang zwischen den elektrischen und wärmetechnischen Grundlagen bei Hochleistungslichtbogenöfen. *Radex-Rdsch.* **1973**, 519–27.
37. Ottmar, H., and D. Ameling, The high-powered electric-arc furnace for steelmaking — definition and limits of productivity (German). Ergebnisse eines 10-jährigen Betriebes. Paper given to a group of visitors from Instituto Latinoamericano del Fierro y el Acero [ILAFA] on 4 June 1973 in Düsseldorf. See *Stahl u. Eisen* **94** (1974), 125–32.
38. Ottmar, H., and D. Ameling, Diskussion und eventuelle Definition des Begriffes 'UHP' für die Anwendung als Elektrowärmeaggregat. Prepared for 'Deutsches Komitee Elektrowärme', Arbeitskreis Perturbation.
39. Ottmar, H., A. Oerter, G. Schmeiduch, D. Ameling and U. Siegers, Über einige besondere Probleme bei der Verarbeitung von Eisenschwamm im Hochleistungs-Lichtbogenofen. Internationaler Eisenhüttentechnischer Kongreß, Düsseldorf 1974 (27–30 May). Vol. 2 (Düsseldorf) 1974. Report 3.1.2.1. 14S.
40. Schoenmaker, O. D., and J. Ph. Guldenmundt, Leistungssteigerung im Elektroofen durch organisatorische und schmelztechnische Maßnahmen. *Radex-Rdsch.* 1973, 528–42.
41. *Leistungsspitzen — Begrenzungs-Automatik.* Brown, Boveri & Co., AG publication No. D. IO 51 115 D.

42. Eßmann, H., Überlegungen zur Auslegung und Planung von Hochleistungs-Lichtbogenöfen unter Berücksichtigung neuerer Betriebserfahrungen. Nachr. Eisenhüttenind. GHH Sterkrade 1975, No. 1, 46–56.
43. Ericson, A., Metallurgische Gesichtspunkte zur induktiven Umrührung von Metallschmelzen. *ASEA-Z.* **16** (1971), 117-18.
44. Hanås, B., Induktive Umrührer, Gesichtspunkte der Konstruktion und Anlagentechnik. *ASEA-Z.* **16** (1971), 123–8.
45. Sundberg, Y., Prinzip und Funktion der induktiven Umrührer. *ASEA-Z.* **16** (1971), 107–16.
46. Linder, S., Das Umrühren von Stahl, ein Mittel zur Beschleunigung metallurgischer Reaktionen. *ASEA-Z.* **16** (1971), 119–22.
47. Brandt, S., Automatisierung von Elektro-Lichtbogenöfen. *Techn. Mitt. AEG-Telefunken* **66** (1976), 299–302.
48. Leu, H. W., and O. Schläpfer, *Hochleistungs-Lichtbogenöfen*. Brown, Boveri & Co. AG-Druchschr. No. CH 511 690 D.
49. Computer based control system for electric arc furnaces. *ASEA-Druckschr. AU 80-101 E.*

Cooksley, C. G., Modern British arc furnace practice. *Electrical Times* No. 4426, 1977, 5, 8; after *Arch. Energiewirtsch.* **31** (1977), No. 12, 1078–81.

Maddever, W. J., R. M. Ňikolic, A. McLean, and R. S. Segsworth, The influence of electrode gas injection on arc furnace steelmaking. *Iron & Steelmaker* **4** (1977), No. 11, 33–41.

Billings, S. A., F. M. Boland, and H. Nicholson, Electric arc furnace modelling and control. *Automatica* **15** (1979), No. 2, 137–48.

Bowman, B., Trends in electrical parameters of steelmaking arc furnaces (German). *Elektrowärme*, Issue B, **37** (1979), No. 2, 80–6.

5.5.2 Wear Problems

1. Schwabe, W. E., Arc heat transfer and refractory erosion in electric steel furnaces. *Proc. Electr. Furn. Steel Conf.*, Electr. Furn. Steel Comm., Iron Steel Div., Amer. Inst. min. metallurg. petrol. Eng. Vol. 20. 1962. 195–206.
2. Ottmar, H., A. Oerter, R. Assenmacher and P. Goldstein, Stahlerzeugungsverfahren im Umbruch — die Zukunft der Herdschmelzverfahren. II. The high-capacity electric arc furnace (German). *Stahl u. Eisen* **89** (1969), 466–71.
3. Kreutzer, H. W., K. W. Langner, H. Lutz, K. Unger and M. Velikonja, The refractory lining of electric arc furnaces in the Federal Republic of Germany (German). *Stahl u. Eisen* **94** (1974), 223–29.
4. Bowman, B., and F. Fitzgerald, Hot spots in arc furnaces. *J. Iron Steel Inst.* **211** (1973), 178–86.
5. Oerter, A., and G. Schmeiduch, Untersuchungen des Verschleißfaktors von E-Ofen-Zustellungen. Stahlwerksausschuß-Sitzung der Eisenhütte Südwest. Jan. 1976. Unpublished report.
6. Großkopf, B., P. Jeschke and H. Naefe, Verschleißmessungen mit Radioisotopen in metallurgischen Gefäßen. *Techn. Mitt., Essen* (1977), 160–7.
7. Schmeiduch, G., Einfluß einiger Betriebsparameter auf den Feuerfest-Verschleiß im Elektrolichtbogenofen. Berlin 1978. (Dr-Ing.-Diss. Techn. Univ. Berlin.)
8. Ottmar, H., A. Oerter, G. Schmeiduch, A. Ameling and U. Siegers, Über einige besondere Probleme bei der Verarbeitung von Eisenschwamm im Hochleistungslichtbogenofen. In *Internationaler Eisenhüttentechnischer Kongreß*, Düs-

seldorf 1974, 27–30 May. Vol. 2. [Düsseldorf] 1974. Report No. 3.1.2.1, 14 pages.

9. Ameling, D., R. Assenmacher, E. Elsner and G. Fuchs, Water-cooled sidewall elements in UHP arc furnaces (German). *Stahl u. Eisen* **98** (1978), 429–34.
10. Brachet, J. P., and G. Sartorius, Entwicklung eines Wärmefluß-Meßfühlers und eines Wandstärke-Meßsystems für Lichtbogenöfen. Erfahrungen über den Verschleißvorgang der Ausmauerung und die Regelung der Phasenstromstärke. Direkte Steuerung des Lastreglers durch die drei Fühler. In *Kongreß für Automation und Prozeßsteuerung in Elektrostahlwerken und Gießereien*, Versailles, May 1975, Report No. I.5.
11. Obst, K. K., W. Münchberg and H. Comes, Betriebsversuche über den Einfluß der Legierungselemente auf die Verschlackung von basischen, feuerfesten Stoffen in Lichtbogenöfen. In *Vorträge of the 16th internationalen Feuerfest-Kolloquium*, Aachen, 25.–26. Oct. 1973. Institut für Gesteinshüttenkunde der Technischen Hochschule Aachen, Forschungsinstitut der Feuerfest-Industrie Bonn; Verein Deutscher Eisenhüttenleute, Düsseldorf. Aachen 1973, 274–313.
12. Bergh, S., H. Sandberg and N. Ståhl, Continuous measurement of lining wear in steel furnaces with radioisotopes. *J. Metals* **21** (1969), No. 2, 19–22.
13. Johansson, R., Die Verwendung von AGA-Infrarot-Thermovision für die quantitative Beurteilung des Verschleißes an feuerfesten Ausmauerungen in Öfen und Pfannen. *Fachber, Hüttenprax. Metallweiterverarb.* **13** (1975), 990, 992–4, 997–8.
14. Ståhl, N., Methoden zur Bestimmung des Futterverschleißes in Stahlöfen. *Fachber, Hüttenprax. Metallweiterverarb.* **14** (1976), 581–3.
15. Bollmohr, H., N. Jansen, R. Solmecke and U. Siegers, Betriebs- und Laborergebnisse mit einem dolomitisch zugestellten Lichtbogenofen-Deckel. *Tonind. Z. keram. Rdsch.* **99** (1975), 29–32.

Jackson, B., Refractories technology and refractories output — more and more in less and less. *Refractories J.* **52** (1977), No. 5, 12–14, 16–17, 19–21, 23–25.

Nagayama, Hiroshi, Anwendung einer CaO-Herdzustellung bei einem Lichbogenofen und der Einfluß auf den Oxideinschlußgehalt im Stahl. [Japanese] *Tetsu to Hagané* **63** (1977), No. 10, 1643–52.

Alcock, S., and D. R. F. Spencer, The application and performance of basic refractories in secondary steelmaking ladles. *Trans. & J. Brit. Ceram. Soc.* **77** (1978), No. 2, 45–57.

Bichlbauer, Erich, Operational experience in the use of refractory material in electric arc furnaces with sponge iron charge. *Metallurg. Plant & Technol.* **1** (1978), No. 1, 53–56, 58–61.

Long life experienced with new type of electric furnace spout runner. *Ind. Heat.* **45** (1978), No. 4, 29–30, 32.

Refractories design in electric furnaces at Ford steel division. *Ind. Heat.* **45** (1978), No. 4, 40.

Padgett, G. C., Mechanical behaviour — theory and practice. *Refractories J.* **53** (1978), No. 4, 13–16, 18–20.

Jordan, G. R., Electrode erosion in electric arc furnaces — the controlling parameters. *Ironmaking & Steelmaking* **5** (1978), No. 4, 177–83.

Houseman, D. H., Current practices in refractories for the steel industry. *Steel Times* **207** (1979), No. 8, 615–21.

Kayworth, P. M., and G. P. Carswell, Governing maintenance of BOS and arc furnaces. *Refractories J.* **54** (1979), No. 4, 11–14, 16, 19–20.

Bowman, B., The relationship between the power programme and refractories consumption in an arc furnace. *Ironmaking & Steelmaking* 1974, No. 4, 212–4.

5.6.1 Technology and metallurgy in scrap charging — basic furnace-linings

1. Ottmar, H., A. Oerter, G. Schmeiduch and U. Siegers, The melt-down of sponge iron pellets reduced with Rhenish lignite in the arc furnace of Röhrenwerke Bous-Saar GmbH (German). *Stahl u. Eisen* **96** (1976), 106–12.
2. Rankin, W. M., Production d'acier au four électrique Usine de Houston, Armco Steel Corporation. *J. Metals* **20** (1968), No. 5, 104–7.
3. Yoshihara, H., K. Nakazawa and T. Haraguchi, Construction and operation of UHP electric furnace. *Trans. Iron Steel Inst. Japan* **11** (1971), (Suppl. I) 309–11.
4. Engledow, D., and F. D. Winter, BSC experience of arc-furnace continuous charging. *Ironmaking & Steelmaking* **3** (1976), 359–65.
5 Eßmann, H. Gesichtspunkte für Auswahl und Bemessung moderner Hochleistungs-Lichtbogenöfen. *Klepzig Fachber.* **80** (1972), 334–7.
6. Ottmar, H., and D. Ameling, The high-powered electric-arc furnace for steelmaking — definition and limits of productivity (German). *Stahl u. Eisen* **94** (1974), 125–32.
7. Schwabe, W. E., Arc heat transfer and refractory erosion in electric steel furnaces. *Proc. Electr. Furn. Steel Conf.*, Electr. Furn. Steel Comm., Iron Steel Div., Amer. Inst. min. metallurg. petrol. Eng. Vol. 20. 1962. 195–206.
8. Ottmar, H., A. Oerter, R. Assenmacher and P. Goldstein, Stahlerzeugungsverfahren im Umbruch — die Zukunft der Herdschmelzverfahren: 2. der Hochleistungslichtbogenofen. *Stahl u. Eisen* **88** (1968), 466–71.
9. Thielker, K. H., Installations for preheating the scrap in electric steel plants. I. Experiences with a preheating plant for a 100-ton electric arc furnace (German). *Stahl u. Eisen* **90** (1970), 526–9.
10. Schoenmaker, O. D., Installations for preheating the scrap in electric steel plants. II. Operational results of a scrap preheating plant with scrap baskets lined with refractories (German). *Stahl u. Eisen* **90** (1970), 530–4.
11. Schmidt, F., Installations for preheating the scrap in electric steel plants. III. Operation and results of scrap preheating plants for 10- and 15-ton electric arc furnaces (German). *Stahl u. Eisen* **90** (1970), 534–7.
12. Ottmar, H., A. Oerter and G. Schmeiduch, Betriebsergebnisse eines 50 t-Lichtbogenofens und die voraussichtliche Entwicklung des UHP-Verfahrens. In *Congrès international sur le four électrique à arc en aciérie*, Cannes, 7–9 June 1971. Organisé par l'Institut de Recherches de la Sidérurgie Française, IRSID, Metz 1972. 349–62.
13. Neumann, F., H. Leu, R. Ptach and U. Brusa, The BBC – Brusa steelmaking process (German). *Stahl u. Eisen* **95** (1975), 16–23.
14. Schermer, K., Erfahrungen mit einer ungewöhnlichen Verfahrenstechnik für die Verarbeitung von Eisenschwamm im Elektrolichtbogenofen. *Radex Rdsch.* 1976, 675–98.
15. Harms, F., and W. Englemann, Melting down in a large electric arc furnace using an oxygen jet as a fourth heat-source (German). *Stahl u. Eisen* **85** (1965), 456–64.
16. Hogan, W. T., New economics of steel — more steel with less capacity. *Iron Steel Eng.* **52** (1975), No. 12, 57–60.
17. Noda, H., H. Furuhashi and H. Ushiyama, Operating experiences of UHP arc furnace. In *Congrès Union Internationale d'Electrothermie*, Warsaw, 1972, Report No. 101.

18. Noda, H., A new 70-ton UHP-arc furnace, its design and operating results. See Ref. 11. 376–83.
19. Gross, O., Einblasen von Kalk-Flußspat-Gemischen mit gasförmigem Sauerstoff in den Elektrolichtbogenofen. *Arch. Eisenhüttenwes.* **44** (1973), 311–15.
20. Voss. H., Das Verhalten des Schwefels im Frischprozeß eines basischen Lichtbogenofens unter Verwendung von Weißfeinkalk zur Schlackenbildung. Clausthal 1963. (Dipl.-Arb. Techn. Univ. Clausthal).
21. Haucke, M., H. Clees, H. J. Kopineck and H. Ottmar, Use of pulverized lime in the electric steel plant. Pneumatic transport as a charging method (German). *Stahl u. Eisen* **83** (1963), 441–9.
22. Ottmar, H., Metallurgische Aspekte bei der Erzeugung hochwertiger Stähle im Hochleistungslichtbogenofen. In *Memoria del 8. Congreso Latinoamericano de Siderurgia*. Lima, Perú, Sept. de 1968. Instituto Latinoamericano del Fierro y el Acero. Santiago/Chile 1969. Report Part 3b 1/6.

Miller, T. E., and B. Guffy, Basic arc melting and related quality. *Iron & Steelmaker* **5** (1978), No. 4, 25–27.

5.6.2 Technology and metallurgy in scrap charging — acid furnace-linings

1. Brown, J. W., A. Bueler and B. P. Courvoisier, Use of arc furnace in foundries. *Elektrowärme,* Issue B, **35** (1977), 264–9.
2. Körber, F., Das Verhalten von Mangan, Silicium und Kohlenstoff bei der Stahlerzeugung. *Stahl u. Eisen* **54** (1934), 535–43.
3. Körber, F., and W. Oelsen, Die Wirkung des Kohlenstoffs als Reduktionsmittel auf die Reaktionen der Stalherzeugungsverfahren mit saurer Schlacke. *Mitt. K.-Wilh. Inst. Eisenforsch.* **17** (1935), 39–61; *Stahl u. Eisen* **56** (1936), 181–208.
4. Dobrowsky, F., Erfahrungen beim Betrieb eines sauer zugestellten 7,5-t-Elektrolichtbogenofens. *Berg- u. hüttenm. Mh.* **106** (1961), 327–33.

5.7 Technology and Metallurgy for Sponge-iron Charges

1. Ottmar, H., H. Schenck and W. Dahl, Technical and economic effects of the substitution of scrap by sponge iron in the arc furnace (German). *Stahl u. Eisen* **97** (1977), 731–41.
2. Rodriguez, F. A., and D. H. Carillo, Concepts relevant to steelmaking with HYL metallized pellets. In *ECE-Seminar on the Utilization of Pre-reduced Materials in Iron and Steel Making*, Bucharest, 24–28 May 1976. Economic Commission for Europe, Steel Committee, [United Nations Organization, UNO]. o. O. 1976. Report No. R. 18.
3. Celada, J., and G. E. McCombs, HyL direct reduction operations. *Iron & Steelmaker* **3** (1976), No. 10, 18–22.
4. Ottmar, H., A. Oerter, G. Schmeiduch and D. Ameling, Use of sponge iron as melting stock in a UHP arc furnace — selected test results (German). *Elektrowärme,* Issue B, **32** (1974), 127–37.
5. Bold, D. A., and N. T. Evans, Direct reduction down under: the New Zealand story. *Iron & Steel Internat.* **50** (1977), 145, 147–52.
6. Schmeiduch, G., Einfluß einiger Betriebsparameter auf den Feuerfest-Verschleiß im Elektrolichtbogenofen. Berlin 1973. (Dr-Ing. Diss. Techn. Univ. Berlin.)
7. Pantke, H.-D., Chr. Queens, Demands on the quality of sponge iron for steelmaking (German). *Stahl u. Eisen* **96** (1976), 652–7.

8. Neumann, F., H. Leu, R. Ptach and U. Brusa, The BBC – Brusa steelmaking process (German). *Stahl u. Eisen* **95** (1975), 16–23.
9. Post, G., and D. Ameling, Charging pellets at Hamburg. *Iron Steel Metallurg.* 1975, Magazine April, 43–51.
10. Sibakin, J. G., G. A. Roeder and P. H. Hookings, Electric arc steelmaking with continuously charged reduced pellets. *J. Iron Steel Inst.* **205** (1967), 1005–17.
11. Celada, J., and R. Quintero, Steelmaking with HyL sponge iron. *Proc. Electr. Furn. Conf.* **32**. 1974. 41–6.
12. Perez Ayala, J. L., G. Rodriguez, E. A. Bryan and H. N. Hughes, Electric furnace steelmaking with sponge iron. *Iron Steel Eng.* **40** (1963), No. 8, 69–77.
13. Antoine, J., and J. Dumont-Fillon, Elaboration d'acier au four à arcs à partir de minerai préréduit chargé en continu. Résultats des essais effectués à la Station pilote de l'IRSID. In *Congrès international sur la production et l'utilisation des minerais réduits,* Evian, 29–31 May 1967, Organized by the Chambre Syndicale de la Sidérurgie Française and others. Metz 1968, 323–33.

Dobson, R., W. A. Mullett and I. G. Nixon, Smelting and refining of sponge iron. *Ironmaking and Steelmaking* **4** (1977), No. 5, 265–75.

Miller, Jack Robert, Use of direct reduced iron ore and balanced integrated iron and steel operation. *Ironmaking and Steelmaking* **4** (1977), No. 5, 257–64.

Barbi, A., The benefits and drawbacks of using sponge iron in electric arc steelmaking. *South East Asia Iron & Steel Inst. Quart.* **6** (1977), No. 3, 39–42.

Bleimann, K. R., M. D. Coward and C. F. Hendrix, An update on electric arc furnace melting with up to 100% sponge iron. *Iron & Steelmaker* **5** (1978), No. 6, 25–8.

Schermer, K., and A. Dalrymple, The significance of Dunswart's rotary kiln direct reduction experience to the steelmaker in the USA. *Iron & Steelmaker* **5** (1978), No. 6, 19–24.

Maschlanka, Walter, Günter Post, and Emil Elsner, Utilization of direct reduced iron on different iron and steel production processes. *Metallurg. Plant & Technol.* **1** (1978), No. 2, 13–14, 16–17, 20.

Brown, J. W., and R. L. Reddy, Electric arc furnace steelmaking with sponge iron. *Ironmaking & Steelmaking* **6**, No. 1, 24–31.

Reddy, Richard L., Electric arc furnace steelmaking with sponge iron. *Canad. metallurg. Quart.* **18** (1979), No. 2, 245–50.

Schermer, K., Improved technology for processing sponge iron in the electric arc furnace. *Ironmaking & Steelmaking* (1975) No. 403, 188–92.

Willars, H. M. and R. C. Madden, The development of a new steelmaking process utilising highly metallised sponge iron. *Iron and Steel International* (1975), 313–321.

Pantke, H. D. and Chr. Queens, Melt-down characteristics of sponge iron in the manufacture of steel. *ECE-Seminar Bucharest 1976* (1976).

Ameling, D., G. Rudolph, E. Elsner and H. Knapp, The use of high-metallized sponge iron in electric steel works. *ECE-Seminar Bucharest 1976* (1976).

Rodríguez, F. A., and D. H. Carillo, Concepts relevant to steelmaking with HyL metallized pellets. *ECE-Seminar Bucharest 1976* (1976).

Post, G., and D. Ameling, Charging pellets at Hamburg. *Iron and Steel Metallurgy* (1975), No. 4, 43–51.

Noda, H., H. Furuhashi, and H. Ushiyama, Operating experience of UHP arc furnace. *Warsaw 1972*, (1972) No. 101.

Kirkby, H. J. A., and R. D. Langman, Arc furnaces and their developing role for melting directly reduced iron for steelmaking. *Warsaw 1972* (1972), No. 102.

Engledoer, D., and F. D. Winter, BSC experience of arc-furnace continuous charging. *Ironmaking & Steelmaking* (1976), No. 6, 359–65.
Schwabe, W. E., Electric furnace problems: design and operating requirements for UHP arc furnaces melting prereduced charge materials. *UIE-Congress, Liège* (1976), No. 8, Ia.
Schermer, K., Experience with unusual technology for the processing of sponge iron in the electric arc furnace. *Radex Rundschau* (1976), No. 2, 675–98.
Elliott, J. F., and J. K. Wright, Metallized ore practices in small arc furnaces. *Ironmaking & Steelmaking* (1976), No. 5, 32–5.

Electric Furnace Steel Production
Edited by E. Plöckinger and O. Etterich

6 Induction Furnaces

Karl-Heinz Brokmeier, Dortmund

Steel in relatively small quantities has been successfully melted in medium-frequency furnaces for several decades. The preferred grades here are high-class, such as rust- and heat-resisting steels, tool steels, high-speed steels, and other special steels which are cast into ingots for rolling or forging and also into steel castings.[1] Mains-frequency, 21 000-kW crucible furnaces of 60-t capacity have been very successfully operated for several years for melting malleable cast iron. We can therefore expect mains-frequency crucible furnaces with capacities of 60–150-t and powers of up to 40 000 kW to be used for making steel. We can recognize their advantages and disadvantages compared with arc furnaces from theoretical and physical considerations and from practical experience which has yet to be quantified and evaluated.

The high-performance, mains-frequency crucible furnace has the advantage that it can be connected to any mains electricity supply of adequate power. The power consumed depends on the furnace charge; the furnace therefore works in quasi-continuous operation at maximum electric power but without peak loads. Because the heat energy is supplied inductively, no graphite or other electrodes are used. Charges of small lump-size, e.g. sponge iron, are stirred well by the powerful bath motion; this accelerates reactions in the bath/slag interface layer and always results in a steel bath which is homogenous in both temperature and composition. There is no local overheating of the melt, and iron and alloy melting losses are low.

With induction furnaces, dust- and gas-emissions depend solely on the nature of the charge and are therefore very low in comparison with all other melting processes.[2–4] The noise level arising from melting is so insignificant that no suppression is required; however, it must be considered in connection with the charging process.

Beside these undeniable advantages, there are certain drawbacks which must not be ignored. Crucible-wall thickness must be kept within certain limits — about 100 mm for smaller furnaces and up to 300 mm for 120–150-t furnaces.[5,6] These limits arise from capital cost concerned with the capacitor

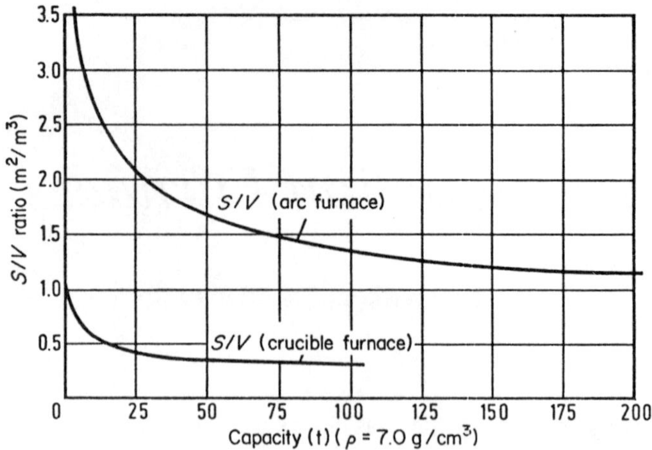

Figure 1. Relationship of bath surface area S to melt volume V (S/V ratio) for arc furnaces and induction crucible furnaces[7]

bank, which increases in size as wall-thickness increases, and to a lesser extent with the reduction in electrical efficiency. Experience with basic refractory linings is limited to furnaces of only about 5–10-t capacity and depends on the working methods; corresponding development work for large furnaces is in progress. Acid-compound-rammed linings in currently used crucible-sizes, offer satisfactory durability provided recognition is given to their limit on temperature-performance and that they are attached by slags containing FeO, MnO, etc.

Metallurgical possibilities are very limited because of the relatively thin crucible walls and the lack of suitable basic linings; however, the vigorous bath motion gives some assistance. Compared with other steel-production methods, the decarburizing process (oxidation) is very limited. On the one hand, the low crucible freeboard promotes the risk of steel splashes and boiling over, and on the other, the specific bath surface area (bath surface area in square metres divided by crucible capacity in tonnes) is relatively small compared with electric arc furnaces (Figure 1).[7] The mains-frequency crucible furnace should preferably be operated with a liquid initial bath; this makes for severe difficulties in steel production with frequent specification changes. Scrap melting from cold, without a liquid heel, requires particular qualities of the scrap. Because of the bath motion, specific power input to mains-frequency furnaces is limited to about 400 kW/t; for this reason, melting-cycle times of under 2 h cannot be achieved.

6.1 APPLICATIONS FOR MEDIUM- AND MAINS-FREQUENCY CRUCIBLE FURNACES

A survey by F. Neumann (Table 1) shows a present-day view of the possible areas of application for larger induction furnaces for steel-production. From what has been explained above, this is the region in which acid-linings provide satisfactory results. Secondary processes, in separate vessels for oxidation and refining, widen the possible applications for induction crucible furnaces appreciably. A separate vessel for alloying enables a quasi-continuous melting process with unalloyed iron to be carried out in a large mains-frequency crucible furnace to produce steels of various compositions. Secondary processing of liquid steel is open to all the usual processes for steel-production, irrespective of whether the liquid steel comes direct from the induction furnace or from a secondary processing vessel.

In the steel foundry, the medium-frequency furnace is a well-known, valuable supplement to the arc furnace. The larger mains-frequency crucible furnace will also find further applications if the problem of the lining can be resolved satisfactorily.

The main application for medium- and large-sized mains-frequency crucible furnaces is at present for the production of grey and malleable cast iron; this will be dealt with in more detail in Chapter 11. Here, modern channel-type induction furnaces (induction-heated holding furnaces and foundry systems) are used in increasing numbers to mechanize and automate the process.

6.1.1 Induction Crucible Furnace Design

In designing induction furnaces, the electrical and physical relationships, technical limitations, and safety considerations must all be borne in mind.[5,6] It is a basic physical law that simultaneous presence of an electric current and a magnetic field acting on the conductor carrying the current lead to the production of a mechanical force, as in the electric motor. The mechanical force acts vectorially at right-angles to the plane formed by the current direction and the magnetic field. In an induction furnace this force always acts towards the middle and leads to vigorous motion in the bath. This motion leads to the top of the melt forming into a dome, the height of which depends on the height of the coil, the specific power, the frequency, and the molten material (see section 4.2.3, Figure 8).

Figure 2 illustrates the range of different furnace voltages used for induction crucible furnaces. For small laboratory furnaces, 200–500 V are used; for medium-sized production furnaces 800–2000 V; for large furnaces up to 3000 V.

The maximum *performance* (i.e. output) obtainable from crucible furnaces depends on the capacity and on constructional, mechanical, and electrical limits. If the refractory lining problems can be overcome (see section 6.1.2.2), then power and hence performance are limited by the maximum permissible furnace coil voltage and current. The voltage limit is determined by the operating conditions, while maximum current is a development question

Table 1. Simplified survey of applications for induction steel melting. (After F. Neumann)

	Aptitude and type of operation[a]		Notes and limitations on scrap quality and type of operation	
	Basic[b]	Acid	Acid crucible lining	Basic crucible lining
Building steel St 32–70	Good	For temperatures always under or equal to 1580 °C and only short superheat to tapping temperature (about 1630 °C)	Dephosphorizing and desulphurizing, oxidation of C, Si, Mn, and Cr not easily possible. Soft steels cause high temperature of lining	Dephosphorizing and desulphurizing possible within certain limits, depending on type of lining and metallurgy
			Pure melting is often sufficient because P and S are within the tolerance. However, soft steels need to be oxidized, and this is only possible with basic linings	
Concrete-reinforcement steel	Good	ditto	P and S limited only for special qualities. Greater analysis tolerances are often admissible	
Non-alloy building steels of special quality	Good	ditto	Higher scrap quality required since P and S must not exceed 0.04%. For soft steels, see above	Dephosphorizing and desulphurizing possible within certain limits, depending on type of lining and metallurgy
Non-alloy tool steels	Good	ditto	High scrap quality required; pure remelting; at higher C content, lower lining temperature	ditto

Premelting charges for VOD, AOD, and other processes	Good	ditto	Suitable for premelting with higher C and lower alloy content and where temperature limits are observed; already practised	ditto
Low alloy building steels (1) General building steels (2) Heat-treated steels (3) Case-hardening and nitriding steels (4) Spring- and roller-bearing steels	Good	ditto Elements such as Cr, Mn, and others attack acid lining; higher-alloy melts of this type not possible	Higher scrap quality required, since P and S must not exceed 0.04%. For soft steels, see above. To No. 4: Relatively low lining temperature with high C content	ditto
High-alloy building steels	Good	ditto	Higher scrap quality required; pure remelting	ditto
High-alloy special purpose steels	Good	ditto	Chemical and thermal loads on lining too high	ditto
Alloy tool steels	Good	not	ditto	ditto

Notes: [a] Mains-frequency crucible furnaces are not suitable for pure charge-processing. If necessary, starting-blocks are used to help.

[b] 'Good' aptitude ratings for basic linings are valid on condition that lining material which can withstand chemical attack and thermal shock is available.

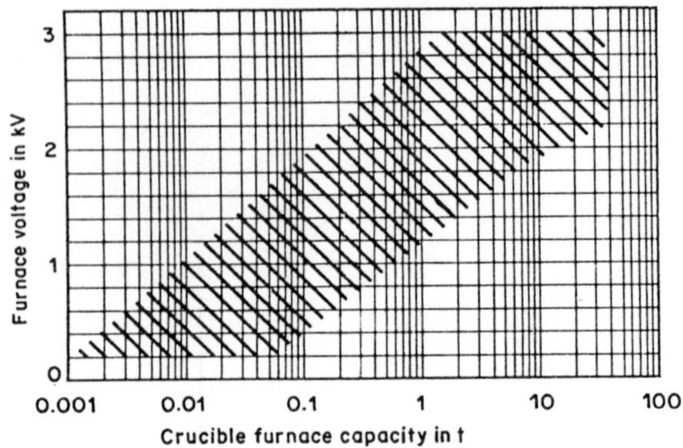

Figure 2. Relationship of furnace voltage to capacity for medium-frequency crucible furnaces[8]

dependent, among other things, on whether there should be water-cooling of the furnace coil laminations.

Furnace *apparent power* is the product of voltage and current, and the active power is the product of apparent power and power factor cos ϕ, measured simultaneously, while both depend on frequency (see section 4.2.3, Figure 10). Because of this relationship, for a given apparent power the maximum active power is attained at the highest cos ϕ, i.e. by mains-frequency furnaces which, having a lower operating frequency and hence lower reactance, transfer coil power with lower apparent power demand from the supply system, i.e. high power factor. However, because power transfer to the furnace coil is proportional to frequency, the medium-frequency coil can transfer higher power to the charge for similar crucible-sizes but at the expense of a lower power factor, i.e. higher apparent power-demand. This leads to the further conclusion that maximum absolute melting output is obtained with mains-frequency furnaces while maximum specific melting output is attained with medium-frequency furnaces.

Melting time is the reciprocal of the specific melting output. The maximum specific output, or minimum melt-down time, is determined by the maximum permissible bath motion. Figure 3 shows these relationships in dependence on mean furnace output, which in turn depends on the initial level to which the crucible is filled (see Figure 13).

The *efficiency* of energy conversion, as is seen in Figure 4, depends only to a small degree on the furnace dimensions; we can expect an induction furnace for steel or cast iron to work at about 80% efficiency.

The following considerations apply for the *power factor*. For a completely absorbed magnetic field, cos $\phi = 0.7$. This would apply for an induction furnace in which the coil surrounded the melt directly without a gap, and where the diameter of the melt is more than 3.5 times its depth in the coil. However,

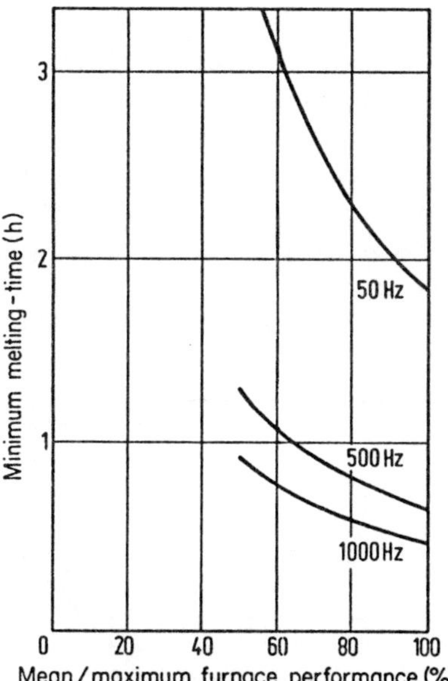

Figure 3. Minimum melting times (without idle periods) for melting steel in a crucible filled to 25% above the top of the coil shown in relation to frequency and the ratio of mean to maximum furnace power

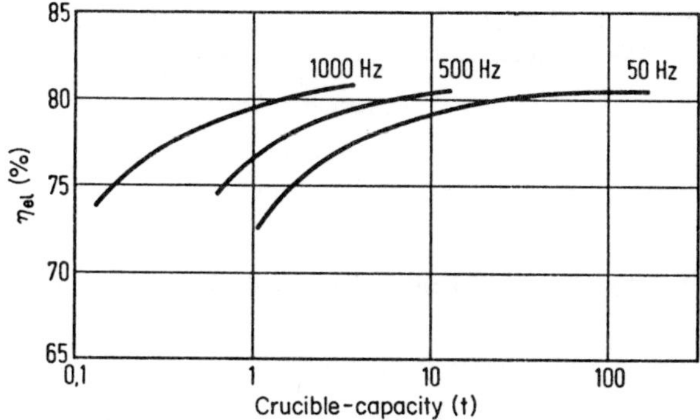

Figure 4. Induction crucible furnace electrical efficiency in relation to furnace-size

since the crucible has to be made from refractory material, the electromagnetic field flows unused in that area. As furnaces become larger, the crucible wall becomes thicker, and so does the unused electromagnetic field. This causes cos ϕ to deteriorate; however, increasing the frequency has the same effect. The capacitor bank required to provide the compensation towards cos $\phi = 1$, which is always required, is inversely proportional to cos ϕ. Therefore, more capacitor power is required for larger furnaces and higher frequencies (see section 4.2, Figure 11). The worsening power factor at higher frequencies is compensated economically because capacitors for higher frequencies cost less than capacitors for lower frequencies; in addition, capacitors for higher frequencies to take up less space per unit. The capacitor requirements of a mains-frequency crucible furnace with a power of 300 kW/t can be ascertained for various operating conditions from Figure 5.

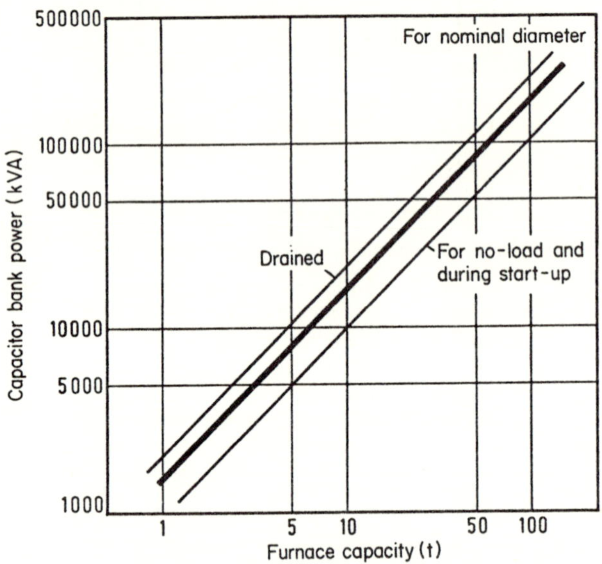

Figure 5. Capacitor requirements for mains-frequency crucible furnaces for steel-melting

6.1.2 Building and Lining Induction Crucible Furnaces

6.1.2.1 Mechanical and electrical equipment

An induction crucible furnace (Figure 6) has a ceramic crucible (1) surrounded by a water-cooled coil (2). Mains-frequency furnaces always have, and newer medium-frequency crucible furnaces usually have, stacks of laminations (3) made from transformer steel to guide the external magnetic field. The furnace structure is made from steel sections (4) arranged and dimensioned so that clamps (5) grip the laminations and the coil, which in turn supports the crucible firmly. Only stable construction and appropriate clamping can guarantee that the inner part of the crucible, made from ramming or bricks, will not crack

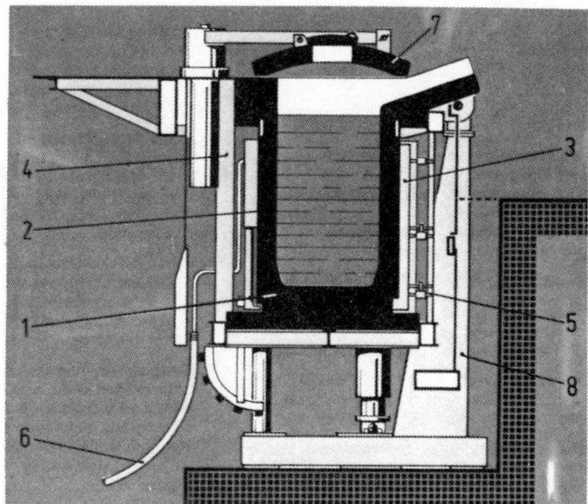

Figure 6. Basic design of an induction crucible furnace[9]

because of stressing. The outer structure also acts as a tilting frame, which enables the whole furnace system to be tilted hydraulically about the spout. If weighing is required, the load-measurement system can be arranged beneath the frame.

The crucible is covered by a roof (7) which can be lifted and swivelled hydraulically. The furnace is clad (8) to protect the coil from dirt and steel splash. The electric current passes through water-cooled cables (6); water for cooling the coil is either supplied via these water-cooled cables or through separate hoses. With proper control, the magnetic load on the laminations is so small that special cooling is unnecessary; water-cooling may only be required for high specific powers or higher frequencies. Waste gases are extracted via a furnace ring, roof, or hood system. The method used for this depends on the way the furnace is charged; noise-suppression may be required in consequence.

Small furnaces for 10–1000-kg tapping weight are mostly used in laboratories and special foundries. Table 2 shows technical details of those commonly used today. Smaller furnaces are usually of open construction, while larger ones usually have laminated stacks. Today, power is supplied almost exclusively via static frequency-converters (500–4000 Hz). The power planned for given capacity depends on specified melting time. Pure melting time is usually 1 h; ½ h is possible at higher frequencies, and 2 h is still economical.

Medium-size furnaces have tapping weights of between 1 and 20-t. For lower weights they operate at medium frequencies; above that, they use mains frequency. The former are nowadays mostly of compact construction and have laminated stacks, as with mains-frequency furnaces; power is supplied via static converters. The smaller furnaces are mostly used for steel melting, while the

Table 2. Guide to smaller, medium-frequency crucible furnaces[9]

Furnace type	Crucible capacity	Frequency	Converter power on MF side	Maximum melting output (no liquid heel)	
				Cast iron (1450 °C)	Steel (1650 °C)
JSM	(t)	(Hz)	(kW)		
0.25	0.25	1000	320	415	365
		1000	320	410	360
0.5	0.5	1000	435	590	525
		1000	535	740	670
		500	380	495	430
0.75	0.75	1000	535	735	660
		1000	760	1090	985
		1000	535	725	645
1.0	1.0	500	645	900	805
		1000	760	1080	975
		500	970	1410	1280
		500	645	885	790
		1000	760	1070	955
1.5	1.5	500	970	1400	1260
		500	1290	1910	1740

larger ones are used almost exclusively for castings; however, the applications for the sizes overlap. Table 3 lists the technical data for large medium-frequency crucible furnaces.

High-capacity furnaces today have melt weights between 20 and 60 t; they are used almost exclusively in the production of grey and malleable iron castings. Larger furnaces of up to 150-t capacity are beginning to be introduced. In order to achieve a high absolute output they have to be operated at mains frequency. Table 4 quotes guide-values for such furnaces. As in Tables 2 and 3, crucible capacities are quoted for several supply power ratings; this allows optimum melting rate adjustment to suit the operation.

Many varieties of *induction crucible furnace installations* are possible and can be made to fit space available. Figure 7 shows an example of a newly erected medium-frequency melting installation. The schematic circuit diagram for a mains frequency induction furnace installation is shown in Figure 8.

Reliable and adequate cooling for the coil is indispensable for induction furnace operation. The transformer, capacitors, and laminations may also be water-cooled to maintain operating temperature within limits. All other components are air cooled. Figure 9 is a schematic diagram of the type of closed-circuit water-cooling system most commonly used today.

Table 3. Guide to large, medium-frequency crucible furnaces[9]

Furnace type ITM	Crucible capacity (t)	Frequency (Hz)	Converter power on MF side (kW)	Maximum melting output (no liquid heel) (kg/h) Cast iron (1450 °C)	Steel (1650 °C)
		1000	760	1090	975
3	2.0	500	970	1440	1290
		500	1290	1970	1780
		500	1610	2490	2270
		500	970	1410	1260
		500	1290	1940	1760
4	3.2	500	1610	2480	2250
		500	1940	3030	2760
		500	2260	3560	3250
		500	1290	1920	1740
		500	1610	2470	2230
5	5.0	500	1940	3010	2740
		500	2260	3540	3230
		500	2560	4010	3620
		500	2900	4580	4140
		500	2260	3520	3210
		500	2560	3990	3600
		500	2900	4560	4120
6	8.0	500	3240	5060	4540
		500	3600	5660	5100
		500	3900	6160	5540

6.1.2.2 Lining of induction crucible furnaces

All refractory linings for steel-production vessels have to fulfil the general mechanical, thermal, and metallurgical requirements and have satisfactory durability; however, with induction crucible furnace linings, the electrical laws of inductive energy transfer have also to be considered.

Mechanical The lining must form a crucible which is tight and strong enough to take the melt and withstand mechanical stresses and thermal shock. It must be able to withstand not only the loads imposed by charging but also erosion caused by the powerful melt motion and slag attack.

Thermal Heat-losses and regenerative heat should be as low as possible. For example, a layer of asbestos between the inside of the coil and the outer wall of the crucible reduces heat-losses. Increasing the thickness of the asbestos layer reduces heat-losses further; however, at the same time the crucible-wall

Table 4. Guide to large mains-frequency crucible furnaces (50 Hz), with melting capacity and power consumption for cast iron and steel at a tapping temperature of 1500 °C[10]

Furnace type IT..	Crucible capacity (t)	Supply power (kVA)	Rated furnace power (kW)	Melting rate (kg/h)	Specific power consumption (kWh/t)
7/1320		1250	1150	2000	605
7/1650	12.5	2000	1800	3350	565
7/1980		2800	2600	5000	545
8/1320		1700	1540	2800	575
8/1650	16.5	2650	2400	4600	545
8/1980		3800	3450	6800	535
9.1/1650		3000	2650	4700	590
9.1/1980	22.0	4250	3800	7000	570
9.1/2200		5100	4700	8750	565
9.1/2450		6300	5800	11000	560
11.1/1650		3750	3450	6200	585
11.1/1980	31.0	5500	4950	9200	570
11.1/2200		6700	6150	11500	560
11.1/2450		8200	7600	14400	555

temperature increases, and insulation which causes very high loads on the crucible lining can lead to premature lining failure. Fairly high heat-losses must be tolerated in order to achieve good durability, especially with basic linings.

Metallurgical According to its composition, the crucible may be attacked directly by the constituents of the melt and by its reaction-products as well. Individual slag-formers and reaction-products attack the crucible, particularly in the slag zone; however, damage caused by infiltration can also occur anywhere else in the crucible. This changes the qualities of the crucible to such an extent that it may no longer be able to withstand the thermal stresses; cracks and spalling then usually lead to rapid failure.

Electrical The insulation of the coil and crucible must be sufficient for the voltages in operation, and the crucible-wall thickness must allow efficient energy-transfer.

The many sharply changing and partly conflicting demands on induction furnace crucibles mean that the main problems in induction melting are the correct choice of materials and crucible dimensions and controlling the operation suitably at every stage.

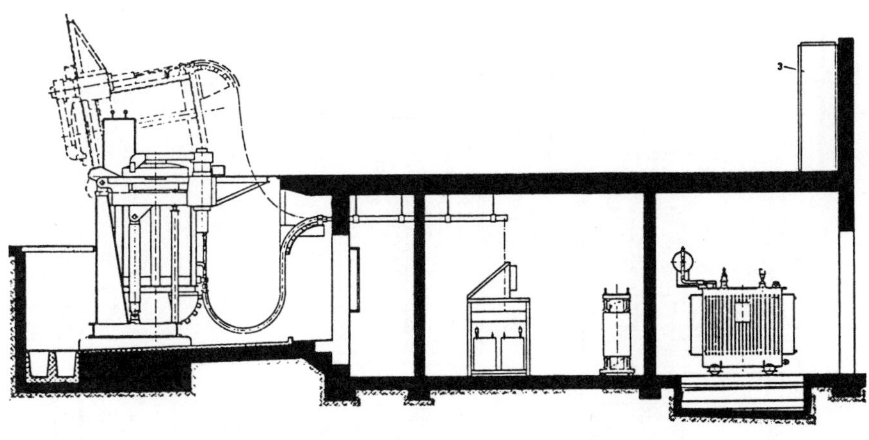

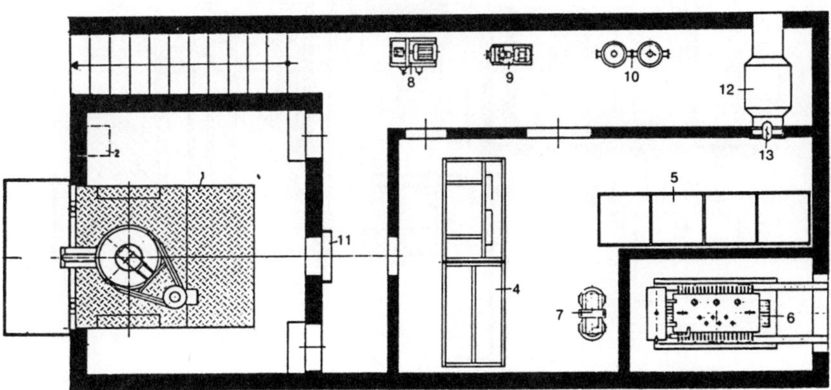

Figure 7. Site-installation plan for a medium-frequency crucible furnace with thyristor convertor.[9] 1 MF crucible furnace; 2 switch-panel; 3 switch-cubicle; 4 capacitor bank; 5 thyristor converter; 6 transformer; 7 reactor; 8 hydraulics; 9 water-pump; 10 water heat-exchanger; 11 cooling-water control; 12 air-filter; 13 room-ventilator fan

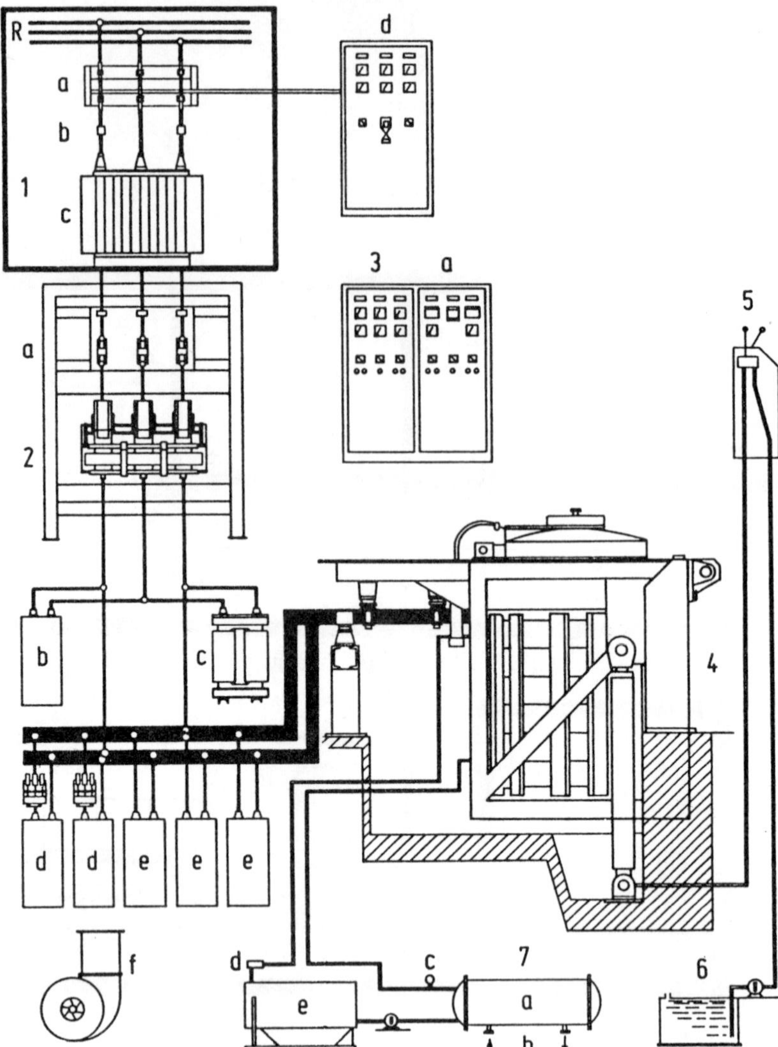

Figure 8. Schematic circuit diagram for a mains-frequency (50-Hz) crucible furnace melting system.[5] 1 High-tension enclosure: (a) circuit-breaker; (b) current-transformer; (c) transformer; (d) instrument panel. 2 Low-tension switchgear: (a) contactors, fuses, and instrument-transformers; (b) balancing capacitor; (c) balancing reactor; (d) fixed capacitors; (e) variable capacitors; (f) ventilation. 3 Control-cubicle: (a) automatic compensation. 4 Mains-frequency crucible furnace. 5 Control-pulpit for tilting. 6 Hydraulic tilting. 7 Indirect cooling system: (a) heat-exchanger; (b) water-supply; (c) pressure-monitor; (d) temperature-monitor; (e) water-tank

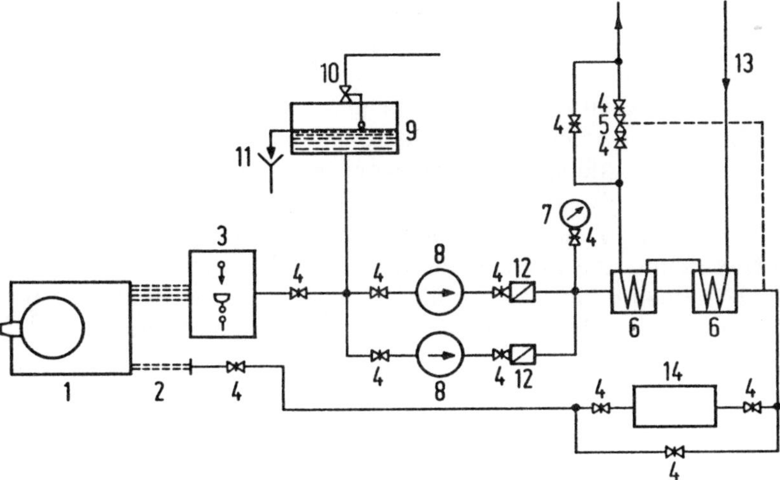

Figure 9. Closed-circuit water-cooling system for an induction crucible furnace installation with heat-exchangers.[11] 1 Furnace; 2 water-cooled cable; 3 cooling-water monitor; 4 stop-cock; 5 temperature-regulating valve; 6 heat-exchanger; 7 manometer (pressure-indicator); 8 pump; 9 header-tank; 10 float-switch (primary make-up water-inlet); 11 tank overflow; 12 non-return valve; 13 secondary water-inlet; 14 water-filter

6.1.2.2.1 Refractory materials for crucibles

Crucible materials for induction furnaces are particularly important since, because of the powerful bath motion, not only the slag but also almost the entire crucible inner surface takes part in the reaction throughout the whole melting process. The crucible materials are mostly constituted from pure, high-melting-point metal oxides, or, in special cases, mixtures of these. The oxides used are SiO_2, Al_2O_3, MgO, CaO, ZrO_2, and Cr_2O_3. These are usually used as ramming compounds, less often in shaped bricks and lastly also for preformed crucibles for purely laboratory use. Up to about 1950 it was up to induction furnace-operators to find ramming compounds suitable for their particular melting programmes. After then, the refractory industry carried out systematic research and found good lining compounds. For acid lining compounds the quality spectrum extends from adhesive sand to high-quality quartzite. Basic compounds and shaped bricks are made from much more varied constituents; at present their application extends to continuous melting of steel in crucibles of about 30-t capacity. The multitude of compounds and furnace sizes, as well as of melting programmes, require precise matching to one another to achieve good crucible-life.

A paper published recently describes the increase in life of rammed basic crucibles for a 3.6-t, medium-frequency furnace. The life was increased from

20 to 61 melts for the same melting programme by increasing crucible wall-thickness from 110 to 130 mm and by using an improved ramming method and increasing density from 2.7 to 2.9 kg/dm^3. Further experiments were aimed at using special metallurgical techniques to strengthen the protective lining of granular periclase which adheres to the inside of the crucible. This layer is about 10 μm thick and is very strong and dense and is heavily enriched with spinels. Carefully controlled additions of aluminium during melting down enable a mean crucible life of 108 melts to be attained. Further refinements of the process enable a mean crucible-life of 125 melts to be reached.[12] Since the end of 1977 attempts have been made to reproduce these results with two 31-t mains-frequency crucible furnaces, newly installed at the same place.[13,14]

6.1.2.2.2 Crucible-lining technique

From what has been said already, it is clear that we can only sensibly make remarks here of a fundamental nature. Depending on the manufacturer and the application, the coils are insulated by surrounding each conductor with fibreglass, plastic strip or lacquer and/or separating the turns with insulators, sometimes cemented in place. Care must be taken that the inner surface of the coil is smooth so that it can make good contact with adjacent refractory material. Pointing and coating the inside of the coil smoothly with special cements has proved particularly effective with basic linings, to the extent that asbestos or fibreglass coats may not be necessary. When first applying a new ramming-compound lining, it is recommended that the manufacturer's notes be followed with regard to the shape of the ramming template, the ramming and drying-out processes, and starting up. It is especially important to produce a dense, sealed crucible, free from cavities or any porous parts, by careful ramming and/or vibration. Post-densification, using rods about 10–12 mm thick, can prolong the life of basic linings if carefully carried out. This can be done after one third and two thirds of the expected life (in charges), during down-time, by access through the upper front face of the ramming in the moveable part of the crucible.

One excellent way of assessing the suitability of a ramming compound is to look at a crucible fragment when wrecking-out a worn lining. Near the surface in contact with the steel, the cross-section of the fragment is densely and heavily infiltrated; after about one third of the remaining thickness the infiltration should cease and the original grain structure should become increasingly recognizable. After about two thirds thickness, it should be possible to rub off some of the grains, and loose compound should be present where the ramming touches the coil. If this generally describes the condition in which the crucible is found, then the compound, the lining technique, and the furnace-operation all conform to the best practice.

6.1.3 Metallurgy in the Open Induction Crucible Furnace

Rudolf Plessing, Kapfenberg

Induction crucible furnaces differ from arc furnaces in ways which are important metallurgically, although both types of furnace are used for making steel from scrap and alloys. These differences are:

(1) High, relatively narrow melting vessel;
(2) Low crucible-wall thickness;
(3) Relatively small area of metal in contact with slag;
(4) Low slag temperature;
(5) No carburizing during melting down;
(6) Powerful bath motion.

Therefore reactions such as oxidation, decarburizing, and dephosphorizing, as well as deoxidizing and desulphurizing, which are usually required in steelmaking, can only be applied to a limited extent or necessity in an induction crucible furnace used for production. Those reactions which take place inside the melt, or at the melt–lining interface, are assisted by the powerful bath motion. On the other hand, the metal–slag reaction suffers because of the smaller contact-area and lower slag temperature. Any attempts at enhancing this reaction by using slags that are more reactive and less viscous are further limited by the requirement for long life and low wall-thickness of the crucible. Induction furnaces are specially suited to the production of alloy steels by the remelting process; this is because the powerful melt motion dissolves alloying additions quickly. For this reason even small induction furnaces can be operated economically.

6.1.3.1 Metallurgical conversion in induction crucible furnaces

As we can presume here that the basic metallurgical processes are known, only the features peculiar to decarburizing in induction crucible furnaces will now be described.[1]

A *decarburizing* or *oxidizing* reaction is only necessary in special cases and is then effected by introducing oxygen. Moreover, with acid linings, decarburizing takes place at higher temperature through a silicon reaction (crucible reaction), according to the following equation:

$$2[C] + (SiO_2) \rightarrow 2\{CO\} + [Si]$$

It is assumed that *dephosphorization* will be effective with basic slags and crucibles. Nevertheless, like decarburization, it should only be resorted to in special cases since the necessary slags, which are rich in iron oxide, attack the crucible lining particularly severely.[1–6]

With regard to *diffusion deoxidation* and its effectiveness, we need only highlight the importance of the lower partial pressure of oxygen with basic linings as compared with acid ones.

Because of the powerful bath motion, conditions are particularly favourable in induction crucible furnaces for *precipitation deoxidation* and the relatively rapid extraction of deoxidation products. Here, as with diffusion deoxidation, more uneven and low oxygen contents may occur in the melt with basically lined crucibles than with acid linings.[7–19] Figure 10 shows the development of oxygen in the metal bath during refining in relation to crucible materials.[8]

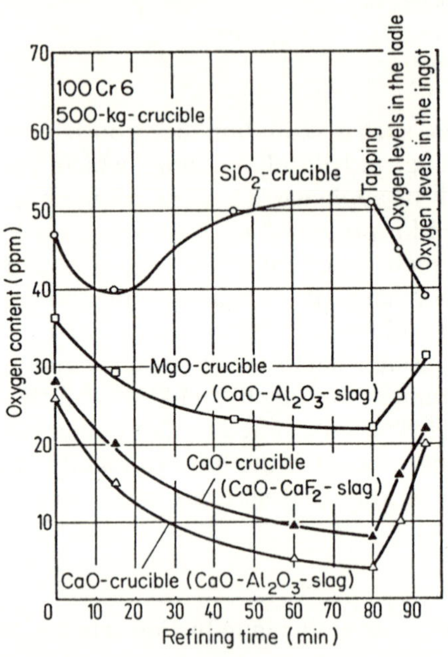

Figure 10. Oxygen-refining history with a 500-kg crucible for a 100Cr6 ball-bearing steel in relation to crucible material[8]

Proper deoxidation of the steel bath is essential for effective *desulphurization* and is only possible in basic lime or magnesite crucibles.[8] Figure 11 shows desulphurization with different slags in a basic crucible.

Gas-absorption and degassing are also of considerable importance in melting in induction crucible furnaces. The reactions in picking up hydrogen from atmospheric moisture, rust, or damp additions, and absorption of nitrogen from the atmosphere into the metallic bath, are well known.[1] Care should be taken to reduce gas-absorption from the atmosphere and from the charge-

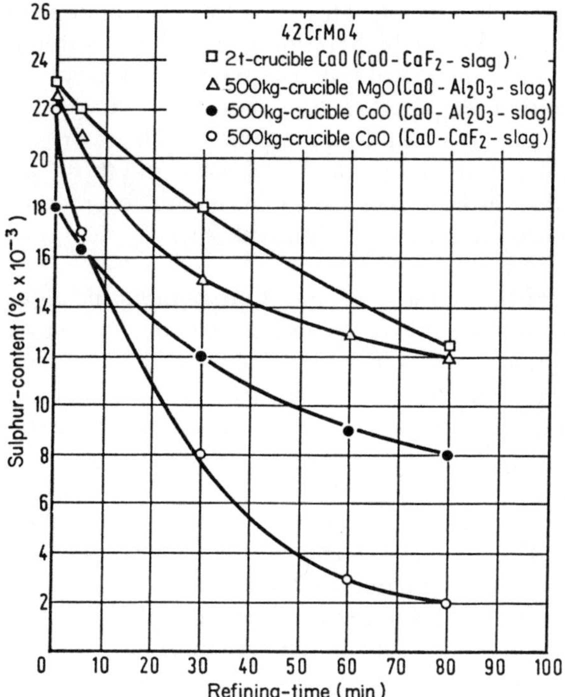

Figure 11. Desulphurization of a heat-treatable 42CrMo4 steel in lime and magnesite crucibles[8]

materials as far as possible by careful selection of raw materials and covering the steel bath as early as possible with suitable slag mixtures. Undesired nitrogen pickup by alloying elements, such as ferro-chromium, having a high affinity for nitrogen, can be avoided to a great extent if these elements are not added until the charge has become partly liquid. The open ladle can be flushed with argon in order to cut hydrogen and free nitrogen-absorption; however, this is so inefficient and requires so much argon that it cannot be used in normal induction-furnace operation.[12–15]

6.1.3.2 Normal metallurgical processing in induction crucible furnaces

From the preceding section it is clear why induction crucible furnaces are preferably used for pure remelting where metallurgical processing is limited to alloying and refining. This requires charges containing sufficiently low levels of phosphorus and sulphur.

The *slag-formers* (1–2% of the metallic charge is sufficient) are most often put on the bottom of the crucible. However, this is not enough to cover the bath where there is a powerful bath motion, as in mains-frequency furnaces. Ferro-molybdenum and ferro-tungsten *alloys* could form volatile oxides during heating up to melting temperature; they should therefore only be added when the charge becomes partly liquid. In the same way, it has proved advantageous to first, if possible, charge only unalloyed scrap and those alloying elements that are nobler than iron. This ensures that oxygen absorbed during melting is effectively reduced into carbon monoxide, which then escapes from the melt.

According to the desired steel composition, *refining* and *alloying* can already be started during melting down. As a rule, a sample for analysis is taken after melt-down is complete; this is then followed by a change of slag. For melting in acid crucibles glass slag is usually applied; for basic crucibles a lime and fluorspar or lime and alumina slag is used, according to the composition; with linings containing Al_2O_3 a magnesite-fireclay slag is used. The slag change is sometimes repeated.

Where low silicon levels are specified with acid-lined crucibles, the melt must not be overheated because of the danger of reducing silicon out of the lining or the slag. This danger is greater with higher carbon and chromium levels. Titanium and aluminium levels in the charge should be as low as possible. The manganese content of the steel is limited to about 2%. The temperature should only be raised to tapping level just before tapping. Aluminium, titanium, and similar elements should only be added just before or, if possible, during tapping.

These precautions are largely superfluous with basic-lined crucibles. In such cases we must take particular care that additions are absolutely dry and free from hydrates; this is because the hydrogen contents of steels from basic-lined induction furnaces are usually higher than in steels produced with acid linings.

6.1.3.3 Iron and alloy melting losses

With both acid and basic linings the iron melting losses are around 2–3%. In induction furnaces the alloy melting losses are relatively small, since there is no local overheating of the charge, and losses due to oxidation are relatively low.

Much has been said in the literature about the melting losses for individual alloying elements in acid and basic induction furnaces.[2,4,17,18] The most comprehensive survey is in the work by R. Rübendorffer and H. M. Kühn.[19] Melting losses are significantly affected by the mode of operation, the nature of the lining, and the composition of the steel to be melted. Most data in the literature concern the relative melting loss which is assessed analytically, while the output of liquid metal is ignored. However, aboslute melting losses form the basis for calculating production costs; these losses can only be ascertained from the materials balance. Table 5 shows the absolute melting-losses for different groups of steels and alloy-levels for charges of 30–50% typical recycling scrap in acid- and basic-lined, medium-frequency furnaces. The table

Table 5. Alloy melting losses in induction crucible furnaces; scatter ranges are determined by:

$$\text{Absolute alloy Melting loss (\%)} = \frac{(\text{total alloy charge (kg)} - \text{total alloy output (kg)}) \times 100}{\text{total alloy charge (kg)}}$$

Steel group and range of alloy content (%)	Absolute alloy melting-loss (%)		Remarks
	Basic and neutral lining	Acid lining	
Low-alloy case-hardening and tool steels			
Cr 0–2.0	3–5	3–5	
Mo 0–0.5	3–4	3–4	
Ni 0–2.0	2–3	2–3	
V 0–0.5	3–5	3–5	
W 0–2.0	3–4	3–4	
Cu 0–1.0	2–3	2–3	
Medium-alloy hardenable and tool steels			
Cr 2.0–5.0	3–5	3–5	
Mo 0.3–2.0	3–4	3–4	
Ni 2.0–5.0	2–3	2–3	
V 0.2–3.0	4–6	4–6	
W 0.5–3.0	3–4	3–4	
Co 0.5–3.0	2–3	2–3	
High-alloy tool steels			
Cr 2.0–15.0	3–5	3–5	
Mo 0.5– 5.0	3–4	3–4	
Ni 2.0– 5.0	2–3	2–3	
V 0.2– 3.0	4–6	4–6	
W 0.5–20.0	3–5	3–5	
Co 0.5– 5.0	2–3	2–3	
High-Mn Steels			
Mn 10–20	9–12	Acid melting impossible	

(*contd.*

Table 5 cont.

Alloy melting losses in induction crucible furnaces; scatter ranges are determined by:

$$\text{Absolute alloy Melting loss (\%)} = \frac{(\text{total alloy charge (kg)} - \text{total alloy output (kg)}) \times 100}{\text{total alloy charge (kg)}}$$

Steel group and range of alloy content (%)	Absolute alloy melting-loss (%)		Remarks
	Basic and neutral lining	Acid lining	
Stainless and heat-resisting steels; Cr and Cr Ni(Mo) steels; superalloys			
Cr 5.0– 30.0	3–6	4–7	Nb melting loss in scrap up to 25%; from added Nb up to 8%
Mo 1.5– 5.0	3–5	3–5	
Ni 5.0–100.0	2–3	2–3	
W 0 – 2.0	3–4	3–4	
Cu 2.0– 5.0	2–3	2–3	
Co 2.0– 50.0	2–3	2–3	
Nb 0 – 5.0	8–12	10–15	
High-speed steels			
Cr 3.5– 5.0	3–5	3–5	For Mo, V, and W: Lower limit of scatter for the lowest and upper limit for the highest-alloy levels
Mo 1.0–10.0	3–5	3–5	
W 1.0–20.0	4–6	4–6	
V 1.0– 5.0	4–6	4–6	
Co 2.0–20.0	2–3	2–3	
Al – Ni–Co magnet steels			
Ni 12.0–25.0	1.5–2		Acid melting not recommended for quality reasons, because required Si levels cannot be attained
Cu 2.0– 5.0	1.5–2		
Co 5.0–35.0	1.5–2		
Al 7.0–15.0	7–9		

was compiled from the author's experience for melt weights from 300 to 2500 kg. It should be remembered here that alloying elements with a higher affinity for oxygen reduce the melting losses of elements with a lower affinity. The various brands of steel in the individual groups are often made up of very different alloy steels, and therefore various scatter ranges are quoted.

The high-alloy and metal outputs, beside the other advantages, can be decisive for choosing the induction crucible furnace, in particular for high-quality steels. Such steels are produced in small quantities and contain high levels of Ni, Co, W, Mo, V, and similar elements. Induction furnaces are also very flexible and present no problems in changing programmes.

6.1.4 Melting Technique for Open Induction Crucible Furnaces

Karl-Heinz Brokmeier, Dortmund

With coreless induction furnaces, the electrical power input during melt-down depends not only on constructional and electrical characteristics but also on the type of scrap (packing density) and the level to which the furnace crucible is filled.

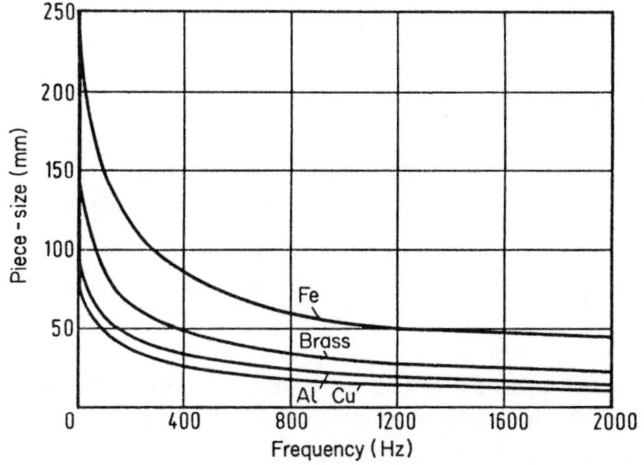

Figure 12. Smallest scrap piece-size in relation to frequency

Figure 12 shows the relationship between the smallest charge piece-size and the frequency at which the furnace works. The bulkier and more difficult to pack the scrap is, the lower is the electrical power input, and the longer is the melt-down time, leading to higher losses and therefore higher electricity costs.

If clean scrap of small piece-size can be used, special conditions apply. F. Bardenheuer[1] has already introduced the 'fill-factor' concept for these conditions. Fill-factor is the ratio of the weight of scrap initially charged to the weight of a hypothetical ingot which would completely fill the crucible. Although a medium-frequency furnace melts down both bulky scrap and that of small piece-size better than a mains-frequency furnace, it is both desirable and more efficient to charge the two types of furnace with scrap of the largest possible piece-size, packed relatively tightly in the crucible.

However, in cold-charge operation a *mains-frequency furnace* with a good charge still only yields 70–75% of the output of a furnace operating

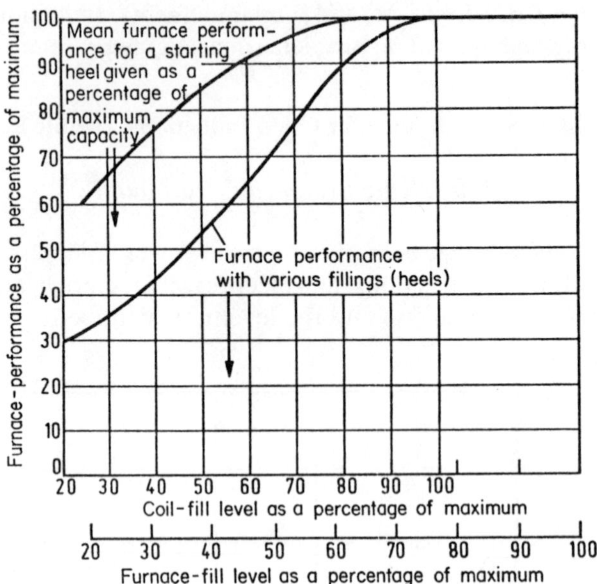

Figure 13. Mean effective output in relation to initial charging, and effective output in relation to percentage of furnace/coil height filled

continuously, with a liquid hot heel. Figure 13 shows the output of a mains frequency furnace in relation to its initial filling with remaining melt. The crucible should always be filled as full as possible so as to utilize the supplied power as efficiently as possible. Operated in this way, the induction crucible furnace is an ideal continuous melter. Quasi-continuous melting operation of a mains-frequency furnace enables a certain amount of molten steel to be removed (practical maximum, 40%)[2] at a desired composition and temperature. The new charge is then added to the remaining melt and a suitable amount of energy is applied. At the same time the necessary alloying materials are added, so that the desired composition is again reached at the end of the new melting cycle. Because of its simplicity, this quasi-continuous mode of operation has proved well suited to foundry work where the melting programme does not change. If this 'liquid heel' operation has to be interrupted because of an alloy-change, then the furnace must be completely emptied before beginning with a new quality of melt. In any event, the dependence of output on the furnace contents and on the start-up curve should be borne in mind. The output of a furnace system should be planned to accommodate variable operating circumstances.

The *medium-frequency furnace* is a better solution for cold-charge operation. It has the advantage of greater flexibility in the size of scrap it can accept.

Even with small furnaces a satisfactory relationship between average and maximum output can be achieved when the furnace is full (up to 90%).

6.1.4.1 Furnace charging

The following scrap-charging methods have proved suitable for mains-frequency furnaces in quasi-continuous operation. For material of small relatively uniform piece-size a shaker conveyer is used; a charging-bucket is used for medium-sized scrap, while a lifting-magnet is used for large pieces. The charging and exhaust-gas extraction systems should be matched to one another. Charging-material for continuous charging and any scrap added must be absolutely dry and as free as possible from rust, scale, coatings, etc. The first charge for each melt in medium-frequency furnaces must be particularly carefully inserted in order to avoid any damage to the crucible inner walls and bottom. This also applies with mains-frequency furnaces after idle times or quality changes. Thin chips and loose scrap steel sheet have proved very suitable as initial charge materials. The order in which individual pieces of charge material are inserted, by charging-basket or by hand, is also important. If the scrap is uniform and small in piece-size it is very prone to form bridges. In such cases the scrap underneath melts, while that above welds together without sinking into the bath and can only be melted down with difficulty. The liquid metal underneath in the crucible is then greatly overheated and attacks the lining. Such a 'bridge' can often be made to melt by tilting the furnace a little; the 'bridge' will then be melted sufficiently by contact with the overheated melt below to enable small pieces of scrap to be carefully inserted through the resulting hole.

Part of the allocation of slag-formers will already have been laid on the bottom of the crucible. In this way, slag can form immediately after the first part of the charge melts, thus lessening the attack on the refractory lining.

6.1.4.2 Furnace operation

What has been said above describes furnace-operation during melt-down. Since the refractory lining (and particularly basic lining) is the main problem with induction furnaces, the whole melting process has to be geared as far as possible to sparing the lining. It follows that the bath temperature should only be raised to tapping level just before tapping and that the slag-composition should always be maintained to a standard that ensures that the lining cannot be attacked, or at least that attack will be minimal, while due regard is paid to the metallurgical requirements.

During the melting and refining periods the furnace power and the insulation-resistance of the crucible lining should be monitored by watching special instruments. These indicate irregularities in wall-wear; in the event of any danger of breakout, a steel connection forms through a crack in the crucible, and causes an alarm signal to sound indicating a connection between the melt and the coil. If this happens, emergency tapping procedure must be carried out at once. Other system components such as the transformer, capacitors, coil-cooling, etc. are only monitored by warning devices.

6.1.4.3 Deslagging

Depending on the nature of the crucible lining and the slag composition, the slag may either be liquid or granulated. Liquid slag is either tapped or drawn off. Granulated slag may be removed with mechanical deslagging devices; the furnace-manufacturers offer appropriate equipment.

6.1.5 Melting Practice and Metallurgy for Sponge Iron in Induction Furnaces

Erwin Dötsch, Dortmund

The characteristic of the induction crucible furnace described in the previous section has two features. These are powerful bath motion and freedom from overheating; both make induction melting appear particularly attractive for processing sponge iron. This is why there has been some vigorous development in this field in the last few years.[1–9]

Without going into details about the results of research, the present state of the art can be described as follows:

(1) Only mains-frequency furnaces are suitable for the melting performance of at least 30 t/h which is required.
(2) Liquid-heel operation, which has to be used, can easily be realized with sponge iron — plain or in briquettes. Charging is continuous and is controlled to suit the furnace output in such a way that a constant bath temperature is maintained (Figure 14).
(3) At present, only acid refractory linings are available for induction furnaces with the required capacity of at least 30 tonnes. This means an acid slag, and so dephosphorizing cannot take place in the melting furnace.
(4) In general, acid linings cannot be used for producing low-carbon steels. However, it may be economical in special circumstances to use the induction furnace to produce a steel-like melt, which could then be further processed in a post-treatment installation.
(5) Controlling the vigorous boiling reaction in the relatively slim melting crucible and processing the slag arising are problems which still have to be solved on a production scale.

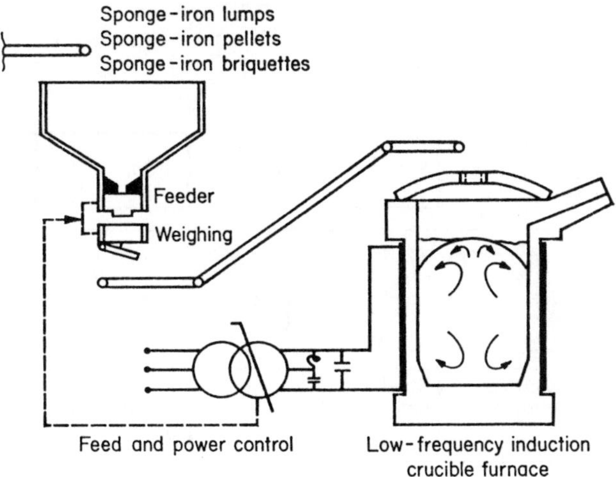

Figure 14. Melting-down sponge iron in an induction crucible furnace[2]

(6) Induction-melting of sponge iron is now available for the production of cast iron.[2,9,10] The boiling reaction here is less vigorous, because most of the residual oxygen is reduced by silicon. The tough slag which arises can be removed from the furnace in a reasonable time using a slag rake. The good economics of the process can be seen by comparing the advantages of sponge iron with the increased silicon-melting loss.

Since 1978, the entry into service of several large induction crucible furnaces for steel-production has significance for sponge iron as a charge for induction furnaces and especially for the development of a suitable basic refractory lining. One process worth watching involves suitable metallurgical methods to form a protective coating on the crucible inside wall. This coating brings about a significant increase in the life of the refractory lining.[11]

6.1.6 The Use of High-capacity Induction Crucible Furnaces in Steel Production

Erwin Dötsch, Dortmund

What has been said so far indicates that larger induction crucible furnaces are now used mainly to produce cast iron. However, we now also have the results of early experience in using large induction crucible furnaces for steel production.

Figure 15 shows three different processes, greatly simplified in the diagram, for using an induction furnace for producing steel (see also Chapter 7, Table 1).[1]

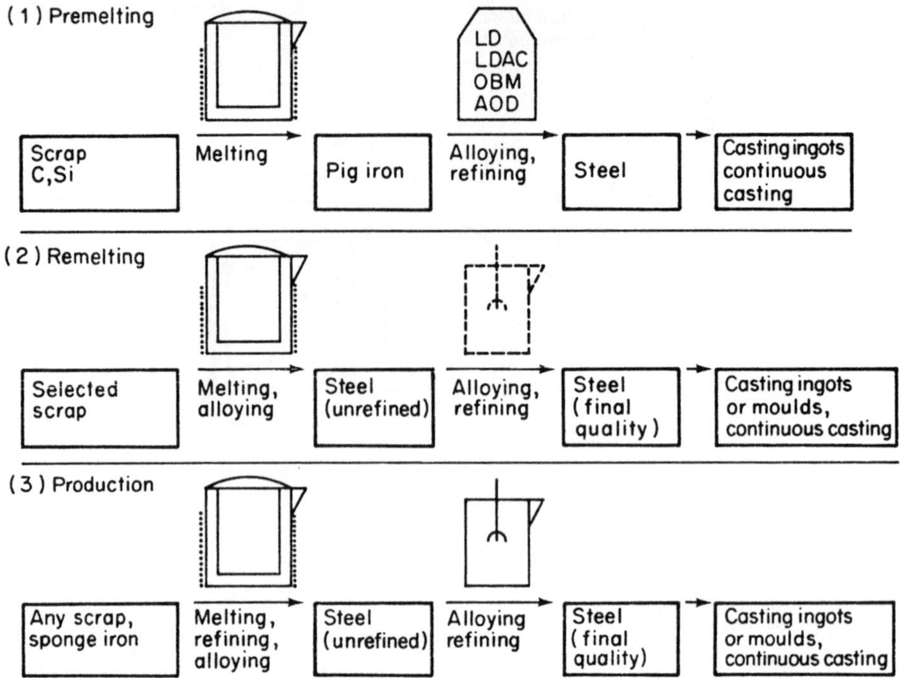

Figure 15. Schematic diagram of methods for using an induction crucible furnace in different steel-production proceses. LD = Linz Donawitz; LDAC = Linz Donawitz (French variant); OBM = oxygen bottom-blown; AOD = argon-oxygen-decarburization

(1) Producing a preliminary melt for further processing in a blast converter. Since the resulting premelt is hardly different in composition from pig-iron, the crucible furnace is used here much in the same way as for pig-iron production.
(2) Remelting from selected scrap. This steel is produced for eventual post-treatment in the ladle, and its composition is governed by the charge material to a great extent. No metallurgical refining to remove phosphorus and sulphur is carried out in the crucible. In this process, use is made of experience in transferring steel melting from medium-frequency furnaces to larger units and the special characteristics of mains-frequency furnaces.
(3) Producing steel from any type of scrap or sponge iron. In this process, the crucible furnace is not only used for melting but as a refining vessel as well.

At the time of writing, three induction crucible furnace installations were in use for premelting with *acid* linings using the first process above.

Since mid-1979, three 63-t induction crucible furnaces have been in use at the Allegheny Ludlam Steel Corp Brackenbridge Works in the USA, producing premelted charges from steel scrap and alloying materials for use in LD (Linz Donawitz) converters.[2] The power supply amounts to 21 MW each at a mains frequency of 60 Hz. Preheater feeders supply the crucible furnaces; these heat the scrap up to 500°C in horizontal vibrator-conveyors.

The melting system at the Fagersta Works in Sweden (also operating since mid-1979) works in a similar way. A 17-MW distributing plant, operating at a mains frequency of 50 Hz, can be connected to either of two crucible furnaces of 65-t capacity each. Here, too, the scrap is pre-heated on vibrator-conveyors. The induction furnaces produce molten charges for a 40-t LD converter.

The first installation in the Federal Republic of Germany has been in service since mid-1980 at Boschgotthardshütte O. Breyer GmbH, Siegen.[4] The crucible furnace capacity there amounts to 43-t; the power supply is 10 MW at 50-Hz mains frequency. Operating with a liquid heel, it produces 15 tonnes of melt per hour at a tapping temperature of 1600°C for further processing in a 15-t AOD converter (argon–oxygen decarburization).

The second process, remelting from selected scrap, has been in operation since mid-1978 in two 32-t mains-frequency crucible furnaces with basic (magnesitic) linings at Stahlwerken Bochum AG.[3,5] The two furnaces are fed in tandem by an 8-MW distributing plant. The crucible furnaces are fed automatically and melt 64 tonnes of steel, ready to cast, in 8 h, using the low-cost tariff. In cold-charge operation, i.e. when the furnace is emptied completely and restarted with scrap for each melt, the basic crucible currently achieves a life of 76 melts; it thus constitutes a significant step forward in the field of basic lining of induction crucible furnaces.

Experience in operation shows that induction furnaces used for steel-production, e.g. for steel castings, can significantly lower melting costs while at the same time being more acceptable in the environmental sense. The only precondition for this is the availability of suitable scrap.

6.1.7 Metallurgy and Melting Techniques in Vacuum Induction Crucible Furnaces

Wolfgang Stawicky, Altena

Rohn already wrote about the principles of melting in induction crucible furnaces back in 1929.[1] However, the breakthrough of this process for large-scale practical application only occurred in the 1960s in the USA.[2,3] The decisive factor was the requirement for high-quality steels and alloys for the aircraft industry and for nuclear reactors and electronics. The electronics industry in particular requires very high purity in connection with the miniaturization of components. Furnaces have been built with capacities of up miniaturization of components. Furnaces have been built with capacities of up to 50 t, and melting capacity has reached 300 000 tonnes per year.[3–5]

6.1.7.1 Application of vacuum melting

The cost of melting in a vacuum-induction crucible furnace is about two to four times that for other melting processes. This limits the process mainly to the production of nickel and cobalt-based alloys and high-alloy steels which have to meet special requirements. The process is also used where other methods are inadequate. Figure 16 illustrates the position occupied by the vacuum induction crucible furnace as a primary melting furnace relative to other processes commonly used today.

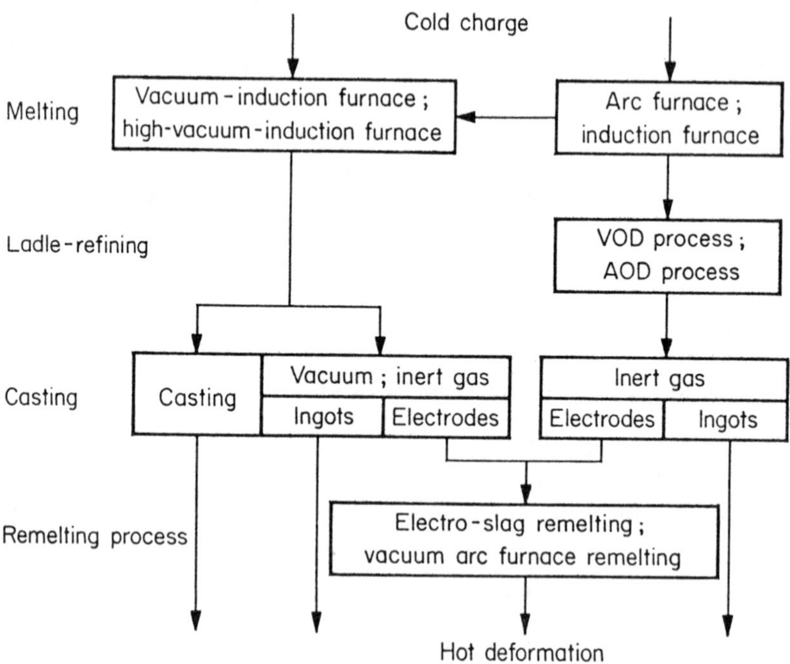

Figure 16. The vacuum-induction furnace as a primary melting system in relation to other methods

The economics of the process have been improved by the development of combined methods and special designs. Here, melting-down, oxidizing, and desulphurizing are taken over by an arc furnace, and any further processing of the melt then takes place in the vacuum-induction crucible furnace.[2,6] Here, we should differentiate between vacuum- and high-vacuum-induction crucible furnaces. In high-vacuum-induction furnaces melting can take place down to a minimum pressure of 10^{-4} torr; tapping is either carried out under vacuum or in an inert gas atmosphere.

Steels for particularly exacting requirements are premelted in a high-vacuum furnace, cast into electrodes, and then remelted in a vacuum arc furnace.

One special development is a combination of a high-vacuum induction furnace with an appended refinement zone using electron-beam heating.[7,8]

6.1.7.2 Metallurgy

It is characteristic of refining in vacuum induction crucible furnaces that the metallurgical process parameters, namely pressure, temperature, and duration of treatment, can be freely influenced to a great extent, and, assisted by the effective bath motion, enable intended metallurgical processing to be carried out quickly.[9] The refractory lining, which usually is based on MgO and Al_2O_3, needs special attention, since there is a possibility of chemical reactions and pick-up of oxygen from the lining at a late stage[10] (see Figure 17). A crystalline lime (CaO) lining creates favourable conditions; however, its peculiarities as a lining and for melting have to be borne in mind.

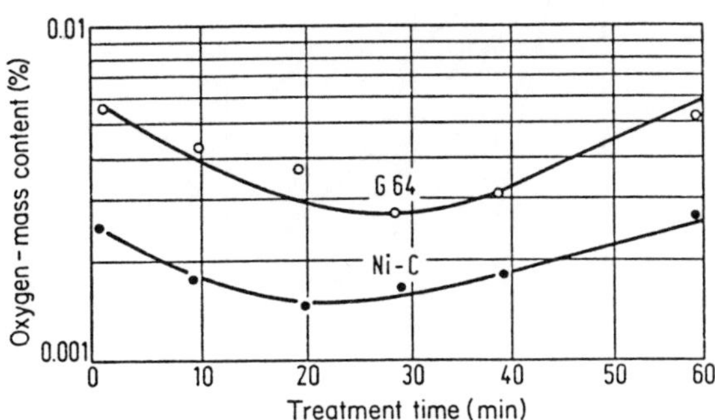

Figure 17. Change of oxygen content in a nickel-based Cr–Mo–V alloy (G64) and in nickel at 0.22%C (Ni–C) in an MgO crucible at a pressure of 1.33×10^{-5} mbar[10]

Because of the pressure relationships in the carbon–oxygen reaction, it is possible to achieve low C and O levels in the steel bath by adjusting the balance while the pressure is dropping. This also hold good for high-chromium melts, in which it is possible with vacuum-refining to obtain the lowest possible carbon levels of 0.01% or less without loss of chromium. The low oxygen levels attainable do not require precipitation deoxidation and allow high degrees of purity to be reached in the steel.

Hydrogen-separation proceeds according to Sievert's Square Root Law, and hydrogen levels of less than 1 ppm can be attained. Nitrogen-reduction follows the same law, but can be delayed by nitride-formers (e.g. Cr, Mn) and even prevented by using strong nitride-formers such as Ti or Al.[11,12]

Freedom from contaminants is especially important in nickel-based alloys. A few parts per million of impurities can be enough to prevent hot-deformation or to impair the required qualities markedly. Prolonged treatment at low pressure can remove Pb, Bi, Te, Cu, and Se; however, As, Sn, and Sb cannot be affected under vacuum. Tin-evaporation is only observed in melting iron[12] (see Figure 18).

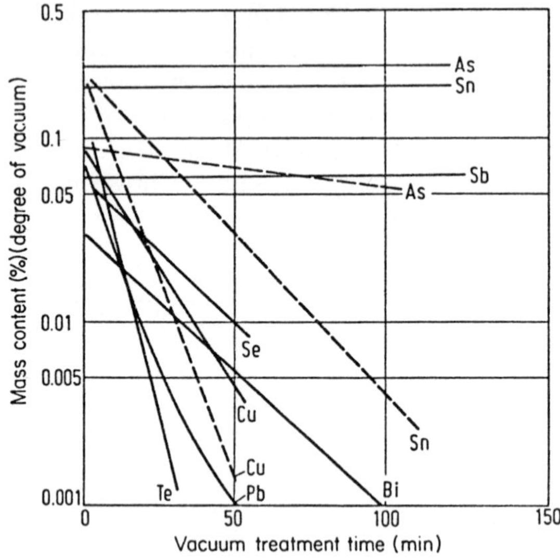

Figure 18. Evaporation of impurities from NiCr 8020 and iron melts under vacuum.[12] —— NiCr 8020 melts 1566°C, 6.63×10^{-3} mbar; -------- iron melts 1600°C, 6.65×10^{-3} mbar

6.1.7.3 Melting technique

Scrap should be selected for vacuum induction furnace processing in the same way as it is for melting in open induction furnaces. In addition, if a boiling reaction is intended, titanium, aluminium, and other elements which have an affinity for oxygen must be removed before vacuum-processing. In addition to careful selection of scrap, high-grade ferro-alloys and the purest metals are essential for the successful production of superalloys in vacuum induction furnaces.[13]

Melting down must begin with metals such as Ni, Co, Fe, and Mo, which have no great affinity for oxygen and nitrogen. Titanium, aluminium, and similar metals are only added after the end of the C–O reaction and the removal of all impurities.

Casting must be done under vacuum in order to preserve the high quality achieved in vacuum-melting. The refractories must then fulfil the same requirements as were described in section 6.1.7.2. The casting technique is conventional, and thus the grain cannot be directionally frozen.

6.2 Channel-type Induction Furnaces and Induction Transfer Ducts

Karl-Heinz Brokmeier, Dortmund

As was mentioned earlier, the old mains-frequency channel furnaces are now only of historical importance in steel-melting. In modern form, as induction-heated holding furnaces and foundry installations, they can be found with cylindrical crucibles, drums with a horizontal axis, with rectangular box shapes, and in many other forms evolved for special purposes in increasing numbers in highly automated foundries. They are also begining to be used as holding furnaces for steel.[1] The main element of the channel furnace is a ceramically lined closed channel which is metal-filled from the furnace bath, and where the metal acts as a secondary winding. The channel is flanged to the furnace vessel. The secondary coil (winding) is surrounded by transformer laminations which act as a core for a primary winding which carries the mains-frequency current. More details can be found in the technical literature.[2]

Induction transfer ducts are used in foundries between melting, holding, and casting units to improve the flow of material and co-ordination of individual production processes.

In steelworks they can be used for metallurgical purposes in counterflow processes. Here the steel is transported uphill, while the slag runs downwards. The counterflow produces large imbalances all over the reaction interface between the steel and the slag; this shortens reaction times to only 10% of the time required for confluent metallurgical processes.[3,4] Stacks of laminations are located under the ceramically clad ducts; these laminations carry individual coils, and the current is controlled in a way that produces a moving magnetic field which is followed by the flow of material.[2]

6.3 LITERATURE REFERENCES

6 Induction Furnaces

1. Hegewaldt, F., Schmelzen von Stahl im Induktionsofen. 19. Internationales Wissenschaftliches Kolloquium, Technische Hochschule Ilmenau 1974. Vortragsreihe A3.

2. Wetzel, K., H. D. Winkler and K. Muckoff, Staubemission eines 5 t-Induktionstiegelofens beim Schmelzen von Gußeisen. *Gießerei* **60** (1973), 603–7.
3. Domres, G., Staubemission von drei 22 t Induktionstiegelöfen beim Gußeisenschmelzen mit Schrottvorwärmung. *Gießereiprax.* **92** (1974), 367–9.
4. Wetzel, K., Möglichkeiten zur Erfassung und Begrenzung der Feststoffemission aus Induktionsschmelzöfen in Eisengießereien. Essen 1975. (Schriftenreihe der Landesanstalt für Immissions- und Bodennutzungsschutz des Landes Nordrhein-Westfalen. H. 35.) 66–8.
5. Brokmeier, K.-H., *Induktives Schmelzen*. Essen, 1966.
6. Hegewaldt, F., Problems in the design of induction melting furnaces (German). *Elektrowärme* **28** (1970), 197–207.
7. Neumann, F., Cast iron melting processes compared (German). *Elektrowärme.* Issue B, **32** (1974), 278–82.
8. Sommer, F., and E. Plöckinger, *Elektrostahlerzeugung*. 2nd edition. Düsseldorf 1964. (Stahleisen-Bücher. Vol. 8.)
9. BBC-Brown, Boveri & Co. A.G. publication D IO 40312 D Dortmund, 1974.
10. BBC-Brown, Boveri & Co. A.G. publication D IO 50265 D Dortmund, 1975.
11. BBC-Brown, Boverie & Co. A.G. works drawing.
12. Wilke, K., Erhöhung der Haltbarkeit einer basischen Tiegelzustellung für einen 3,6 Mittelfrequenzofen durch gezielte metallurgische Maßnahmen beim Erschmelzen von unlegiertem und legiertein Stahlguß. Aachen 1976. (Dr-Ing.-Diss. Techn. Hochsch. Aachen.)
13. Dötsch, E., Steelmaking in mains frequency coreless induction furnaces — present status and significance for the foundry industry (German). *Elektrowärme*, Issue B, **36** (1978), 154–9.
14. Wilke, K., Großinduktionstiegelöfen. Paper at Internationalen Kongresses für Hüttentechnik, 18.–20. June 1979 in Düsseldorf.

Three induction melting furnaces will provide hot metal for BOF at Brackenridge works of AL. *Ind. Heat.* **45** (1978), No. 10, 26.

Swannell, B. E., Practical aspects of bulk melting with induction furnaces: experiences with a modern induction melting plant in a European foundry. *Brit. Foundrym.* **71** (1978), No. 11, 254–60.

Wardle, Bob, Allegheny Ludlum turns on to induction melting. *Metal Producing* **17** (1979), No. 4, 51–3.

Harrison, Wm. Leonard, Engineering and installation of coreless induction furnace plant. *Elektrowärme,* Issue B, **37** (1979), No. 4, 209–14.

6.1.3 Metallurgy in the open induction crucible furnace

1. Sommer, F., and E. Plöckinger, *Elektrostahlerzeugung*. 2nd edition. Düsseldorf 1964. (Stahleisen-Bücher. Bd. 8.)
2. Bardenheuer, P., and W. Bottenberg, Die Entphosphorung und Entschwefelung im kernlosen Induktionsofen. *Mitt. K.-Wilh.-Inst. Eisenforsch.* **15** (1933), 49–53.
3. Goedecke, W., Die metallurgischen Möglichkeiten des Induktionsschmelzofens im Vergleich zu anderen Ofensystemen. *Gießerei* **41** (1954), 405–10.
4. Wever, F., and H. Neuhauß: Über die Metallurgie des eisenlosen Induktionsofens. *Mitt. K.-Wilh.-Inst. Eisenforsch.* **8** (1926), 171–9.
5. Zotos, J., and D. A. Colling, Optimum desulphurization and dephosphorization of steels in air induction furnaces. *Proc. Electr. Furn. Steel. Conf.*, Electr. Furn. Steel Comm., Iron Steel Div., Amer. Inst. min. metallurg. petrol. Eng. **18** (1960), 273–78.

6. Polzguter, F., Über die Stahlerzeugung im kernlosen Induktionsofen größerer Bauart. *Stahl u. Eisen* **51** (1931), 513–20.
7. Fischer, W. A., and M. Wahlster, Untersuchungen über die Abschiedungsgeschwindigkeit primärer Desoxydationsprodukte aus Eisenschmelzen. *Arch. Eisenhüttenwes.* **28** (1957), 601–9.
8. Schöberl, A., W. Holzgruber and E. Kahler, Der Einfluß der Tiegelzustellung auf den Ablauf metallurgischer Reaktionen im kernlosen Induktionsofen. *Berg- u. hüttenm. Mh.* **111** (1966), 531–41.
9. Etterich, O., Kristallkalk, die höchstbasische Betriebszustellung für Induktionsöfen und ihre neuen metallurgischen Möglichkeiten. Aachen 1963. (Dr-Ing.-Diss. Techn. Hochsch. Aachen.)
10. Fischer, W. A., and H. Engelbrecht, Die gleichzeitige Entschwefelung und Desoxydation von Stahlschmelzen. *Stahl u. Eisen* **75** (1955), 70–5.
11. Fischer, W. A., and Th. Colmen, Der Einfluß des Kohlenstoffgehaltes auf die Entschwefelung von Eisenschmelzen durch einen Kalk-Flußspat-Tiegel im Hochfrequenzofen. *Arch. Eisenhüttenwes.* **21** (1950), 355–66.
12. Sperl, H., R. A. Weber and H. Wiemer, Betriebsergebnisse mit der Spülgasbehandlung von Stahl in der Gießpfanne. *Stahl u. Eisen* **96** (1976), 1056–60.
13. Speith, K. G., and O. Steinhauer, Verfahren zur Spülgasbehandlung von Roheisen und Stahl in der Pfanne. *Stahl u. Eisen* **83** (1963), 75–80.
14. Bächtold, H., and W. Waldvogel, Erfahrungen in der Stahlgießerei beim Spülen von Schmelzen mit Argon in der Gießpfanne. *Gießerei* **56** (1969), 313–18.
15. Dewshap, P., and G. Hoyle, Jet degassing of molten steel. *J. Iron Steel Inst.* **203** (1965), 988–94.
16. Pölzguter, F., Über die Bauart und Anwendung des kernlosen Induktionsofens im Elektrostahlbetrieb. *Stahl u. Eisen* **55** (1935), 773–9.
17. Hessenbruch, W., Zur Kenntnis des Hochfrequenz-Induktionsofen IV. Weitere Beiträge zur Metallurgie des eisenlosen Induktionsofens. *Mitt. K.-Wilh.-Inst. Eisenforsch.* **13** (1931), 169–81.
18. Weitzer, H., Die Abbrandverhältnisse im kernlosen Induktionsofen. *Stahl u. Eisen* **59** (1939), 1353–8.
19. Rubensdörffer, F., and H. M. Kühn, Metallurgische Arbeitsweise beim Schmelzen von Stahl im Induktionsofen. *Gießerei* **50** (1963), 41–51.

6.1.4 Melting technique for open induction crucible furnaces

1. Badenheuer, F., Bau und Betrieb großer kernloser Induktionsöfen. *Stahl u. Eisen* **55** (1935), 821–8.
2. Dötsch, E., Steelmaking in mains frequency coreless induction furnaces — present status and significance for the foundry industry (German). *Elektrowärme*, Issue B, **36** (1978), 154–9.

6.1.5 Melting practice and metallurgy for sponge iron in induction furnaces

1. Pantke, H.-D., and Ch. Queens, The processing of sponge iron in foundaries (German). *Stahl u. Eisen* **94** (1974), 1114–20.
2. Dötsch, E., Use of sponge iron in the crucible induction furnace (German). *Elektrowärme*, Issue B, **32** (1974), 273–7.
3. Brokmeier, K.-H., and F. Neumann, Leistungsgrenzen von Induktionstiegelofen. In *Internationaler Eisenhüttentechnischer Kongreß*, Düsseldorf 1974 (27–30 May) Vol. 2. [Düsseldorf] 1974. Report No. 3.1.2.8. 23 pages.

4. Dötsch, E., and F. Hegewaldt, Use of the crucible-type induction furnace for steelmaking (German). *Elektrowärme*, Issue B, **33** (1975), 248–53.
5. Wenzel, W., F.-R. Block and V. Grumbrecht, Melt-down of sponge iron in the induction furnace (German). *Stahl u. Eisen* **95** (1975), 496–502.
6. Swietalski, J., and E. Mazanek, Stahlerzeugung im Hochfrequenzofen aus Eisenschwamm. *Metalurg. i Odlewn.* **1** (1975), 239–50.
7. Frohberg, M. G., M. L. Kapoor, H.-D. Pantke and Ch. Queens, Das Verhalten von Phosphor beim Einschmelzen von Eisenschwamm. Arch. Eisenhüttenwes. **46** (1975), 695–700.
8. Pantke, H.-D., and Ch. Queens, Demands on the quality of sponge iron for steelmaking (German). *Stahl u. Eisen* **96** (1976), 652–7.
9. Geck, H.-G., and W. Maschlanka, Schmelzen von Midrex-Eisenschwamm in Induktions- und Kupolöfen. *Foundry Trade. J.* **141** (1976), 969–70, 973–5, 978, 981–2, 984–6, 989.
10. Kowalke, H., and W. Mainz, Einsechmelzversuche mit brikettiertem Eisenschwamm in Induktionstiegelöfen zur Herstellung von Gußeisen mit Kugelgraphit. *Gießerei* **64** (1977), 525–30.
11. Wilke, K., Erhöhung der Haltbarkeit einer basischen Tiegelzustellung für einen 3,6 Mittelfrequenzofen durch gezielte metallurgische Maßnahmen beim Erschmelzen von unlegiertem und legiertem Stahlguß. Aachen 1976. (Dr-Ing.-Diss. Techn. Hochsch. Aachen.)

6.1.6 The use of high-capacity induction crucible furnaces in steel production

1. Dötsch, E., Stahlerzeugung im Netzfrequenz-Induktionstiegelofen und Bedeutung für die Gußerzeugung. *Elektrowärme* Issue B. **36** (1978), 154–9.
2. Wardle, B., Allegheny Ludlum turns on to induction melting. *Metal Producing* **17** (1979), No. 4, 51–3.
3. Dötsch, E., Induktiv beheizte Anlagen zum Einschmelzen, Speichern und Gießen von Stahl und Gußeisen. *Elektrowärme* Issue B. **38** (1980), 229–30.
4. Glasmeyer, U., and E. Biener, Die Stahlherstellung auf dem Verfahrensweg Netzfrequenz-Induktionsofen — AOD-Konverter — Horizontal-Stranggießanlage bei der Boschgotthardshütte in Siegen. Eisenhüttentag 1980, Düsseldorf.
5. Wilke, K., Basisch zugestellte Induktionstiegelöfen mit einem Fassungsvermögen von 3,6 t und 32 t für die Erschmelzung von Stahl, *METEC 79*, Vol. II, 129/46.
6. Dötsch, E., and B. Schulze Zumhülsen, Netzfrequenz-Induktionstiegelöfen einer Großstück-Stahlgießerei. *BBB-Nachrichten* 1980, No. 3, 110–16.

6.1.7 Metallurgy and melting techniques in vacuum induction crucible furnaces

1. Rohn, W., Technische Eigenschaften vakuumgeschmolzener Metalle. *Z. Metallkde.* **21** (1929), 12–18.
2. Schlotter, R., Technologie der Vakuum-Schmelz- und Umschmelzverfahren für Superlegierungen und hochwarmfeste Stahle. *Gießerei* **61** (1974), 75–85.
3. Wahlster, M., and H. Spitzer, Technology and metallurgy of the processes for the production of special steels and comparable materials (German). *Stahl u. Eisen* **92** (1972), 961–72.
4. 60,000 pound vacuum induction furnace with many innovations in operation at cyclops plant. *Ind. Heat.* **37** (1970), 1487–1501.

5. Schlotter, R., Fortschritte im Vakuumschmelzen von Werkstoffen für die Luft- und Raumfahrt. *Vakuumtechn.* **18** (1969), No. 2, 31–40.
6. Stawicky, W., Schmelzen von nichtrostenden Stählen unter Vakuum. *Gießerei* **53** (1966), 229–37.
7. Hunt, Ch. d'A., and C. V. Harrison, Airco's facility for steel refining and casting with induction furnaces and electron beams (German). *Iron Steel Eng.* **48** (1971), No. 8, 85–8.
8. Harrington, Th. H., The electron-beam continuous hearth refining process and its products. *Canad. metallurg. Quart.* **10** (1971), No. 2, 137–45.
9. Winkler, O., The theory and practice of vacuum melting. *Metallurg. Rev.* **5** (1960), 1–117.
10. Snape, E., and P. R. Beely, The effect of crucible-melt-reactions on the oxygen content of vacuum induction melted nickel-base-alloys. *Trans. Amer. Foundrym. Soc.* **76** (1968), 183–8.
11. Kraus, Th., Einführung in die theoretischen Grundlagen der Vakuumbehandlung von Metallschmelzen. *Neue Hütte* **17** (1972), 581–5, 738–43.
12. Holt, R. T., and W. Wallace, Impurities and trace elements in nickel-base-super-alloys. *Internat. metals rev.* 1976, March, 1–16.
13. Schlatter, R., Vacuum induction melting. *J. Metals* **24** (1972), No. 5, 17–25.

Szekely, J., C. W. Chang, and W. E. Johnson, Experimental measurement and prediction of melt surface velocities in a 30,000 lb inductively stirred melt. *Metallurg. Trans.* **8B** (1977), No. 3, 514–17.

Ohno, Reiichi, Kinetics of evaporation of manganese, copper and sulfur from iron alloys in vacuum induction melting. *Trans. Iron Steel Inst. Japan* **17** (1977), No. 12, 732–41.

Suzumoto, Teiichi, Toru Takahashi, Kazuo Karashima and Sigenori Kawai, Auslegung eines 25-t-Vakuumraffinationsofens (Japanese). *Tetsu to Hagané* **63** (1977), No. 13, 2126–33.

Cleave, David A. van, The 26 Cr–1 Mo stainless that used to be produced in an electron beam furnace is now vacuum induction melted. *Iron Age* **221** (1978), No. 13, 47–50.

6.2 Channel-type Induction Furnaces and Induction Transfer Ducts

1. Bonis, P., and Ch. Shivdusani, Channel induction furnaces for holding, superheating and stirring steel. In *VIII Internationaler Elektrowärme-Kongreß,* Liege 1976, Ic.4.
2. Gerbig, H. E., State of engineering and trends of development for induction heated holding furnaces and casting equipment (German). *Elektrowärme,* Issue B, **35** (1977), 269–75.
3. Schenck, H., The effect of type of process on the technical operating efficiency of reactions between two phases, particularly slag and metal (German). *Stahl u. Eisen* **84** (1964), 311–26.
4. Schenck, H., M. G. Frohberg and D. Papamantellos, Ergänzende Ableitungen zu Stoffaustausch zwischen Schlacke und Metall bei permanent und bei transitorisch wirksmen Phasen. *Arch. Eisenhüttenwes.* **37** (1966), 13–19.

Dowsing, R. J., Channel-furnace superheat can help cut steelmaking costs. *Metals & Mater.* 1978, No. 6, 33–5.

Electric Furnace Steel Production
Edited by E. Plöckinger and O. Etterich

7 *Secondary Steelmaking*

HANS WILHELM KREUTZER, DÜSSELDORF

It can be seen from earlier chapters that the arc furnace is a very versatile melting system which can be used to produce a wide range of steels from the simplest types to the most highly alloyed grades. This is because reducing conditions can be established in the electric furnace, while with oxidizing melting and refining processes (Siemens–Martin open hearth and basic oxygen-converters) reduction can usually only be done in the ladle.

This disadvantage of basic oxygen-converters and open-hearth furnaces was one of the reasons for the intensive development of ladle-refining, in which the arc furnace also took part but with a different aim. With ultra-high-power arc furnaces the aim is to shorten the tap-to-tap time by transferring metallurgical work out of the furnace.

With many post-treatment ladle-refining processes today it is immaterial whether the steel originates from a basic oxygen-converter, an open-hearth furnace, or an arc furnace.

It is only possible to take full advantage of the huge power-capability and to achieve rapid amortization of the cost of the capital investment in a modern arc furnace, including the outlay on environmental protection, if the furnace is operated with long power-on times and at the highest average power possible. It is assumed that the time after melt-down, when boiling and deoxidation take place at much-reduced power, and the time when the power is off are both kept as short as possible.

With the conventional 'classic' two-slag refining process often only the heat loss is compensated by energy supplied during these times. Long tap-to-tap times and high temperatures during the refining period lead to heavy wear and long fettling times. The ladle-refining processes described below enable these time-consuming oxidizing and reduction stages to be performed outside the furnace. Thus ultra-high-power furnaces with specific powers of up to 800 kVA/t can be used exclusively for melting down and superheating, and tap-to-tap times of under 2 h can be achieved. At tapping weights of 130 to 140 tonnes the practical limitation of the 2-h furnace nowadays is the load capacity

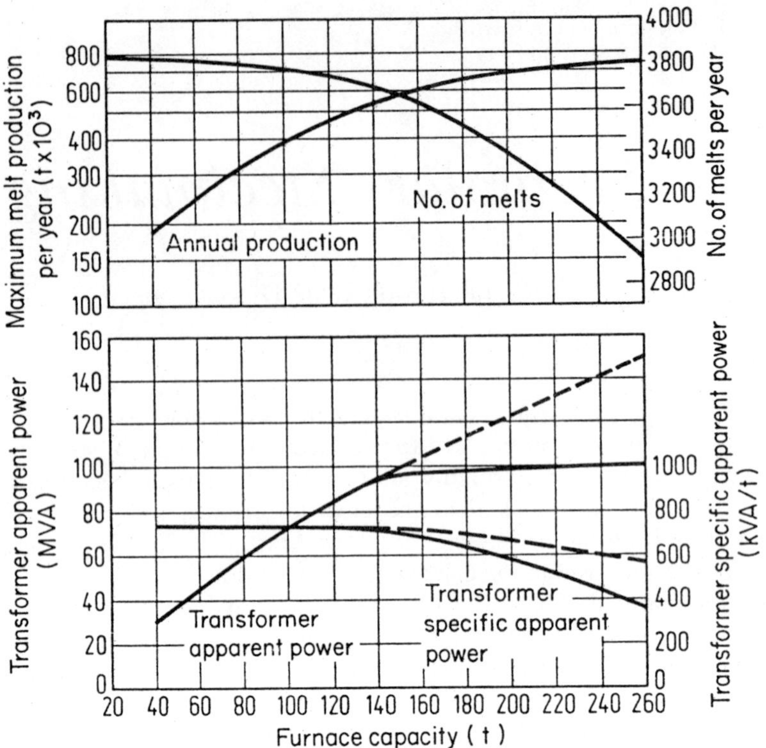

Figure 1. The relationship between specific transformer power, annual production, and the greatest number of melts attainable annually with modern arc furnaces (average power-consumption for melt-down 425 kWh per tonne of charge; time from end of melt-down to restart 1 h; 355 operating-day availability per year; utilization time 88% of available operating time; time for re-lining, repairs, etc. 12% of available operating time)

of the graphite electrodes. The annual outputs attainable with modern arc furnaces are shown in Figure 1.

The choice of which parts of the metallurgical processes to be transferred from the furnace to after-treatment units depends on actual technical and economic conditions. The type of programme, the quality required of the end-product, the energy supply, the raw material, and alloy bases as well as the casting system all affect the choice of process combination in any steelworks. The course of developments is leading towards the use of UHP furnaces solely as 'melting machines'.

The same developments are to be expected with induction furnaces when the problems of basic refractory linings in large furnaces are solved.

7.1 RESIDUAL FURNACE-REFINING WITH LADLE POST-TREATMENT

The melting vessel need not be confined to melting down scrap if it can also be used for other specified process stages which do not involve extra time. In addition, all the process stages that involve a great deal of energy are carried out in the melting vessel as far as possible. It is an advantage to melt down most of the ferro-alloys, especially chrome and nickel, in the electric furnace since ladle-refining is carried out with minimum energy-consumption or none at all. However, this requires precise conservation and control of furnace heat throughout all the post-treatment stages up to tapping, and the steel must eventually be at the right temperature for tapping from the melting vessel.

In general, no more than 3% by weight of alloy materials are added in ladle-refining without additional heat; otherwise, the tapping temperature would be excessive, especially with small furnaces.

7.1.1 Refining with Arc Furnaces

As has been said before, we can establish different oxygen partial pressures in arc furnaces. Oxidizing conditions are determined by charging with solid oxygen-carriers (ore, metal oxides, rolling scale) or by blowing in gaseous oxygen. In this way unwanted companion elements, such as aluminium and silicon, which enter the charge with scrap or alloying materials, can be oxidized during melt-down.

A non-alloy or low-alloy charge can be dephosphorized during melting down.[1–6] Powdered lime can be added to the liquid phase; lump lime can be fed continuously through the furnace roof, or lime or limestone can be included with the main charge.[5,7,8] If the amount of carbon and/or carbon carriers in the charge is raised, then more phosphorus and sulphur can be removed before the end of melt-down by feeding in extra oxygen during the melting-down period.[9] In any case, a reactive, lime-rich slag must be provided early during melt-down. The powerful electrodynamic forces in the high-current arcs, together with increased evolution of carbon monoxide and boiling limestone,[5] all invigorate the mixing of the slag with the bath and speed up the reactions.

The oxidizing conditions during melt-down limit the extent of desulphurizing and dephosphorizing. Figure 2 shows how the degree of desulphurization, measured in the distribution of sulphur in the steel bath and the slag, also depends on the amount of iron oxide in the slag.[10,11] The sulphur distribution is

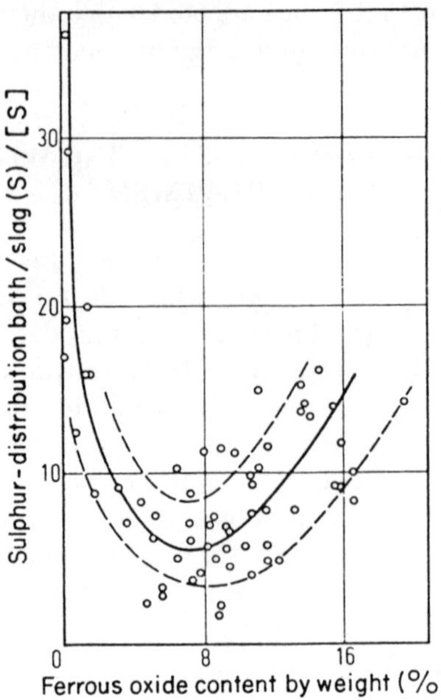

Figure 2. Relationship of sulphur distribution to the level of ferrous oxide in the slag in an electric arc furnace (after H. Schenck and E. Steinmetz[10])

unfavourable at first while the iron-oxide level is rising; however, the distribution coefficient rises again with high levels of FeO. This increasing sulphur distribution is connected with the saturation level in the $CaO - FeO_n - SiO_2$ slag system. The equilibrium values show that with oxidizing slags the sulphur distribution rises very quickly with FeO_n levels above 30% in the slag.[11] In addition, as with the oxygen blast process, blowing oxygen in during melt-down leads to the sulphur dissolved in the slag oxidizing into sulphur dioxide, which is released as gas[12] together with the dust. In arc furnaces a sulphur content of, say, 0.07% imported with the charge can be reduced by about half with oxidizing slags. In special cases sulphur levels of less than 0.01% can be achieved in the furnace using black slags, although only with several changes of slag.[13]

When sponge iron is fed continuously into an arc furnace and melted down, the conditions are not the same as when melting scrap. The residual oxygen content in the sponge iron necessitates the addition of carbon unless this is already present in the sponge iron because of a direct reduction process.[14] The amount of carbon required depends on the degree of metallization of the sponge iron. If the metallization is low then with high power from the transformer the amount of CO evolved can limit the addition of carbon and with it the melting performance.[14] The residual oxygen and the large quantity of slag resulting from the mostly acid gangue in the ore lead to effective dephosphorization. Sponge iron produced by the gas-reduction process is low in sulphur. The vigorous boiling action and slag frothing prevent nitrogen pickup; it is thus possible to produce deep-drawing steels in the arc furnace since these steels need very low nitrogen contents.[15] Another advantage of using sponge iron is the low level of trace elements. If there is no need for dephosphorizing and desulphurizing because of the type of charge and the specified contents of the finished product, then there is no need for any slag work, and dolomitic lime can be used to protect the refractory brickwork.[16] This also applies if it is customary to use an after-treatment for desulphurizing.

When producing high-chromium steels, most of the chrome is included as chrome-rich scrap or low-cost, high-carbon/high-silicon ferro-alloys added to the charge. If much carbon arrives with the charge, then the additional cost of pre-decarburization by blowing in oxygen near the end of melt-down is tolerated in order to relieve secondary refining units. The temperature must not be too low, and carbon content must not drop below about 0.40% in order to avoid much chromium loss. Slag-reduction is necessary before tapping unless the slag is transferred with the melt into a post-treatment converter and reduced there. No dephosphorizing is possible with this type of process where chrome content is high.

Before ladle-refining finally broke through, there were attempts to speed up the time-consuming work of reduction, alloying, desulphurizing, and the removal of deoxidation products in the arc furnace under white, lime-based slags, low in heavy metal oxides. However, the main problem with this is the small specific area of reaction between the steel and the slag.[17] Electromagnetic stirring coils[18] were more effective than manual or mechanical methods; however, these are seldom used today. The principal difficulty still lies in changing over from oxidation to reduction in just *one* vessel. Oxygen pickup from contaminated refractories and from residual slag delays the establishment of reducing conditions and degrades the attainment and reproducibility of equilibrium. Only by tapping can complete separation between oxidizing and reducing conditions be achieved. Transfer of slag rich in heavy metal oxides into the ladle can be avoided by careful design of the spout and other equipment as well as by controlling the slag.[19]

7.1.2 Refining with Induction Furnaces

Induction furnaces have been used for a long time for the production of medium- and high-alloy steels by remelting plain and alloy scrap which is low in phosphorus and sulphur, together with ferro-alloys. This is characterized by the avoidance of any carburization and low alloy losses (see Chapter 6). The possibilities for metallurgical slag work are very limited because the slag is only indirectly heated above the metal bath and can only be moved indirectly by the bath motion. As Table 1 shows, applications for the induction crucible furnace are limited by the type of refractory lining used.[20–23] While an acid lining can be used in case A (Table 1), as in the foundry industry, this has only limited possibilities in case B, and a basic lining is essential for case C. These applications have already been partially realized.

However, ladle-refining can only be used economically for high hourly outputs and larger melt weights; the specific heat losses with small weights are too high. The refractory problems with basic-lined crucibles of over 20 tonnes are still largely unresolved today. We therefore have the variant in case A in Table 1 in which a premelt high in carbon and silicon is prepared in an acid crucible and is then oxidized and refined in an appended secondary treatment process.

The great advantages offered by induction furnaces and the large-scale efforts being made in many places allow us to hope that it will be possible in future to use larger furnaces with basic linings.[23] This would also open up further possibilities for melting sponge iron continuously; as in arc furnaces, carbon-monoxide evolution is limited here because of the small specific area of the bath surface. However, decarburizing rates of 1 kg of carbon per square metre of bath surface per minute have been measured with experimental basic crucibles.[24,25]

The pronounced bath turbulence obtained with induction furnaces can be used very effectively in conjunction with an appropriate basic lining and a lime-based slag to obtain steel of high purity with low oxygen content. Tests have been carried out with crucibles of up to 1 tonne size,[26] made from fused lime, using lime/fluospar and lime/alumina slags.[27,28] Here, too, application in larger crucibles and large-scale production falls down because of problems with the refractory lining which so far remain unsolved.

When all these as yet unanswered questions have been dealt with, wide-ranging application of the possibilities of secondary refining should open up a large and interesting field of application for induction furnaces.

7.2 SECONDARY REFINING SYSTEMS

After-treatment systems were developed over the last thirty years (Figure 3)[29] in parallel with melting and oxidizing vessels which are designed for the highest power and output; each can operate at optimum only for certain refining process stages and is not very suitable or may be totally unsuitable for others. Figure 4 shows the various refining-process stages from melting up to attainment of the exact temperature for casting as well as the suitability of melting and oxidizing methods for the various steel-production processes. The

Table 1. Three possible processes for making steel in mains-frequency induction furnaces with acid or basic linings:

A: Producing a 'pig iron' premelt followed by oxidation in a converter
B: Remelting from selected scrap and alloys
C: Melting any scrap or sponge iron

Charge	Mains-frequency induction crucible furnace	Melt when tapped	Post-treatment 'ladle-refining'	Final steel	End-product
A Scrap, Carbon Carriers, Ferro-silicon, Ferro-alloys (carbon-rich)	Melting Acid lining	'Pig iron' premelt	Oxidizing, Refining, Alloying in converter (LD, OBM, AOD) or in ladle (VOD)	High-alloy and Low-alloy steels	Ingots, Continuous casting, Castings
B Selected clean scrap	Remelting and alloying Acid lining	Limited tapping temperature and types of steel	Desulphurizing, Deoxidizing, Alloying, Purging — according to circumstances	All steel compositions	Ingots, Continuous casting Castings
C Any scrap Sponge iron	Melting Alloying Refining Basic lining		Desulphurizing, Deoxidizing, Alloying, Purging, — according to circumstances	All steel compositions	Ingots, Continuous casting, Castings

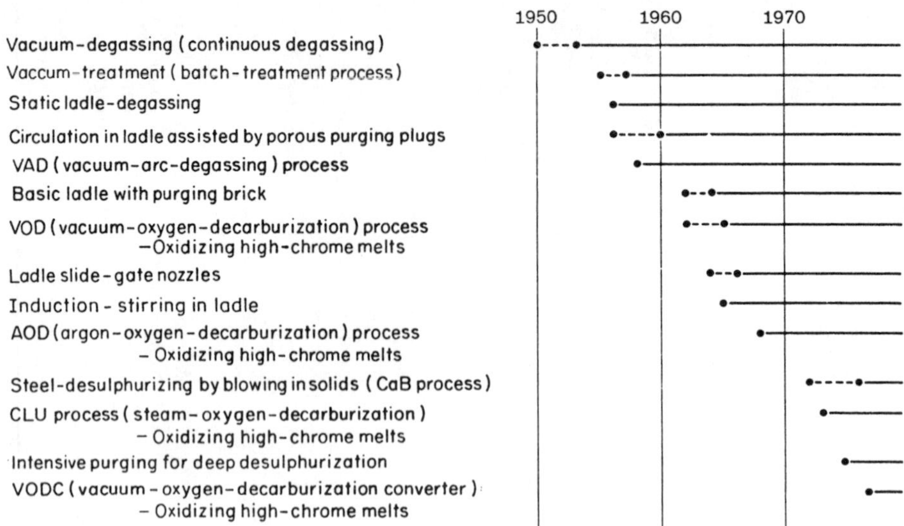

Figure 3. Chronological development of significant secondary refining processes[29]

Legend: ■ Very well suited; ▩ Well suited; ▨ Moderately suited; □ Ill suited

	Oxygen-refining processes		Hearth-melting processes	
	LD/LDAC converter	OBM/LWS converter	Siemens-Martin (open-hearth) furnace	UHP arc furnace (single slag)
Melting solid charges				
scrap	Moderately suited	Moderately suited	Very well suited	Very well suited
directly reduced iron ore	Moderately suited	Moderately suited	Moderately suited	Very well suited
ferro-alloys	Ill suited	Ill suited	Moderately suited	Very well suited
Decarburizing				
non-alloy melts	Very well suited	Very well suited	Well suited	Moderately suited
non-alloy melts (finely decarburized)	Ill suited	Very well suited	Ill suited	Ill suited
high-Cr-alloy melts	Well suited	Well suited	Ill suited	Well suited
high-Cr-alloy melts (finely decarburized)	Ill suited	Moderately suited	Ill suited	Ill suited
Dephosphorizing				
non-alloy melts	Well suited	Well suited	Moderately suited	Moderately suited
Desulphurizing				
coarse desulphurizing	Moderately suited	Moderately suited	Moderately suited	Moderately suited
fine desulphurizing	Ill suited	Moderately suited	Ill suited	Ill suited
Alloying				
low-alloy steels	Moderately suited	Moderately suited	Moderately suited	Moderately suited
high-alloy steels	Ill suited	Moderately suited	Ill suited	Moderately suited
narrow chemical composition	Ill suited	Moderately suited	Ill suited	Moderately suited
Degassing				
hydrogen-removal	Ill suited	Moderately suited	Ill suited	Ill suited
lowest nitrogen contents	Well suited	Well suited	Moderately suited	Ill suited
Deoxidizing				
carbon deoxidation	Ill suited	Moderately suited	Ill suited	Ill suited
precipitation deoxidation	Ill suited	Moderately suited	Ill suited	Moderately suited
Narrow casting-temperature range	Moderately suited	Moderately suited	Moderately suited	Moderately suited

Figure 4. Grading steelmaking processes according to metallurgical possibilities[30]

UHP arc furnace working in oxidizing conditions under a slag is best for melting from scrap, directly reduced ore and many ferro-alloys. It is also fairly suitable for decarburizing, dephosphorizing, and rough desulphurizing.

As has been said before, it is ill suited for all other process stages for the following reasons:

(1) It is not possible in an arc furnace to change over quickly and completely to reducing conditions after melting down and oxidation. Subsequent pickup of oxygen from the slag, from refractories contaminated with heavy metal oxides, and from the atmosphere cannot be avoided completely.
(2) All reactions between a steel bath and a refining slag (deoxidation, refining, desulphurizing) only proceed at optimum rates if matter can be exchanged between the slag and the steel over the shortest diffusion paths and the most intimate phase-contact possible. In arc furnaces, the specific interface areas between steel and slag are small.[17] Conditions can be improved by positive mixing using mechanical, pneumatic, or electromagnetic methods, but only at a relatively great cost in time.
(3) Reactions which are assisted by partial vacuum (decarburizing and degassing) can only be carried out incompletely and to limited effect in arc furnaces.

Steel after-treatment processes can be divided into three groups, according to purpose:

(1) Oxidizing post-treatments for decarburizing at reduced carbon monoxide partial pressure for:
    (a) Unalloyed steels to obtain the lowest levels,
    (b) High-chrome steels to reduce chromium oxidation.
(2) Reducing treatments for:
    (a) Deoxidizing,
    (b) Separating out deoxidation products,
    (c) Alloying,
    (d) Desulphurizing,
    (e) Homogenizing (temperature and chemical composition).
(3) Vacuum treatments for:
    (a) Lowering gas contents (hydrogen and nitrogen),
    (b) Improving purity by vacuum deoxidation (carbon deoxidation),
    (c) Adding elements with a high affinity for oxygen,
    (d) Cutting down reoxidation (vacuum ingot casting).

These treatments may be carried out in ladles, converters, or vacuum installations. Oxidizing post-treatments for decarburizing are carried out in converters or ladles by adding oxygen while lowering the partial CO pressure.

This requires either very intensive mixing with appropriate reactive gas mixtures or vacuum treatment with forced circulation. The second group of reduction post-treatments can be carried out in ladles, converters, and vacuum installations with appropriate additives. The very numerous vacuum-treatment processes enable either reducing or oxidizing conditions to be used and can thus be employed in various ways for a very broad production programme.

Post-treatment installations should eliminate the drawbacks of arc furnaces enumerated above as well as of other steel-production processes. They should enable a very high degree of matter-interchange between steel and the reactive medium (oxygen, slags, deoxidation media, desulphurization media, etc.). With all post-treatment systems for steel in the static state, the relationship with bath volume of the motionless reaction interface between the slag and the steel bath is significantly more unfavourable than with arc furnaces. It is therefore necessary to intensify the interchange of matter between steel and slag, i.e. between steel and reactive medium, so as to increase turnover and improve utilization. In addition, spurious reactions must be avoided. Such reactions involve pickup of oxygen from the slag and from refractories or the atmosphere, particularly during reducing post-treatments.

Mechanical mixing can be used for desulphurizing pig iron.[13] In steel after-treatment the temperatures are around 300°C higher; this has so far made such mixing systems impracticable because the life of refractory parts is limited. Pneumatic mixing via purging bricks, lances, or jets in the bottom or wall can greatly improve the exchange of matter between the steel bath and the slag;[31] however, this requires treatment vessels with enough freeboard and high specific volume (cubic metre of vessel volume per tonne of steel). Specific treatment volumes for ladles, converters, and vacuum units are shown in Table 2. Homogenizing and light purging require the lowest specific volume with a freeboard from 200–300 mm. Intensive purging requires a freeboard of up to 1 m. Converters have a working volume of 0.4 m^3 per tonne of steel; this is three times the volume of the steel to be treated. Small AOD converters have a specific volume of up to 0.7 m^3 per tonne of crude steel. The greater the specific volume, the more turbulence and specific gas throughput can be achieved in blowing; in AOD converters this can amount to up to 1 m_n^3 per tonne per minute.

7.2.1 Ladles

It is desirable in all applications that ladles for ladle-treatment processes should be basic-lined; in very many cases it is absoultely essential. Acid-lined ladles can be used for fine adjustment of chemical composition, for homogenization within narrow temperature limits, or for limited desulphurization. Purity can also be improved significantly by moderate purging in acid ladles after tapping from the electric furnace but keeping back the black slag there. However, if the

Table 2. Characteristic values for after-treatment in ladles and converters

	Ladle treatment					Refining high-Cr melts		
	Homogenizing		Intensive purging		Lancing	Vacuum ladle	Converter	Vacuum converter
	Atmosphere	vacuum	Atmosphere	vacuum		VOD	AOD, CLU	VODC
Specific volume (m^3/t liquid steel)	0.17		0.22			0.22	0.4–0.7	0.5
Specific gas throughput (m^3/t min)								
Purging Brick Ar, N_2	0.001	0.0005	0.01	0.008		0.0005/10		0.01
Jets O_2							0–0.75 } 0.6 to 1.0	
Ar, N_2, H_2O							0.15–0.5 }	
Lancing O_2 Ar, N_2						0.2/0.7	1.2 (0.2)[a]	1.2

[a] Blown in via side-jets.

free oxygen content in the steel must be brought down to very low levels, as is necessary with extreme desulphurization down to below 0.003%, then ladles in bauxite, magnesite, or dolomite must be used.[32] Magnesite or dolomite linings cannot be avoided for after-treatments involving high steel-temperatures held for long periods, as with the VOD process for example; this is because of the problem of lining durability.[33–6]

Basic linings will be dealt with in Chapter 8, section 8.2. Acid refractory linings hardly change their volume at temperatures above 800°C. On the other hand, the expansion coefficients of basic materials remain constant right up to the highest temperatures; therefore these materials are very prone to cracking with changes of temperature (see also Chapter 6, section 6.1.2.1). Therefore basic-lined ladles can be used where marked temperature changes can be avoided. This is only possible if the ladle is not allowed to cool down after the treatment when the steel has been tapped off. A slide-gate nozzle on the ladle is essential here.[37–38] If there is a long waiting time before refilling the ladle it is absolutely essential to keep the ladle hot by using burners. However, if the UHP furnace tap-to-tap time, the after-treatment time, and the time for teeming into a continuous casting system are optimized in relation to one another, it is then possible to dispense with intermediate heating for the ladle. It is becoming increasingly common practice to fit the ladle with a cover after filling and then either to remove the cover just before tapping or even to tap through an opening in the cover. Figure 5 shows how the ladle has developed in recent times from a simple pouring ladle into a treatment ladle and then into a reactor with several purging bricks and increased freeboard.[29,39] Alloys are added through a small hatch or hole in the middle; lances for blowing in solids are also introduced in this way.[40,41]

Use of an overhanging, vacuum-tight cover has made today's ladle into a vacuum reactor. For tapping, the ladle is increasingly carried on a ladle bogie instead of by crane. A tilt mechanism on the bogie enables tapping to be carried out without splashing the ladle side-wall; it also facilitates deslagging or slag-changing.[42] A single central purging brick is sufficient in most cases for intensive purging under vacuum; this is also better for the bath/slag reaction than a purging brick arranged off-centre.[43] Table 2 contains data on gas flow for purging via lances and bottom bricks.

7.2.2 Converters

Converters for secondary treatment of steel were developed from the bottom-blown, air-refining Thomas and Bessemer converters and from top or bottom-blown oxygen converters. These converters are used for changing molten pig iron into steel by oxidation. They must have high specific volumes, i.e. large free converter volumes, of 0.7–1.0m^3 per tonne in order to attain high gas-flow rates and high conversion rates.

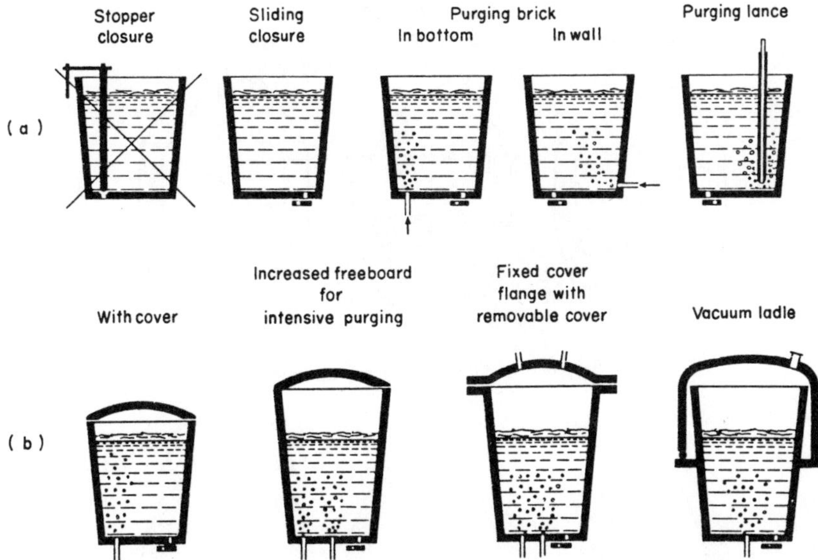

Figure 5. Development of the ladle from pouring-ladle to metallurgical reactor[29]

Converters for secondary treatment of molten steel or ferro-alloys are also characterized by specific volumes that are significantly greater than those of secondary treatment ladles (Table 2). Figure 6 shows schematic representations of three converter processes for oxidizing melts high in chromium.[44,50] A reaction gas, consisting of a mixture of oxygen, argon, or nitrogen with steam, is blown in through jets in the converter wall or bottom. However, these converters can also be used purely as mixers for intense bath/slag reactions and for degassing;[51] in that case, the driving or impulse gas is argon or nitrogen, depending on the requirements for the steel. Because the specific gas flow rates are very much higher (Table 2), the metallurgical conditions may be better than for ladle-treatments, and the treatment times may be shorter.

The durability of bottom-jets is not sufficient to permit blowing with pure oxygen, and so a cooling gas (argon, nitrogen, or steam) is blown through the outer ring of a ring-jet. With high-carbon, high-chromium melts, the oxidizing rate can be raised considerably in the high-carbon region up to around 0.5%C by top-blowing at a high specific oxygen-flow rate by means of a lance.[52,53]

Three developments of the converter process can be distinguished. With AOD converters the process gas is injected through several jets (3 to 5) in the lower part of the converter wall. The name Argon–Oxygen-Decarburization indicates the original gas composition and the aim of the process, which is to oxidize high-chromium melts. The process was developed in the laboratories of the Linde Division of the Union Carbide Corporation in the USA; it was

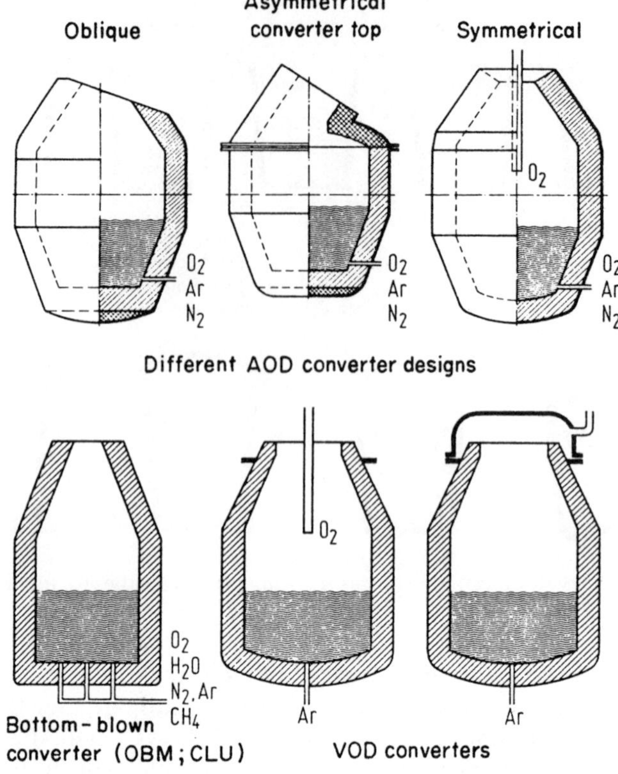

Figure 6. Converter processes for oxidizing and treating non-alloy and high-chromium alloy premelts[29]

brought to maturity for production with a 15/20-t converter in 1968 in collaboration with Joslyn Stainless Steels at their Fort Wayne works, Indiana.[44] Apart from its purely metallurgical functions yet to be discussed, the argon lowers the temperature at the jet-nozzle. These days, argon is largely replaced by nitrogen on economic grounds if this is metallurgically possible or desirable (nitriding). At the time of writing, over a hundred converters were in operation in around seventy works throughout the world, and a further seven firms intended to install such converters in 1980.[54] The smallest converter takes a melt weighing 3 t; the largest, 150 t. Because of their versatility, AOD converters will be used in increasing numbers for low melt-weights in foundries.[54]

The CLU process was developed jointly by the firms Creuzot Loire (France) and Uddeholm (Sweden). The first converter of this type came into operation in October 1973 at Uddeholm's Degerfors steelworks in Sweden. In comparison with the AOD process, the main feature of the CLU process is the use of an oxidizing gas mixture consisting of oxygen and steam. The steam takes the place of most of the argon used in the AOD process; argon is expensive and is not available everywhere. The gas mixture is blown in through ring-jets in the bottom of the converter. The steam dissociates completely into hydrogen and oxygen in the process. Apart from this endothermic dissociation reaction, which is important for the life of the bottom-jets, the steam has the same metallurgical purposes as argon in the AOD process. The endothermic dissociation-reaction cuts down the amount of scrap that has to be added for cooling purposes. Up to now, the CLU process has attained only limited importance. The OBM (Oxygen Bottom-blown Maxhütte) process, which is used almost exclusively for oxidizing pig iron that is either low or high in phosphorus, can basically also be used as an oxidizing reactor for high-carbon ferro-alloys or for melts rich in chromium and carbon.[55]

The last variant to be developed was the VODC or VODK process.[50] Vacuum-Oxygen-Decarburization in converters was developed in 1977 by Thyssen Edelstahlwerken at their Witten works, using an oxygen-blast converter and experience with the VOD process. It makes use of the high oxidation rates obtained with oxygen top-blowing combined with the advantages of vacuum-oxidation. After oxidation with a lance under atmospheric pressure, a vacuum cover is placed on the converter; further decarburizing is then carried out by blowing in argon and a reaction between the carbon-laden bath and the oxygen in the chrome oxide-laden slag, while the slag is reduced 'of its own accord'.

Another process involving oxidation under vacuum has also been tried out, this variant is preferred for new installations.

Converter design has taken various forms in the course of development (Figure 6); however, symmetrical converters are almost the only type built today.

Development of the AOD process was troubled in a way similar to that of the LD (Linz Donawitz oxygen top-blowing) process and that of continuous casting. Today most of the chrome and chrome-nickel steels for strip-mill production are processed in AOD converters. The AOD converter would have become established sooner if the problems with the refractory lining had been overcome in the first few years.[56] It has only been possible to bring about a clear reduction in refractory costs because of the arrival of improved refractory materials and the use of a better-balanced and controlled process with optimized control of bath temperatures. Linings of ceramically bonded dolomite bricks are now common, especially in European converters, as well as chrome–magnesite linings.[57,58]

7.2.3 Vacuum-treatment Systems

Vacuum-treatment of steel on a large scale began because of the need to make large forgings for power-station rotors from alloy steels of which the hydrogen content had to be low so as to avoid failures due to flake-formation. Vacuum-treatment was one of the first forms of secondary refining processes for steel (Figure 3).[59–61] While continuous degassing processes were still intended to be used exclusively for degassing, batch-treatment installations enabled refining to be extended to improve not only deoxidizing but also decarburizing and purity generally.[62] Some of the improvements were originally attributed to the vacuum treatment; however, they are more strongly influenced by enforced motion of the melt. With the adoption of basic ladles and cessation of the use of crucible reactions and the development of slide-gate nozzles for ladles (for high-temperature ladles) as well as of bottom-purging bricks, all metallurgical treatments could be carried out under vacuum even in existing ladles.[33,37] Most vacuum-treatment reactions still involve enlargement of volume; in the vacuum-treatment of steel this means that oxidation of carbon, hydrogen, and nitrogen are brought about by lowering the partial pressure of the gas concerned.

Figure 7 provides a survey of vacuum-treatment processes in common use today.[29,63] Most are in three fields of application.

(1) Decarburizing high-chrome melts down to the lowest levels of carbon and fine decarburizing of non-alloy melts to carbon levels below 0.01%.

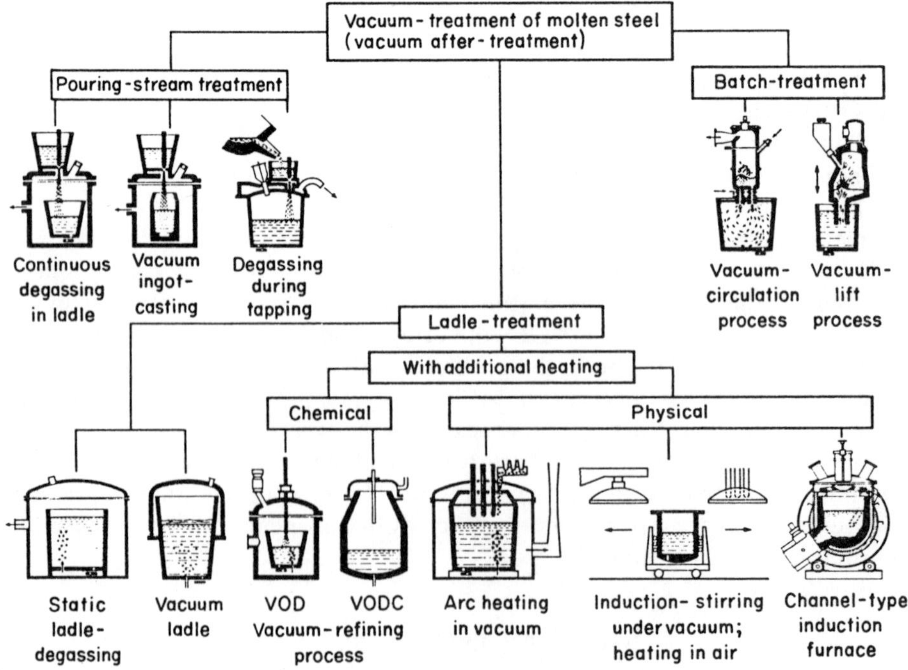

Figure 7. Survey of the most common vacuum-treatment processes used in the production of steel and castings[29]

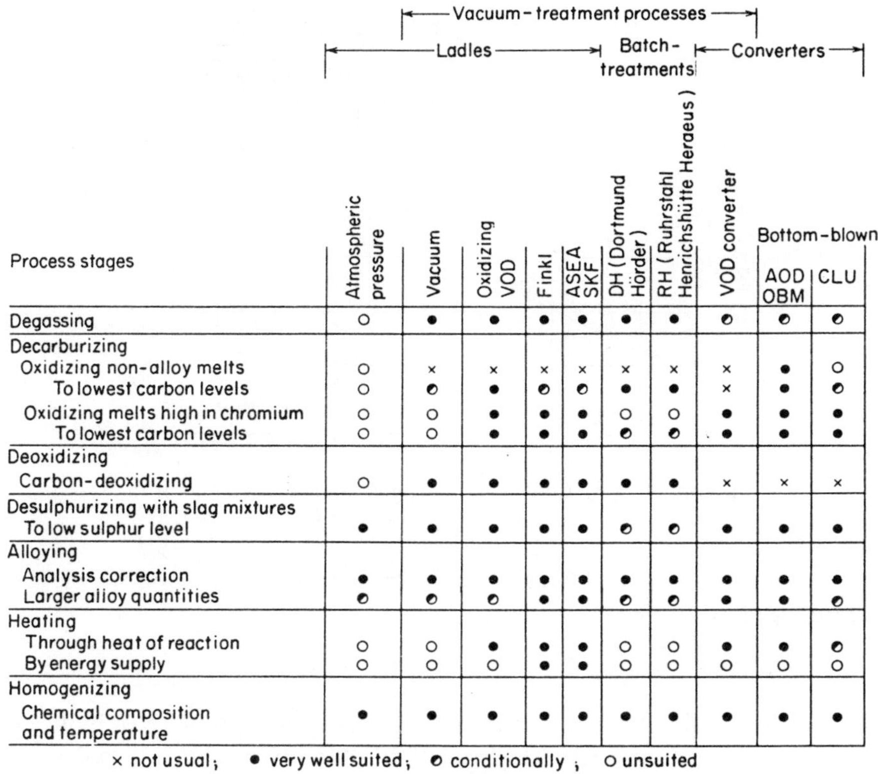

Process stages	Atmospheric pressure	Vacuum	Oxidizing VOD	Finkl	ASEA SKF	DH (Dortmund Hörder)	RH (Ruhrstahl Henrichshütte Heraeus)	VOD converter	Bottom-blown AOD OBM	Bottom-blown CLU
	Ladles (atmospheric)	Ladles — Vacuum-treatment processes				Batch-treatments		Converters		
Degassing	○	●	●	●	●	●	●	◐	◐	◐
Decarburizing										
Oxidizing non-alloy melts	○	×	×	×	×	×	×	×	●	○
To lowest carbon levels	○	◐	●	◐	◐	●	●	×	●	◐
Oxidizing melts high in chromium	○	○	●	●	●	○	○	●	●	●
To lowest carbon levels	○	○	●	●	●	◐	◐	●	●	●
Deoxidizing										
Carbon-deoxidizing	○	●	●	●	●	●	●	×	×	×
Desulphurizing with slag mixtures										
To low sulphur level	●	●	●	●	●	◐	◐	●	●	●
Alloying										
Analysis correction	●	●	●	●	●	●	●	●	●	●
Larger alloy quantities	◐	◐	◐	●	●	◐	◐	●	●	◐
Heating										
Through heat of reaction	○	○	●	●	●	○	○	●	●	◐
By energy supply	○	○	○	●	●	○	○	○	○	○
Homogenizing										
Chemical composition and temperature	●	●	●	●	●	●	●	●	●	●

× not usual; ● very well suited; ◐ conditionally suited; ○ unsuited

Figure 8. Survey of feasibility of after-treatment systems for individual refining-process stages[29]

(2) Carbon deoxidation; i.e. removal of dissolved oxygen via the carbon in the bath.
(3) Reducing the level of atomic hydrogen and nitrogen dissolved in the steel.

Vacuum-treatment is also beneficial when adding alloying elements with a high affinity for oxygen. In addition, vacuum-treatment, as shown in Figure 8, can be used universally for all metallurgical refining. This is also why it is particularly used where the production programme has numerous variations and there are very many different melting and casting requirements. Certain types of steel and certain steel characteristics can only be produced by means of vacuum-treatment.

7.3 DEGASSING

Degassing was the starting-point for ladle- or secondary refining (see Figure

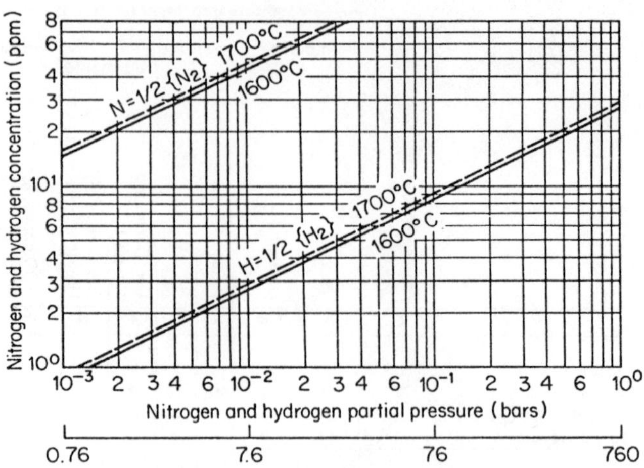

Figure 9. Solubility of hydrogen and nitrogen in iron melts at 1600 and 1700°C[63]

3[59–62]). The pickup of dissolved atomic hydrogen and nitrogen follows Sievert's square root law:

$$(\%H) = K \cdot \sqrt{(P_{H_2})} \quad \text{or} \quad (\%N) = K \cdot \sqrt{(P_{N_2})}$$

i.e. it is proportional to the square root of the partial pressure of hydrogen/nitrogen in the atmosphere (Figure 9).[63] Solubility is also affected by the alloying elements. The nature of nitrogen fixation in iron and steel melts is the subject of extensive research.[64–6] Nitrogen-solubility falls with increasing silicon and carbon contents, but it rises with increasing percentages of elements such as manganese, chromium, vanadium, and titanium, which all have an affinity for nitrogen.

Molten iron dissolves about 0.045% nitrogen and about 0.0025% hydrogen at partial pressures of 1 bar. Before post-treatment, commercial steels contain about 3–8 ppm hydrogen (Figure 10). The solubility of hydrogen in melts drops somewhat with rising temperature. The solubility of atomic hydrogen jumps sharply during solidification; if the initial level is high, expansion can cause bubbles of hydrogen to form as the metal freezes.

Solubility drops further during cooling, and so atomic hydrogen may become segregated in solid form. If the strength of the steel is exceeded this can lead to internal cracking and flaking. In order to avoid this trouble and prolonged cooling or homogenizing, the hydrogen contents in sensitive steels must be below 2 or 3 ppm; this is particularly important for large blooms and objects that are bulky when finished. The hydrogen partial pressure must be below 0.004 bar (3 torr) to obtain hydrogen levels of 2 ppm. Such low partial pressures call for modern vacuum-treatment installations.[67–9]

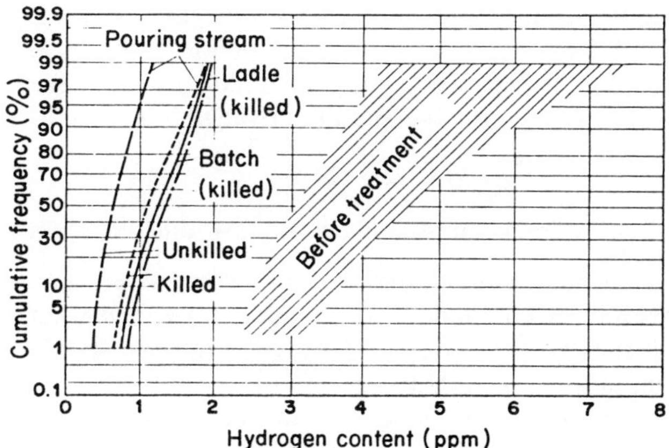

Figure 10. Hydrogen levels in non-alloy steels before and after various vacuum-treatments[63]

It is also important to avoid pickup of hydrogen from damp refractories which can produce a raised partial steam pressure.[63] The success with which hydrogen is removed in a vacuum-treatment system depends on many factors.[66] Optimum conditions occur when an unkilled steel melt is poured into a vacuum (pouring-steam degassing, Figure 10). In batch-treatment processes hydrogen content drops with time, as with RH process (Ruhrstahl Henrichshütte/Heraeus), and with the number of lifts, as with the DH process (Dortmund Hörder) and is again the lowest when unkilled melts are treated. Static degassing processes provide the most unfavourable conditions, but here, too, carbon monoxide evolution assists the removal of hydrogen from unkilled melts. Killed melts require extreme agitation by stirring with argon or inert gas so that steel from the bottom of the melt is moved up to the top layers for degassing.

Degassing processes are very complex. Solubility coefficient varies with concentration, but, in addition, the movement of matter from the interior of the melt to the boundary layer, the matter-transfer conditions, the dynamic viscosity, the flow conditions, and nucleation are all important for degassing.[66] Thus it is possible, even in AOD converters, to effect degassing which is adequate for most purposes by using a high specific flow of purging gas.

7.4 DECARBURIZING

In arc furnaces the charge generally consists of collected or recycled scrap in which the carbon content is quite low. It is cheaper to oxidize molten pig iron in converters. The premelt only has higher carbon contents when solid carbon carriers (pig iron or cast-iron scrap) or expensive high-carbon ferro-alloys, particularly ferro-chromium, are processed.

Other conditions can occur in induction furnaces. Until the problems with basic linings in large mains-frequency induction furnaces (over 10 t) are completely solved, it can make sense to remelt a high-carbon, high-silicon premelt in an acid-lined crucible and then oxidize it in a secondary-treatment unit.[20–22]

7.4.1 Decarburizing Plain and Low-alloy Steel Melts

There is no problem in oxidizing high carbon contents in arc furnaces by blowing in gaseous oxygen via hand-lances or through water-cooled wall- or roof-lances. It is true that the geometrical shape of the arc furnace and the technical and economic limitations of exhaust-gas removal prevent the attainment of oxidation rates as high as in a basic oxygen furnace; however oxygen flow rates of up to 1.5 m_n^3 per tonne per minute can still be used.[67,68]

In all hearth-melting processes, as in the basic oxygen steelmaking process, it is difficult to achieve low carbon levels (under 0.03%); indeed it is only possible at the cost of very high iron-melting losses. However, the OBM process is an exception in which lower carbon monoxide pressures can be achieved in a way similar to the AOD and CLU processes.

For secondary decarburizing, use is made of the dependence on pressure of the carbon–oxygen reaction.[72] This kind of after-treatment can also be carried out in principle in a bottom-blown converter using mixed or inert gas. However, secondary decarburizing is generally carried out in a vacuum-treatment unit.[69–73] The areas of application are production of very low-alloy soft iron, annealed dynamo steel, and special deep-drawing steels containing less than 0.01% carbon. The process of decarburizing is shown schematically in Figure 11.[63]

Starting with unalloyed, unkilled melts with carbon contents above 0.04%, the desired carbon level of under 0.01% can be achieved directly by lowering the carbon monoxide pressure p_{co} to below 0.01 bar.[69–71] If the initial carbon content is higher, extra oxygen will be required (Figure 11, B). It is customary today to carry out this secondary decarburization in batch-treatment installations used in basic oxygen steelmaking plants with high converter tapping weights.

7.4.2 Decarburizing High-chromium Alloy Melts

Industrial production of high-chrome steels which resist rust, acid, and heat began in 1920. Direct decarburizing of a bath containing chromium with ore was limited to low chrome levels. This process was characterized by low output, high energy-consumption, and the need to use soft, expensive types of ferro-chromium to obtain the right composition.

Only with the introduction of oxygen-refining from 1945 was it possible in the Federal Republic of Germany to raise steel-bath temperature quickly in oxidizing by making use of exothermic reactions and to shift part of the

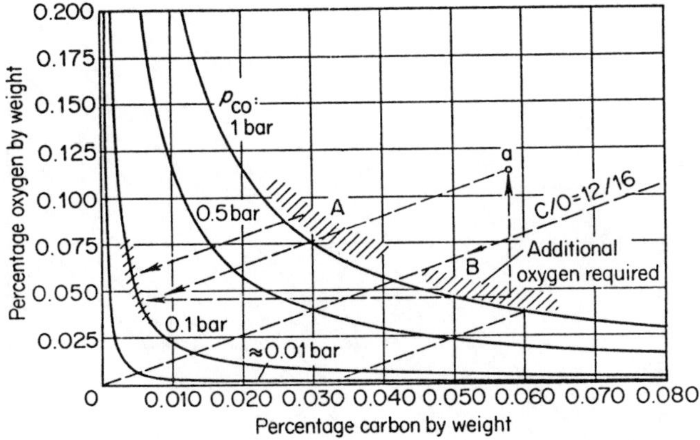

Figure 11. The oxygen–carbon equilibrium in relation to carbon monoxide pressure p_{co} and the course of carbon and oxygen contents in vacuum-decarburizing of iron melts containing carbon[63]

decarburizing process to take place before chromium-oxidation. Oxygen-refining was only applied to metallurgical processing later, after industrial oxygen-refining of high-chrome melts. Fundamental research by D. C. Hilty[74] and other scientists[75–8] into the dependence on temperature of chromium–carbon–oxygen equilibrium and the reduction equilibrium between steel and slag has supported further development.

The increases in oxygen flow rate enabled oxidation rates to be raised. At the same time, the conditions for mass-production of chrome steels were improved by the development of large arc furnaces charged through the roof and equipped with electromagnetic stirring as well as by improvements in refractory hearth-linings and temperature-measurement techniques and the speeding-up of the analytical process.

As the demand for high-chrome steels increased, there was a drop in the relative amount of recycling scrap available for steel-production, and so the amount of chromium fed in as ferro-chromium had continually to be raised.

Oxygen-refining in arc furnaces enabled cheap grades of ferro-chromium, rich in carbon, to be charged. The share of the market taken by high-carbon ferro-chromium increased at the expense of more costly refined and 'carbon-free' grades of ferro-chromium (see Chapter 3, section 3.4). Thus the average carbon content in the charge kept on rising; this trend has stabilized today in different ways for different kinds of steel.

Figure 12 shows the relationship between chromium–carbon equilibrium and temperature.[79] Chromium-oxidation behaviour with temperature is different from that of carbon because the free-formation enthalpy for carbon monoxide falls with increasing temperature while the opposite is true for metal oxides. Final carbon levels of under 0.03% can be achieved in an arc furnace

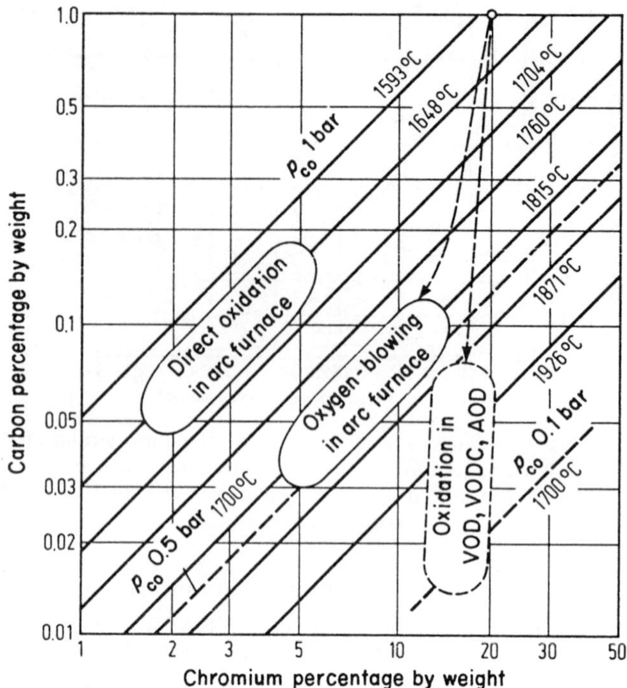

Figure 12. Final contents for carbon–chromium temperature equilibria for decarburizing high-chromium melts[79]

from initial levels of about 1% C and 15% Cr (see Figure 12) but only by rapidly raising the bath-temperature to over 1900°C. This rise in temperature is conveniently achieved partly by electrical means at first and then, without adding further energy, from the exothermic oxidation of silicon, manganese, carbon, iron, and chromium. The chromium–carbon–oxygen equilibrium for p_{co} = 1 bar and 1800°C is shown in Figure 13 after E. Steinmetz.[78,80]

During oxidation the amount of oxygen dissolved in the melt rises along the curve while the carbon level decreases. When the equlibrium line with the chromium-oxide phase is reached, chromium oxidation, i.e. chromium slagging, takes place instead of the preferred oxidation of carbon. It is essential to reduce the chromium in the slag for economic reasons. With arc furnaces and especially those of large capacity the unfavourable ratio of the bath volume to the area of the reactive interface between the bath and the slag complicates reduction of the chromium. Re-ladling, i.e. pouring steel and slag from the furnace into the ladle and then back into the furnace, does not produce sufficient improvement.

Economical production of high-chromium alloy steels on a large scale requires the following conditions:

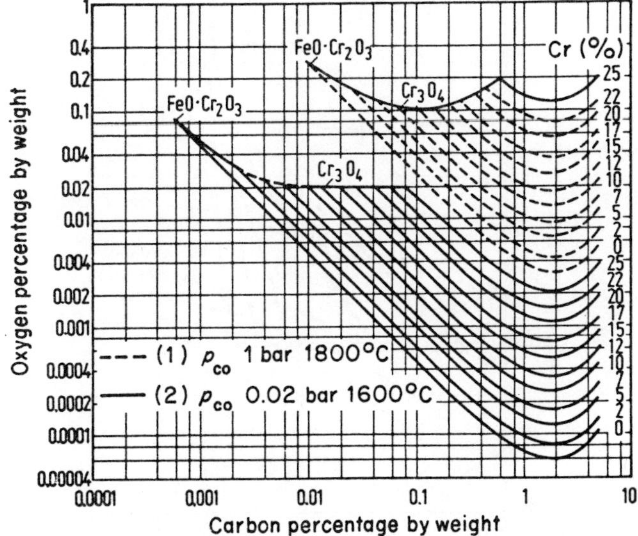

Figure 13. Equilibrium between carbon and oxygen in iron-chromium melts at 1800°C with p_{co} of 1 bar and at 1600°C with a p_{co} of 0.02 bar[80]

(1) High productivity (high rate of output plus high rate of oxidation);
(2) Use of cheap ferro-chromium alloys containing high levels of carbon and silicon;
(3) Low consumption of reduction media for slag-reduction;
(5) High chromium output, even with low final carbon levels; and
(5) Low process-temperature to protect the refractory lining.

Increasing quality-requirements adds the following as well:

(6) Low gas contents (nitrogen and hydrogen);
(7) Lowest possible carbon levels;
(8) Low sulphur levels.

While the first group of requirements is of an economic nature, the last three requirements cannot be fulfilled in an arc furnace on its own.

The dependence on temperature of the chromium–carbon equilibrium is plotted in Figure 12 for p_{co} of 1 bar as well as 0.1 bar and 0.5 bar for 1700°C. For example, at bath temperatures of 1700°C and 10% Cr, a p_{co} of 0.1 bar results in an equilibrium carbon level of 0.01%, while at p_{co} of 1 bar the equilibrium carbon level is about 0.20%. As can be seen in Figure 13, lowering p_{co} at unchanging carbon level in the bath not only reduces the level of oxygen dissolved in equilibrium but also displaces the equilibrium line for the

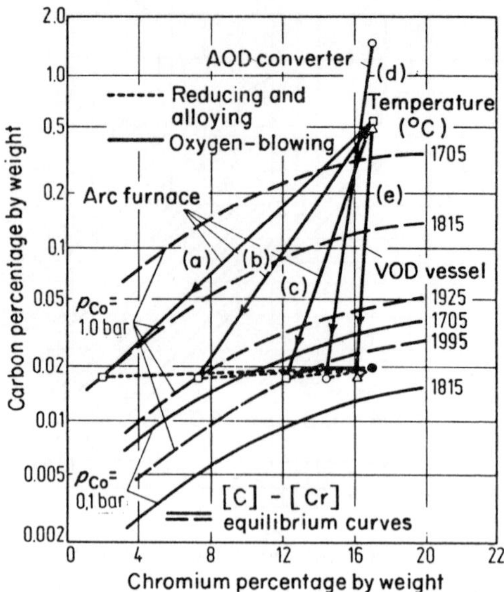

Figure 14. Change in chemical composition during oxidation of steels containing large percentages of chromium in arc furnaces and in AOD and VOD processes[45]

chromium-oxide phase. It is therefore possible to obtain lower carbon levels in oxidizing high-chrome steel melts by lowering the partial CO pressure before the chromium reacts with the slag in preference.

Figure 14 schematically illustrates the three processes for oxidizing high-chrome steels in arc furnaces (the oldest method) and in AOD converters and VOD installations.[45] The charges initially contain 17% chromium and 1.8% carbon on average. The chromium–carbon line for AOD converters lies between those for conventional production in arc furnaces and for vacuum-oxygen installations. In arc furnaces at $p_{co} = 1$ bar it is only possible to obtain low carbon levels with tolerable chromium oxidation at the expense of high bath-temperatures. These high temperatures arise in the course of oxygen-refining because of the exothermic oxidation of steel companion-elements and iron at low carbon levels, and especially through oxidation of chromium. The high bath-temperatures have to be reduced by, say, adding high-chrome, carbon-free scrap of the same composition as the melt or of non-alloy scrap and soft, expensive ferro-chromium. Chromium is reduced from the slag by using ferro-silicon, silico-chromium, or aluminium. The high proportion of scrap used for cooling is paid for indirectly with expensive, energy-intensive silicon.

With the VOD process, where pressure drops steadily down to 0.1 bar, carbon is oxidized with almost no chromium loss so long as care is taken to homogenize the steel bath by vigorous motion. Appropriate process-control can enable the heat emitted during oxidation of carbon, silicon, and chromium to suffice to compensate heat-losses and, as the carbon level drops, to raise the bath-temperature by around 80–150°C up to the required tapping-temperature. The amount of reduction-material consumed is small, and scrap for cooling is only needed in small quantities, if at all.

With the AOD process it is not possible to work at such low p_{co} pressures as with the VOD process. If we want to limit the rise in temperature during oxidation, then we have to put up with heavier oxidation of chromium, which in turn requires an increase in material consumed for reduction. A constant bath-temperature, which is desirable, can be achieved by adding cooling material as often as possible. An energy balance-sheet for the AOD process shows that about 40% of the total energy 'credit' comes from the melt while about 60% is emitted by the reactions. The melt accounts for 55% of the total energy-losses; the cooling-material contributes 13.4% of this together with 21.3% for chromium-reduction.[79]

If we start with a charge of unalloyed scrap and a high-carbon grade of ferro-chromium (e.g. charge chrome) for an 18%-Cr steel, then the charge may contain more than 2% carbon. There is no problem in an arc furnace in oxidizing the carbon down to about 0.5% without significant oxidation of chromium; however, time is required for oxidation and perhaps for a brief supplementary slag-reduction period. Whether preoxidation is carried out in the arc furnace, and if so to what extent, depends on the timing of the operation of the arc furnace and of the supplementary AOD converter or VOD installation as well as on the amount of carbon present. With the AOD process the slag from the arc furnace is generally transferred to the converter, regardless of whether there has been any preoxidation. With the VOD process slag-reduction always takes place and the ladle is transferred without slag into the VOD installation. The oxidation process in both AOD and VOD systems is controlled to achieve the best thermodynamic and economic conditions, according to the nature of the charge and the final composition required. Thus, with high initial levels of carbon, there may be a brief preoxidation period in the AOD converter using an oxygen-lance with a high specific delivery rate.[52,53] When oxidizing via jets in the side or bottom, the ratio of oxygen to inert gas must be lowered in three to five stages as the carbon level drops; this will lower the p_{co} pressure continuously in the region required at the time from 0.8 to 0.2 bar.[45,46]

In the VOD process the pressure in the system, and with it the p_{co} pressure, are reduced continuously as the carbon content drops. As vacuum-pump performance falls off with reducing pressure, the pumps should be designed for adequate delivery according to experience. On the other hand, with high levels of carbon in the bath, if the pressure is not much reduced high oxygen-delivery rates and rapid oxidation can give rise to vast amounts of waste gas. This means that with sufficiently powerful vacuum pumps the initial level of carbon processed in a VOD installation may be as high as 1.8%, and high oxidation rates and short process-times can be reached with specific oxygen flow rates of up to 0.7 m_n^3 per tonne per minute (Tables 2 and 3).[43,51,80–84]

In the VODC process important stages of the AOD and VOD processes are carried out under optimum conditions using appropriate equipment.[50] It remains to be seen to what extent this latest process becomes established. With high suction-pump performance it should be possible to use lances for oxidizing under vacuum down to low carbon levels in large converters with short oxidation times and little slagging of chromium. The amount of inert gas required for bath motion is small (Table 3).

The lowest carbon levels possible with all three processes can be achieved by very intensive purging with inert gas at very low p_{co} pressures at the end of the process (boiling-out).

Control of the VOD process has been much improved by exhaust gas analysis. The CO/CO_2 ratio is measured continuously using a solid-state electrolytic cell.[84] The times for the start of oxidation and of intensive decarburizing and for the end of oxidation at about 0.08%C can be determined precisely for the type of installation. This method can also be used to avoid overoxidation of the bath.

To avoid overoxidation in the AOD process as well, the oxygen-utilization coefficient and the quantity of oxygen required are computed using a statistical model. In order to optimize the efficiency with which oxygen is used, this must be related to temperature and to the composition of the bath (C, Si, Cr, Ni, Mn) at the time as well as to the amount of slag and the quantities of oxygen and inert gas blown in during the various blowing stages. Models in use take account of starting conditions and monitor p_{co} changes in the bath at any time together with changes in composition and temperature and compare them with the p_{co} pressure of the reaction gas.[46,79] The aim is precise control of carbon oxidation and calculation of the quantities of reduction materials and slag-formers required.

A blow-diagram for a VOD melt is shown in Chapter 14, Figure 6. Corresponding diagrams for AOD, VOD, and VODC processes are shown here in Figures 15 and 16.[43,45,46]

After slag-reduction, steel baths are generally subjected to a reduction-treatment during which sulphur content is lowered and purity as regards oxide content is improved via the slag phase accompanied by intensive agitation of the bath. The AOD process has the additional advantage of enabling the slag to be changed quickly and substituted by a fresh secondary slag for desulphurizing. The ratio (%S)/[%S] of sulphur in the slag to that in the melt amounts to about 100 for both processes, whereas it can reach a value of 200 for AOD secondary slag.

In practice, approximate equilibrium can only be achieved in an acceptable

Table 3. Guide to consumption and process values for oxidizing chrome steels in the AOD, VOD, and VODC processes

		Process		
		VOD	AOD	VODC
Initial C content	(%)	max. 1.8, usually <1	1.5–3.0	1.5–3.0
Starting temperature	(°C)	about 1600	about 1520	1520
Max. oxidizing rate dc/dt	(%C/min)	0.006–0.030	0.04 or 0.08[a]	0.08
Oxygen flow rate	(m_n^3/t min)	0.2–0.70	0.3–0.75 or 1.2[a]	1.2
Slag basicity	(%CaO/%SiO_2)	1.2–2.0	1.2–1.5	1.2–1.5
and for secondary slag			2	2
Sulphur distribution	(%S)/[%S]	about 100	about 100	about 100
and for secondary slag			200	200
Treatment temperatures	(°C)	*1600*/1700	1550/*1700*	1600/*1700*
Treatment time from fill to end of tapping	(min)	about 140	about 90	about 90
Oxygen consumption	(m_n^3/t fl)	about 15–20	about 30	about 30
Oxygen efficiency (CO)	(%)	20–60	30–90	50–80
Gas consumption				
— argon and nitrogen	(m_n^3/t fl)	about 0.8	15–25	1
Reduction silicon	(kg/t fl)	about 5–8	10–18	about 10
Lime	(kg/t fl)	20–35	70–100	40–50
Chromium output	(%)	98	98	98

[a] With lancing.

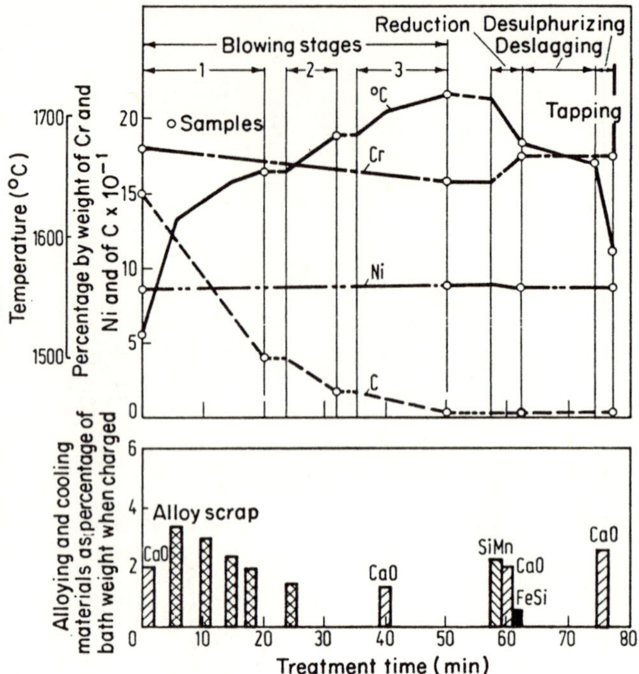

Figure 15. Changes in temperature and chemical composition during oxidation of an austenitic Cr–Ni melt in an AOD converter[45]

time if care is taken to ensure very good contact between phases and balanced concentration by means of vigorous bath motion.

Table 3 contains consumption and process values for producing austenitic chrome-nickel steels (X 5 CrNi 18 10) containing about 0.04% carbon. The greater consumption of argon and reduction materials in the AOD process should be compared with the cost of the vacuum plant for the VOD process. The oxygen efficiency (carbon oxidation) in the VOD process depends on the lancing conditions (jet characteristics and distance) and on the stirring intensity as well as the ladle geometry, the slag (quantity and composition), and the carbon levels at the start and finish. Bottom-blown AOD converters are very oxygen-efficient at high carbon levels, but their efficiency drops when very low carbon levels are required.

The decision as to which process to use depends on what quantities are to be produced and on the quality requirements and what raw materials are available as well as the types of production system used before and after treatment. The AOD converter is better suited for carbon levels above 1.5% in the charge; in normal production for a strip mill it can economically turn out large quantities of steel that is not especially low in carbon. The VOD process is more versatile

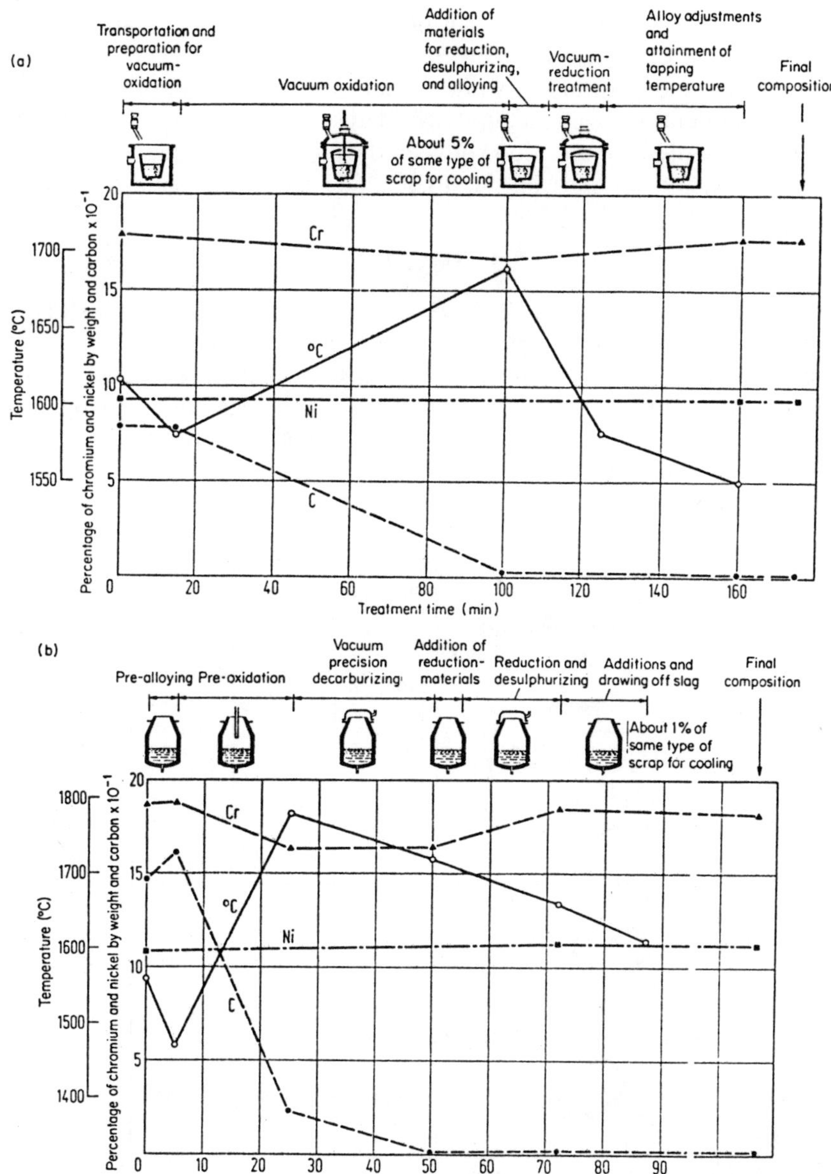

Figure 16. Changes in temperature and chemical composition during treatment of a high-chrome melt in (a) the VOD process and (b) the VODC process

and can be used to produce steels ranging from high-quality steels low in chromium to large ingots for forging. Rust-resistant ELC steels (extra-low carbon) can be produced economically at low p_{CO} pressures. Ferritic steels with nitrogen contents below 50 ppm can be produced with the VOD process if oxidation is started when the carbon level is above 1% and the bath is vigorously agitated with argon while no air is allowed to enter the vacuum vessel. Subsequent nitrogen pickup is avoided by enclosed pouring direct from the treatment-ladle. Competition between the two processes leads to continual improvements, and comparison should be made in every possible application.

7.5 Deoxidizing

When making steel in ultra-high-power furnaces it is customary nowadays to tap from under a black slag which is rich in iron oxides, unless a high-chrome charge is deoxidized in the furnace before tapping so as to recover chromium. Development is aimed at completing the oxidation stage with tapping and to eliminate the need for additional reduction of melts which must be deoxidized before casting; this requires tapping to be carried out as free from slag as possible. Figure 17 illustrates a few possibilities for this.[19]

Tapping methods	Run of slag		
	Before	During	After
(1) Without holding slag back 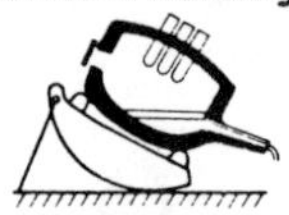	5 to 10%	40 to 70%	20 to 80%
(2) Plugging the spout 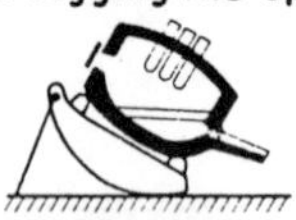	No	Yes	Yes
(3) Low-set spout (siphon) 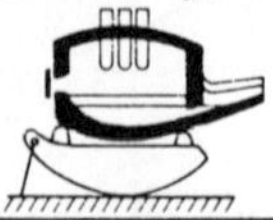	No	Little	Little
(4) Sliding-gate or stopper 	No	Little	Little
(5) Deslag before tapping 	No	No	No

Figure 17. Possible ways of keeping back black slag rich in iron oxides when tapping arc furnaces

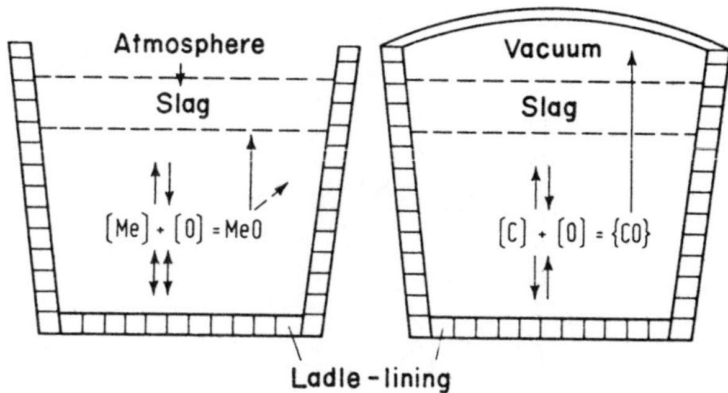

Figure 18. Reactions during deoxidation and vacuum-deoxidation in the ladle[19]

Effective deoxidation in the ladle is only possible if further oxygen supply is stopped as soon as possible (Figure 18).

A deoxidized melt can still pick up oxygen from the atmosphere and from the slag and the refractory lining. The system aims to balance the partial pressures of the oxygen in the deoxidized steel bath with those in the slag and the refractory lining, especially during vigorous purging in the ladle using lances or via one or more porous purging bricks. It is therefore essential to ensure that no slag rich in iron oxides gets into the ladle or else that it is removed before deoxidizing begins.

The development of direct electrochemical measurement of oxygen activity in molten steel by means of EMF probes (see Chapter 14) constituted an important step forward in the understanding and control of deoxidation processes.[85] Figure 19 illustrates the dependence of oxygen activity in iron melts on the amounts of various deoxidation elements present in the melt. Equilibrium here is just as important for the reaction between the molten iron and oxygen in the lining. Because oxygen activity is even more dependent for carbon, with normal quantities oxygen is picked up from the unstable acid lining of the ladle. This is also valid for melts deoxidized with aluminium. It is thus possible to over-purge an aluminium-killed melt in an acid ladle; i.e. continuous purging can lower the aluminium content and give rise to a steady increase in oxygen content. If a very low oxygen content is required then basic ladle-linings provide an answer; however, the possibility of 'poisoning' by an iron-oxide rich slag must be eliminated. Figure 20 clearly illustrates an example of the effect of different ladle-linings on the oxygen activity in the melt as well as the level of oxygen throughout a cross-section of the ladle.[19] The pressure dependence of the C–O reaction accelerates oxygen pickup from acid ladle-linings during intensive purging under vacuum.

If a high degree of purity is required in the end-product, then not only must we prevent any new formation of deoxidation products through oxygen pickup but, in addition, existing oxides must be eliminated as far as possible.[86]

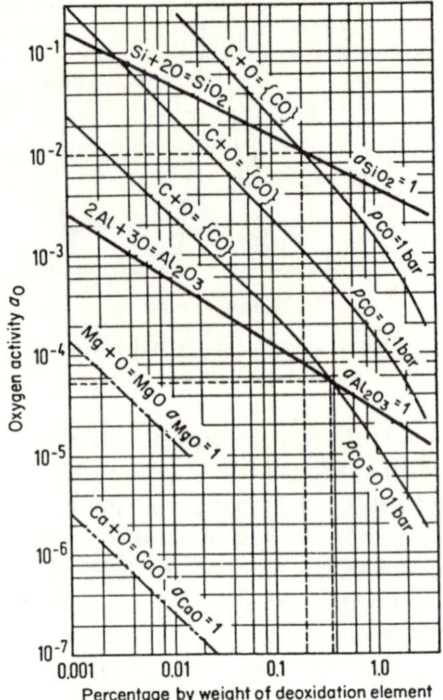

Figure 19. Oxygen activity in iron melts in relation to the content by weight of deoxidation elements and the dependence on pressure of the deoxidation of carbon at 1600°C[19]

Measurement of the amount of dissolved oxygen cannot provide a reliable indication of the degree of oxidic purity. The rate of separation depends on the mechanism of formation of the oxide particles, on their ability to agglomerate, and on their rate of ascent through the melt as well as their wettability with the slag or the refractory lining.[86,87] Blowing in solids, especially calcium compounds and lime-bearing fluxes, not only assists desulphurization but can also have a beneficial effect on the level of oxide inclusions in the steel. The same is also true of intensive purging under a slag of lime, alumina, and fluorspar. Vigorous bath motion assists agglomeration and the segregation and elimination of oxide particles.[19,40,41,88]

7.5.1 Vacuum-deoxidizing

Vacuum-deoxidizing makes use of the dependence on pressure of the C – O reaction already discussed in section 7.4.1 on vacuum-decarburizing (Figure 11). However, in precision decarburization we start with low carbon and high

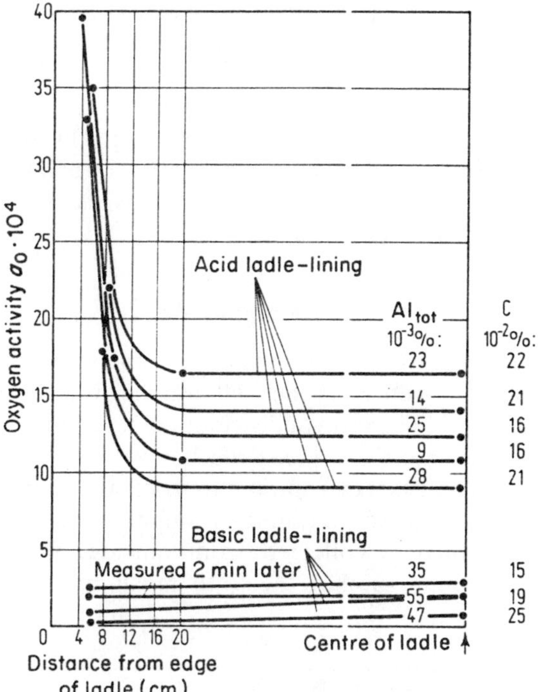

Figure 20. Measurement of oxygen activity on the edge and in the middle of basic and acid-lined ladles after intensive purging of low-alloy structural steels under vacuum (after J. Otto and G. Pateisky[19])

oxygen levels and stoichiometrically reduce the carbon to very low levels, possibly with the addition of oxygen. In vacuum-deoxidizing we start with an unkilled melt and reduce oxygen content stoichiometrically, leaving carbon dissolved in the melt. Because the initial oxygen level has already been lowered, the ensuing precipitation deoxidation results in fewer deoxidation products. A further beneficial effect can be obtained by adding carbon to the melt which is already low in carbon, in the vacuum installation.[19,20]

Particular mechanical properties can be obtained for heavy forging blooms for the power industry and for special wire-products if the usual deoxidation elements, i.e. silicon and aluminium, are not used. As shown in Figure 19, with a partial CO pressure of 1 bar at a temperature of 1600°C a carbon level of 0.2% reduces oxygen activity to exactly the same extent as an equal amount of silicon. If the p_{co} pressure is lowered to 0.01 bar then the oxygen activity with 0.35% carbon corresponds to that with an equal amount of aluminium. Such a melt can be cast in the killed state. However, the required oxygen activity levels can only be achieved at these p_{co} pressures with a basic ladle-lining which contains no silicon. A two-stage process is sometimes used for producing heavy forging blooms. After tapping from the arc furnace the oxygen level of the unkilled melt is lowered far enough by vacuum ladle-treatment for desulphurization to be carried out under a lime–fluorspar slag. This is then followed by final deoxidation during vacuum-casting of the ingot, when the dependence on pressure of the C – O reaction comes fully into its own; any additional reoxidation is also effectively prevented.[19]

7.6 Desulphurizing

All hearth-melting processes operate exclusively under oxidizing conditions in which a little sulphur is given up by the melt into the air while a limited amount is absorbed by the slag.[89] The ratio of distribution of sulphur between black arc furnace slag and the bath (%S)/[%S] reaches values of from 3 to 8; with white slag low in iron oxides it can amount to from 50 to 100.[10]

It is difficult to control the amount of sulphur imported with scrap — particularly scrap which is collected from outside. Figure 21 illustrates possible ways of limiting sulphur levels to maximum values specified for finished steel.[89]

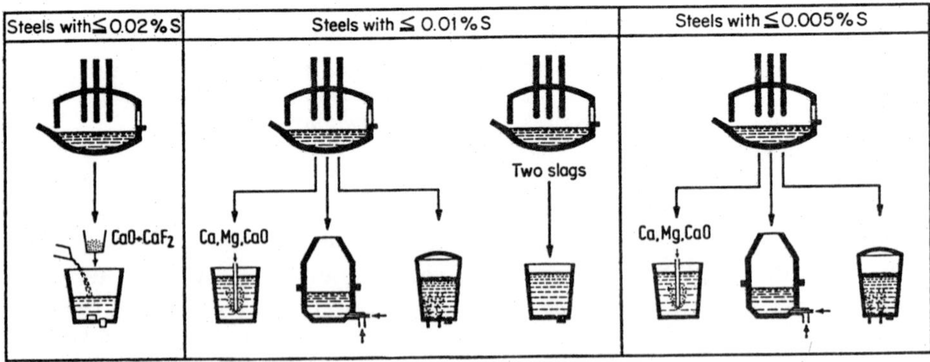

Figure 21. Processes for producing low-sulphur steels in UHP arc furnaces without a reducing slag[13]

It is not possible to achieve very low sulphur levels by adding desulphurizing materials to the ladle during tapping. Defined, low sulphur levels can only be achieved by special treatment of steel that has already been thoroughly deoxidized. The same conditions apply for desulphurizing as for deoxidizing, since desulphurizing reactions are very closely associated with oxygen activity and with the free oxygen content of the bath.[41,88] If desulphurizing is carried out in the tapping ladle, no slag high in iron oxides must get into the ladle. Ways of holding back slag when tapping an arc furnace were indicated in section 7.5. Any oxygen picked up complicates extraction of sulphur; thus the lowest sulphur levels are obtained in basic-lined ladles under lime–alumina–fluorspar slags low in iron oxides. Covering the ladle or working in a vacuum to prevent atmospheric oxidation is a further help. The lower the oxygen activity, the more effective is the desulphurizing process. Recent research has shown that the following relationship is valid for particular operating conditions and specified steel products:

$$[\%S] \approx 30a_{[O]}$$

where [%S] is the percentage by weight of sulphur in the bath and $a_{[O]}$ is the oxygen activity level.

The difficulties that arise when, for quality reasons, it is necessary to specify vacuum-deoxidizing or deoxidizing without aluminium or silicon were indicated in section 7.5. Certain precautions must be taken during desulphurizing with lime-fluorspar slag under atmospheric pressure so as to attain low hydrogen levels.

Table 4 summarizes the large number of desulphurizing processes. Pneumatic processes must be extended by intensive purging (Figure 5).[19,41] The ladle-desulphurizing processes which are most economical and most often used today involve blowing in desulphurizing materials containing magnesium, calcium, or lime and fluorspar, using immersion-lances and intensive purging.[13,40,88,90,91,92]

When desulphurizing with an immersion-lance the powdered desulphurizing material should have a grain-size of less than 1 mm and must be blown as deeply as possible into the ladle (Figure 22); the action must be completely smooth, without any pressure-pulsation. Calcium vaporizes instantly on entering the melt because of its high vapour-pressure; this causes calcium bubbles to rise through the melt. The dwell-times of the calcium bubbles or of the lime mixture in the melt affect the efficiency with which the desulphurizing materials are used.

During desulphurizing in an appropriately equipped ladle (Figure 5), the degree of sulphurization depends on the composition of the slag. It improves as the slag number rises.[40,41] This number is:

$$\frac{\%CaO}{\%SiO_2} / Al_2O_3$$

Sulphur levels of 0.020% can be lowered to below 0.002% in 10 min purging time with low oxygen levels and a basic-lined ladle. EMF measurements have helped to clarify these relationships.[85,87]

The desulphurizing stage in the AOD or VOD processes for producing high-chrome melts follows the same rules. Here, too, it is essential to mix the steel intensively at low oxygen-activity levels using a slag with a good ability to

Table 4. Processes for desulphurizing steel

Group	Process	Means of Desulphurization	Desulphurization (%)
Steel-Production Processes	Oxygen-blast process	CaO	30–50
	Hearth-melting process with one slag	CaO	30–50
	Electric arc furnace with two slags	CaO + CaF_2	<80
	AOD	CaO + CaF_2	95
Pouring-stream	Molten steel (Perrin)	$CaO–Al_2O_3$	40–60
Mixing processes	With solid slag formers	$CaO–CaF_2$	45–70
Pneumatic processes	CAB process	Ca/Mg	95
	MW process	$CaO–CaF_2$	95
Electro-magnetic Processes	ASEA–SKF	rare-earth metals	60–90
	Counter-current channel	CaO + CaF_2	<90

Characteristic

Practicability	Ladle-lining	In service	Treatment-time
Simple	Normal	Yes	n/a
Simple	Normal	Yes	n/a
Simple, two-slag	Normal	Yes	20–40 min
Medium, two-stage process	Normal	Yes	5–10 min
Simple	Normal	Yes	On tapping
Simple	Normal	Yes	On tapping
Blowing system	Basic	Yes	8–20 min
Blowing system	Basic	Yes	8–20 min
Complex, three-stage process	Basic	Yes	1–2 h
Electromagnetic channel		No	On tapping

Outlay

Addition (kg/t)	Means of stirring	Desulphurization system		Additional energy	
		Capital cost	Second ladle	Ladle	Melt
50–60	n/a sometimes N_2 or Ar	High	No	No	n/a
40–60	n/a	High	No	No	n/a
20–40	n/a	High	No	No	n/a
35	Argon	High	No	No	+50°C
25–50	n/a	Medium	Sometimes	Sometimes	+50°C
20–30	n/a	n/a	Sometimes	Sometimes	+50°C
1–2	Argon	Low	Yes[a]	Yes	+50°C
3–5	Argon	Low	Yes[a]	Yes	+50°C
1–2.5	Electric power	High	Yes[a]	Yes	Arc
5–10	Electric	Medium	–	Yes	Induction

[a] With ladle covered.

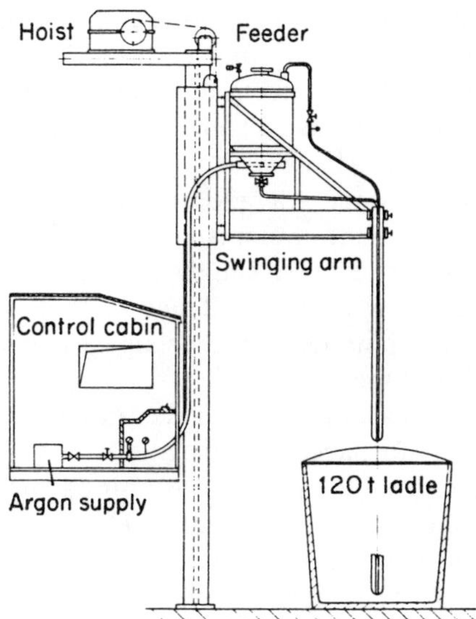

Figure 22. Solid-material feed system for desulphurizing steel using the CaB process[13]

absorb sulphur in order to attain a low final sulphur level. Clearly, the AOD converter can be used just as well as or even better than a ladle for desulphurizing structural and tool steels.[51]

7.7 Alloying and Temperature-adjustment

After-treatment processes have made it possible to adjust the chemical composition of final products to within very fine limits. It is thus possible to meet the wishes of steel-users and manufacturers who specify a final product with very narrowly defined physical characteristics and which can be worked economically in manufacturing-works where there is a high degree of automation. The alloying process used depends on the manner of deoxidation and on the amounts of alloying ingredients to be contained in the steel. For low- and medium-alloy steels all the alloying elements can be added after deoxidizing. This is common, for example, after precision-decarburizing of electrical steels for dynamo and transformer laminations.[39,67,68] However, it is also possible to carry out fine corrections to the composition while tapping after pre-alloying or after oxidizing (AOD or VOD treatment).

Alloying to within fine limits of composition requires good bath-circulation to homogenize the melt;[31,93] all after-treatment processes offer this facility. It is also necessary that the alloying elements added are absorbed into the melt without melting loss. Vacuum-treatment units and especially batch-treatment systems are especially effective in this respect. These requirements are particularly important for the production of micro-alloyed steels and when adding alloying elements which have a high affinity for oxygen.

Alloying elements added to a ladle which operates in atmospheric conditions can be fed in via the purging inlet. All modern vacuum-treatment installations, and especially batch-treatment systems and AOD converters, too, are nowadays equipped with very extensive high-level storage-bunker installations and fully automated weighing, dosing, and feed-systems which enable alloying to be carried out to fine limits using the cheapest and most metallurgically suitable ferro-alloys every time.[42,46,69,70]

Many steelworks (both blast furnace and electric) use a standard low carbon level at which they tap furnaces for specified steel quality groups. The advantage of this is that it enables a time-consuming sample test to be dropped and simplifies the melting or oxidizing process. The carbon level is subsequently adjusted to fine limits in a vacuum-installation or ladle by adding carbon or carbon-rich ferro-alloys.[70,94,95] The levels of carbon and other alloying elements must be set accurately, particularly when several melts of identical quality are poured in succession into a continuous casting system (sequence-casting).[70]

When large quantities of alloying materials are added, their cooling factor must be taken into account. However, with elements which have a positive heat of solution, such as silicon metal, this can be omitted or even overcompensated. The Finkl or ASEA–SKF processes are particularly suitable where large quantities of alloying material have to be charged for small tapping-weights (Figure 7), since these processes enable the melt to be heated with an arc during or between after-treatments.[96,99] All other processes, especially oxidation treatments, make do without this additional heating if the melt is superheated to a sufficient degree before tapping from the furnace to enable heat lost during subsequent treatment to be compensated.

Short tap-to-tap times and very rapid ladle-turnover as well as the covering of ladles (Figure 5) all help to conserve the high heat-content of basic ladles and to reduce temperature-loss from the melt. In this way, a very narrow band of temperature can be maintained throughout casting from the start to the end of pouring operations. This is absolutely essential for long casting times in continuous-casting plants producing slabs or billets, where the quality requirements are high, and also particularly for sequence-casting. If these conditions cannot be met it is first necessary to achieve a stable relationship between the ladle and the melt by lengthy stirring and possibly also by additional cooling.

7.8 LITERATURE REFERENCES

1. Ottmar, H., Performance capabilities of a 50-ton arc furnace (German). *Stahl u. Eisen* **86** (1966), 201–7.

2. Goldstein, P., and K. Dziggel, Possibilities of increasing the efficiency of a 50-ton electric arc furnace (German). *Stahl u. Eisen* **88** (1968), 1189–92.
3. Schöberl, A., W. Holzgruber and H. Raisky, Maßnahmen zur Leistungssteigerung von Lichtbogenöfen. *Stahl u. Eisen* **82** (1962), 1335–45.
4. Durrer, R., and G. Heintze, Refining liquid pig iron in electric arc furnaces. *J. Iron Steel Inst.* **192** (1959), 13–25.
5. Liestmann, W. D., and H. E. Wiemer, Mode of operation and results of a 45-ton high-power electric arc furnace (German). *Stahl u. Eisen* **93** (1973), 217–21.
6. Harma, F., and G. Zingel, Erschmelzen von niedriglegierten Stählen im Lichtbogenofen ohne Feinungszeit. *Stahl u. Eisen* **84** (1964), 1187–97.
7. Haucke, M., H. Clees, H.-J. Kopineck and H. Ottmar, Verwendung von Feinkalk im Elektrostahlwerk. *Stahl u. Eisen* **83** (1963), 441–9.
8. Fenne, M., Das Einblasen von pulverisierten Stoffen in Lichtbogenöfen. Fachber. Hüttenprax. *Metallweiterverarb.* **15** (1977), 215–16, 218–19.
9. Harms, F., and W. Engelmann, Melting down in a large electric arc furnace using an oxygen jet as a fourth heat-source (German). *Stahl u. Eisen* **85** (1965), 456–64.
10. Schenck, H., and E. Steinmetz, Considerations on the transfer of material between liquid metal and slag elucidated by the example of desulphurization (German). *Stahl u. Eisen* **88** (1968), 818–23.
11. Steinmetz. E., and F. Oeters, Metallurgical fundamentals of the behaviour of sulphur in liquid iron (German). *Stahl u. Eisen* **97** (1977), 373–81.
12. Neuhaus, H., H. J. Langhammer, H. Kosmider and H. Schenck, Möglichkeiten zur Entschwefelung von Stahl über die Gasphase beim Frischen mit Sauerstoff. *Stahl u. Eisen* **82** (1962), 1279–87.
13. Kalla, U., H. W. Kreutzer and E. Reichenstein, Technological possibilities to obtain defined sulphur contents in steel (German). *Stahl u. Eisen* **97** (1977), 382–93.
14. Ottmar, H., H. Schenck and W. Dahl, Technical and economic effects of the substitution of scrap by sponge iron in the arc furnace (German). *Stahl u. Eisen* **97** (1977), 731–41.
15. Rigaud, M., A. H. Marquis and T. E. Dancy, Electric arc furnace steelmaking with prereduced pellets. *Ironmaking & Steelmaking* **3** (1976), 366–72.
16. Ameling, D., J. Sittard, W. Resch and J. Geiseler, Use of dolomite lime in the arc furnace (German). *Stahl u. Eisen* **98** (1978), 1281–6
17. Zingel, G., Der UHP-Ofen in der Edelstahlerzeugung. *Radex-Rdsch.* 1973, 550–5.
18. Speith, K. G., H. v. Ende and W. Schneider-Milo, Die Wirkungsweise einer elektro-induktiven Badbewegung auf die metallurgischen Reaktionen im Lichtbogenofen. *Stahl u. Eisen* **78** (1958), 215–20.
19. Schäfer, K., The behaviour of oxygen during steelmaking from the melting and refining vessel to solidification (German). *Stahl u. Eisen* **99** (1979), 412–20.
20. Dötsch, E., Stahlerzeugung im Netzfrequenz-Induktionstiegelofen und Bedeutung für die Gußerzeugung. Paper at the 9th BBC-Stampfmassentagung 10–11 May 1978 in Dortmund.
21. Dötsch, E. and F. Hegewaldt, Schmelzen von Stahl in Großrauminduktionstiegelöfen. *Fachber. Hüttenprax. Metallweiterverarb.* **15** (1977), 429–33.
22. Kreutzer, H. W., 9. BBC-Stampfmassentagung 10/11 May 1978 in Dortmund. *Stahl u. Eisen* **98** (1978), 721.
23. Wilke, K., Erhöhung der Haltbarkeit einer basischen Tiegelzustellung für einen 3.6-t-Mittelfrequenzofen durch gezielte metallurgische Maßnahmen beim Erschmelzen von unlegiertem und legiertem Stahlguß. Aachen 1976. (Dr-Ing.-Diss. Techn. Hochsch. Aachen.)
24. Pantke, H. D., and C. Queens, The processing of sponge iron in the crucible induction furnace (German). *Stahl u. Eisen* **94** (1974), 1114–20.

25. Pantke, H. D., and C. Queens, Anforderung an die Qualität von Eisenschwamm für die Stahlerzeugung. *Stahl u. Eisen* **96** (1976), 652–7.
26. Etterich, O., Kristallkalk, die höchstbasische Betriebszustellung für Induktionsöfen und ihre neuen metallurgischen Möglichkeiten. Aachen 1963. (Dr-Ing.-Diss. Techn. Hochsch. Aachen.)
27. Schöberl, A., W. Holzgruber and E. Kahler, Der Einfluß der Tiegelzustellung auf den Ablauf metallurgischer Reaktionen im kernlosen Induktionsofen. *Berg-u. hütten. Mh.* **111** (1966), 534–41.
28. Fischer, W. A., and O. Etterich, Operational results of melting special steels in the medium frequency furnace with crucibles of fused lime (German). *Stahl u. Eisen* **87** (1967), 28–34.
29. Kreutzer, H. W., J. Otto and K. Schäfer, Sekundärmetallurgie. 1st Internationaler Kongreß für Hüttentechnik – Verfahren und Anlagen. Düsseldorf, 18–20 June 1979. Report 7.6.
30. Baum, R., K. Schäfer, H. W. Kreutzer and H. Sperl, Dislocation of metallurgical process steps of steelmaking to subsequent plants (German). *Stahl u. Eisen* **95** (1975), 973–81.
31. Maas, H., Fortschritte bei der Anwendung technischer Gase in der Metallurgie. Anwendung technischer Gase in der Vakuummetallurgie. *Techn. Mitt., Essen.* **70** (1977), 75–9.
32. Metzing, J., and W. Landt, Einsatz von Stahlgießpfannen mit hochwertigen feuerfesten Stoffen im Stahlwerk der Thyssen-Henrichshütte AG in Hattingen. *Stahl u. Eisen* **99** (1979), 13–17.
33. Comes, H., and H. Wagemann, Operating experience with basic ladles at the after treatment of steel (German). *Stahl u. Eisen* **94** (1974), 386–90.
34. Baum, R., H. Zörcher, B. Großkopf, H. Naefe and M. Oberbach, Investigation of the wear of the refractory lining in metallurgical vessels by measuring the distribution of radioactivity (German). *Stahl u. Eisen* **96** (1976), 572–7.
35. Münchberg, W., Wear in VOD ladles (German). *Stahl u. Eisen* **99** (1979), No. 23.
36. Walter, M., and H. Zörcher, Lining of ladles for vacuum refining of high-chromium melts produced using the VOD process (German). *Stahl u. Eisen* **99** (1979), No. 23.
37. Muth, S., The use of slide-gate nozzles in ladles for bottom pouring (German). *Stahl u. Eisen* **91** (1971), 173–81.
38. Einsatz von Schieberverschlüssen in Stahlwerken. Düsseldorf 1975. (Fachausschußbericht des Vereins Deutscher Eisenhüttenleute. No. 2.012.)
39. Betriebsergebnisse mit der Spülgasbehandlung von Stahl in der Gießpfanne. *Stahl u. Eisen* **96** (1976), 1056–9.
40. Gruner, H., F. Bardenheuer, H.-W. Rommerswinkel and H. Schulte, Steel desulphurization subsequent to the melting process (German). *Stahl u. Eisen* **96** (1976), 960.
41. Gruner, H., H.-E. Wiemer, F. Bardenheuer and W. Fix, Metallurgische Maßnahmen und Bedingungen zur Stahlentschwefelung über das Schlackenreaktionsverfahren. *Stahl u. Eisen* **99** (1979), No. 14.
42. Planung und Inbetriebnahme eines Hochleistungslichtbogenofens mit Prozeßsteuerung für die Edelstahlerzeugung. *Fachber. Hüttenprax. Metallweiterverarb.* **16** (1978), 780–7.
43. Meyer, H., M. Walter, R. Baum, and H. Zörcher, Vacuum refining of high-chromium melts. *Stahl u. Eisen* **99** (1979), 1315–18.
44. Labee, C. J., Argon–oxygen steelmaking marks major advance. *Iron Steel Eng.* **47** (1970), No. 6, 112–14.
45. Gorges, H., H. Graf, H. Lutz, P. G. Oberhäuser and H. Mülders, Experiences with the AOD process for the manufacture of stainless steels (German). *Stahl u. Eisen* **96** (1976), 1251–8.

46. Behrens, K. F., E. Köhler and K. Unger, The production of stainless, acid and heat resistant steels in the works Krefeld of Thyssen Edelstahlwerke — commissioning of an AOD plant (German). *Stahl u. Eisen* **99** (1979), No. 23.
47. Leroy, P., J. Morlet and J. Saleil, Le procédé C.L.U. d'affinage des aciers inoxydables. *Rev. Métallurg.* **72** (1975), 499–505.
48. Brasilis, J., J. Rouzier, P. Leroy, J. Morlet and J. Saleil, Elaboration des aciers inoxydables au convertisseur C.L.U. (Creusot-Loire/Uddeholm). *Rev. Métallurg.* **73** (1976), 441–6.
49. Öberg, K. E., Das C.L.U.-Verfahren in Degerfors. *Jernkont. Ann.* **161** (1977), 38–9.
50. Bauer, H., K. F. Behrens and M. Walter, Vacuum refining of high-chromium steels in the converter (German). *Stahl u. Eisen* **97** (1977), 938–44.
51. Glasmeyer, U., Four years of AOD operation with dolomite as refractory material. *Metallurg. Plant & Technol.* **2** (1978), 21–3.
52. Gorges, H., W. Pulvermacher, W. Rubens and H. A. Dierstein, Development and advantages of top blowing in combination with the AOD-process at Fried. Krupp Hüttenwerke AG. *Iron & Steelmaker* **5** (1978), No. 7., 30–3.
53. Pulvermacher, W., and H. A. Dierstein, Preliminary oxidation with the aid of a top blowing lance in the AOD converter of Fried. Krupp. (German). *Stahl u. Eisen* **99** (1979), No. 23.
54. Meyer, H., and M. Walter, Vacuum refining of high-chromium melts in the converter using the VODK-process. *Stahl u. Eisen* **99** (1979), 1325–8.
55. Knüppel, H., K. Brotzmann and H. G. Faßbinder, The oxygen bottom blowing converter, a new process of steelmaking (German). *Stahl u. Eisen* **93** (1973), 1018–24.
56. Münchberg, W., and K.-H.- Obst, Überblick über die Entwicklung der Zustellung von AOD-Konvertern in der Bundesrepublik Deutschland und in der Welt. *Stahl und Eisen* **99** (1979), No. 23.
57. Unger, K.-D., and E. Miecke, The lining of the AOD converters at Thyssen Edelstahlwerke AG in Krefeld (German). *Stahl u. Eisen* **100** (1980).
58. Münchberg, W., and K.-H. Obst, Wear in VOD ladles (German). *Stahl u. Eisen* **99** (1979), No. 23.
59. Tix, A., Betriebliche Anwendung der Stahlentgasung im Vakuum, besonders bei großen Schmiedestücken. *Stahl u. Eisen* **76** (1956), 61–8.
60. Harders, F., H. Knüppel and K. Brotzmann, Verfahrensgrundlagen der großtechnischen Entgasung von Walz und Schmiedestahlen nach dem Vakuumheber-Verfahren. *Stahl u. Eisen* **79** (1959), 267–72.
61. Thielmann, H., and H. Maas, Stahlentgasung nach dem Umlaufentgasungs-Verfahren. *Stahl u. Eisen* **79** (1959), 276–82.
62. Knüppel, H., A. Drevermann and F. Oeters, Die Entfernung des Sauerstoffs bei der Behandlung von Stahlschmelzen in Vakuum. *Stahl u. Eisen* **79** (1959), 414–9.
63. Kreutzer, H. W., Vacuum treatment of modern steel (German). *Stahl u. Eisen* **92** (1972), 716–24.
64. Tsu, Y., and T. Saito, Theoretische Untersuchung der Wechselwirkungsenergie zwischen gelösten Atomen und Stickstoff in flüssigem Eisen. *Trans. Japan Inst. Metals* **14** (1973), 339–46.
65. Kunze, H.-D., Diffusion des stickstoffs in flüssigen eisenreichen Mehrstofflegierungen. *Arch. Eisenhüttenwes.* **44** (1973), 173–9.
66. Kraus, Th., Einführung in die theoretischen Grundlagen der Vakuumbehandlung von Metallschmelzen. *Neue Hütte* **17** (1972), 581–5, 738–43.
67. Petersen, K., and H. Kahnwald, Direct collection of the dust-containing fumes of electric-arc furnaces during the production of high chromium steels and their de-dusting with the aid of cloth filters. *Stahl u. Eisen* **93** (1973), 910–14.

68. Scriven, D., Lichtbogenofenbetrieb mit Einblasen von Sauerstoff. *Ironmaking & Steelmaking* **1** (1974), 193–200.
69. Delhey, H.-M., L. Fiege and H. Vorwerk, Concept and operation of the DH vacuum plant in the LD steelmaking plant of Fried. Krupp Hüttenwerke AG, works Rheinhausen (German). *Stahl u. Eisen* **99** (1979), 18–23.
70. Pflipsen, H. D., and R. A. Weber, The operation of vacuum degassing plants in the oxygen steelworks Bruckhausen and Ruhrort of Thyssen AG (German). *Stahl u. Eisen* **98** (1978), 747–53.
71. Köhler, E., W. Rudack, H.-D. Schöler, R. A. Weber and W. Weitkämper, Leistungssteigerung von Oxygen-konvertern und Umlaufentgasungsanlage bei der Herstellung von Qualitäts- und Edelstählen. In *Internationaler Eisenhüttentechnischer Kongreß,* Düsseldorf 1974 (27–30 May). Vol. II. Düsseldorf 1974. Report 3.2.2.8. 17 pages.
72. Schenck, H., E. Steinmetz and R. Thielmann, Die Kinetik der Reaktion zwischen Kohlenstoff und Sauerstoff in flüssigem Eisen bei 1600°C. *Arch. Eisenhüttenwes.* **42** (1971), 79–84.
73. Weber, R. A., H. Schicks, G. Pietzko, K. Nürnberg, E. Köhler and T. Kootz, Investigation into decarburization during the vacuum circulation process (German). *Stahl u. Eisen* **92** (1972), 664–78.
74. Hilty, D. C., Relation between chromium and carbon in chromium steel refining, *J. Metals. Trans.* **185** (1949), 91–5.
75. Hilty, D. C., W. D., Forgeng and R. L. Folkman, Oxygen solubility and oxide phases in the Fe–Cr–O system. *J. Metals. Trans.* **7** (1955), 253–68.
76. Healy, G. W., and D. C. Hilty, Influence of oxygen in the decarburization of chromium steel. *J. Metals. Trans.* **9** (1957), 695–707.
77. Tesche, K., Gleichgewichte zwischen Kohlenstoff, Sauerstoff und den Metallen Chrom sowie Mangan in Abhängigkeit von der Temperatur beim Frischen mit Sauerstoff im basischen Lichtbogenofen. *I. Chrom. Arch. Eisenhüttenwes.* **32** (1961), 437–50.
78. Steinmetz, E., Physikalisch-chemische Grundlagen. Gleichgewichte in homogenen und heterogenen Systemen. *Techn. Mitt., Essen.* **66** (1973), 337–41.
79. Watanabe, T., Extrafurnace refining of stainless steel — refining characteristics of VOD and AOD. Unpublished report. Nov. 1978.
80. Baum, R., R. Hentrich, Z. Yun and H. Zörcher, The refining of stainless steels in a vacuum-oxygen-refining plant (German). *Stahl u. Eisen* **95** (1975), 8–11.
81. Schmidt, M., O. Etterich, H. Bauer and H.-J. Fleischer, Production of high-alloy special steels in the basic oxygen converter. I. Plants and processes (German). II. Metallurgische Grundlagen der Oxydation hochchromreicher Eisen-Kohlenstoff-Legierungen. *Stahl u. Eisen* **88** (1968), 153–68.
82. Bauer, H., O. Etterich, H.-J. Fleischer and J. Otto, The reducing and oxidizing vacuum treatment of melts of high-grade steel (German). *Stahl u. Eisen* **90** (1970), 725–35.
83. Schlager, W., A. Eggenhofer and G. Kaiser, Erfahrungen bei der Herstellung rost-, säure- und hitzebeständiger Stahle mit einer Vakuum-Frischanlage. *Berg-u. hüttenm. Mh.* **119** (1974), 1–9.
84. Otto, J., G. Pateisky and H.-J. Fleischer, Concentration variation of high chrome steels in vacuum oxidising and process control by means of continuous oxygen partial pressure measurement in the waste gas (German). *Stahl u. Eisen* **96** (1976), 939–45.
85. Pluschkell, W., I. Present state and development problems of the electrochemical oxygen measuring techniques (German). II. Anwendungsbeispiele für den Einsatz elektrochemischer Sauerstoffsonden im Stahlwerk. *Stahl u. Eisen* **99** (1979), 398–411.

86. Oeters, F., H.-J. Selenz and E. Förster, Metallurgical fundamentals of the deoxidation of steel (German). *Stahl u. Eisen* **99** (1979), 389–97.
87. Förster, E., Gemeinschaftsarbeit, 'Desoxidieren, Gießen und Erstarren von Stahl' des Vereins Deutscher Eisenhüttenleute (VDEh). *Stahl u. Eisen* **98** (1978), 445–51.
88. Förster, E., W. Klapdar, H. Richter, H. W. Rommerswinkel, E. Spetzler and J. Wendorff, Deoxidation and desulphurisation by blowing of calcium compounds into molten steel and its effects on the mechanical properties of heavy plates (German). *Stahl u. Eisen* **94** (1974), 474–85.
89. Kalla, U., H. W. Kreutzer and E. Reichenstein, Technological possibilities to obtain defined sulphur contents in steel (German). *Stahl u. Eisen* **97** (1977), 382–93.
90. Spetzler, E., and J. Wendorff, Das Einblasen von Erdalkalien in Stahlschmelzen und ihre Auswirkungen auf die Gebrauchseigenschaften von Stahl. *Radex-Rdsch.* 1976, 595–608.
91. Haastert, H.-P., W. Meichsner, H. Rellermeyer and K. H. Peters, Tauchlanzenverfahren. Wirtschaftliche Roheisenentschwefelung durch Einblasen von Carbidgemischen. *Verfahrenstechn.* **10** (1976), 476, 478, 482, 484.
92. Holzgruber, W., and D. Hengerer, Die Entschwefelung von Stahlschmelzen durch Einblasen pulverförmiger Feststoffe in Ofen and Pfanne. *Berg- u. hüttenm. Mh.* **123** (1978), 278–5.
93. Maas, H., and F.-J. Hahn, Methoden zur Vakuumbehandlung von LD- und E-Stählen mit dem RH-Vakuumverfahren. *Vakuum-Techn.* **25** (1976), No. 1, 17–20.
94. Köhler, E., W. Rudack, H.-D. Schöler, R. A. Weber and W. Weitkämper, Leistungssteigerung von Oxygenkonvertern und Umlaufentgasungsanlagen bei der Herstellung von Qualitäts- und Edelstählen. In *Internationaler Eisenhüttentechnischer Kongreß*. Düsseldorf 1974, (27–30 May) Vol. II. Düsseldorf (1974). Report 3.2.2.8. 17 pages.
95. Schrott, R., and H. Oppenhoff, Herstellung von Aufkohlungsschmelzen mit Kohlenstoffgehalten bis 0.8% C in einer Vakuumheberanlage. In *Internationaler Eisenhüttentechnischer Kongreß,* Düsseldorf 1974, (27–30 May). Vol. II. Düsseldorf (1974). Report 4.1.2.8. 18 pages.
96. Tiberg, M., T. Buhre and H. Herlitz, Das 'ASEA-SKF'-Verfahren zum Feinen von Stahl. *ASEA-Z.* **11** (1966), No. 3, 47–50.
97. Finkl, C. W., Vacuum arc degassing process at A. Finkl and Sons Co. *Iron Steel Inst.* **209** (1971), 33–6.
98. Eggenhofer, A., G. Kaiser and F. Bauer, Betriebserfahrungen mit der Vakuum-Beheizungsanlage für 50-t-Schmelzen. *Berg- u. hüttenm. Mh.* **119** (1974), 338–43.
99. Bailey, W. H., Herstellung von Stahl nach dem Finkl-VAD-Verfahren im Werk River Don von BSC. *Ironmaking & Steelmaking* **4** (1977), 209–15.

Secondary steelmaking. Proceedings of a conference organized by Activity Group Committee V (Ironmaking and Steelmaking) of The Metals Society, and held at the Royal Lancaster Hotel, London, on 5–6 May 1977. The Metals Society. Book No. 190. 130 pages, many illustrations and tables.

Pinder, E., A review of secondary steelmaking processes. *Trans. & J. Brit. ceram. Soc.* **76** (1977), No. 6, 148–54.

Olette, M., Considération de type fondamental sur les opérations métallurgiques hors du four (French). *Circ. Inform. techn.* **34** (1977), No. 9, 1969–92.

Brotzmann, K., Secondary steelmaking: review of the present situation in Europe. In *Secondary Steelmaking*. The Metals Society Book No. 190. London 1978, 27–31.

Susuki, Y., and T. Kuwabara, Secondary steelmaking: review of present situation in Japan. *Ironmaking & Steelmaking* **5** (1978), No. 2, 80–8.

Orehoski, M. A., Secondary steelmaking — refining processes in the USA. In *Secondary Steelmaking*. The Metals Society. Book No. 190. London 1978, 14–26.

Housemann, D. H., Refractories and the newer steelmaking processes — 23: Secondary steelmaking processes — 1. *Steel Times* **205** (1977), No. 8, 833, 836–7, 840–1.

Houseman, D. H., Refractories and the newer steelmaking processes — 26: Secondary steelmaking processes — 4. *Steel Times* **206** (1978), No. 2, 205–6, 209–11.

Houseman, D. H., Refractories and the newer steelmaking processes — 25: Secondary steelmaking processes — 3. *Steel Times* **205** (1977), No. 10, 1005, 1008–10, 1014, 1016, 1044.

Houseman, D. H., Refractories and the newer steelmaking processes — 24: Secondary steelmaking processes — 2. *Steel Times* **205** (1977), No. 9, 939, 942–5.

Deily, Richard, Round-up of AOD furnaces. *Iron & Steelmaker* **5** (1978), No. 7, 27–9.

Netoskie, R. R., D. A. Heckel and J. P. Wiencek, Metallurgical aspects of the AOD. *Iron & Steelmaker* **5** (1978), No. 7, 15–20.

Koncsics, David M., Optimizing AOD product cost. *Iron & Steelmaker* **5** (1978), No. 7, 34–7.

Williams, A. G. and I. Ludlam, The AOD process and equipment design. *Steel Times Intern.* **206** (1978), No. 12, 51–2, 54–6, 58–9.

Venne, L. J., and C. Z. Oldfather, Enhancement of physical properties by argon oxygen decarburization. *Trans. Amer. Foundrym. Soc.* **86** (1978), 139–42.

Kaufman, John W., and Carlos, E. Aguirre, Characteristics of refractory wear provides basis for new practice for AOD process. *Ind. Heat.* **45** (1978), No. 4, 12–14.

Kaufman, John W., and Carlos E. Aguirre, Characterization of refractory wear provides basis for new practice for AOD process. *Ind. Heat.* **45** (1978), No. 5, 32–4.

Johnsson, O., and K. Öberg, Economic aspects of stainless steel production in the CLU converter process. In *Secondary Steelmaking*. The Metals Society. Book No. 190. London 1978, 88–91.

Leach, J. C. C., A. Rodgers and G. Sheehan, Operation of the AOD process in BSC. *Ironmaking & Steelmaking* **5** (1978), No. 3, 107–20.

Vacuum metallurgy. Papers presented at the proceedings of the 1977 Vacuum Metallurgy Conference held in Pittsburgh, Pa., June 20–22. Ed. R. C. Krutenat. Princeton, NJ: Vad Science Press 1977. VI, 568 pages, many illustrations and tables.

Lorenz, A., Methods to produce vacuum in large vacuum metallurgical plants. In *Proceedings of the 5th Intern. Conf. on Vacuum Metallurgy and Electroslag Remelting Processes*. Munich, 11–15 Oct. 1976. Hanau 1977, 29–32 and 4 tables.

Krayzel, M., and Z. Motloch, Vacuum treatment outside the furnace in Czechoslovakia. In *Secondary Steelmaking*. The Metals Society. Book No. 190. London 1978, 64–8.

Kajioka, Hiroyuki, Yoshihiro Irie, Aribumi Niida, Kazushige Umezawa and Yoshihiro Okamura, Refining of quality steel using BOF-ladle furnace process. *Nippon Steel techn. Rep.* No. 11, 1978, 85–97.

Ushiyama, H., G. Yuasa and T. Yajima, Ladles furnace (LF) process in Japan. In *Secondary Steelmaking*. The Metals Society. Book No. 190. London, 1978, 101–16.

Baum, R., and H. Zörcher, Stahlwerke Südwestfalen Geisweid vacuum oxygen refining process. In *Secondary Steelmaking*. The Metals Society. Book No. 190. London 1978, 69–72.

Wayne, Ted J., The VAD process. *Proc. Electr. Furn. Conf.* **36** (1978), 82–9.

Tivelius, B., T. Sohlgren and C. Wretlind, The processing of Z-steel in an ASEA–SKF ladle furnace. In *Secondary Steelmaking*. The Metals Society. Book No. 190. London, 1978, 54–8.

Sundberg, Yngve, Mechanical stirring power in molten metal in ladles obtained by induction stirring and gas blowing. *Scand. J. Metallurg.* **7** (1978), No. 2, 81–7.

Ford, H., A. Byrne, D. M. Clinton and S. R. Laird, Manufacture of special steel tubes by ASEA process at Clydesdale Works of BSC. In *Secondary Steelmaking*. The Metals Society. Book No. 190. London, 1978, 92–5.

Cook, F. T., Low-alloy billets by the Finkl process at Brymbo. In *Secondary Steelmaking*. The Metals Society. Book No. 190. London, 1978, 96–100.

VAD process being expanded throughout the world. *Iron Steel Eng.* **56** (1979), No. 10, 65–6.

Pflipsen, H. D., and R. A. Weber, Installation of vacuum degassing plant at the Bruckhausen and Ruhrort Works of August Thyssen Hütte AG. In *Secondary Steelmaking*. The Metals Society. Book No. 190, London, 1978, 37–43.

Kurita, M., H. Ichikawa and Y. Tozaki, DH degassing of slabs for plate and sheet at Kashima Steel Works. In *Secondary Steelmaking*. The Metals Society. Book No. 190. London, 1978, 44–9.

Suzuki, Y., and T. Kuwabara, RH vacuum treatment in Nippon Steel Corporation. In *Secondary Steelmaking*. The Metals Society. Book No. 190. London, 1978, 32–6.

Scaninject I. 1st International conference on Injection Metallurgy, Lulea, Sweden, 9–10 June 1977. Preprints organized jointly by Jernkontoret, Lulea, 1977.

Scaninject II. 2nd International conference on Injection Metallurgy, Lulea, Sweden, 12–13 June 1980. Preprints organized jointly by Jernkontoret, Lulea, 1980.

Scott, Jr, W. W., and R. A. Swift, Advantages of ladle injection of calcium and magnesium reagents for steel desulfurization. *Proc. Electr. Furn. Conf.* **36** (1978), 38–51.

Aoki, T., Bullet shooting: an improved method of Al and Ca addition. *Iron & Steel Intern.* **51** (1978), No. 5, 307, 309, 311–12, 315–17.

Hater, M., W. Recknagel, and M. Haneke, Experience in making special high-tensile grades by improved metal desulphurization and DH vacuum process. In *Secondary Steelmaking*. The Metals Society. Book No. 190. London, 1978, 51-3.

Arbeletche, C., and B. F. Mehring, Injection of powders into liquid iron and steel. *Steel Times* **207** (1979), No. 8, 605, 607–13.

Oberg, Karl-Erik, and Frank, J. Weiss, The Scandinavian lancers ladle injection metallurgy system. *Proc. Electr. Furn. Conf.* **36** (1978), 54–60.

Spetzler, E., K. Nürnberg and R. Bruder, Influence of alkaline earth on steel metallurgy. *Trans. Amer. Foundrym. Soc.* **85** (1977), 311–16.

Nürnberg, K., E. Spetzler and W. Klapdar, The TN process and its effect on materials properties and on shop practice. In *Secondary Steelmaking*. The Metals Society. Book No. 190. London, 1978, 59–63.

Tivelius, Bertil, and Tomas Sohlgren, Secondary Steelmaking by the ASEA/SKF- and the TN-Process: a comparison. *Iron & Steelmaker* **5** (1978), No. 11, 30–9.

Electric Furnace Steel Production
Edited by E. Plöckinger and O. Etterich

8 Tapping and Casting Steel

FRIEDHELM REGNITTER, VÖLKLINGEN, HEINZ SPERL, DÜSSELDORF, HEINZ JUNG, VÖLKLINGEN, DIETER RUBENSDÖRFFER, VÖLKLINGEN, AND JOACHIM MEYER ZU HERINGDORF, BOCHUM

Steel which has been melted in an arc furnace, and in some cases subjected to secondary treatment in special vessels, is tapped into casting ladles where it often undergoes final treatment and is then finally poured into ingot moulds, continuous-casting systems, or steel-casting moulds. There are also a few cases of special processes such as centrifugal casting,[1–3] pressure casting,[4–5] and others; however, these will not be discussed in this book.

8.1 TEMPERATURE AND CASTABILITY

The necessary tapping temperature depends on the level of casting temperature desired, bearing in mind all the temperature losses which occur during tapping and waiting in the ladle right up to the end of pouring. The weight of the melt, the size of the ladle, the extent of any ladle preheating, the nature of any ladle-treatment, and the time taken for casting all have significant influence.

The quantity of greatest importance in determining casting temperature is the liquidus temperature which depends on the composition of the steel (cf. Chapter 14, Table 1). Ever since the introduction of casting flux and an effective way of protecting the stream of molten metal during pouring, it has been possible to pour steels containing higher percentages of elements such as chromium, aluminium, and titanium, which oxidize easily, at superheat temperatures that are hardly any higher than for non-alloy steels. However, the nature of any treatment, such as degassing, purging with gas, or the application of various deoxidizers, to which the steel is subjected in the casting ladle, has an important effect on the tapping-temperature required in order to keep to the desired casting temperature.

Individual processes and temperature histories during tapping and pouring, deoxidizing in the ladle, and the solidification of fully killed, half-killed, or unkilled steels can all be studied in the comprehensive literature quoted in the appropriate references[6–9] if they are not covered in other sections of this book.

8.2 LADLES AND LADLE-LININGS

8.2.1 Technical Requirements

The principal dimensions and technical requirements for steel-casting ladles with capacities from 10 to 400 tonnes are laid down in an iron and steelworks leaflet published by VDEh, the Association of German Iron and Steel Engineers.[10] This makes the interchange of steel-casting ladles much easier and provides for standardization of other system-components such as ladle transporter-wagons, tilting-devices, and similar equipment around the casting shop. The technical requirements lay down specifications for materials and non-destructive testing of ladle-construction.

In 1942 and 1943 the various steelworks came to an agreement with the refractory industry about the standardization of brick forms for ladle-linings. This standard for refractory brick formats and casting materials was published by the VDEh steelworks committee as Report 397, and has since frequently been brought up to date. The latest edition contains data on ladle bricks, stopper rod bricks, perforated bricks, nozzles, and runner bricks.[11]

8.2.2 Wear-processes

The durability of ladle working-linings depends on a number of wear-processes, some of which overlap each other, and we can break them down into chemical, physical, and mechanical wear-processes.

Chemical wear is affected by slag basicity.[12] The oxides CaO, MgO, FeO, and MnO promote slagging of acid fireclay linings,[13–15] while with Al_2O_3, TiO_2, Fe_2O_3, and P_2O_5 this is much less, or there may even be a slag-inhibiting effect.[16] On the other hand, these oxides all react with basic magnesite and dolomite linings. Direct reactions between the molten metal and the refractory lining also play a part.

Physical wear is affected by slag-viscosity. Depending on the porosity of the refractory materials, decreasing viscosity leads to heavier slag-infiltration and subsequent consumption of the lining. Lining-wear is also affected by the amount of slag and the temperature of the melt. Countermeasures include stiffening the ladle slag,[17] tapping with as little slag as possible, capping the steel bath with a covering agent, and, in some cases, water-cooling of acid ladles.

Mechanical wear of the lining is caused by the tapping stream impinging on the lining and by flow over the lining caused by bath motion during gas-purging or ladle-refining and when removing skulls.

Because of these effects, the ladle has to be lined in a way that anticipates the entire wear-spectrum by shaping the lining appropriately or using different types of refractory material.[18,19]

8.2.3 Refractory Wall-linings

A wall-lining normally consists of a backing-lining and a working-lining. The main reason for incorporating the backing-lining is to act as a safety-barrier to prevent steel-breakout. When a ladle is lined with basic materials and high-alumina refractories having high thermal conductivity, the backing-lining also provides thermal insulation.[20–21] Thin plates or mats of insulating material are used so that the insulating lining does not take up too much of the ladle volume; products based on aluminous silicate have proved particularly suitable for this. When a ladle-lining is being changed from acid to basic, it must be remembered that the density will be considerably greater (see Table 1 in Chapter 3, Section 3.6). The insulating lining also helps replacement of the working-lining. The thickness of the backing-lining and the bricking techniques used depend on operating conditions such as the size of the ladle, the type of production programme, and the length of time the steel is expected to dwell in the ladle.

The introduction of acid working-linings brought about a radical change in lining technique which also proved to be economical; this was the change from ramming to centrifugal application. The main benefit from this was an improvement in lining-life, but it also resulted in significant reduction in the financial outlay on refractories and interim repairs.[22–6] Brick linings are still used in a few steelworks; the reasons for this include the size of the steelworks, a small number of ladles, the capacity of the melting vessels, and the type of production programme as well as the high cost of installing modern slinging-machines in old steelworks.

Research is still in progress into the effects of rammed and centrifugally applied refractory ladle-linings on the purity of steel. Acid-linings, rammed or centrifuged, are not suitable for pouring certain types of steel. The sands most commonly used for rammed linings originate from natural sources, and their usefulness in steelworks has been proved empirically. The grain structure of the sand[17] and the chemical and mineral composition[7,27–9] as well as the moisture content all serve as measures of the suitability of a sand.

Centrifugal application of linings for steel casting ladles was introduced in the 1960s. This technique offers the advantage of improved and more regular packing of the sand, a shorter time required for application of the lining, improved lining life, and improvements in working conditions. Sand for slinging must meet the same requirements as ramming compound, except that the grain size of slinger sand must not exceed 3 mm and the water content must be between 8% and 9% in order to prevent premature wear of the slinger-head.[30]

Transfer of metallurgical refining, such as dephosphorizing, desulphurizing, deoxidizing, and alloying, into the ladle results in increased wear because of chemical reactions between the lining, the slag, and the molten metal.[31] In addition, some refining processes such as gas-purging and vacuum-treatment stress the lining even more because of the increased time the steel remains in the ladle and because the temperature is usually higher. In such cases, dolomite and magnesite linings are very durable in large-scale industrial use, with lives of up to over 100 melts, and have proved themselves, even though the material costs can be many times greater.[19,32–44] However, it has only been possible to use basic bricks and compounds since the introduction of the sliding-gate nozzle, because ladles with basic linings must not be cooled and must most definitely not be quenched.[44–51] It is worth using sequences that are extended as long as is practicable in order to achieve the longest possible lives for basic-lined ladles. If interruptions in utilization cannot be avoided, a drop in temperature should be limited as far as possible by covering the ladle and/or using supplementary heating. This is particularly important, because the thermal expansion coefficient of basic linings is greater than that of acid ones above 1000°C (see Chapter 3, section 3.6, Table 1). One should be particularly wary of damage which is then inevitable with basic linings because of spalling, cracking, undercutting, etc.

It is important that the steelworks engineer should achieve an even ladle wall-wear profile, so as to avoid high material-consumption caused by the need for early replacement of the wall-lining because a few spots wear more than the rest.[18] For this reason, many steelworks use higher-grade, costlier refractory materials in places, such as the slag area or where the tapping stream hits the wall, where wear is heaviest. The quality of firebricks normally used, which contain around 22–30% alumina, is improved by the addition of sillimanite, corundum, and bauxite; however, marked improvement can only be achieved with alumina contents above 60%.

With acid-linings, high wear-rates in any gaps can be avoided by pointing them with a high-alumina mortar or one based on zirconium silicate.

8.2.4 Refractory Linings for the Ladle-bottom

The refractory linings of ladle-bottoms pose a particular problem, since it is hardly ever possible to equalize the lives of the wall and bottom, and even then, special measures are required. Bottom-wear is caused by impingement of the tapping stream and removal of residual steel as well as erosion, to some extent, in the vicinity of the outlet nozzle.

Ladle-bottoms are usually covered with rectangular bricks 155–250 mm thick. The bottom lining extends under the wall-lining as well as covering the floor. Bottom bricks are mostly made from burnt, semi-acid fireclay and are formed dry or half-dry; plastic-formed bricks are less often used. All qualities of mortar are used up to high-alumina and zirconium mortars.

Nozzle bricks incorporated in the ladle-bottom are also subjected to mechanical and thermal shock during nozzle-changes. Therefore the average life of nozzle-bricks is less than that of the rest of the bottom, and nozzle-bricks have to be changed from one to six times in the course of a ladle lining-life. Nozzle-bricks are usually formed semi-dry from fireclay of normal commercial quality, but high-alumina bricks and those containing zirconium are also used.

Perforated bricks for gas-purging are usually made of higher-grade material than outlet nozzle-bricks and mostly contain zirconium, bauxite, and corundum.

8.2.5 Drying-out Steel Casting-ladles

It is particularly important that rammed and centrifugally applied linings in steel casting-ladles are dried out, because any moisture in the sand and binder will be driven out and yet the wall-lining must have a sound ceramic bond for the first charge. If drying-out is insufficient, residual moisture in the lining can destroy the wall immediately or lead to uncontrollable temperature drop and pickup of hydrogen by the steel. The time required for drying-out does not depend on wall-thickness or ladle-size, but is usually less than 24 h. Holes of about 4 mm diameter evenly distributed over the ladle surface simplify and speed-up the drying-out process.

8.2.6 High-temperature Protective Layers

Several attempts have recently been made to apply an additional protective layer over the usual ladle-lining by spraying on a compound which withstands high temperatures. This is best done by flame- or plasma-spraying. The material applied may consist of fire-resistant oxide or compound in powder or fibre form, e.g. Al_2O_3, $Zr_2O_3 \cdot SiO_2$, and Cr_2O_3.

However, we have as yet no operational experience with this technique. It may also be open to question whether the method is at all suitable for practical use since up to now the costs have been too high for industrial application, and the success of basic linings has overtaken these endeavours.

8.2.7 Casting-ladle Closures

Secondary treatment of steel in the casting-ladle calls for elevated tapping-temperatures, extended dwell-times, and powerful motion of the steel and slag as well as a rapid turnover, all of which led, in ladles fitted with stoppers, to difficulties which could only be overcome by using sliding-gate nozzles. Continual improvement in sliding-gate-nozzle technology and the development of special refractory materials particularly for the region where surfaces rub against each other, have led to increasing conversion of old, stoppered ladles and fitting all new ladles with sliding-gate-nozzle systems.[45–51]

In addition to the advantages already mentioned, sliding-gate nozzles also enable pouring operations to be controlled more easily and accurately and incur less cost in replacing the shutters and in routine maintenance as well as making operation simpler in general and improving safety in the casting bay. When it is necessary to pour steel at very different rates, two sliding-gate nozzles with different outlet-diameters can be used in one ladle. This system, using two slide-nozzles in the ladle-bottom, enables the time for which a ladle can be operated continuously to be doubled. It is essential that the equipment is set up very carefully if all these advantages are to be fully exploited.

Four types of sliding-gate nozzle system have proved themselves in industrial use so far. Three of these are very similar in construction; their main differences are in the ways the sliding-plates are held together to open and close the nozzles. In one of these, springs are used to press the plates together, while in both the others bolts are used. The fourth system is quite different; here the plates are moved by an electric motor instead of hydraulically, and the plates are rotated instead of moved linearly. In addition, the flow of molten steel can be regulated via two pouring outlets of different diameters; this is important for keeping the flow rate as steady as possible when pouring a melt.

8.3 INGOT-CASTING

8.3.1 Ingot Geometry and Moulds

The size and shape of an ingot are determined by metallurgical and economic criteria, according to the application, as well as by steel quality, melt weight, and the possibilities for shaping the ingot.

In addition, the way the steel solidifies is of decisive importance; there is a limit to ingot size for highly alloyed steels which are inclined to segregation during crystallization. Square cross-sections are usual for elongated products or strip; rectangular sections are used for sheet, and round ingots are used for tube, while ingots for forging are usually polygonal in section.

Slenderness ratio is defined as the ratio of the height of an ingot to its mean diameter.[52] This ratio is smaller for reverse-tapered ingots because this is metallurgically advantageous for the crystallization process (Table 1).

Increasing slenderness ratio results in a higher solidification rate for a given ingot weight and pouring rate.

The moulds are slightly tapered to facilitate ingot-removal (stripping) but, above all, also to obtain a fine grain structure with fully killed steel. The literature usually defines this conicity C as follows:

Table 1. Guide to normal ingot formats

Type of steel	Format	Ingot Weight (t)	Mould-orientation Normal	Mould-orientation Inverse	Slenderness ratio H/D_{mean}	Conicity $(D_{max}-D_{min})/2H$	Crop-end (%)
Ordinary, unkilled steel	Square-section ingots	0.5–12	x		2.0–4.1	1.0–1.9	–
	Slabs: broad side	4.0–50	x		1.4–2.1	1.5–2.0	–
	Slabs: narrow side				3.0–3.6		
Ordinary, killed steel	Square-section ingots	0.5–12	x		2.0–3.2	1.0–2.5	9–13
	Slabs: broad side	4.0–50	x		1.4–2.1	1.5–2.0	11–17
	Slabs: narrow side				3.0–6.0		
Quality steel, killed	Square-section ingots	0.5–10	x	x	2.7–3.2	2.0–3.0	9–13
	Slabs: broad side	4.0–50	x	x	1.4–2.1	1.5–2.0	11–17
	Slabs: narrow side				3.0–6.0		
Special high grade steels: Structural steel (stainless)	Square-section ingots	0.5–7.0		x	3.0–3.8	2.0–4.0	4–12[a]
	Polygonal forging ingots	1.5–500 (1.5–100)		x	1.5–3.4	3.0–6.0	8–15[a]
	Slabs: broad side	4.0–50		x	1.4–2.1	1.5–2.0	5–14[a]
	Slabs: narrow side	4.0–25			3.0–6.0		
Special high grade steels: Tool steel	Square-section ingots	0.5–3.5		x	2.8–3.6	2.5–6.0	4–15[a]
	Polygonal forging ingots	0.5–80		x	1.4–3.0	4.0–7.0	10–20[a]
Special high grade steels: High-speed steel	Square-section ingots	0.5–2.5		x	1.5–2.9	2.5–6.0	4–15[a]
	Polygonal forging ingots	0.5–5.0		x	1.5–2.9	4.0–8.0	10–20[a]

[a] The size of the crop end is very dependent on the type of process (exothermic hood-lining, ingot-top heating, etc.).

$$C\% = \frac{(\text{large diameter} - \text{small diameter})}{2 \times \text{ingot height}} \times 100$$

High conicity is a disadvantage during rolling and forging in that too much time is needed to make the sides parallel; on the other hand, it is metallurgically advantageous because there is less segregation with fully killed steels in reverse-conical moulds (normal-conical or taper = small end up; reverse conical or taper = larger diameter on top).

The gradient of normally tapered moulds, preferred for unkilled or ordinary killed steels, is 1–2.5% for blooms; it goes up to 3% for reverse-taper moulds for rolling ingots and can be still higher for forging ingots.

Normally tapered moulds help the ingot top to solidify and make it simple to strip the ingot. On the other hand, the increased top solidification rate does cause a risk of secondary piping. This danger can be alleviated, however, by reducing taper.[53] Reverse-tapered moulds are generally used in the production of high-quality, special steels. This means killed steels, which are poured under a cap (hot-top). The cap keeps the ingot-top hot so that molten steel can flow while the inside of the ingot solidifies progressively. Ingots cast in reverse-tapered moulds have hardly any secondary shrinkage cavities, unlike ingots cast in normally tapered moulds; instead they only have a somewhat porous structure in the centre. Working procedure in the casting-pit is more complicated, and more expensive, too, since the hood-frame must also be stripped. It is essential that work in the casting-pit be carried out with particular care when very slender, untapered ingot moulds are used. Ingots have been made economically in this way in recent years for ring-production and for use as electrodes for the remelting process.

The ingot mould is a casting, often still made from grey cast iron (hematite), into which the molten steel is poured and solidified. As the shape of the ingot mould not only influences the quality of the ingot but also affects working procedure in the casting-pit, a sensible compromise has to be made between the conflicting demands of the rolling mill, the forge, and the steelworks.

The literature contains various views on the choice of the right composition for iron castings; this is equally true for the new ingot-mould materials, while there are many factors of the most varied kind which affect the life of an ingot mould (constructional details, type of steel, the length of time the steel stays in the mould, and the mutual effects alloying elements in the cast iron have on one another). The effects of the individual alloying elements and impurities on the cast iron are generally known. The saturation intensity concept, introduced by E. Heyn and corrected by Piwokarsky, is a valuable aid for establishing the composition of cast iron.[54] According to data in the literature, saturation intensity should be less than unity in order to achieve good ingot-mould life.[55]

The following analyses are recommended (among others) in the literature:[9]

Author	C%	Si%	Mn%	P%	S%
Eisenhütte	3.4–3.6	1.6–2.0	0.4–0.6	0.1	0.08
Pieper	3.5–3.6	1.5	0.5–0.7	0.1	0.1
Cerkasov	3.78	1.3	1.5	0.085	0.037
Stadler	3.5	2.0	0.7	0.1	0.1

Steelworks in the USA, which have their own ingot-mould foundries and suitable raw materials, have been working for a long time with ingot-moulds made from steel. Such moulds and others made from nodular cast iron have been introduced in West Germany over the last ten years with increasing success, i.e. longer life and better ingot-surfaces. These moulds have the following compositions:

	C%	Si%	Mn%	P_{max}%	S_{max}%
Steel moulds	4.1–4.5	1.0–1.5	0.8–1.2	0.08	0.04
Nodular c.i. moulds	3.6–3.9	1.7–2.2	max 0.3	0.03	0.008

After the success with casting slab ingots in steel moulds, great progress was also made with enclosed steel ingot-moulds (blind moulds with bottom holes) for group (uphill) casting of special, high-grade steels. Nodular cast iron also did well for open ingot-moulds with cross-sections that are square, almost square, round, and polygonal as well as for components such as castings for top discards, which are subject to a high level of mechanical stress. In the Federal Republic of Germany about 20% of all ingot moulds are of hematite, while about 10% are of nodular cast iron; the remaining 70% approximately are steel (this includes figures for manufacturers that produce moulds only for their own use).

Ingot-moulds for square-section ingots usually used to weigh about 1.16 times the ingot weight (for slabs this was about 1.3); today the figures are 0.9 and 1.20.

8.3.2 Ingot-casting Processes

Steel may either be poured from the top or uphill (Figure 1). The design and arrangement of the moulds, etc. can be referred to in the relevant literature.[58,59]

Top-pouring of fully killed steels into reverse-tapered moulds provides good conditions for crystallization with a minimum of shrinkage cavities and removes oxidic impurities (deoxidation products) as well. All the same, some non-metallic impurities inevitably become trapped in the bottom third of the ingot; this is especially true for large ingots which take a long time to solidify. The

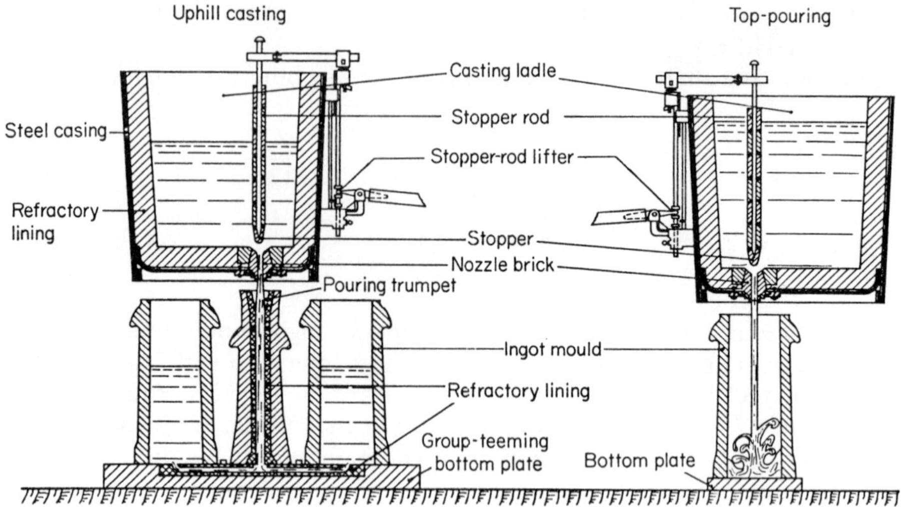

Figure 1. Schematic illustration of top-pouring and uphill casting of ingots[59]

various segregation processes responsible for this are described in the literature.[60,61] One disadvantage of top-pouring is that a good ingot surface can only be obtained with great difficulty.

The bottom-pouring or uphill process makes it possible to cast several ingots at the same time, thus significantly shortening the pouring time. The main advantage of this technique over top-pouring is, however, that less heat is lost during teeming. The rising melt causes less stress on the mould and therefore mould-life is prolonged. It is also easier to observe the process and adjust the pouring rate; this makes it possible to achieve much better ingot surfaces, especially if a suitable casting flux is used.

The improved ingot-surface quality saves costs of machining raw ingots and billets and compensates the extra expense in the casting-pit and the somewhat lower output caused by sprue i.e. steel trapped in the trumpet and gates. However, it is still not possible with uphill casting completely to eliminate precipitation of non-metallic impurities in the bottom third of the ingot.

With top-pouring each ingot is usually cast direct from the casting-ladle (ordinary steel) or via a funnel. The fit between the mould and its bottom plate and cap must be good enough to prevent steel entering the joints or escaping. Any steel that gets into gaps between the mould and its cap or bottom plate is trapped when the ingot shrinks, and this causes cracks across the ingot.

When high-grade steels are top-poured it is advisable to use a protective atmosphere of argon or nitrogen, especially when casting forging ingots. Ingot surface-faults such as oxide skin and shell can largely be eliminated;

purity is usually improved, too, by avoiding pickup of oxygen from the atmosphere into the molten steel.

Opinions differ as to what the optimum rate is for pouring. Steel may either be poured at a relatively high ladle-temperature (70–120°C above liquidus) and a low liquid level rise-rate or else it is poured faster with less superheating. In general, higher pouring rates require less superheating in order to produce ingots with no cracks and a low level of segregation.

With uphill casting, particularly with large groups of moulds, teeming must start fast in order to compensate the initial substantial loss of heat from the steel in the trumpet and runner-bricks, and to prevent premature solidification there. The teeming rate is only reduced to normal casting rate after the steel has started to rise in all moulds in the group. In uphill casting the surface of the molten steel in the moulds is covered with flux; this absorbs ingot-scum and prevents it adhering to the walls of the mould. The introduction of casting flux in the early 1960s made work easier in the casting pit and brought about a significant improvement in ingot surface quality. These advantages also led to increasing application of the technique to the casting of high-grade ingots for forging. In group casting with capped moulds or open moulds, a sandwich plate is used between the moulds and the base. In uphill casting of forging ingots a similar system is used, as can be seen in Figure 2; here a base plate incorporating a rounded chamber is set between the bottom plate with its channels and the special mould. The pouring operation is lengthy; this means that very high-grade refractory materials are required for the trumpet and runner bricks. The amount of casting flux added depends on ingot surface area and amounts to about 2 kg per tonne of ingot for normal ingots for the rolling mill. The type of casting flux used should be chosen to match the steel quality and the pouring temperatures.

Problems that can occur with uphill casting include missing out an ingot and steel breakout from the trumpet or runner bricks or from between the mould and the bottom plate. Common reasons for missing out one or more ingots are low steel temperature or obstruction in the runner bricks. The length of time an ingot has to stand to solidify completely depends mainly on the ingot diameter and the composition of the steel. It can be calculated approximately using Nelson's formula:[62]

$$t = \left(\frac{D}{2}\right)^2$$

where t is the time to solidify in minutes and D is the ingot top diameter in inches. Thus a 4-t hooded ingot takes about 2 h to solidify while a 7-t one takes about 2½ h.[63]

8.3.3 Ingot-top Treatment

High-grade steels are normally poured into inverse-conical moulds equipped with caps which hold molten steel to enable any shrinkage cavity in the top to be

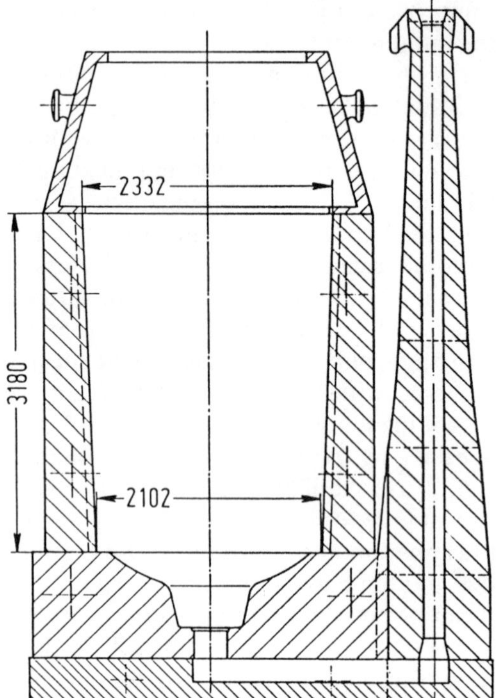

Figure 2. Casting a 139-t forging ingot by the uphill method. (Works drawing by Thyssen Giesserei AG, Mülheim a.d., Ruhr)

filled. Prolonging the time during which the steel in the hot-top remains molten enables non-metallic impurities to rise into the dead-head which is cut off in hot-working later on and scrapped. There are a number of ways of keeping the steel hot in the ingot. These include insulated tops with insulated exothermic linings and an insulated top used in conjunction with electric heating of the ingot head.[57,64–7] Whereas classical electric heating of the ingot head has now lost almost all importance, the BEST process (Boehler-Electro-Slag-Topping), which was developed about ten years ago, has become established in the production of heavy forging ingots. This process enables the weight of the crop head to be minimized while definitely preventing non-metallic impurities from collecting in the bottom third of the ingot and going a long way towards avoiding ingot segregation altogether.[68] With all the ingot hot-topping methods mentioned, the outlay on heating and insulation should be matched to practical improvements in output and ingot quality.

8.4 CONTINUOUS CASTING

8.4.1 The Development of Continuous Casting

Continuous casting was developed very rapidly after the Second World War.[69,70] Nowadays about a quarter of all crude steel made throughout the world is produced in continuous-casting systems. Steel-producers are today generally convinced that continuous casting is at least as economical as ingot production and can match the quality of the latter across much of the production spectrum for high-quality steels. Continual development of the technique aimed at improved steel characteristics is leading to increasing adoption of the process in works producing special high-grade steels.[71] The reasons for continuous-casting systems are:

(1) Lower investment outlay compared with that for a blooming train (mini-steelworks);
(2) About 10% more productivity than with conventional ingot-casting;[72]
(3) High degree of consistency of steel composition along the whole length of the strand; better core quality, especially with flat strands; high inherent surface quality, leading to savings on an otherwise expensive surfacing process;
(4) High degree of automation;
(5) Friendlier to the environment;
(6) Better working conditions.

8.4.2 Types of Installation

The first continuous-casting plants were aligned vertically; however, with larger cross-sections, increasing strand-length, and, above all, with increasing pouring-rates this type of construction leads to unreasonable building-heights. These factors also lead to a considerable increase in the length of the liquid phase which has metallurgical effects. The length of the liquid phase in a continuously-cast strand is determined by the following formula:

$$L = \frac{D^2}{4\chi^2} V_c$$

Where D = strand thickness (mm)
χ = solidification characteristic (mm/min$^{1/2}$)
These values amount to 26–33 for the whole cooling length.
V_c = casting rate (m/min).

Efforts to reduce building-height first led to continuous-casting systems in which molten metal passed into a vertical mould and solidified completely before being bent or where the strand was bent in the liquid phase and later to the bow-type installation which has a curved mould (Figure 3) and is the system most used today. Vertical systems and those in which the strand is bent when completely solidified have long straight liquid phases and can lead to unacceptably high capital outlay. However, these systems have metallurgical advantages because of their geometry, and purely vertical installations also have advantages from the point of view of maintenance. A vertical system in which the strand is bent while still in the liquid phase has the advantage that the building need not be as tall as when the strand is bent after solidification; however, the liquid-phase bending system requires higher initial capital outlay and greater maintenance costs. The bow-type system represents a compromise between the costs of capital outlay and of maintenance and what can be achieved metallurgically.

Continuous-casting is suitable for the production of almost any cross-section imaginable; square, rectangular, polygonal, round, and oval sections are all available.[73] There are also some instances of preliminary sections for tubes and rails being formed in continuous-casting systems.[74] Various systems produce slabs, blooms, and billets. Sections with a breadth/thickness ratio greater than 1.6 are normally described as slabs. Billet-machines produce square or nearly-square, round, or polygonal cross-sections up to 160 mm across. Larger sections and those with a breadth/thickness ratio less than 1.6 are cast in bloom-machines.[75] Billets nowadays normally produced in continuous casting plants measure 90–250 mm round; blooms produced in this way range from 80 × 80 to 300 × 300 mm, and slabs are 50–350 mm thick and 300–2500 mm wide.

Continuous-casting output-rates have risen sharply, especially in the last few years.[76] This is essentially because of increase in the breadth of the strand and in casting rate. The following outputs have been exceeded per section per minute:

Slabs	5 tonnes
Blooms	1 tonne
Billets	350 kg

Finally, we should mention horizontal continuous-casting systems which are already used for non-ferrous metals and cast iron[77] and which are being further developed for steel.[78,79a,79b] R. Thielmann and R. Steffen have produced a comprehensive report about the state of development of horizontal continuous-casting systems for producing billets from unalloyed and alloy steels; it includes a detailed survey of the literature as of early 1980.[79c] Horizontal continuous-casting systems have three important advantages over conventional continuous-casting systems:

(1) Low height and cost of building;
(2) Simple means of protecting the melt against reoxidation;
(3) No strand deformation because the ferrostatic pressure is much lower.

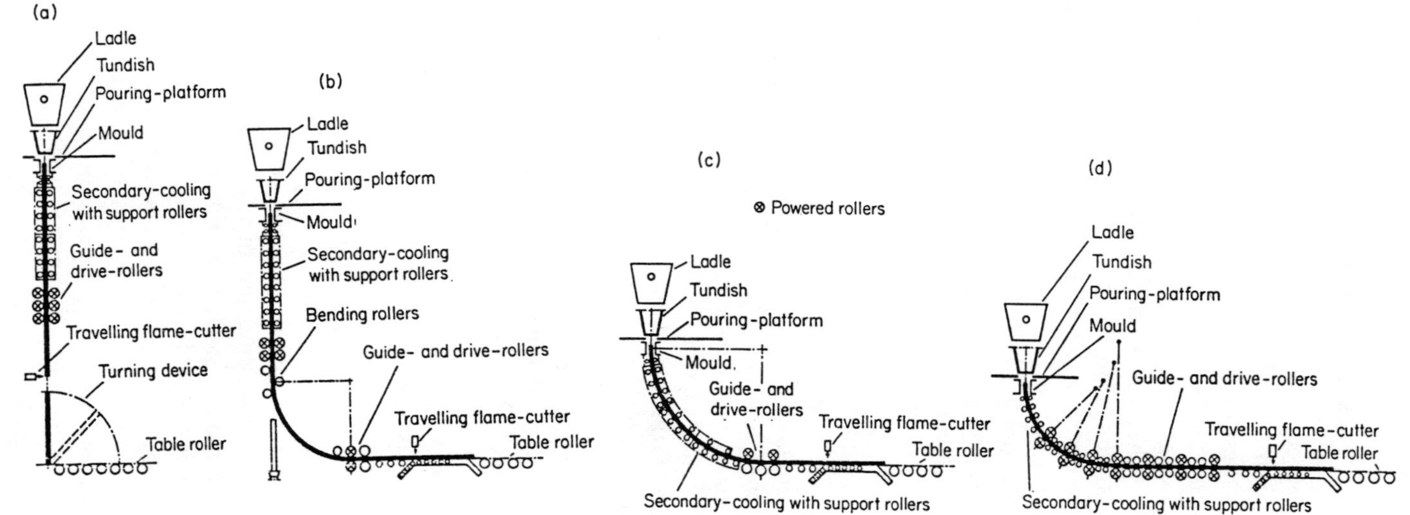

Figure 3. Schematic diagrams of different types of continuous-casting system.[59] (a) Vertical system; (b) deflected system; (c) bow-type installation; (d) oval-bow system

8.4.3 Casting Technique

Molten steel is poured from a casting ladle via a tundish into an open, water-cooled copper mould. At first the bottom of the mould is closed off by a starting-bar, which then leads transport of the hot strand from the mould into the continuous withdrawing rolls. The strand, which starts to solidify in the mould, passes through a cooling system before it finally reaches the withdrawing rolls, whereupon the hot strand takes over transport. The starting-bar is separated from the hot strand before or after it reaches the parting device. The latter, which may either be a flame-cutter or hot shears, moves at the same rate as the hot strand and cuts it into the lengths required.

The purpose of the tundish is to feed a defined quantity of molten steel into one or more moulds. This can be done by using nozzles controlled by stoppers, slide-gates, or other means. The tundish may initially be cold, warm, or hot, according to the nature of its refractory lining.[80,81] Where difficult steels are processed, the pouring stream is protected against oxidation between the tundish and the mould by using an inert atmosphere, submerged pouring, or submerged boxes. The mould not only forms the strand section but also extracts a defined quantity of heat, so that the strand shell is strong enough for transport by the time it reaches the mould-outlet. The mould may be made from copper tube or hardenable copper alloy, depending on the shape and size of the strand to be cast. As a rule, tubular moulds are used for smaller sections. The interior surface of the mould may be coated with chrome or molybdenum to reduce wear and to suit heat-transfer from the alloy being cast. The mould is tapered to match steel-shrinkage and casting-rate and the type of steel concerned.[82] Moulds used today range from 400 to 1200 mm in length overall, but their usual length is between 700 and 800 mm. The problem of steel adhering to the mould-sides is usually countered by oscillating the mould sinusoidally relative to the strand and by adding lubricant (oil or casting flux) in an attempt to cut friction between the mould and the steel. The lubricant, particularly casting-flux, has an additional metallurgical function. The choice of lubricant depends on the qualities required and the casting conditions; it is particularly important that casting-flux should be chosen to match the quality-programme precisely.

The level of the steel in the mould may be controlled manually or by an automatic system. Either method may be used to keep the level constant or to match the incoming molten steel, i.e. to accommodate variations in casting rate. Manual control is effected via the stopper in the tundish or by varying the output rate. An automatic control system may meter radioactivity or infra-red radiation or measure temperature via a probe in the mould wall to determine the steel-level and compensate any changes by actuating the stopper-mechanism (for constant pouring rate) or controlling the speed of the withdrawing rolls (varying casting rate).

The type of starting-bar used for continuous-casting depends on the type of installation. Rigid starting-bars can be used in vertical systems, while articulated dummy bars or flexible strip have to be used in bowed installations. The starting bar can be connected to the hot strand in different ways. One is by welding the fluid steel using a jointing element (flat slab, screw, or fragment of rail) which is soluble in the starting-bar; another is by casting the connector in a specially shaped head in the dummy bar in a way that enables it to be released by unlatching.[83]

The thickness of the solidified strand shell on leaving the mould depends first of all on how long the steel is in contact with the mould, but it also depends on the specific thermal conductivity of the mould and on the amount of superheat that steel has when it enters the mould. It can be determined with fair accuracy using the following parabolic formula:

$$C = \chi\sqrt{t}$$

where C is the thickness of the strand shell (mm)
χ is the solidification characteristic (mm/min$^{1/2}$)
t is the solidification time (min)

The solidification characteristic in and near the mould lies between 20 and 26, depending on the operating conditions; for the secondary cooling-area the figure is 29–33. The thickness of the solidified strand shell on leaving the mould is about 8–10% of the strand-thickness, depending on casting rate. A secondary cooling-area under the mould speeds up completion of the solidification process. The coolant usually is water but a water/air mixture or compressed air are also sometimes used. The secondary cooling area is divided into several zones to suit coolant flow rates. The necessary quantity of water is sprayed over the entire strand by spray-bars.[84] The ferrostatic pressure may be so high in relation to the strand cross-section and the casting rate that the strand has to be supported to prevent buckling. The equipment for this is expensive in plants producing blooms and especially slabs.[85] The literature contains details about other parts of these systems such as transport, guidance, and strand-cutters.[83]

8.4.4 Quality in Continuous Casting

Continuous-casting is used nowadays for almost all non-alloy steels and many low-alloy, killed structural steels as well as a large proportion of high-alloy quality steels, particularly austenitic chrome-nickel steels. In continuous-cast steels primary crystallization approximately corresponds to that in ingots having the same section, usually with more pronounced columnar crystallization caused by the relatively rapid cooling. Figures 4 and 5 illustrate cross-sections through an alloy structural steel strand measuring 120 × 140

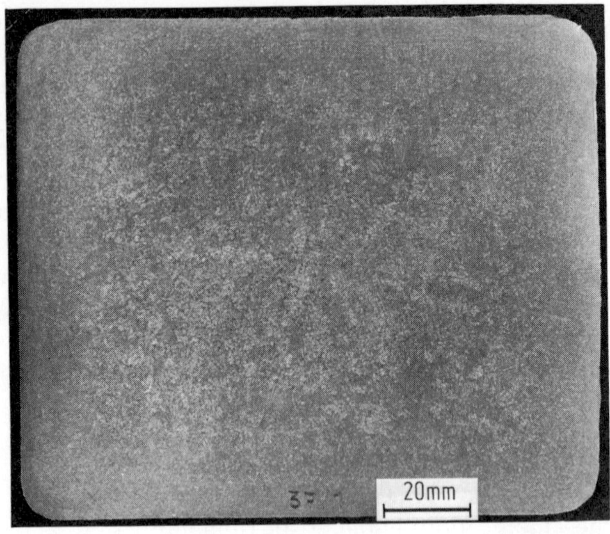

Figure 4. Grain-structure in a 120 × 140 mm strand of 36CrNiMo4 steel[86]

Figure 5. Grain-structure in a strand of X8Cr17 steel measuring 480 × 140 mm[86]

mm and a ferritic, stainless steel strand measuring 480 × 140 mm. Figures 6(a) and 6(b) show the characteristic grain-structure obtained in continuous-cast slabs of austenitic chrome-nickel steels measuring 1000 × 165 mm.

Because of their cast structure and core porosity, continuous-cast steels must undergo at least five to ten stages of hot-formation, depending on the use for which they are intended. When it is properly employed the continuous-casting process can produce many types of steel of which the internal and surface qualities are superior to those obtained with ingots.[88] Electromagnetic stirring has proved to be an important aid to improving

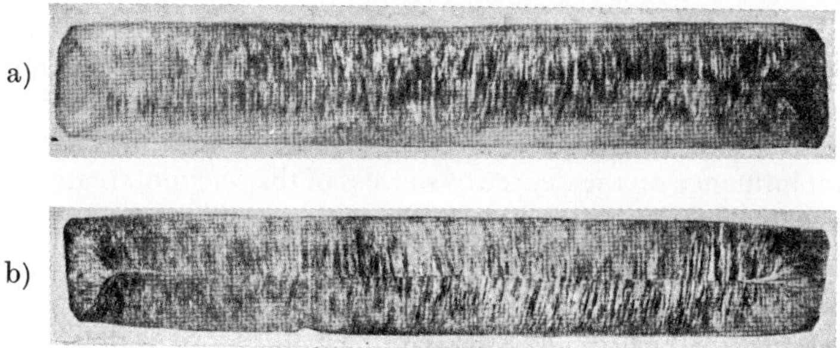

Figure 6. Cross-sections through two continuous-cast austenitic Cr–Ni steel slabs measuring 1000 × 165 mm: (a) not stabilized; (b) titanium-stabilized[87]

strand quality.[89] The relevant literature gives details of faults obtained in continuous casting.[90]

8.4.5 Designing Continuous-casting Systems

The design of a continuous-casting system is basically determined by the melt weights and melting-sequence time used in the steelworks as well as by the desired strand-dimensions and the types of steel to be cast. It must be borne in mind that maxima for the number of strands, charge time, and casting rate cannot be exceeded for given cross-sections and grades of steel.[83,84] In the planning stage it should be remembered that minimizing set-up time and limiting the range of cross-sections promote high system-availability; sequence-casting should also be considered.[91] The quality requirements for high-grade and special steels can be met by using special measures such as protecting the pouring stream between the ladle and the distributer, electromagnetic stirring coils in the mould and/or the secondary cooling section,[89] and well-built strand-guides in bloom-production units.

8.5 VACUUM INGOT-CASTING

While ingots for rolling and forging are normally cast in air after the steel has previously been treated under vacuum, heavy forging ingots are often cast in a vacuum, in which case the melt must first be degassed if very low hydrogen levels are specified.

In vacuum ingot-casting the mould stands inside an evacuated vessel. The vacuum vessel is sealed off from the ladle containing the melt by a steel closure. When pouring starts, the stream of molten steel breaks through the closure and is broken down into droplets by the action of gas expelled from the molten metal. The increase in surface area and the shortened diffusion paths resulting help the process of degassing. With higher levels of oxygen in the steel the intensity of degassing is increased by the formation of carbon monoxide bubbles, so that not only is there a reduction in non-metallic inclusions but also very low levels of hydrogen (around 1.5 ppm) can be attained. The pressure in the vacuum vessel and the distance the pouring stream falls in the vacuum have a significant influence on the degree of success of this vacuum treatment, as has the extent of deoxidation.[92]

Vacuum ingot-casting has a further application in melting steel in vacuum induction furnaces in order to provide protection against pickup of oxygen and hydrogen by a melt which has been thoroughly degassed (cf. Chapter 6, Section 6.1.6).

8.6 LITERATURE REFERENCES

8 Tapping and Casting Steel

1. Simoneit, K., and W. Rädeker, Stahlhohlguß als Vormaterial. *Stahl u. Eisen* **68** (1948), 419–26.
2. Lückerath, W., Schleudern von Stahlblöcken größter Abmessungen. *Stahl u. Eisen* **70** (1950), 209–18.
3. Barthelemy, Ph., P. Delbey, P. Peytavin, J. Dedieu, J. L. Gatelais and L. Babel, Entwicklung des kontinuierlichen Schleudergießverfahrens. *Rev. Métallurg.* **71** (1974), 437–52.
4. Drever, J. R., G. R. Lohman and J. Woodburn jr., Pressure pouring of stainless steel slabs. *Iron Steel Eng.* **40** (1963), No. 12, 123–8.
5. Weymueller, C. R., Outlook for pressure casting. *Metal Progr.* **87** (1965), No. 5, 113–14, 116, 118, 120, 122, 124–7.

Gießen und Erstarren von Stahl. Informationstagung, Luxembourg, 29 November–1 December 1977. Vol. 1. (Also deals with the process in brief. Compiled by H. Jacobi. Organized and published by the EEC Commission.) Düsseldorf: Verl. Stahleisen 1977. 423 pages, many illustrations and tables. 8°. = EUR 5903 d, c.f.

Jacobi, Hatto, Gießen und Erstarren von Stahl. Zusammenfassender Forschungsbericht (über die EGKS-Verfahren 1969–1972). = Part I. Forschungs-Vertrag No. 6210–50 of the European Coal and Steel Community. Published by the EEC Commission general directorate for scientific and technical information management, Luxembourg, 1977. 194 pages, 120 illustrations, 22 tables. 4°. = EUR 5861 d, e, f. Part I.

Sheffield International Conference on Solidification and Casting, Sheffield University, 18–21 July 1977. Vol. 1–3 + supplement. (Paper No. 1–81). Preprints. Sponsor: Sheffield Metallurgical and Engineering Association and The Department of Metallurgy of the University of Sheffield in association with The Metals Society.

Kemlo, Kenneth Garry, Reduction of ingot butt losses. *Iron & Steelmaker* **6** (1979), No. 3, 17–23.

Sweetman, D., Ingot weight control for yield improvement. *Steel Times* **207** (1979), No. 6 (Supplementary Intern. Iss.), 91–5.

8.1 Temperature and castability

6. *Die physikalische Chemie der Eisen- und Stahlerzeugung*. Verein Deutscher Eisenhüttenleute. Düsseldorf, 1964.
7. *Gießen und Erstarren von Stahl.* Verein Deutscher Eisenhüttenleute. Düsseldorf, 1967.
8. Knüppel, H., *Desoxydation und Vakuumbehandlung von Stahlschmelzen.* Vol. 1. Düsseldorf, 1970.
9. Wagner, C., *Herstellung von Stahlbrammen durch Blockgießen und Walzen.* Düsseldorf, 1975.

8.2 Ladles and ladle-linings

10. *Stahl-Eisen-Betriebsblatt 330010-77.* Issue Jan. 1977.
11. Vereinheitlichung feuerfester Baustoffe für Gießpfannen und Unterguß. *Stahl u. Eisen* **98** (1978), 32–6.
12. Schwiete, H. E., and N. Neises, *Untersuchungen über die Verschlackung von Schamotte-Pfannensteinen.* Köln/Opladen, 1964. (Forschungsberichte des Landes Nordrhein-Westfalen. No. 1299.)
13. Abratis, H., Zur Kinetik der Reduktion von Kieselsäure durch kohlenstoffhaltige Eisenschmelzen. *Arch. Eisenhüttenwes.* **43** (1972), 277–82.
14. Abratis, H., Reaktionen feuerfester Stoffe mit Stahlschmelzen. *Arch. Eisenhüttenwes.* **44** (1973), 329–36.
15. Abratis, H., Reaktionen feuerfester Stoffe mit Schlackenschmelzen. *Arch. Eisenhüttenwes.* **44** (1973), 691–6.
16. Maas, H., and H. Abratis, Reduktion hochtonerdehaltiger feuerfester Stoffe in Eisen-Kohlenstoff-Schmelzen. *Arch. Eisenhüttenwes.* **40** (1969), 153–7.
17. König, G., and G. Horn, Experiences gained in lining pouring ladles with ramming mixes (German). *Stahl u. Eisen* **89** (1969), 1092–6.
18. Cherubim, W., and W. Resch, Operating experience with the refractory lining of steel pouring ladles (German). *Stahl u. Eisen* **95** (1975), 582–92.
19. Baum, R., F. Taake, H. Zürcher, W. Münchberg, J. Stradtmann and G. Zingel, Dolomitisch zugestellte Gießpfannen in einem Edelstahlwerk. In *Vorträge des 20. internationalen Feuerfest-Kolloquiums,* Aachen, 13–14 Oct. 1977. Institut für Gesteinshüttenkunde der Technischen Hochschule Aachen; Forschungsinstitut der feuerfest-Industrie, Bonn; Verein Deutscher Eisenhüttenleute, Düsseldorf. Aachen, 1977, 40–67.
20. Knop, K., H. Richter, H.-W. Rommerswinkel and J. Wendorff, Wärmetechnische Überlegungen und Betriebsergebnisse an basisch zugestellten Stahlpfannen. *Tonind.-Ztg. keram. Rdsch.* **98** (1974), 26–9.
21. Thorildsson, S., Preheating of ladles. *Scand. J. Metallurg.* **2** (1973), 277–81.
22. Hergenhahn, R., Experience with and comparison of linings in rammed and bricked ladles (German). *Stahl u. Eisen* **88** (1968), 974–8.
23. Ludwig, H. W., and H. Schneider, Development of rammed ladle linings in open-hearth and basic oxygen steels plants (German). *Stahl u. Eisen* **89** (1969), 1096–9.
24. Hartmann, T., Introduction of rammed pouring ladles in a basic converter steel plant (German). *Stahl u. Eisen* **89** (1969), 1100–2.
25. Bauer, G., and J. Koenitzer, Contribution to the problem of ladle life in a basic oxygen steel plant with large converters (German). *Stahl u. Eisen* **89** (1969), 1102–5.

26. Abele, K. H., G. Bauer, E. Hoffken and G. Klages, Refractory lining of steel pouring ladles (German). *Stahl u. Eisen* **94** (1974), 213–18.
27. Berens, L. W., A. Wijkström and E. Klinger, *Gieß.-Prax.* **82** (1964), 239–48.
28. Cigler, V. D., and V. L. Bulach, Verwendung kieselsäurehaltiger feuerfester Steine und Stahlgießpfannen. *Ogneupory* **31** (1966), No. 12, 32–4.
29. Sojfer, V. M., and V. F. Martynenko, Zustellung von Stopfenpfannen mit einem chemisch härtenden Gemisch. *Stal in Deutsch* **6** (1966), 752–4.
30. Kleeschulte, H., Experiences with the use of a slinger for ladle lining (German). *Stahl u. Eisen* **89** (1969), 859–63.
31. Baum, R., K. Schäfer, H. W. Kreutzer and H. Sperl, Dislocation of metallurgical process steps of steelmaking to subsequent plants (German). *Stahl u. Eisen* **95** (1975), 973–81.
32. Allcock, S., D. Newbound and D. R. F. Spencer, The application and performance of basic refractory materials in steel ladles in the United Kingdom. See Ref. 19. p. 1–39.
33. Mönchberg, W., J. Stradtmann and W. Deilmann, Verschleißverhalten von basischen, feuerfesten Steinen auf Dolomitbasis beim Einsatz in Stahlgießpfannen im diskontinuierlichen Betrieb. See Ref. 19. p. 68–87.
34. Forchheimer, O. L., and D. A. Brosnan, USA dolomite brick practice in steel ladles. See Ref. 19. p. 88–121.
35. Deilmann, W., and W. Zednicek, Versuchsergebnisse bei Gießpfannenzustellungen mit Magnesitprodukten und Steinen auf Basis Mag/Dol-Simultansinter. See Ref. 19. p. 122–48.
36. Mizuno, Y., N. Shimada, H. Kyoden and Y. Namba: The performance results of basic ladle lining. See Ref. 19. p. 149–72.
37. Störmann, J., and S. Chaudhuri, Beitrag zur Verwendung basischer Steine in Stahlgießpfannen, besonders von neuentwickelten Spinellsteinen für die sekundäre Stahlerzeugung. See Ref. 19. p. 173–84.
38. Jeschke, P., U. Muschner, M. Oberbach and J. Pirdzun, Die Verwendung von magnesitischem Feuerfestmaterial in der Pfannenmetallurgie und im gesamten Verfahrensablauf beim Stranggießen von Stahl. See Ref. 19. p. 185–217.
39. Lakin, J. R., The effect of process variables on the performance of the linings of secondary steelmaking units. See Ref. 19. p. 218–27.
40. Bauer, G. E. Hees, E. Eisermann and G. König, Moderne Gießpfannenzustellungen. See Ref. 19. p. 228–52.
41. Resch, W., H. Bunse and P. Artelt, Slingern von basischen feuerfesten Massen in Stahlgießpfannen für die metallurgische Nachbehandlung. See Ref. 19. p. 285–303.
42. Piret, J., and J. Claes, Amélioration de la résistance des sables argileux à l'attaque des scories de poches à acier au moyen d'oxyde de chrome. See Ref. 19. p. 304–22.
43. Bönschen, W., S. Kaspar, G. Spiess and M. Dott, Zustellung von 100-t-Stahlgießpfannen mit bauxitischen Slingermassen. See Ref. 19. p. 261–77.
44. Comes, H., and H. Wagemann, Operating experience with basic ladles at the aftertreatment of steel (German). *Stahl u. Eisen* **94** (1974), 386–90.
45. Aliprandi, G., P. Ghirotti and R. Ricci, The present situation of the sliding gate nozzle casting system in Italy. See Ref. 19. p. 352–78.
46. Weidemüller, Ch., O. Krause and H. Jammernegg, Der Einsatz von Schieberkeramik für Gießpfannen. See Ref. 19. p. 379–403.
47. Schweinsberg, H., M. Glaveris, W. Krönert and P. Dietrichs, Zur Haltbarkeit von Schieberverschluß-Systemen in Stahlgießpfannen; Zusammenhänge zwischen

mineralogischem Aufbau und physikalischen Eigenschaften von Korundplatten. See Ref. 19. p. 404–30.
48. Mutsaarts, Ph., M. Avaert, M. Equeter and R. Golinski, Expérience européenne de la "Busette Rotative" en poche de traitement et de coulée d'acier. See Ref. 19. p. 431–65.
49. Deilmann, W., and B. Grabner, Weiterentwicklungen von basischen Keramikteilen für Schieberverschlüsse. See Ref. 19. p. 466–87.
50. Russell, C. K., and J. C. Huntington, Studies of the properties and wear characteristics of slidegate refractories. See Ref. 19. p. 488–509.
51. Otto, J., and S. Muth, Untersuchungen der thermischen Beanspruchung von Pfannenschieberverschlußplatten beim Gespannguß von Stahl. See Ref. 19. p. 510–32.

Hargreaves, J., The evolution of ladle lining refractories. *Refractories J.* **53** (1978), No. 6, 11–15, 17.
Zörcher, Heinz, and Manfred Walter, Zustellung von Pfannen für das Vakuumfrischen hochchromhaltiger Schmelzen nach dem VOD-Verfahren. *Stahl u. Eisen* **99** (1979), No. 23, 1318–21 (Stahlwerksaussch. 1032).
Carswell, G. P., M. P. Crosby and D. R. F. Spencer, Development of high performance ladle linings. *Refractories J.* **55** (1980), No. 1, 9–12, 14–16, 18–20.

8.3 Ingot-casting

52. Lecompte, H., *Cours d'aciérie.* Paris 1962.
53. Smrha, L., E. Janosch, J. Brábnik, B. Idzikovsky and J. Uherek, *Hutnik, Praha,* **24** (1974), 373–6.
54. Piwowarski, E., *Hochwertiges Gußeisen (Grauguß) seine Eigenschaften und die physikalische Metallurgie seiner Herstellung.* 2nd edition. Berlin/Göttingen/Heidelberg, 1951.
55. Sakwa, W., Haltbarkeit von Stahlwerkskokillen. *Gieß.-Praxis.* **19** (1967), 361–4.
56. Uhlitzsch, H., *Die konstruktive Gestaltung von Stahlwerkskokillen.* Berlin, 1955. Deutsche Akademie der Wissenschaften in Berlin. Class A, paper 1.
57. Gathmann, E., Neuere amerikanische und englische Kokillen. *Iron Trade Rev.* **85** (1930), No. 2, 25–7.
58. Leitner, F., and E. Plöckinger, *Die Edelstahlerzeugung.* 2nd edition. Vienna/New York, 1965.
59. Jellingshaus, M., *Was der Elektrostahlwerker von seiner Arbeit wissen muß.* Part 1. Lichtbogenofen. Düsseldorf 1972. (Stahleisen-Schriften. H. 7.).
60. Emi, T., On the origin and removal of large non-metallic inclusions in steel. *Scand. J. Metallurg.* **4** (1975), 1–8.
61. Flemings, M. C., Principles of control of soundness and homogeneity of large ingots. *Scand. J. Metallurg.* **5** (1976), 1–15.
62. Nelson, L. H., Einflußgrößen bei der Seigerung und Erstarrung von Stahlblöcken. *Amer. Inst. Min. Metallurg. Engr., Techn. Publ.* No. 802, 7, *Met. Technol.* **4** (1937), No. 3.
63. Pochmarski, L., and H. Hiebler, Untersuchungen an beruhigt vergossenen Blöcken zur Verminderung ihrer Stehzeiten in Kokillen und Tieföfen. *Berg- u. hüttenm. Mh.* **121** (1976), 172–9.
64. Knüppel, H., A. Diener and R. Scheel, Lunkerausbildung und Kopfseigerungen beim Gießen in Kokillen mit ausgekleideten Hauben. *Arch. Eisenhüttenwes.* **35** (1964), 603–12.

65. Hopkins, R. K., The Kellog-electric-hot-top-process. *Proc. Electr. Furn. Steel Conf.* Electr. Furn. Steel Comm. Iron Steel Div., Amer. Inst. min. metallurg. Engrs. **6** (1948), 106–32.
66. Johannson, E., and E. Helin, Elektrische Blockkopfbeheizung von Schmiedeblöcken. *Stahl u. Eisen* **75** (1955), 1756–65.
67. Martin, W., and E. Thon, Die Einwirkung der elektrischen Blockkopfbeheizung auf das Blockinnere schwerer Schmiedeblöcke. *Stahl u. Eisen* **75** (1955), 1765–74.
68. Plöckinger, E., Die Entwicklung und Anwendung des B.E.S.T.-Verfahrens für die Herstellung großer Schmiedeblöcke. *Berg- u. hüttenm. Mh.* **122** (1977), 466–73.

Williams, John, Preservation of cast iron items in iron and steel plant casting operations. *Steel Times* **205** (1977), No. 10, 1017–18, 1021–2, 1024–5, 1027, 1042–4.

Vincent, D., Development of hot topping systems to maximise yields. *Steel Times* **206** (1978), No. 1, 73–5, 78–81, 83–4.

Robertson, S. R. and E. F. Fascetta, An analytical technique for the determination of the thermal contact resistance between a solidifying metal and mold. *Metallurg. Trans.* **8B** (1977), No. 4, 619–24.

Lee, S. J., Winds of change in the casting pit. *Refractories J.* **53** (1978), No. 2, 13–21.

Unglamorous but costly ingot stools get a new lease of life from refractory protection. *Metal Producing* **16** (1978), No. 8, 42–3.

Kelly, B. F., A. G. Goedegebuure, and M. J. McCarthy, Plant studies of air entrainment and nitrogen during bottom pouring. *Steel Times* **206** (1978), No. 9, 98–102.

Hennessy, Robert J., Factors influence mold and stool life. *Proc. Nat. Open Hearth and Basic Oxygen Steel Conf.* **60** (1977), 402/4.

Hurtuk, D. J., R. Sobolewski and R. A. Mosser, Thermal phenomena and microstructural effects in ingot molds. *Proc. Nat. Open Hearth and Basic Oxygen Steel Conf.* **60** (1977), 320–37.

Nakagawa, Yoshitaka, Herstellverfahren großer Schmiedeblöcke (Japanese). *Tetsu to Hagané* **64** (1978), No. 12, 1771–87.

8.4 Continuous casting

69. Continuous casting: A global view of steelmakings former 'enfant terrible' *Magazine Metals Produc.* **13** (1975), No. 10, 42–9.
70. Alberny, R., Etat actuell de développement de la coulée continue. *Circ. Inform. techn.* **34** (1977), 913-23.
71. Nemoto, H., Development of continuous casting operation. *Trans. Iron Steel Inst. Japan* (1976), 51–65.
72. Petersen, U., Continuous casting of steel-actuell and future aspects. *Trans. Iron Steel Inst. Japan* **11** (1971), 67–75.
73. Ushijima, K., Working process and resulting properties of continuously casting steel. *Trans. Iron Steel Inst. Japan* **15** (1975), 380–92.
74. Vogt, G., and K. Wünnenberg, Stranggießen und Auswalzen von Vorprofilen aus Stahl für die Trägerfertigung. *Klepzig Fachber.* **82** (1974), 375–9.
75. Bauer, K. H., H. v. Ende and F. Regnitter, Present state of the process of continuous casting of steel (German). *Stahl u. Eisen* **93** (1973), 233–42.
76. Vogt, G., and K. Wünnenberg, Continuous casting of steel at a high rate (German). *Stahl u. Eisen* **94** (1974), 462–73.

77. Mandle, D., and T. Wertli, Horizontales Stranggießen. *Blech Rohre Profile* **24** (1977), 157–62.
78. Hoffmann, D., Ein neues Stranggießverfahren für Metalle mit geringer Wärmeleitfähigkeit. *Metallwiss. u. Techn.* **29** (1975), 27–33.
79[a] Stanek, V., and J. Szekely, A mathematical model of the closed mold (Watts) Horizontal Continuous Casting Process. *Metallurg. Trans.* **7B** (1976), 619–30.
79[b] Horizontal continuous casting of steel. *Metallurgia, Redhill* **47** (1980), No. 2, 80.
79[c] Thielmann, Rainer, and Rolf Steffen, Entwicklungsstand horizontaler Stranggießanlagen zur Herstellung von Knüppeln aus unlegierten und legierten Stählen. *Stahl u. Eisen* **100** (1980), No. 7, 401–7.
80. The MH-high-temperature-tundish. *Concast-News* **15** (1976), No. 2.
81. Cold tundish lining: what can it do for you? 33. *Magazin Metal Produc.* **15** (1977), No. 4, 39–42.
82. Singh, S. N., and K. E. Blazek, Heat transfer and skin formation in a continuous casting mold as a function of steel carbon content. *J. Metals* **26** (1974), No. 10, 17–23, 26–7.
83. Wasmuth, J. Th., *Das Stranggießen von Stahl.* Düsseldorf, 1975. (Stahleisen-Schriften. H. 8.)
84. Weinreich, W., Beitrag zur Kenntnis der Kühlungs- und Erstarrungsbedingungen beim Stranggießen von Stahl. Berlin 1968. (Dr-Ing.-Diss. Techn. Univ. Berlin.)
85. Baumann, H. G., *Stahlstrang-Gießanlagen.* Düsseldorf, 1976.
86. *Werkbilder 'Vereinigte Edelstahlwerke Aktiengesellschaft'* (VEW), Werksgruppe Kapfenberg.
87. Plöckinger, E., and B. Tarmann, Breitband aus Stranggußbrammen. *Berg- u. hüttenm. Mh.* **107** (1962), 133–44.
88. *Continuous Casting of Steel.* London, 1977. (Metals Society Book. No. 184.) 107–67.
89. Alberny, R., and J. P. Birat, Brassage électromagnétique et qualité des produits de coulée continue. *Circ. Inform. techn.* **34** (1977), 925–44.
90. Baumann, H. G., and W. J. Löpmann, Stahlstrangfehler. *Drahtwelt* **59** (1973), 469–76, 526–36.
91. Baumann, H. G., J. Czikel and G. Schäfer, Beitrag zur Planung von Stahlstrang-Gießanlagen. *Klepzig Fachber.* **81** (1973), 542–5.

Etienne, A., Réflexions sur le contrôle du refroidissement secondaire des installations de coulée continue (French). *Circ. Inform. techn.* **34** (1977), No. 9, 1929–42.

Irving, W. R., and A. Perkins, Basic parameters affecting the quality of continuously cast slabs. *Ironmaking & Steelmaking* **4** (1977), No. 5, 292–9.

Schwabe, Hermann, The long arm of concast. *Steel Times* **205** (1977), No. 11 (Intern.-No.), 90–1, 93, 97–9.

D'Avenia, M., and A. Guiga, Starting production with new continuous casting plants. *Ironmaking & Steelmaking* **4** (1977), No. 6, 338–9.

Duke, J. D., and R. K. Ozeki, Casting slabs for plate application at US Steel's Texas works. *Iron & Steelmaker* **4** (1977), No. 10, 21–7.

Harbold, L. M., and J. F. Emig, Caster production and quality — can both be accomplished? *Iron & Steelmaker* **4** (1977), No. 10, 11–20.

Nakano, Y., Y. Noguchi, F. Hoshi and Y. Muranaka, Continuous casting of stainless steel slabs. *Ironmaking & Steelmaking* **4** (1977), No. 6, 361–7.

Wyckaert, A., Control of the parameters in continuous-casting practice in influencing the surface finish and internal cleanness of the steel. *Ironmaking & Steelmaking* **4** (1977), No. 6, 340–9.

Deily, Richard, L., Casting raw steel — USA. *Iron & Steelmaker* **4** (1977), No. 12, 31–3.
Kyriakides, S. K., and J. M. Hambright, Product quality of strand cast steel at inland steel company. *Iron & Steelmaker* **4** (1977), No. 12, 17–25.
Continuous casting of steel. Proceedings of an international conference organized by The Metals Society, London, and L'Institut de Recherches de la Sidérurgie Française, IRSID, and held in Biarritz, France, on 31 May–2 June 1976. The Metals Society. London 1977. 324 pages, many illustrations and tables. 4°. = The Metals Society. Book No. 184.
Marr, H. S., Technological problems in continuous casting. *Iron & Steel Intern.* **51** (1978), No. 2, 87, 89, 91, 93–4, 97–100.
A study of the continuous casting of steel. Past trends, current technology and statistical survey. Future impact of the process. International Iron and Steel Institute, Committee on Technology. Brussels 1977. 288 pages, many illustrations and tables.
Corlis, R. G., and K. A. Ströbele, Meltshops logistics for series casting of steel billets or slabs. *Iron Steel Eng.* **55** (1978), No. 4, 25–31.
Schmid, Markus, and Dieter Rubensdörffer, High speed casting of billets. *Metallurg. Plant & Technol.* **1** (1978), No. 1, 29, 32–4.
Siegel, R., Shape of two-dimensional solidification interface during directional solidification by continuous casting. *Trans. ASME, Ser. C, J. Heat Transfer* **100** (1978), No. 1, 3–10.
Adams, James, S., and Carl L. Valdiserri, Thinking behind the concept of the jumbo caster at Great Lakes. *Iron Steel Eng.* **55** (1978), No. 5, 31–4.
Jauch, Rudolf, Hubert Löwenkamp, Friedhelm Regnitter, Klaus Fischer, Herman Schroer, Rolf Wilhelm Simon and Erwin Jericho, The quality of continuously-cast slabs and billets. *Metallurg. Plant & Technol.* **1** (1978), No. 2, 24–33.
Ichikawa, Hiroshi, Isao Yamazaki, Tomohiko Kimura and Yasuyuki Tozaki, Progress of quality steel slab casting at Kashima steelworks. *Proc. Nat. Open Hearth and Basic Oxygen Steel Conf.* **60** (1977), 243–56.
Diserens, M., Conspal for cleaner steel. *Concast News* **17** (1978), No. 3, 5–6 (AKS E46.8).
Wiesinger, Horst Alois, Design and layout planning of billet casting machines. *Mettalurg. Plant & Technol.* **1** (1978), No. 4, 5–6, 8, 10, 12, 14–16.
Caster squeezes into small space. *Iron Age Metalworking Intern.* **17** (1978), No. 11, 16MP19.
Samarasekera, I. V., and J. K. Brimacombe, The continuous-casting mould. *Intern. Metals Rev.* **23** (1978), No. 6, 286–300.
Siegel, Robert, Analysis of solidification interface shape during continuous casting of a slab. *Intern. J. Heat & Mass Transfer* **21** (1978), No. 11, 1421–30.
Schrewe, H., and K. Wünnenberg, Advantages and drawbacks of new developments for strand casters. *Iron & Steelmaker* **5** (1978), No. 12, 15–20.
Komma, Gerhard, and Bernhard Krüger, Designing slab casting machines for high productivity. *Iron & Steel Intern.* **51** (1978), No. 6, 373, 375–6, 377–80.
Shah, Raymond, New ideas improve continuous casting. *Iron Age Metalworking Intern.* **18** (1979), No. 1, 40MP10–12, 40MP16, 40MP19, MP23.
Maringer, Robert E., and Carroll E. Mobley, Direct casting of wire, filament and fiber. *Wire J.* **12** (1979), No. 1, 70–4.
Marr, H. S., Electromagnetic stirring: stepping stone to improved continuously cast products. *Iron & Steel Intern.* **52** (1979), No. 1, 29–31, 33, 35–7, 39–41.
Improved system for measuring molten steel level in mold of continuous casting

machine. *NKK News* **19** (1979), No. 3, 1, 3 (AKS E46.26).
Weinberg, F., Continuous casting. *Metals Technol.* **6** (1979), No. 2., 48–55.
Young, K. P., R. G. Riek, and M. C. Flemings, Structure and properties of Thixocast steels. *Metals Technol.* **6** (1979), No. 4, 130–7.
Flick, Max, and Wolfgang W. John, Concast continuous caster at National Steel Corporation producing the widest cast slabs in the world. *Metallurg. Plant & Technol.* **2** (1979), No. 3, 16, 20, 24.
Grill, A., Cooling system to prevent centreline cracks in continuously cast steel billets. *Ironmaking and Steelmaking* **6** (1979), No. 2, 62–7.
New continuous caster and cooling line concept developed. *Iron Steel Eng.* **56** (1979), No. 5, 72–3.
Nettelbeck, Klaus, and Ernst Bachner, High-speed continuous casting of slabs. *Iron Steel Eng.* **56** (1979), No. 5, 45–8.
Ooi, H., and Y. Iida, High quality bloom manufacture: continuous casting scores another success. *Iron & Steel Intern.* **52** (1979), No. 3, 135, 137, 139, 141, 143.
Kosarski, Z. Z. J., Some aspects of the continuous casting process. *Brit. Foundrym.* **72** (1979), Special Number, June, 97, 99, 100–2, 104–5.
Meinhold, Ulrich, and E. A. Planck, Latest technology in continuous casting of hollow billets or rounds for seamless pipe. *Iron Steel Eng.* **56** (1979), No. 6, 61–2.
Palmaers, A., A system for controlling molten steel level in the tundish of a continuous casting machine. *Metallurg. Rep.* CRM No. 54, 1979, 41–2.
Florchak, J. V., Straight- v. curved-mould continuous casters for slab production. *Ironmaking & Steelmaking* **6** (1979), No. 3, 123–30.
Singh, S. N., A practical solution to the problems of alumina build up in nozzles during continuous casting of aluminium-containing steels. *Iron & Steelmaker* **6** (1979), No. 6, 40–6.
Indyk, B., and R. Hadden, Continuous casting of small cross-section billets. *Metallurgia, Redhill* **46** (1979), No. 7, 444, 446, 448–50.
Cuscino, Thomas A., High-tonnage bloom and billet casting facilities. *Iron Steel Eng.* **56** (1979), No. 8, 46–51.
Application of electromagnetic stirring to continuous casting of steel. *Steel Times* **207** (1979), No. 10 (Suppl.), S47, S49–S52.
Etienne, A., and B. Mairy, Heat transfer in continuously cast strands. *Metallurg. Rep. CRM* No. 55, 1979, 3–13.
Hatzenbichler, H. Recent developments in continuous steel casting. *Steel Times* **208** (1980), No. 4, 284, 286, 288, 290, 295.
Kollberg, Sten, Contribution to the theory and experience of electromagnetic stirring in continuous casting. *Iron Steel Eng.* **57** (1980), No. 3, 46–54.

8.5 Vacuum ingot-casting

92. Coupette, W., *Die Vakuumbehandlung des Flüssigen Stahles*. Wiesbaden, 1967.

Electric Furnace Steel Production
Edited by E. Plöckinger and O. Etterich

9 Remelting Processes

Erwin Plöckinger, Kapfenberg

9.1 CLASSIFICATION OF REMELTING PROCESSES IN STEELMAKING TECHNOLOGY

Today, remelting processes have become indispensable for secondary refining in the production of high-grade steels and alloys with much higher standards of purity and technical properties compared with metals produced conventionally. The customer's quality specification nowadays often calls for the remelting process to be used in order to guarantee attainment of the qualities he seeks.

All these processes are based on the principle of remelting solid charge materials, preferably from sacrificial electrodes, into water-cooled moulds in which the remelted steel solidifies continuously, mostly as a directionally solidified ingot or casting. One of a number of different types of remelting furnace is used, according to the qualities desired of the final product. In selecting the type of remelting furnace, the costs of conversion, which can sometimes differ markedly among the types of furnace used, must be borne in mind.

Remelting improves quality in the following ways:

(1) The remelting furnace results in an ingot which has virtually no internal faults and has less micro-segregation and no ingot-segregation, secondary piping, or discontinuities. Hot-shaping can therefore be effected with less working, and in special cases working can even be dispensed with altogether.
(2) Remelting removes harmful impurities from the steel. Here gases, tramp elements, and, also alloying elements having a high vapour-pressure are preferentially removed by the vacuum remelting process, while in the ESR (electro-slag remelting) process non-metallic inclusions (oxides, sulphides) are removed by metallurgical reaction between the steel and the remelting slag.

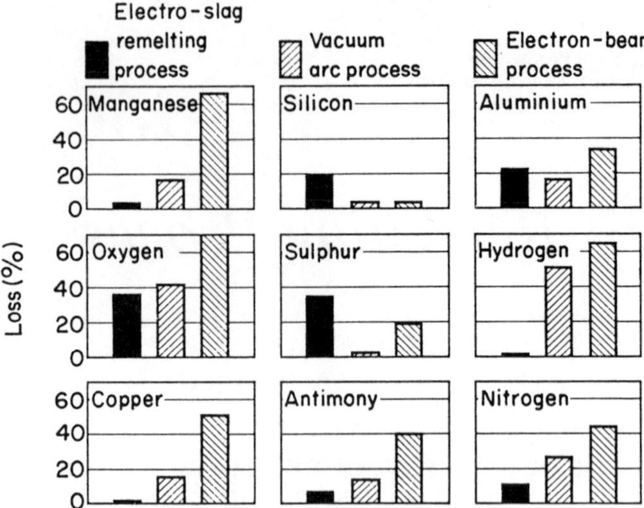

Figure 1. Melting losses with different remelting processes[1]

Figure 1 shows comparisons between the changes in chemical composition which can be expected in remelting for the three most important processes used in production today: vacuum-arc, ESR, and electron-beam furnaces.[1] It should be borne in mind here that corrections to the chemical composition, such as compensation for unwanted melting losses, are possible only in the ESR and electron-beam processes and then only to a limited extent. The scientific possibilities of each individual remelting process are the reason why the different processes can co-exist.

Finally, it should be noted that the starting-material for a remelting process must be prepared to suit the process to be used. For instance, for the ESR process the hydrogen content must be low enough while the sulphur content must be low for the vacuum remelting process. Bearing these preconditions in mind, the primary melting process used to make the remelt electrodes is of secondary importance from the point of view of quality, this process is largely determined by the melting system available and by economic considerations.

9.2 VACUUM ARC FURNACES

Manfred Wahlster, Hanau

The vacuum arc furnace was developed in Germany as long ago as 1905 for remelting tantalum.[1] Later, it was also used for producing metals such as molybdenum, titanium, or zirconium, which have a high melting point or an

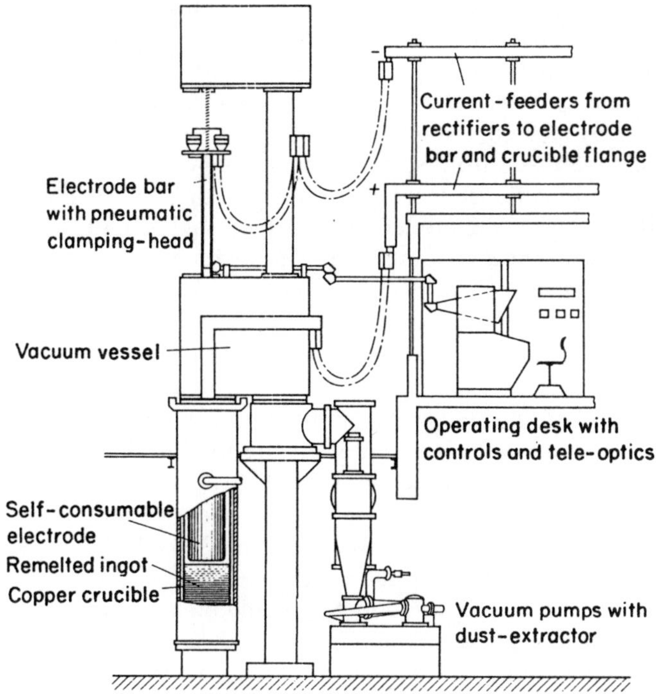

Figure 2. Arrangement of a vacuum arc furnace[1]

affinity for oxygen.[2] Success in remelting these metals and their alloys led to its application for remelting medium- and high-alloy steels and nickel and cobalt-based alloys, which, together with titanium alloys, form the principal areas of application for vacuum arc furnaces.[3–9]

In this process an electrode made of the specified alloy is connected as a cathode via a dc arc to a water-cooled copper crucible and melted down under a vacuum of 10^{-2} to 10^{-3} torr. This method precludes any contamination of the melt by crucible material or atmospheric reaction.

9.2.1 Mechanical and Electrical Design of Vacuum Arc Furnaces

The layout of a vacuum arc furnace is shown schematically in Figure 2.[10] Essentially, it consists of a vacuum vessel with a water-cooled copper crucible flanged at the lower end. A bar with a consumable electrode attached is introduced into the vacuum vessel via a vacuum-tight seal. The electrical and mechanical connections are usually made via a pneumatic clamping head. The vacuum vessel is connected to a pump via a dust-extractor. The control cabin with the operating desk and tele-optics is situated on the working platform, which is usually level with the shop-floor since the water-cooled remelting crucible is usually located under the floor.

The moulds are made from seamless or welded copper tubes of 15–30 mm wall-thickness, with a flange welded to the top end. The melting current must be fed into this tube as symmetrically as possible so as to avoid lateral deflection of the arc which can cause inhomogeneity in the remelted ingot. The inside of the mould must be kept clean to ensure smooth ingot surfaces. Lives of up to 1000 melts or more without repairs can be achieved by careful handling.

A closed water-circuit is convenient for cooling the mould, and continuous temperature-monitoring enables even and adequate cooling to be maintained. Production installations require around 1000 l of cooling water per minute.

Oil diffusion pumps with Roots blowers are used to maintain the vacuum. Both types of pump are often used in parallel.

The melting process is observed via the tele-optics system; an image of the arc is displayed on a glass screen at the operating desk. This is especially helpful in starting up and finishing off the melting process.

The risk of an explosion caused by the mould melting through and ensuing water ingress can be as good as eliminated in steel-remelting furnaces by expert servicing and the warning and shut-down systems which are almost universally used today. Nevertheless, it is still advisable to install the mould below floor level for safety reasons. However, unlike with titanium melting, safety walls are not necessary to protect the steel-melt operators.

Nowadays the preferred d.c. current source is silicon rectifiers because of their low losses and robustness. These have a no-load voltage of about 60–80 V which drops to between 30 and 36 V under full load; the arc voltage is a few volts lower.

The optimum current for remelting steel depends on the type of steel and usually has to be determined empirically. Moreover, it is not proportional to the electrode cross-section area. The following can be taken as approximate guide-values for remelting carbon steels: 7000 A for 400-mm diameter ingots; 24 000 A for 1000-mm and 32 000 A for 15 000-mm. This assumes standard distances between the donor electrode and the crucible wall of 50 mm for a 400 mm-diameter ingot and 75 mm for 1500 mm.

Very little has changed in the external appearance of the vacuum arc furnace and auxiliary equipment since its application on a large scale for the production of remelt ingots of high-purity alloys based on iron, nickel, or cobalt. Melting systems are nowadays usually equipped with two remelting stations in order to increase productivity; in this way solidification of one remelted ingot can take its course without interfering with remelting of the next ingot. Figure 3

Figure 3. 7-t vacuum arc furnace (Leybold-Heraeus GmbH & Co. KG, Hanau)

illustrates the arrangement of a 7-t vacuum arc furnace of the type preferred for the production of superalloys.

9.2.2 Metallurgical Possibilities and Melting Operations

9.2.2.1 Metallurgical processes in vacuum-remelting

The reactions which take place during remelting in a vacuum can lead to exceptional purification of the metal being remelted. In addition to the diminished micro-segregation and improved ingot structure caused by controlled solidification, vacuum-remelting also brings about a decided improvement in purity.

One of the most important processes is the vacuum extraction of hydrogen down to a level well below the limit required to prevent flaking. Most of the hydrogen departs from the thin molten film formed on the electrode and from the metal droplets thrown into the mould between the electrodes and the melt in a vacuum of from 10^{-1} to 1 torr.[6]

Oxygen-reduction takes place by floating off the oxides in the presence of carbon and also through the formation of carbon monoxide. It is advisable, however, to minimize the C–O reaction. Thus the type of deoxidation reaction has a marked effect on the oxygen-content of the remelted ingot. In particular, a deoxidation process using aluminium, adjusted to the type of steel concerned, has proved a good way of achieving optimum purity.[6,9]

The oxides remaining in the steel and those which separate during solidification are evenly and finely dispersed throughout the ingot, so that much greater microscopic purity is achieved in comparison with conventionally cast ingots.

The nitride and carbonitride content of steels such as titanium-stabilized stainless steels cannot be improved to the same degree as the oxygen content. One reason for this is the greater density of these inclusions compared with oxides (5.0 as against 3.5 g/cm^3); the other reason is based on their relatively small mean diameter of 10 μm. However, even with the much-reduced rate of ascent and separation thereby caused, vacuum-melting still leads to a clear improvement in purity.

Nitrogen-removal is only possible to a limited extent and is heavily dependent on its activity in the melt. On the other hand, it is possible to produce nitrogen alloy steels by melting in a defined gaseous atmosphere (e.g. nitrogen) in a vacuum arc furnace.

As for other incidental and trace elements, only those which have a relatively high vapour pressure can be affected by vacuum-remelting. It is important in practice that the degree of evaporation of manganese depends on the melt temperature, the pressure in the furnace, and the activity coefficient of the manganese.[11] The loss of chromium is generally negligible; the same goes for the change in carbon content. There is no significant reduction in sulphur level; however, there is a notable improvement in the sulphide-distribution. For this reason, the donor electrode must already have the low level of sulphur (and phosphorus) specified.

Condensates of the undesirable trace elements, lead, arsenic, antimony, and tin are always found in the fume-extractor; however, the amounts concerned are so small that a change in their concentration in the steel can only be determined analytically when their initial level is high. Electrodes for nickel alloys having high strength at elevated temperatures are therefore usually prepared in a vacuum induction furnace to ensure that they have the smallest possible scatter and lowest possible levels of these harmful trace elements.

9.2.2.2 Remelting technique

Remelt electrodes are usually cast, rolled, or forged bars, most often with a circular cross-section. Bearing in mind the vacuum-remelting processes discussed in the previous section, the electrodes must be prepared with compositions appropriate to that of the remelt ingot to be produced. The electrode surfaces are cleaned of casting skin and scale by means of special processes such as shot-blasting or pickling. In most cases, machining is not necessary. The electrode diameter fills the inside of the mould except for a small annular gap, which ranges from 20 to 25 mm for small ingot-diameters up to 75 mm for large ingots.

Before melting starts, the furnace is evacuated; in this process all traces of moisture, which tends to collect on water-cooled elements, must be removed. After the pressure has dropped to 10^{-2} torr, an arc is struck between the electrode and a metal plate of the same material on the bottom of the mould.

The choice of remelting conditions is crucially important for the quality of the remelt ingot. As with all remelting processes, the aim is to obtain the flattest liquid phase possible by using low current values and thereby to obtain an evenly solidified ingot with very low segregation.[12,13] The lower limit of the remelting current is determined by the quality of the ingot-surface; it is essential for low currents that the electrode should have clean surfaces and low gas contents.[6] Low gas content is normally achieved by vacuum treatment of the melt before casting the electrode. The optimum melting rate depends on the material and is determined empirically.

At the end of the remelting process the bath-depth is decreased by steadily reducing the process current; however, sufficient material has to be melted off the electrode to compensate for volume-reduction in solidification and to avoid formation of a shrinkage cavity on the top of the ingot. After the ingot has solidified completely, the furnace is returned to atmospheric pressure, the crucible is opened, and the ingot is removed.

The ingot-surface must be cleaned or machined before further processing; this is especially important for crack-sensitive materials. It is particularly necessary to do this if the ingot-surface is spotted with metal droplets or inclusions caused by splashes on to the crucible wall during remelting. When remelting materials containing high levels of volatile elements which condense on the wall of the mould, the ingot-surface layer can become enriched with these elements which must then be removed by machining. The hot ductility of remelted ingots is generally better than that of the alloys from which they are made; remelted ingots can therefore generally be worked at lower temperatures than would be necessary for conventionally cast ingots.

9.2.2.3 Applications and trends in development

The improvement in material quality attainable through remelting under vacuum has opened up a large field of application for the vacuum arc furnace. It will always be needed for the production of steels and alloys based on iron, nickel, and cobalt, when the lowest possible gas contents are required and volatile trace elements are to be removed.

Vacuum arc furnaces used now in production are extensively mechanized and automatically controlled; they can be simple to operate and trouble-free in service. With few exceptions, remelted ingots seldom exceed 1000 mm in diameter or 15 tonnes in weight. Increasing ingot diameter above 1500 mm and weight over 50 t has not proved worthwhile in practice.

9.3 ELECTRO-SLAG REMELTING FURNACES

Helmo Jäger and *Gert Kühnelt, Kapfenberg*

In the electro-slag remelting (ESR) process, a metal electrode is melted down in a metallurgically active slag. The heat necessary for the remelting process is produced by passing an electric current through the molten slag which acts as a resistance heating element. The electrode metal is refined by melting it down in droplets through the slag; it is then solidified in a water-cooled mould to form an ingot which preferably has its grain parallel with its longitudinal axis.[1–7]

9.3.1 Design of ESR Installations

The following equipment (Figure 4) is required to operate an ESR installation:

(1) A water-cooled mould in which the ingot is formed;

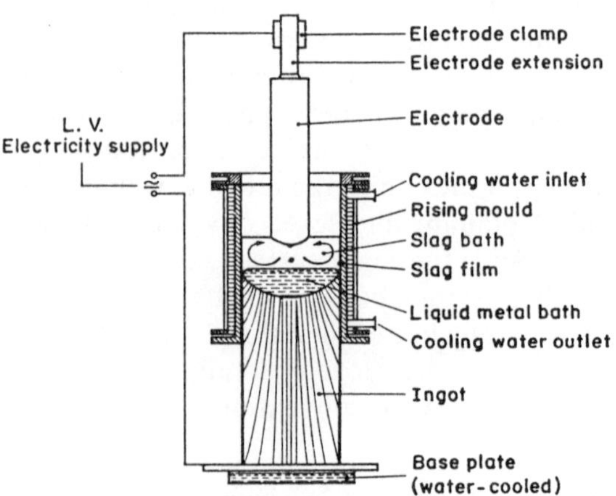

Figure 4. Schematic representation of electro-slag remelting

(2) A water-cooled base plate beneath the mould;
(3) An electrode holder to clamp each donor electrode via an extension-piece. The electrode-holder is mounted on a carriage mounted on a column;
(4) Power is supplied from the mains via one or more transformers. The high-current circuit comprises the copper leads to the electrode clamp, the electrode with its extension, the molten slag, the ESR ingot, the base-plate, and the return conductors back to the transformer.

There are single-electrode installations, with one electrode and one furnace column, and other installations with electrode changing facilities (Figure 5) which predominate today. In the latter type, two electrode trolleys on two columns enable several electrodes (including short ones) to be melted down in

Figure 5. ESR installation with rising mould and electrode-changer for ingots of up to 35 t with maximum ingot-diameter of 1150 mm (VEW Vereinigte Edelstahlwerke, Kapfenberg Steelworks, Austria)

succession to form one remelted ingot. ESR installations with one electrode usually work with a fixed mould.[8,9] In another design, a short mould is used[4–6] which encloses only the molten slag and the top of the ingot where solidification takes place. Either the mould is raised or the base-plate is lowered with the ingot as the ingot grows. The mould is made from steel plate, copper, or copper alloy.[1,2,4,7,10] The mould walls are water-cooled.

The movements required during the remelting process are powered hydraulically (e.g. the clamp system) or electrically via cables, shafts, gears, etc. This mainly concerns raising, lowering, and swivelling the electrode carriage and the mould and movement of the base-plate and auxiliaries (e.g. slag furnace). These movements are usually controlled from a central console which also contains control and monitoring devices for the water-cooling, hydraulics, and the controls for regulating the remelting process.[6,7]

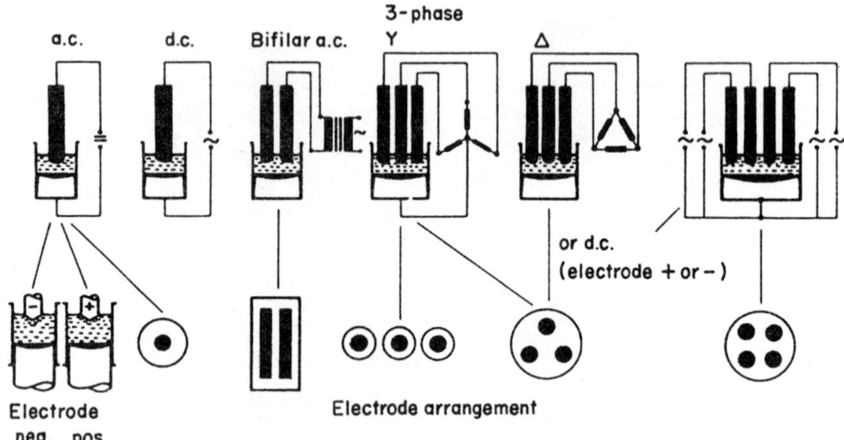

Figure 6. Possible ESR circuits

ESR installations may be powered by a.c.[2,4,6] (single or polyphase) or by d.c. with cathodic or anodic polarity (Figure 6).[6] In bifilar installations, usually used for producing slabs, the current flows through the slag between two electrodes.[9,12] Large ESR installations are also powered by thyristor-controlled low-frequency a.c. in order to minimize electrical losses in the supply cables at the high current levels required.[5,13] The remelt parameters (voltage, current, power, slag resistivity, slag bath height, and electrode-to-ingot diameter ratio) must be suitably matched in order to attain constant melting and ingot-growth with a uniform grain-structure.[4,14–16]

The depth of immersion of the electrode-tip is kept at optimum by automatic voltage or current control. An approximately constant melting rate can be maintained by continuously measuring electrode weight or length and controlling the power.[2,7]

Extensive development is in hand to automate the process. This work concerns monitoring the melting operation, keeping the melting rate constant, and control of the whole process including starting, electrode-changing, and hot-topping.[4,17,18]

ESR furnace operation requires the use of the following auxiliaries. The electrodes are fixed to their stubs by a device which enables them to be electro-slag welded, flash-butt welded, or arc-welded by hand.[1,9] Many ESR installations are equipped with slag pre-melting furnaces, which are single-phase arc furnaces with a copper, graphite, or MgO crucible. Three-phase furnaces or induction furnaces with graphite crucibles are less common.[1,4,5,19]

When producing an ESR ingot from several electrodes, the tip of each new electrode must be preheated so that the slag bath temperature will not drop excessively during electrode-changes and to avoid electrode-cracking due to thermal shock.[14,20] Preheating systems used for this are usually vertical tube furnaces and are either gas-fired or are fitted with electric resistance-heating. Other auxiliaries are required, i.e. for preparation of start-up plates, cutting equipment for cropping electrodes or sectioning long ingots and equipment to feed in deoxidizers, alloying elements and slag materials.

Unlike vacuum arc furnaces, in which only round ingots can be produced, a multitude of different shapes can be produced by electro-slag remelting. As well as round and polygonal sections, ESR systems can produce square sections, rectangular ingots of various breadth-to-thickness ratios, and mouldings.[1–4]

ESR slab furnaces may be operated with single electrodes and single-phase power, or they may use two or four electrodes in bifilar or double-bifilar arrangement; otherwise, they may use three-phases with three electrodes.[10,12,21] Existing production systems can make slabs of up to 2400 × 510 mm cross-section.

In recent years various methods have been developed for producing hollow ESR ingots.[4,19] At the time of writing one Russian-type hollow-ingot system was operating industrially in Sweden. There are also a few semi-industrial systems for producing straight and curved tubes, pressure vessels, etc.[22,23]

Electro-slag casting has been developed for producing special forms such as drop-forging dies, crankshafts, valve bodies, pressure vessels, and others where the crucible form is close to that of the final product.[1,4] One special application is the production of hot or cold work rolls, using various techniques involving fixed and rising moulds; the journals may be cast or welded on.[24,25]

Worn hot or cold rolls may be resurfaced with a wear-layer deposited by the ESR process. This may be applied by a system using special tubular or tube-segment shaped electrodes or bar electrodes arranged within the annulus.[24]

With this technique a T-shaped mould can also be used to provide ample space for the electrodes to be arranged in the slag zone.[1,2,4] Such T-shaped moulds can also be used for producing ingots of small diameter (about 150 mm) and for producing multiple ingots.[9,25]

In the CESPM or CESM (continuous electroslag powder melting) process an unalloyed strip electrode is fed into the slag bath; this electrode carries the required alloying ingredients in powder form, held in place magnetically on the strip.[26,27] Another process has been developed in which an ingot is first produced by conventional casting, then its core is removed and the space is refilled by electroslag remelting.[28,29]

9.3.2 Metallurgical Possibilities and Melting Operations

9.3.2.1 Metallurgical processes in remelting under slag

Three different interfaces are available for metallurgical reactions between the molten metal and the slag. These are the film of liquid metal on the electrode face, the surfaces of metal drops falling through the slag, and the surface of the molten pool on the ingot.[3] The main part of the reaction takes place on the face of the electrode,[4,5,30,31] where slag continuously flows past a thin metal film of about 30–100 μm.[32] The metal–slag reactions proceed rapidly because of the high slag temperatures — 1700 to 1900°C — and the large reaction interface, almost to equilibrium.[6]

In addition to the physical and chemical transformations, absorption processes are also involved. One such process purifies the metal by absorbing non-metallic inclusions into the slag.[1,4,31] Apart form this, there are reactions with the gas atmosphere at the surface of the slag.

The course of the required metallurgical reactions depends on the type of slag used. The remelting slag should fulfil the following requirements: stability at processing temperature (low constituent vapour pressure), lower liquidus temperature than that of the metal being remelted, suitable electrical resistivity, metallurgical suitability for the reactions desired (e.g. desulphurizing), avoidance of unwanted reactions (e.g. alloy melting losses), low viscosity, low surface tension, high interfacial tension between slag and metal, the ability to absorb non-metallic inclusions from the metal being remelted, suitable specific heat and thermal conductivity, absence of harmful constituents, availability, and low cost. These characteristics are discussed in the extensive literature.[1–3,7,33]

The slags most frequently used in practice belong to the $CaO–Al_2O_3–CaF_2$ system.[3] In certain cases MgO, SiO_2, and TiO_2 are added to ESR slags. Table 1 shows a survey of the slags generally used for steel-remelting.[34]

Sulphide and oxide inclusions are separated from the metal during remelting by means of two processes: absorption of inclusions by the slag and physical

Table 1. Typical slag compositions for remelting alloy tool steels[34]

Composition						Application
CaF_2	CaO	Al_2O_3	MgO	SiO_2	TiO_2	
70	30					General-purpose slag
70	15	15				For high-speed steel; good desulphurization
55	15	25	3	2		For hot-working steels
40	30	30				Low-melting-point slags for high-carbon tool steels
30	30	30		10		
50		25			25	For sulphur-bearing tool steels
35	15		5	10	35	Conductive at room temperature (for cold-starting)

and chemical reactions between metal and slag with participation by the gaseous atmosphere above the slag.

Under highly basic slags ($CaO/SiO_2 > 5$) desulphurization in remelting amounts to 50–80%.[1,2,7] Remelting in air assists desulphurization because sulphur in the slag is oxidized into SiO_2 which then escapes into the gas phase.[3,5] The oxygen level which can be reached in the ESR ingot depends first of all on the action of deoxidizers in the metal and on the activity of the oxides of these deoxidizers in the slag. In addition, low oxygen partial pressure in the slag is also essential for the attainment of low oxygen levels in ESR ingots.[35] Aluminium and silicon are the most important deoxidizers in electro-slag remelting. The effect of the CaO/SiO_2 ratio in the slag on the oxygen content in the remelted steel is illustrated in Figure 7; it is an indication of silica activity. It is necessary to deoxidize the slag continuously in order to avoid a reduction in its basicity caused by a rise in the level of silica in the slag; this can be done by adding aluminium. Aluminium addition will also remove oxygen absorbed from the atmosphere and from the scale on the electrode surface. When remelting steels containing alloying elements such as titanium and aluminium, which have a high affinity for oxygen, experience shows that melting losses can be reduced by lowering the oxygen partial pressure above the slag by an inert atmosphere of argon or nitrogen and by coating the electrodes (e.g. with an aluminium-based coating) to inhibit scaling.[4,36,37] The carbon content does not change during remelting under basic slags, provided that the carbon level in the slag is low.[2,3]

During remelting under a basic slag about 30–60% of the silicon is burnt

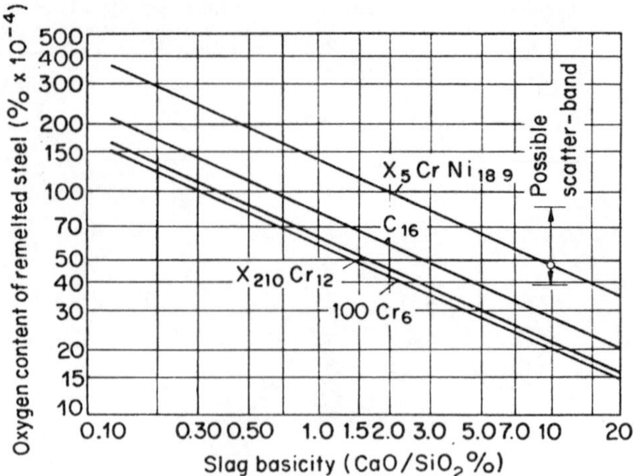

Figure 7. Effects of slag-basicity of average oxygen contents of various steels after remelting[5]

off.[3,5] However, this can be very much reduced by continuous deoxidation using aluminium.

When using basic slags the manganese melting loss amounts to a maximum of 5–10%.[3,5] Normally there is no change in phosphorus content during remelting.[2] According to Medovar,[1] it is possible to remove phosphorus by using CaF_2–BaO slags.

In ESR steel remelting there is no change in the levels of vanadium, chromium, molybdenum, tungsten, nickel, tin, copper or of harmful trace elements such as arsenic, antimony, lead, and bismuth.[1,2,7]

When remelting under aluminous slags having a low silica level, the silicon will reduce the alumina to a certain extent. This can lead to an increase of 0.01–0.02% in the aluminium level of the ESR ingot compared with the electrode. Special measures are required if low aluminium levels (below 0.010%) have to be maintained in the ESR ingot.[38] There is an appreciable aluminium melting-loss at aluminium levels of a few tenths of a per cent and over; this can be compensated by adding aluminium continuously during remelting.[4,7]

As with all elements having a high affinity for oxygen, there will be a high titanium melting-loss; this can be significantly reduced by adding TiO_2 to the slag and by efficient deoxidation of the slag, using inert gas and protectively coating the electrodes.[4,37] A special process has been evolved for remelting aluminium or titanium alloy steels.[4,39]

The steel nitrogen content does not change during remelting, except when the initial level in the electrode metal is low.[1,3] It is possible to produce ingots with high nitrogen levels by using a special process of remelting under increased pressure.[40]

Remelting under a basic slag can lead to a considerable increase in the hydrogen level brought about through damp raw materials for the slag, rust on the electrodes, moisture condensed on the mould and the base-plate, leaks in the mould and base-plate, or high atmospheric humidity.[1–3]

During remelting the hydrogen level in the molten steel depends mainly on the moisture in the air above the slag bath; however, the hydrogen content of the electrode and the chemical composition of the slag (especially the CaO content) also play a part.[38] The following precautions are effective in avoiding pickup of hydrogen from the atmosphere:

(1) Using lime-free slags such as those containing Al_2O_3 and CaF_2;[4,41]
(2) Covering the slag with a dry inert gas.[4,6,38]

ESR aims at producing a grain structure largely oriented along the length of the ingot; this is brought about by maintaining the right shape and depth of the molten-metal pool which in turn depends on the melting rate. The smaller the volume of the molten pool, the less segregation can take place on a macro scale. This affects large-scale ingot-segregation as well as stratification.[16,42]

The macrostructure, the dendrite arm-spacing, and the degree of micro-segregation are affected by the local solidification rate. These are influenced by the time spent in the two-phase liquidus solidus region and by the temperature gradient behind the solidification front.[6,15]

Micro-segregation, too, cannot be completely suppressed in the ESR ingot; however, its extent is much less than in conventionally-cast steels, even for larger ingot cross-sections.[43,44]

ESR installations can be operated on a.c. or d.c. of either polarity. With d.c. remelting, polarization phenomena occur at the interfaces between the slag and the tip of the electrode and between the slag and the molten metal.[45,46]

When the electrode is positive, negative ions (anions) accumulate on the electrode-tip. Anion complexes, such as of $Al_2O_5^{5-}$, produce a high electrical resistance, cause overheating, and thus increase the melting rate. Other undesirable anions, such as O^{2-}, S^{2-}, and OH^-, also accumulate; these are discharged and dissolved in the metal film on the electrode-tip. A corresponding accumulation of cations (Ca^{2+}, Mg^{2+}) occurs at the interface, of opposite polarity, between the slag and the liquid metal on the ingot, and the reactions described above take place in reverse. As the polarization reactions are very dependent on current-density, the effects on melting output and purity are varied.[8,11] These phenomena have severely limited the application of the d.c. variant of the process. More than 90% of all ESR systems are powered by a single- or polyphase a.c.[4,13] However, it is possible to remove undesirable ions from the remelting system by discharging them on to non-consumable auxiliary electrodes biased with d.c. superimposed on the normal a.c. circuit.[46]

9.3.2.2 Process parameters

The heat required for electro-slag remelting is supplied by applying electrical energy to the slag bath. Only 25–45% of this heat is used for preheating and melting the electrode; the rest is lost through conduction and radiation to the mould, base-plate and atmosphere.[2,5,16,47] Figure 8 shows a typical heat balance for a 220-mm square ingot.[5]

Thermal balance sheet

	kcal/h	%
Heat removed in cooling water	122 500	66.0
Radiation	46500	25.0
Ingot (400°C)	17000	9.0
Total heat-input	186000	100.0

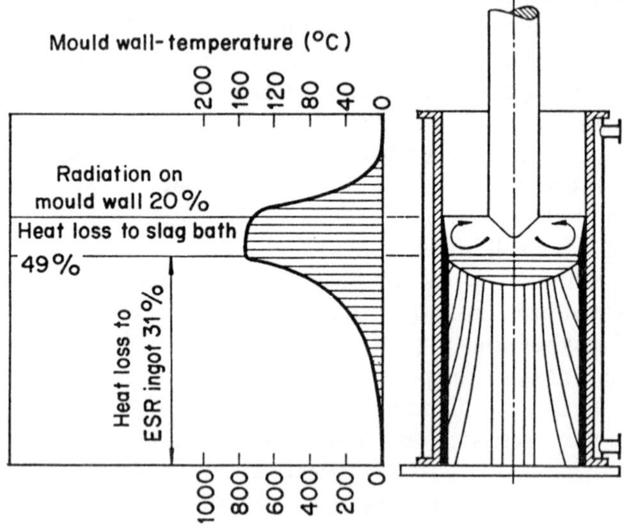

Figure 8. Thermal balance sheet for a 220-mm square ingot produced by ESR process[5]

As is schematically illustrated in Figure 2 in section 4.3, an a.c.-powered ESR system actually consists, from the electrical point of view, of a series of reactances and resistances. The slag bath represents a resistance R_s; the various conductors represent a resistance R_v, and an inductive reactance X_L.

Data on adjusting the most important process parameters and the way they affect each other can be found in the extensive literature.[1,2,4,5,7,16,48] Basically, attention should be paid to the following parameters:

(1) The power supplied to the slag bath. An increase in the power applied raises the melting rate and the depth of the molten crater. However, for optimum solidification the depth of the crater should not be more than half the diameter of the ingot.[2,5] This will be achieved in rough approximation if the following equation is followed:[4,5,16]

$$r = 0.8 - 1.0 \times D \quad (1)$$

where r is melting rate (kg/h)
and D is ingot diameter (mm)

(2) The voltage-drop in the slag bath. In ESR this lies in the region of 30–60 V.
(3) Current. Progressively greater current values are needed as ingot-diameter increases.
(4) Slag-resistivity. For remelting steels, slags of resistivity between 0.2 and 0.5 Ωcm are used.
(5) Electrode-diameter (d). For single-electrode installations with a cylindrical rising mould this is between 0.4 and 0.75 times the ingot diameter (D).[2] With T-shaped rising moulds and fixed moulds the d/D ratio may be greater.
(6) Electrode-distance for ESR systems the distance between the melt-down electrode and the melt is 60–250 mm.

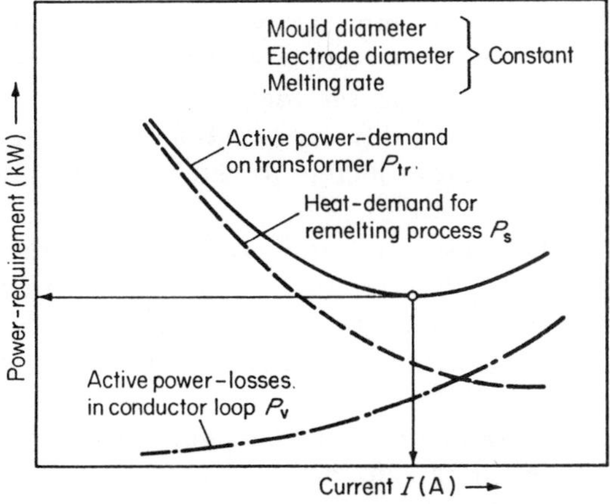

Figure 9. Optimum electrical conditions for electro-slag remelting[16]

Figure 9 shows optimum electrical conditions for particular mould- and electrode-diameters for a single-phase a.c. system.[16] As current increases, both the electrode distance and the required slag-bath depth will decrease. Thus the heat-demand for remelting decreases because the heat-losses to the mould-wall also decrease. On the other hand, the active power-losses in the conductor loop increase with the square of the current. The sum of the active power-loss and the heat-demand for remelting can be plotted as a curve for the active power-demand on the transformer; the latter gives the minimum power-demand required and the associated current value. Power-demand optimized according to these principles leads to ESR systems with a specific consumption of between about 800 and 1600 kWh/t.

9.3.2.3 Operation of the remelting process

The electrodes for remelting can be cast by any suitable system. Steels containing higher levels of easily oxidizable alloying elements will inevitably lose a proportion of these in remelting in air; this should be anticipated by arranging for a correspondingly higher alloy content in the electrodes. There is no single view on the best way of deoxidizing the electrode metal. In any event, a sufficiently elevated silicon content (0.2% or more) is advantageous for thorough deoxidation and desulphurization during melting. If low hydrogen levels must be attained in the ESR ingot, then the electrode must be made from vacuum-degassed metal.

As a rule electrodes are cast either in static moulds or by continuous casting.[2] The weight of the electrode for ESR systems without provision for changing electrodes is determined by the weight of the ingot; for systems with electrode-changing facilities the individual electrodes may be prepared in shorter lengths of, say, 2000 mm. Only the smallest possible amount of material should be left over at the end so that a good ingot-to-electrode-yield ratio is attained. After burning off the inevitable electrode stub, a new electrode is descaled (preferably by sand-blasting) and welded to the extension-piece. In some cases rolled or forged electrodes are used; discarded rolls can also be used as electrodes.

Remelting slags may be used in a premelted or presintered state; otherwise, the individual ingredients are mixed first. Lime and lime-rich slag materials in particular must be stored in airtight packs in order to avoid a high moisture content. Some works preheat the slag materials to reduce their moisture before remelting.

Slags for producing ESR ingots of small diameter (under 500 mm) are usually added to the mould in solid state; for larger ingot-diameters slags are premelted and poured into the mould in the liquid state.[2,7]

When using a rising mould or retracting base-plate, the mould is cleaned and placed on a starting plate which rests on the base-plate. The mould-bearing surface is electrically insulated and sealed with asbestos. After clamping the electrode into its holder and turning on the cooling water, which is pre-warmed to above the dew-point, the system is ready to begin remelting. When starting to arc on a solid slag, a molten pool can be facilitated by adding exothermically reactive mixtures.

If the slag is already liquid, then no arc is necessary to start the process, and the electrode-tip is dipped into the slag-bath. With fixed moulds the slag can be fed to the mould by means of a siphon-like system. Using a liquid slag has the advantage that the base of the ingot will contain less hydrogen.

During remelting the melting rate should, if possible, be kept constant along the length of the ingot. Small adjustments to the voltage and current will be necessary, especially for long ingots, if there is any change in the composition and hence the resistivity of the slag. Any change in the inductive reactance of the high current circuit must also be compensated.

In installations with a rising mould or retracting base-plate, if excessive cooling of the bottom of the ingot is undesirable, insulation-segments can be applied to the emerging ESR ingot.

In systems with electrode-changers,[14] when the electrode has melted down to a stump about 10 mm long this is raised and swung away and a new electrode is dipped into the slag-bath. In modern systems this change is usually done automatically. The change-time is between 30 and 45 s.

At the end of the remelting process the solidification front is slowly raised by progressively reducing the voltage and current in order to achieve an ingot-top which is flat and free from porosity and shrinkage-cavities.[2] Finally, the power supply is cut, the slag is allowed to harden, and the mould is removed.

9.3.2.4 Ingot faults and remedies

Ingot-surface faults Grooved or flaky surfaces can be caused by low slag-temperatures or incorrect slag-composition (high liquidus temperature, low interface tension between steel and slag)[2] if a coating of slag several millimetres thick forms between the ingot and the mould. In extreme cases this can lead to metal breakthrough, since inadequate conduction of heat from the molten metal pool inhibits formation of a strong ingot skin.[41] The remedy is to raise the power supplied and/or use a different slag.

In units with a rising mould or retracting base-plate ingots may exhibit transverse flaws which are caused by excessive friction between the ingot-surface and the mould just under the slag-zone. These problems can be prevented by resetting the mould or changing the mould-geometry (taper).

Internal ingot faults Pores, gas-blisters, and gas-bubble segregation can occur in ESR ingots, caused by high hydrogen levels or inadequate deoxidation of the metal; however, they can confidently be avoided by appropriate measures to reduce hydrogen and oxygen levels.[49] Faults in the very top of the ingot, such as open or closed cavities or large slag-inclusions, are caused by incorrect ingot hot-topping. Large internal faults in ESR ingots can be caused by electrode-fragments. For this reason, care must be taken to ensure that electrode-surfaces are acceptable (free from cracks, skins, or seams) and that electrode-hardness is low enough; where electrodes are made from steel susceptible to heat-treatment cracking, the electrode-tip should be preheated. Foreign metallic inclusions, which are usually in the shape of drops, originate from molten welding rod or from the extension to which the remelt electrode was attached.

9.3.2.5 Typical applications for electro-slag remelting

If the remelting conditions are correctly established, the metal purification and directional solidification in the process produce the following quality improvements over the initial material:

(1) High degree of purity. The incidence of non-metallic inclusions in the ESR ingot is very low because the oxygen and sulphur contents are kept to the lowest possible levels. The remaining slag particles are very small and are particularly evenly distributed.[1,5–7]
(2) High homogeneity. The mainly vertical solidification process completely eliminates ingot-segregation and the dendrite-spacing in the core is less than in conventionally cast ingots of the same size.[43] Moreover, ESR ingots have an especially high density.[41]
(3) In consequence of their high purity and homogeneity, remelted steels are extremely tough and exhibit little anisotropy in the deformed state.[1,41] This is achieved after less deformation than with conventionally cast ingots.[4,25] It is specially worth mentioning that ESR metal takes a very high polish[51] and shows little distortion when hardened. Many steels also show improved workability after remelting.[1,4,7]

The quality-improvement brought about by electro-slag remelting is exploited today for many types of steel and also for superalloys and non-ferrous metals.[1–4] This process is preferred for preparing blanks for highly stressed forgings, usually of low-alloy structural steel for the electrical power industry,[41,50] for aircraft components,[52] for cold- and hot-working steels,[34,43,53,54] for cold-rolls,[24] highly-stressed ball-bearings,[1,44] and other applications. High-alloy ferritic and austenitic steels are also remelted by the ESR process where the specification calls for high purity, homogeneity, polishability, and corrosion-resistance.

9.3.2.6 Plant-size and output

ESR production has grown very sharply over the last few years.[6,20,55] Table 2 is a survey of ESR production in various countries in the Western world. ESR production can be expected to double within the next five years. At present,

Table 2. Survey of ESR installations in the Western world capable of producing ingots of over 500 mm diameter (1977)[4,55]

Country	No. of installations capable of producing ingots of dia. over 500 mm	Estimated annual production 1977 (t/y)
(1) Austria	4	12 000
(2) Belgium	2	8 000
(3) Brazil	1	4 000
(4) France	3	9 000
(5) Britain	15	44 000
(6) Italy	2	8 000
(7) Japan	11	30 000
(8) Luxembourg	1	4 000
(9) Sweden	9	24 000
(10) Spain	1	7 400
(11) USA	33	125 000
(12) West Germany	8	30 000
Total	90	305 400

the largest ingot which can be produced is 2.3 m in diameter, and the heaviest ingot weighs 120 tonnes.[9,17]

9.3.2.7 Trends in development

The ESR process today is counted among the most important remelting processes for special steelworks. However, a great deal of development work is still required for special applications. Above all, reliable technology must be developed for economical production of hollow ingots. There is also a definite need for long slabs with a high breadth-to-thickness ratio; for rolls, and especially for cold-rolls which can be used as-cast. Such cold-forming rolls should also be covered with an armoured, wear-resistant coating.[4,24]

Another development being pursued is an ESR process using a liquid charge with the addition of scrap simultaneously heated by a plasma-burner.[26,27]

It remains to be seen whether production of large forging ingots via the ESR process or one of the special processes described in section 8.3.3 will go ahead in the future. This will depend not only on technical success but also on the economics of these processes.[4,6]

9.4 ELECTRON-BEAM REMELTING FURNACES

Manfred Wahlster, Hanau

The electron-beam melting furnace, known for over sixty years,[1] appears to be one of the most flexible processes because of its very wide applicability. This is mainly because the metallic charge can consist of pieces of scrap, granulate, sponge iron, or swarf; whereas electro-slag remelting and vacuum arc processes require electrode-charges.[2] The most important requirement for the large-scale industrial application of this vacuum-melting process was the development, in the mid-1950s, of a powerful electron-gun.[3] Since about 1960 the furnace sizes and outputs attainable have been rising and are becoming increasingly attractive for steel-production. At the same time, the design of the electron-beam sources has become varied and has led to the appearance of different types such as ring-beam, flat-beam, and circular-beam guns and the multi-chamber gun developed in East Germany.[4–7] It is now a practical proposition to produce ingots of up to 18-t weight and 800-mm diameter.[8]

9.4.1 Mechanical Construction and Electrical Equipment

As well as the changes in the design of electron-guns, there are many ways of using the guns, according to the chosen design of the electron-beam furnace. Three types developed for steel-remelting are schematically illustrated in Figure 10.[9] The upper-right diagram shows melting by means of a vertical, dripping electrode. This method guarantees even melting rate, axisymmetrical power-distribution throughout the molten bath, and constant solidification conditions in the crucible by employing several electron-guns, evenly spaced. It has the additional advantage that, usually, the electrode diameter is not important; for instance, when remelting the usual type of cast ingot it can be larger than the diameter of the final remelt ingot.

The horizontally fed remelt electrode arrangement shown in the upper-left diagram is most often used in Eastern Europe. Here, too, the ratio of energy used purely for remelting to that required for maintaining the temperature at the crucible surface can be varied widely. Multi-chamber electron-beam furnaces are a further development of this type of furnace; the layout is illustrated in Figure 11.[10] the most important characteristics of three such furnaces are listed in Table 3.[8] This type of furnace typically has one electron-gun and magnetic lenses to guide the beam through narrow, cooled tubes (the latter are used to restrict flow) and to adjust the diameter of the beam. Just before it enters the melting space, the electron-beam passes through a magnetic deflection system which allows the beam to be held steady or periodically deflected. The beam is focused, according to the state of the control parameters, on the consumable electrode and the melt-bath or the surface of the melt in the water-cooled copper mould. The vacuum in the electron-gun space is, to a great extent, decoupled from that in the melting space, so that the pressure in the melting space can rise to 10^{-2} torr.

If in electron-beam remelting, it is intended to separate remelting and refining from tapping the molten steel, then the overflow melting system

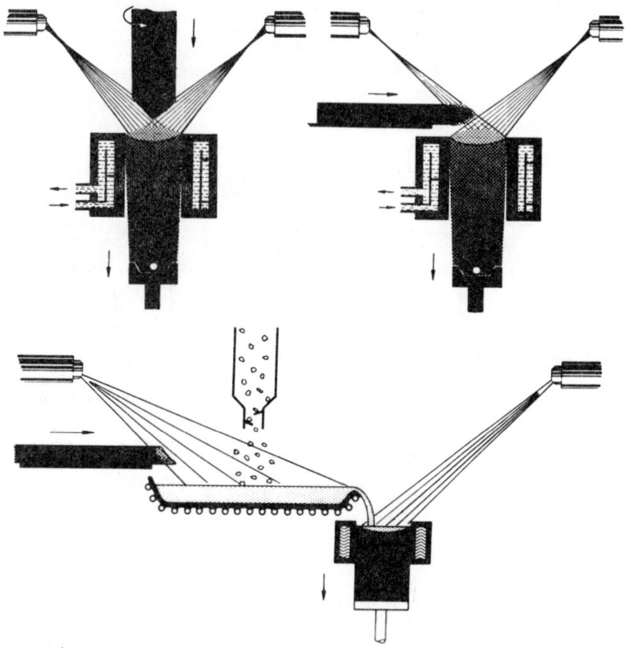

Figure 10. Different types of electron-beam furnace[9]

illustrated in the lower part of Figure 10 is used. This sharp division removes any connection between the type and shape of the starting-material and the form and size of the end-product. By distributing the electron-beam energy evenly over the melt surface, it is possible to limit the bath-depth so that the liquid pressure is low so that volatile, harmful trace-elements can evaporate in the vacuum and distil out.

9.4.2 Metallurgy and Melting Operations

Remelting in electron-beam furnaces occurs at a much lower melting-space pressure than in vacuum-arc furnaces. This pressure lies between 10^{-2} and 10^{-5} torr, depending on the gases given off by the material being remelted. Degassing is therefore efficient; however, high evaporation-losses can be expected. Although this is desirable for the removal of harmful trace-elements which have a high vapour pressure, it does complicate the melting process for some alloys if relatively high losses of elements such as manganese and chromium have to be contended with. Loss by evaporation of

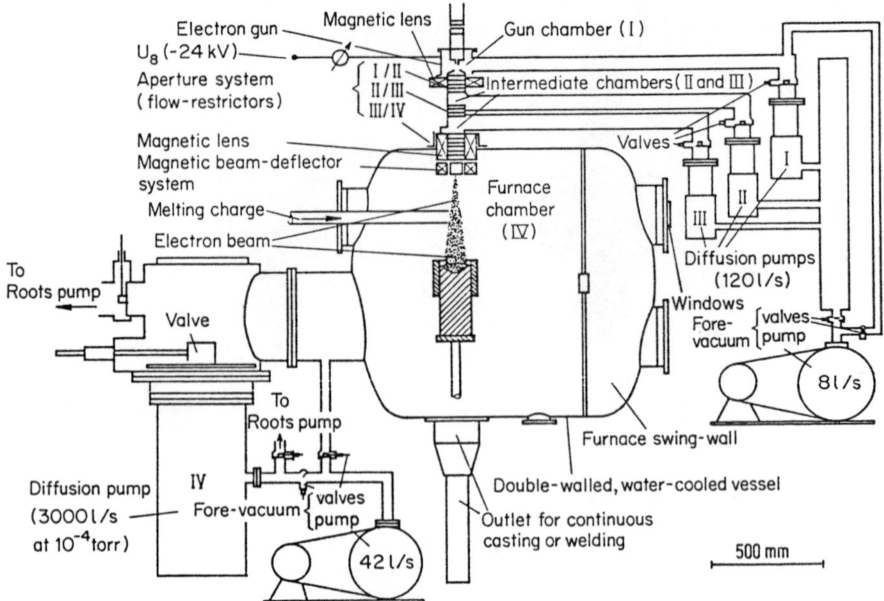

Figure 11. Schematic diagram of 60-kW, multi-chamber electron-beam furnace[10]

base and trace-elements and alloying elements increases as remelting rate decreases (see Figure 12).[9] As with all remelting processes, here, too, the concurrence of both desirable and undesirable metallurgical reactions forces technical and economic compromises to be made.

As with remelting in vacuum arc furnaces, the chemical composition of the electrode must be matched in advance to the desired composition of the remelted ingot; this includes low levels of phosphorus and sulphur. Casting skin and scale must be carefully removed. A high degree of purity, a dense, non-segregated grain structure, and a good surface finish can all be achieved in the remelted ingot by careful adjustment of the remelting conditions.

The application of electron-beam remelting furnaces today is not centred so much on iron- and nickel-based alloys but lies far more with reactive or refractory metals.[11] However, interesting results from the quality point of view are being achieved with the former group and use of the process in special cases appears completely justified.

9.5 PLASMA REMELTING FURNACES

Erwin Plöckinger, Kapfenberg

In addition to the remelting processes discussed in the preceding sections, it appears that the remelting of electrodes into ingots by means of plasma torches

Table 3. Characteristics of multi-chamber electron-beam furnaces (EMO)[8]

Quantity	Units	EMO 60	EMO 250	EMO 1200
Total electricity requirement	KVA	about 110	450	1600/500
	V	380	380	10 000/380
	Hz	50	50	50/50
Electron source (nominal)	kW	60	250	1200
Acceleration voltage	kV	20	30	35
Beam current	A	3	8.4	35
Melting space vacuum unit				
Oil diffusion pump	l/s	8000	30 000	2 × 80 000
Roots pump	m^3/h	600	3600	4 × 3600
Rotary vane pump	m^3/h	60	350	4 × 350
Electron source vacuum unit				
Oil diffusion pump	l/s	2 × 500	2 × 2000	2 × 2000
Rotary vane pump	m^3/h	30	150	150
Charging air-lock vacuum unit				
Roots pump	m^3/h			2 × 1800
Rotary vane pump	m^3/h			2 × 150
Mould diameter				
max.	mm	150	280	max. 800
min.	mm	25	50	
Bar length	mm	500	I no valve 1200	max. 3000
	mm	800	I with valve 800	
	mm		II no valve 1600	
Melting rod size				
Length	mm	max. 1000	I 1300	2000
Diameter	mm	100	110	270/290
	mm		II 2200	
	mm		170	
Cooling-water	l/min	about 150	400	about 2500
Building height above platform	mm	2400		
Platform height	mm		3500	
Bar length				
500 mm	mm	1500		
800 mm	mm	1750		
Unit height above floor	mm			about 9300
Depth of continuous-casting channel	mm			about 8500

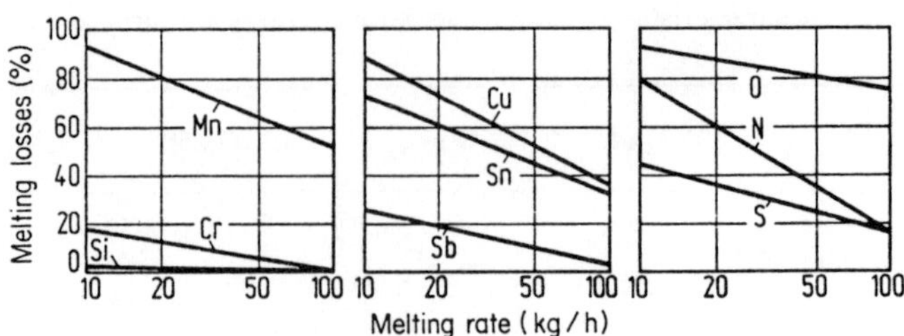

Figure 12. Effect of melting rate on melting losses in electron-beam furnaces[9]

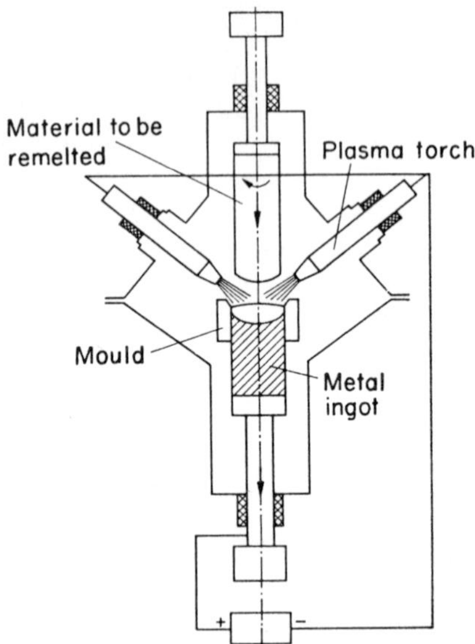

Figure 13. Scheme for a plasma arc remelting furnace (Paton Institute of Welding Technology, Kiev, USSR)[3]

is also gaining importance.[1,2] This process was developed at the Paton Institute in the USSR; its principle is illustrated in Figure 13. The electrode is melted by plasma torches arranged symmetrically and radially around the electrode; these torches heat the molten bath in the water-cooled copper mould at the same time. The ingot produced is continuously withdrawn downwards from the bottom of the mould.[3] The electrode, plasma torches and mould, are all situated in a melting chamber where the plasma gas is circulated in closed circuit. Melting can take place at either reduced or elevated pressure. As a rule, argon is used as the plasma gas. Nitrogen-alloy steels can also be produced if nitrogen is used at elevated pressure. The most important

Table 4. Characteristics of plasma remelting furnaces[4]

Characteristic	Furnace type		
	U-461	U-467	U-600
Max. ingot weight (kg)	30	460	5000
Ingot diameter (mm)	100	250	630
Ingot height (mm)	500	1200	2200
Total power of plasma torch (kW)	160	360	1800
Type of electricity supply	d.c.	a.c.	a.c.

characteristic data about plasma-remelting furnaces available today are shown in Table 4.[4] This table shows the maximum ingot weight to be 5 tonnes with a diameter of 630 mm. Another type of furnace uses a hollow melt-down electrode which also serves as a consumable plasma torch.[5]

The remelt process is metallurgically very adaptable. As well as enabling melting to take place in various gas atmospheres at different pressures as already mentioned, it also allows metal–slag reactions to be carried out as in electro-slag remelting. The argon atmosphere promotes relatively effective degassing and also inhibits selective evaporation of alloying elements. Otherwise, the properties of plasma-remelted ingots are similar to those of ingots produced in vacuum-arc electro-slag or electron-beam furnaces.

In the USSR, plasma remelting furnaces are also used for remelting non-ferrous metals such as platinum and palladium as well as for the production of high-temperature alloys and nitrogen steels.

9.6 LITERATURE REFERENCES

9.1 Classification of Remelting Processes in Steelmaking Technology

1. Wahlster, M., and H. Spitzer, Technology and metallurgy of the processes for the production of special steels and comparable materials (German). *Stahl u. Eisen* **92** (1972), 961–72.

9.2 Vacuum Arc Furnaces

1. Bolton, W. von, Das Tantal, seine Darstellung und seine Eigenschaften. *Z. Elektrochem.* **11** (1905), 45–51.
2. Winkler, O., The theory and practice of vacuum melting. *Metallurg. Rev.* **5** (1960), 1–117, 8 tables.
3. Böhmer, S., and K. H. Werner, Entwicklungsstand des Umschmelzens von Edelstählen nach Sonderverfahren. *Freiberg. Forsch.-H., Series B*, No. 122, 1966, 9–36.

4. Ludwigson, D. C., and F. R. Morral, A summary of comparative properties of air melted and vacuum melted steels and superalloys. Battelle Memorial Institute. Columbus, Ohio, 1960, (DMIC report. 128.)
5. Bungardt, K., and K. Trömel, Über das Schmelzen im Lichtbogen-Vakuumofen mit selbstverzehrender Elektrode. *Arch. Eisenhüttenwes.* **35** (1964), 725–37.
6. Peter, W., and H. Spitzer, Steel elaborated in the vacuum arc furnace for forgings (German). *Stahl u. Eisen* **86** (1966), 1383–93.
7. Lüdemann, K.-F., H.-J. Eckstein and S. Böhmer, Veränderungen verschiedener mechanischer und Gebrauchseigenschaften von Stahlen durch das Umschmelzen nach Sonderverfahren. *Freiberg. Forsch.-H.*, Series B, No. 126, 1967, 9–20.
8. Eckstein, H.-J., Das Umschmelzen von Stahl unter Vakuum und schützenden Medien. *Neue Hütte* **12** (1967), 449–59.
9. Spitzer, H., and P. W. Bardenheuer, Qualitätskontrolergebnisse von umgeschmolzenen Edelstählen. *DEW Techn. Ber.* **10** (1970), 1–14.
10. Wahlster, M., and H. Spitzer, Technology and metallurgy of the processes for the production of special steels and comparable materials (German). *Stahl u. Eisen* **92** (1972), 961–72.
11. Schmitz, J. A., Verdampfungsvorgänge beim Umschmelzen von Stahl im Vakuumlichtbogenofen. *Arch. Eisenhüttenwes.* **41** (1970), 373–9.
12. Sperner, F., and G. Persson, Vorgänge beim Umschmelzen von Stahl im Vakuum-Lichtbogenofen. *Stahl u. Eisen* **82** (1962), 1099–1105.
13. Chasin, G. A., F. I. Sved, D. P. Dolinin, L. Savenok, and G. D. Beksler, Der Einfluß der Stromführung auf die Erstarrungsbedingungen des Metalls beim Umschmelzen mit Lichtbogen im Vakuum. *Izvestija Vysšich Učebnych Zavedenij, Černaja Metallurgija* **8** (1965), No. 1, 43–9.

State-of-the-art advantage boost vacuum art remelting performance. *Metal Producing* **16** (1978), No. 3, 46–8.

9.3 Electro-slag Remelting Furnaces

1. Lataš, J. V., and B. I. Medovar, In *Das elektroschlacke-Umschmelzen*. Moscow, 1970, 38–75.
2. Duckworth, W. E., and G. Hoyles, *Electroslag Refining*. London, 1969.
3. Nafziger, R. H., The electroslag melting process. Washington 1976. (*Bulletin Bureau of Mines No. 669.*)
4. *Electroslag Remelting and Plasma Arc Melting*. National Academy of Sciences, Washington, 1976.
5. Holzgruber, W., and E. Plöckinger, Metallurgical and technological fundamentals of electroslag remelting of steel (German). *Stahl u. Eisen* **88** (1968), 638–48.
6. Wahlster, M., Das Elektroschlacke-Umschmelzverfahren — Wo steht es heute? *Radex-Rdsch.* 1975, 517–30.
7. Hoyle, G., P. Dewsnap, D. J. Salt and E. M. Barrs, Electro-slag refining technology. *BISRA Report MG/A/416/66.*
8. Petrman, I., and I. Kašík, Das Abschmelzen der Elektrode und die Bildung des Rohblocks beim Elektroschlacke-Umschmelzen. In *Teorie a praxe elektrostruskového pretavováni oceli a slitin.* SNTL, Prague, 1966, 17–33.
9. Paton, B. E., B. I. Medovar, Ju. V. Lataš, O. P. Bondarenko, V. M. Baglaj and A. G. Gobacenko, Die Herstellung großer Brammen nach dem Elektro-Schlack-

en-Umschmelzverfahren unter Verwendung von zwei Elektroden. *Freiberg, Forsch.-H.*, Series B, No. 151, 1969, 73–7.
10. Irving, R. R., Plate steels score breakthrough in quality. *Iron Age* **210** (1972), No. 15, 50–2.
11. Holzgruber, W., and K. Peterson, Die metallurgischen Vorgänge beim Elektroschlacke-Umschmelzen mit Gleich- und Wechselstrom. In *Symposium Wissenschaftliche Probleme des Schweißens und der speziellen Elektrometallurgie,* Kiev, May, 1970, Kiev 1970, Proc. Chapt. 4. pp. 90–105.
12. Little, J. H., T. J. Queen and I. M. Mackenzie, Metallurgical study of electroslag-refined plates. In *Electroslag Refining,* London, 1973, (ISI-Publ. 157.) pp. 43–53.
13. Machner, P., and W. Holzgruber, Das Elektroschlacke-Umschmelzen mit niedrigfrequentem Wechselstrom. In *VII. Internationaler Elektrowärmekongreß,* Warsaw, 18–22 Sept. 1972. Warsaw 1973, S.-A.N. 168. pp. 1–6.
14. Holzgruber, W., Ch. Kubisch and H. Jäger, Einfluß der Umschmelzbedingungen auf die Makro- und Mikrostrukur electroschlacke-umgesechmolzener Blöcke unter besonderer Berücksichtigung der Verhältnisse beim Elektrodenwechsel. *Neue Hütte* **16** (1971), 606–10.
15. Mitchell, A., and S. Joshi, The thermal characteristics of the electroslag process. *Metallurg. Trans.* **4** (1973), 631–42.
16. Machner, P., Die Hauptparameter des ESU-Prozesses und deren Einfluß auf den Blockaufbau und die Wirtschaftlichkeit des Verfahrens. *Berg- u. hüttenm. Mh.* **118** (1973), 365–72.
17. Leader in consumable electrode melting systems. Consarc Corporation. Rancocas/NJ, 1971.
18. Efroimovich, Yu. E., V. E. Pirozhnikov, A. F. Kablukovskii and V. M. Vinogradov, Programmed control system for the electroslag melting process. *Metallurgist* 1966, 459–61.
19. Medovar, B. I., Electroslag remelting of alloy steels and alloys. *Metallurgist* 1970, 252–5.
20. Holzgruber, W., Möglichkeiten und Grenzen der Beeinflüssung des Erstarrungsgefüges legierter Stähle beim Elektroschlacke-Umschmelzen. *Radex-Rdsch.* 1975, 409–21.
21. Nishiwaki, M., T. Yamaguchi, M. Koba, K. Okohira and N. Sato, Operation of large bifilar ESR furnace for slab production and quality of slabs and heavy plates produced. In *Proceedings of the 5th International Conference on Vacuum Metallurgy and Electroslag Remelting Processes. Munich,* 11–15 Oct. 1976. Hanau 1977, pp. 197–200.
22. Mullins, P., ESR plant casts 17-t hollow ingots. *Iron Age Metalworking Internat.* **15** (1976), No. 9, 41–2.
23. Ujiie, A., S. Sato, S. Sakai and J. Nagata, Development of new dynamic casting for application of cylindrical products. Production of high temperature alloy tube by newly developed 'YOZO' technique. In *Proceedings of the 4th International Symposium on Electroslag Remelting Processes.* Tokyo, 7–8 June 1973. Publ. by the Iron and Steel Institute of Japan, Tokyo, 1973, pp. 168–82.
24. Jäger, H., and Ch. Kubisch, Improved cold rolls through electroslag remelting. See Ref. 21. pp. 209–12.
25. Paton, E. O., *Electroslag Casting*. Kiev, 1976.
26. Levaux, J., and J. Sunnen, Continuous electroslag powder melting for the production of large ingots. *Iron Steel Eng.* **49** (1972), No. 3, 72–6.
27. Parsons, R. C., Production of tool steels by electroslag powder melting. In *Proceedings of the 3rd International Symposium on Electroslag and other Special*

Melting Technologies, 8–10 June 1971, Pittsburgh, P. I, Pittsburgh 1971. pp. 179–93.

28. Austel, W., and Ch. Maidorn, Metallurgical and technological properties of forgings made by the MHKW-process. See Ref. 21. p. 241–2.
29. Austrian patent 269933 dated 15 July 1968.
30. Paton, B. E., B. I. Medovar, Yu. G. Emelianenko, G. A. Boiko and V. A. Tihonov, Investigation methods, transformation and removal of non-metallic inclusions during ESR. In *Proceedings of the 4th International Symposium on Electroslag and other Special Melting Technologies*, Pittsburgh, 16–18 Oct. 1974. pp. 433–48.
31. Moldavskij, O. D., *Mechanismus der Auflösung der Oxydeinschlüsse im Elektroschlacken-Flußmittel.* Moscow, 1968. pp. 124–33.
32. Klujev, M. M., and Ju. M. Mironov, Über die Größe der Reaktionsoberfläche beim Elektroschlacken-Umschmelzverfahren. *Stal in Deutsch* **1967**, 964–8.
33. Winterhager, H., R. Kammel and A. Gad, Elektrische Leitfähigkeit, Dichte und Oberflächenspannung fluoridhaltiger Schlacken für das Elektroschlacke-Umschmelzverfahren. Opladen 1970. (Forschungsberichte des Lands Nordrhein-Westfalen, No. 2115).
34. Schlatter, R., Electroflux remelting of tool steels. *Metals Engng. Quart.* **12** (1972), No. 1, 48–60.
35. Miska, H., and M. Wahlster, Verhalten des Sauerstoffs beim Elektro-Schlacke-Umschmelzen. *Arch. Eisenhüttenwes.* **44** (1973), 19–25.
36. Medovar, B. I., J. V. Lataš and L. M. Stupak, Mögliche Ursachen des Sauerstoffzutrittes beim Elektroschlacken-Umschmelzen und Methoden zum Schutz des Metalls vor Oxydation. *Avtomaticeskaja svarka* 1961, No. 11, 47–52.
37. Fontaine, P. I., and D. J. Palmer, Electroslag remelting of 18% nickel maraging steel. *Iron & Steel* **45** (1972), 286–92.
38. Jäger, H., G. Kühnelt and H. Straube, Investigations regarding the control of hydrogen and aluminium contents in ESR ingots. See Ref. 30. pp. 306–22.
39. Pateisky, G., H. Biele and H. J. Fleischer, The reactions of titanium and silicon with Al_2O_3–CaO–CaF_2 slags in the ESR process. In *Special Electrometallurgy*. P. II. Reports of the International Symposium on Special Electrometallurgy, Kiev, June 1972. Kiev 1972. pp. 110–21.
40. Kubisch, Ch., and W. Holzgruber, Results obtained with the Böhler pressure ESR-process. See Ref. 27. P. III, pp. 267–84.
41. Jauch, R., A. Choudhury, H. Löwenkamp and F. Regnitter, Production of large ingots using the electroslag remelting process (German). *Stahl u. Eisen* **95** (1975), 408–13.
42. Schumann, R., and C. Ellebrecht, The application of electroslag remelting for the production of heavy forging ingots. *Electric Furn. Proc. Steel Conf.* Vol. **33**, 174–80.
43. Kroneis, M., E. Krainer and F. Kreitner, Ultrahochfeste Stähle unter Berücksichtigung der Erschmelzung nach dem Elektroschlacke-Umschmelzverfahren. *Berg- u. hüttenm. Mh.* **113** (1968), 416–25.
44. Krucinski, M., A. Nowak and W. Strama, Einfluß der Elektroschlackenfeinung auf die Güte des Stahles LH15. *Hutnik, Katowice,* **39**, (1972), 339–44.
45. Tokuda, M., and M. Ohtani, Electrochemical study on slag-metal reaction. In *Chemical Metallurgy of Iron and Steel,* Iron and Steel Institute, London 1973. pp. 93–5.
46. Jäger, H., Steuerung des Reaktionsablaufes beim Elektroschlacke-Umschmelzen durch Anwendung der Schmelzflußelektrolyse, Leoben 1973 (Diss. Montanuniv. Leoben.)

47. Medovar, B. I., V. L. Sevcov, G. S. Marinskij, A. G. Bogacenko, V. I. Sagan and Ju. P. Stanko, Untersuchung der wärmephysikalischen Charakteristiken beim ESU flacher Blöcke bei Industriebedingungen. *Probleme der Sonder-Elektrometallurgie* H. 1, 1975, pp. 9–14.
48. Bungardt, K., R. Diederichs, H. Preisendanz, E. Schürmann and H. Valentin, Beitrag zur rechnerischen Abschätzung der Schmelzbadgeometrie beim Elektroschlacke-Umschmelzverfahren in Abhängigkeit von den Wärmezu- und -abfuhrbedingungen. *DEW Techn. Ber.* **12** (1972), 157–77.
49. Mukherjee, T., Spot segregates in ESR high carbon steels. See Ref. 27. P. II, pp. 215–41.
50. Zimmermann, E., E. Königer and W. Poettering, Verbesserung der Eigenschaften von hochwertigen Schmiedestücken durch Anwendung des Elektroschlacke-Umschmelzverfahrens. *Radex-Rdsch.* 1972, 563–76.
51. Eckstein, H.-J., and H.-J. Spies, Der Einfluß des Umschmelzens auf die Polierfähigkeit von Stählen. *Neue Hütte* **11** (1966), 286–93.
52. ESR for landing gear. *Iron Age Metalworking Internat.* **11** (1972), No. 1, 25.
53. Randak, A., A. Stanz and W. Verderber, Properties of tool steels and antifriction bearing steels made according to special melting processes (German). *Stahl u. Eisen* **92** (1972), 981–3.
54. Fiedler, H., H. J. Spies, H. Müller and Ch. Weiss, Erfahrungen bei der Erzeugung von Schnellarbeitsstahlen nach metallurgischen Sonderverfahren. *Neue Hütte* **24** (1976), 328–34.
55. *Taschenbuch für Stahlindustrie 1978.* Verein Deutscher Eisenhüttenleute. Düsseldorf, 1977.

Erdmann-Jesnitzer, F., and V. Prozens, Influence of nucleic-forming agents upon the formation of crystalline structure and segregation in the ES remelting of steel. In *Proceedings of the 5th International Conf. on Vacuum Metallurgy and Electroslag Remelting Processes,* Munich, 11–15 Oct. 1976, Hanau 1977, pp. 123–9 and 5 tables.

Medovar, B. E., V. M. Baglai, and L. V. Chekotilo, Electroslag casting of large size high pressure pipes. In *Proceedings of the 5th International Conf. on Vacuum Metallurgy and Electroslag Remelting Processes*, Munich, 11–15 Oct 1976, Hanau 1977, pp. 195–6 and 1 table.

Medovar, B. I., V. F. Demchenko, A. G. Bogachenko, I. Tarasevich and Yu. P. Shtanko, Temperature fields of large ESR slab ingots. In *Proceedings of the 5th International Conf. on Vacuum Metallurgy and Electroslag Remelting Processes,* Munich, 11–15 Oct. 1976, Hanau, 1977, pp. 153–6 and 3 tables.

Electroslag remelting facility of large capacity features flexibility in processing techniques. *Ind. Heat.* **44** (1977), No. 9, 24, 26.

Dilawari, A. H., and J. Szekely, Calculation of current-voltage relationships and heat-generation patterns in the electroslag refining process. *Ironmaking and Steelmaking* **4** (1977), No. 5, 308–12.

Still in its infancy, ESR is maturing, bolstered by bigger mould designs. *Metal Producing* **16** (1978), No. 4, 50–2.

Allibert, M., J. F. Wadier, and A. Mitchell, Use of SiO_2-containing slags in electroslag remelting. *Ironmaking & Steelmaking* **5** (1978), No. 5, 211–16.

Irving, Robert R., ESR raises the quality of plate steels, hollow rounds. *Iron Age* **221** (1978), No. 41, 43–6.

Schwerdtfeger, K., Some aspects on the kinetic of reactions occurring in the ESR process. In *Proceedings of the 5th International Conf. on Vacuum Metallurgy and Electroslag Remelting Processes.* Munich, 11–15 Oct. 1976. Hanau, 1977, pp. 133–9 and 3 tables.

Proceedings of the 5th International Conf. on Vacuum Metallurgy and Electroslag Remelting Processes, Munich, 11–15 Oct. 1976. Under the auspices of the Österreichische Gesellschaft für Vakuumtechnik; Eisenhütte, Österreich; Schweizerische Gesellschaft für Vakuum-Physik und -Technik; Deutsche Arbeitsgemeinschaft Vakuum; Verein Deutscher Eisenhüttenleute in cooperation with the American Vacuum Society and International Union for Vacuum Science, Techniques and Application. Carried out by Manfred Wahlster. Leybold-Heraeus GmbH & Co. KG. Hanau 1977. XIII, 263 pp., many illustrations and tables.

Ballantyne, A. S., R. J. Kennedy and A. Mitchell, The influence of melting rate on structure in VAR and ESR ingots. In *Proceedings of the 5th International Conf. on Vacuum Metallurgy and Electroslag Remelting Processes,* Munich, 11–15 Oct. 1976, Hanau 1977, pp. 181–3 and 4 tables.

El Gammal, T., and M. Hajduk, The use of CaF_2-free slags in the electroslag remelting process. In *Proceedings of the 5th International Conf. on Vacuum Metallurgy and Electroslag Remelting Processes.* Munich, 11–15 Oct. 1976, Hanau 1977, pp. 141–6 and 1 table.

Paton, B. E., B. I. Medovar, V. Y. Saenko, and B. V. Degtyaryov, New combined process LDS and ESR. In *Proceedings of the 5th International Conf. on Vacuum Metallurgy and Electroslag Remelting Processes,* Munich, 11–15 Oct. 1976. Hanau, 1977, pp. 191–3.

Rehák, B., I. Kašik, and M. Karnovský, A contribution to the study of the behaviour of non-metallic inclusions in the electroslag remelting process. In *Proceedings of the 5th International Conf. on Vacuum Metallurgy and Electroslag Remelting Processes,* Munich, 11–15 Oct. 1976, Hanau 1977, pp. 147–52.

Schlatter, R., Direct current electroslag remelting of specialty alloy steels. *Ind. Heat.* **45** (1978), No. 10, 19, 22–3, 25.

Mitchell, A., Electroslag hot-topping of heavy ingots. *Ironmaking & Steelmaking* **6** (1979), No. 1, 32–7.

Jauch, R., A. Choudhury, F. Tince and W. R. Lemor, Electroslag remelting process at Röchling – Burbach for heavy forging ingots of 2300 mm diameter. *Ironmaking & Steelmaking* **6** (1979), No. 2, 75–83.

Lane, David, Electro-slag refining. *Steel Times* **207** (1979), No. 6, 425–7.

Republic steel to install electroslag remelt furnace. *Iron Steel Eng.* **56** (1979), No. 7, 59.

Heymann, Hans, and Willi Austel, The MHKW process – a new type of technology for manufacturing heavy forgings. *Metallurg. Plant & Technol.* **2** (1979), No. 5, 58, 60–1.

9.4 Electron-beam Remelting Furnaces

1. DRP 188466 dated 18. Feb. 1905.
2. Stephan, H., Einsatzmöglichkeiten und Entwicklungstendenzen von Elektronenstrahl-Schmelzöfen. *Neue Hütte* **16** (1971), 650–7.
3. Gruber, H., Das Schmelzen von Metallen mit Elektronenstrahlen. *Z. Metallkde.* **52** (1961), 291–309.
4. Hensel, A., Aufbau und Inbetriebnahme des Vakuumstahlwerks in Edelstahlwerk Freital. *Freiberg. Forsch.-H.,* Series B, No. 122, 1966, 37–56.
5. Ardenne, M. v., S. Schiller, G. Jäger, P. Lenk, H. Schönberg, P. Stein, K. Schmidt and G. Barsikow, Ein Elektronenstrahl-Mehrkammerofen mit einer Leistung von 1700 kW. *Freiberg. Forsch.-H.,* Series B, No. 122, 1966, 57–66.

6. Hentrich, R., Der Einsatz eines Elektronenstrahl-Schmelzofens in einem Edelstahlwerk. (Use of an electron-beam melting furnace in a special steel producing plant). VIth International Congress on Electro-heat. 13–18 May 1968, Brighton. Organized by L'Union Internationale d'Électrothermie and the British Committee for Electro-Heat. Paris 1969. N. 136., 8 pp.
7. Sperner, F., Electron-beam melting of metals (German). *Internat. Z. Elektrowärme* **24** (1966), 33–41.
8. Elektronenstrahl-Mehrkammeröfen der VEB Lokomotivbau-Elektrotechnische Werke 'Hans Beimler'. *Stahl u. Eisen* **94** (1974), 112–13.
9. Wahlster, M., and H. Spitzer, Technology and metallurgy of the processes for the production of special steels and comparable materials (German). *Stahl u. Eisen* **92** (1972), 961–72.
10. Ardenne, M. v., S. Schiller, W. Küntscher, H. Thiel and L. Meyer, Hochvakuumschmelzen von Stahl im 60-kW-Elektronenstrahl-Mehrkammerofen. *Neue Hütte* **6** (1961), 198–211.
11. Wahlster, M., and F. Sperner, Anwendung und Technologie wichtiger reaktiver Metalle. In *Proceedings 8th Plansee-Seminar,* 23–6 May 1977, Vol. 1. No. 4. Reutte 1977. 15 pp.

Kalugin, A. S., Veränderung der Zusammensetzung von Legierungen auf der Grundlage von Eisen, Nickel und Kobalt beim Elektronenstrahl-Umschmelzen (Russian). *Izvestija Akademii Nauk SSSR, Metally,* **19**, (1977), No. 5, 121–7.

9.5 Plasma Remelting Furnaces

1. Paton, B., and others, Das Plasma-Umschmelzverfahren. In 3. *Stahltagung.* 20–22 May 1971, Potsdam/DDR.
2. Paton, B., and V. Lakomski, *Plasmabogenumschmelzen von Metallen und Legierungen.* Schweißen in der UdSSR, 1970, pp. 9–10.
3. Roßner, H.-O., Plasmaverfahren in der Schmelzmetallurgie. *Stahl u. Eisen* **95** (1975), 61–4.
4. *Plasmalichtbogenschmelzen von Metallen und Legierungen im wassergekühlten Kupferkristallisator.* Informationsschrift Licensintorg, Moscow, 1977.
5. Paton, B. E., V. I. Lakomski, G. M. Grigorenko and G. F. Torkhov, Plasma arc remelting means high quality metal. *Licensintorg Journal.* Moscow, 1977. pp. 22–4.

McCombe, C., Plasma-arc — a joint USSR/GDR development in electric steelmaking. *Steel Times* **206** (1978), No. 9, 86–7.
Esser, Fred, Detlev Klöpper, Gerhard Scharf and Walter Lachner, Zur Entwicklung großer Plasmastahlschmelzöfen — Simulationsuntersuchungen zur Prozeßextrapolation. *Neue Hütte* **34** (1979), No. 12, 441–4.
Fiedler, J., and G. Scharf, Stahlerzeugung im 10- und 30-t-Plasmaofen. *Freiberg Forsch.-H.,* Series B, No. 212, 1979, 91–5.

Electric Furnace Steel Production
Edited by E. Plöckinger and O. Etterich

10 Modern Arc Furnace Steelworks

KLAUS DZIGGEL, DORTMUND, AND HERMANN LÜNIG, MOERS

10.1 THE PLANNING OF ARC FURNACE STEELWORKS

Arc furnace steelworks were first built to integrated plans in the 1960s; now the concept is used worldwide, and the many underlying reasons stimulate progressive development and continual change. The aim in building any new furnace is always to create a basis for steel-production which will be technically progressive, reliable, and economical. Metallurgical and technical progress, extensive use of automation, the increasing pressure of competition, and particularly the ever-growing demands of environmental protection and health and safety at work all make total-system planning increasingly essential. If the circumstances which determined the design of electric furnace steelworks in the past are examined, then it is found that individual situations are hardly ever identical. For example, smaller arc furnaces were used for steel castings or for a very broad-based programme for making special steels and ingots, while larger units were used for continuous-casting of reinforced concrete steel or tube steel or solely for continuous-casting of slabs for steel plate or strip.

Nowadays, the arc furnace steelworks, charged with scrap or sponge iron, is the only alternative to the blast furnace converter steelworks that offers extraordinary flexibility in the number and size of furnaces which can be used to suit particular production requirements. For this reason, planning today need not be confined to arc furnace design and layout alone but should pay much more attention to the allied requirements for raw and additional materials and energy supply, etc., together with further processing, and should encompass materials flow and the cost of the finished product and of marketing, all in relation to one another. Besides all the technical and material requirements, parts of which are dealt with in great detail in the various sections of this book, the most suitable site must be selected, having regard these days for the costs of transporting raw materials and finished products, possible effects on the nearest residential areas, and the availability of a workforce with suitable qualifications. At the same time, it is necessary to ensure that there is sufficient know-how for planning and operating the intended works. Thus, planning consists of the following essential tasks:

(1) Market analysis;
(2) Analysis of the steel-production process;
(3) Analysis of the energy supply;
(4) Collecting the necessary production data;
(5) Establishing what charge-materials are to be used and drawing up a materials flow chart;
(6) Preparing plans of the works with details of the installations;
(7) Costing;
(8) A PERT chart for timing and progressing the project;
(9) An economic evaluation.

These days, project planning is accompanied more and more by dynamic modelling and computer analysis, so that all the limiting quantities can be investigated, together with the effect of market-fluctuations.

Most suppliers now offer standardized equipment for all systems and installations. Besides arc furnaces this includes cranes, ladle-heating systems, transporter vehicles, continuous-casting systems, etc., so that there is no problem in putting functional systems together.

Concepts such as 'mini-steelworks' for producing about 0.1–1 million tonnes per annum[1–3] 'ultra-high-power' arc furnaces for installations with high-power electricity supplies, and 'high-capacity arc furnaces' with a vessel diameter over about 6 m are still in use today. However, these are already characteristic of a bygone era of development.

Since then all arc-furnace builders in the world have standardized their systems and, because of worldwide competition, have matched each other. Thus, they now only offer furnaces for high electrical power supplies and a maximum possible sequence of melting cycles, almost to the exclusion of all other types. Even if the available electricity supply cannot provide the power required, installations are still usually constructed with high-power transformers which can be utilized fully later on.

Aids, such as graphical representations of the course of work during a particular time-band, are used for planning to depict the interlinking of the various processes with respect to time.

Up to now, many capacity-calculations were based on the consideration of bottlenecks. Here the time used by equipments where bottlenecks could occur was equated to the time-balance for the whole system. Normally, that part of the system which has the longest operating time is taken to be the bottleneck.

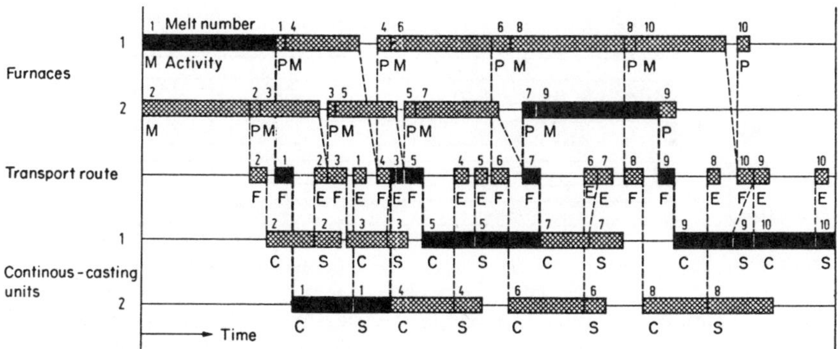

Figure 1. Bar chart showing processes in an electric furnace steelworks[4]

The performance figure is ascertained by dividing the material throughput by the time used by the bottleneck unit. However, this method is not accurate for a system using different processes simultaneously, where some processes may be time-critical while the others are not. For this reason, all processes today must initially be considered as having equal weighting.[4] Then, every time-part of a process is considered to be critical or non-critical, new ways of increasing performance can be considered. Figure 1 is a bar chart which schematically represents one process in an electric furnace steelworks. This scheme can be extended to apply to other parts of the system, such as the scrapyard, for example. Here it is represented by five time-bars. The time used by any production medium consists of active times and waiting times; for the furnaces these comprise melting (M), patching (P) and waiting (W — here shown as blanks). For continuous-casting, they are casting (C), set-up time (S), and waiting. Transport media in motion are either full (F) or empty (E); here again, waiting time is blank. In addition, one sequence of activities involving all elements of the system is represented by heavy black bars. The total elapsed times for the process can only be reduced along this critical path, which involves only critical activities.

Figure 2 is a time-breakdown for a complete system and its various elements. The timing for the complete system is derived by summarizing the timings for the critical system elements. Nowadays, special network analysis and simulation techniques are used to display and calculate the logic and time-behaviour of production processes.

Both methods illustrate work flow with the help of mathematical models. We can use the results to make better recommendations about possible ways of increasing the performance of new installations. Even if an installation is to be planned on the basis of a particular production programme, the total flow of all materials should still be worked out using models for several production programmes and combinations.[5] In this way the adaptability of a steelworks installation can be assessed.

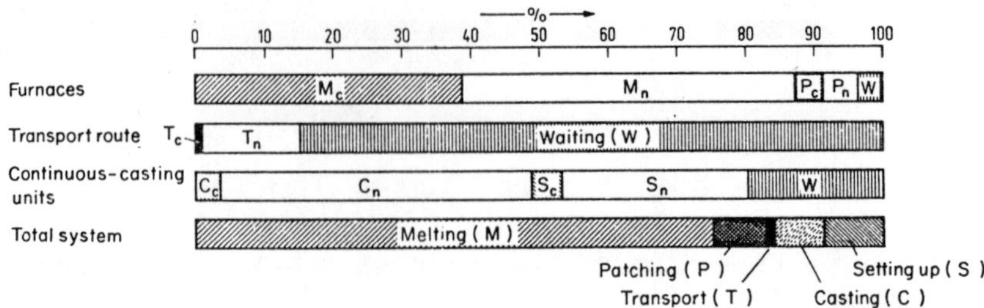

Figure 2. Time-breakdown for a whole electric furnace steelworks system and its elements[4]. Subscript c = critical; subscript n = non-critical

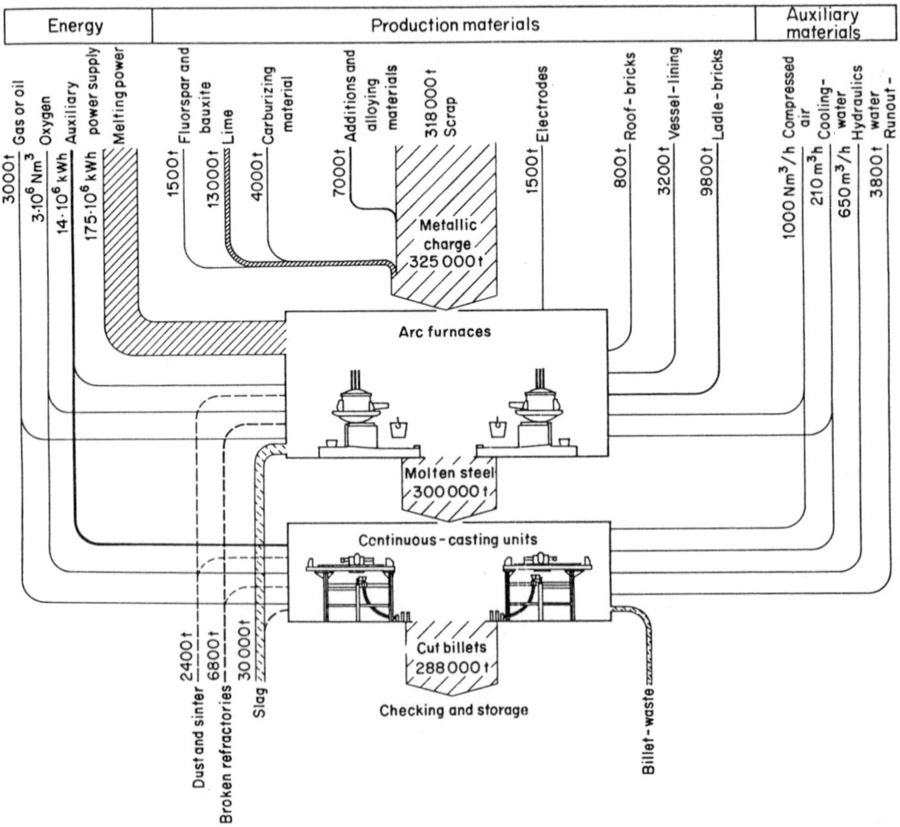

Figure 3. Flow of materials and energy for a steelworks model with an annual output of 300 000 tonnes from two 50/60-t high-performance arc furnaces with two supplementary quadruple continuous billet-casting units[6]

Figure 3 illustrates the flow of energy and materials in an electric furnace steelworks and can be used as the basis for developing simulation models.[6] Another way of representing behaviour in time is to use PERT (Programme Evaluation and Review Technique).

Table 1. Costs of installation and buildings for an electric furnace steelworks using two 120-t high-performance arc furnaces

	Aspect	Costs as percentage of total cost
(1)	Complete arc furnaces	23.5
(2)	Foundation and transformer house	4.5
(3)	Cranes	18.0
(4)	Workshops and steel construction	17.0
(5)	Foundations and building work	9.5
(6)	Steelworks equipment, e.g. ladles, baskets, instrumentation, stackers, pneumatics, electrode stands	11.5
(7)	Energy supply, e.g. low-tension electricity supply, compressed air, water and oxygen supplies	2.5
(8)	Casting system, e.g. casting installation, railway tracks, trucks, winches and moulds	8.0
(9)	Dust-extraction system	3.5
(10)	Bunkerage	2.0

Table 1 illustrates an example of the percentage breakdown of costs for an arc furnace steelworks producing ingots using two 120-t arc furnaces. Naturally, the division of costs is markedly affected if, instead of casting individual ingots, the system output is by continuous casting.

10.2 PRINCIPAL TYPES OF ARC FURNACE STEELWORKS

The construction and operation of electric furnace steelworks are covered comprehensively by the literature.[8–21] Planners and operators are advised to acquaint themselves with the experience which has been collected with these systems before beginning work on a new plan so that they can consider the advantages and disadvantages of applying the different concepts.

If we try to classify the arc furnaces operating throughout the world systematically according to *charging conditions* then we arrive at the following groupings:

Group 1: arc furnace steelworks using *scrap* as the charge-material and sponge iron in place of scrap for quality reasons or during times of scrap-shortage or to take advantage of low prices.

Group 2: arc furnace steelworks using *sponge iron* as the main charge-material and recycling scrap or bought-in scrap when prices are favourable.

Fundamental classification of electric furnace steelworks according to charge-material seems opportune, because the whole system concept for integrated sponge-iron production is considerably more complex than the total concept for steelworks based on scrap.

Another way of classifying arc furnace steelworks, which to a certain extent parallels the development of the arc furnace, is based on the *production programme*. This system results in:

(1) *Classical high-grade steelworks* having highly varied melting programmes, ranging from non-alloy to highly-alloyed constructional steels and tool steels; corrosion-, acid-, and heat-resisting CrNi and Cr steels; also high-speed steels and the like which for various reasons have to be cast into ingots for rolling and forging. Further processing is carried out in the works' own forge and/or rolling mill, and the intermediate products must conform to rigid specifications.

(2) *Steelworks for large-scale production* of high-quality and special constructional steels and special steels produced in large quantities, which can be cast continuously but are often made into rolling or forging ingots for further processing on-site. The quality requirements are generally high.

(3) *Works for mass production* of simple carbon steels in quantities ranging from medium to the highest levels for continuous-casting into billets or slabs which are then worked in special rolling mills into wire, rod, special profiles, sheet, tube, etc. Here the quality requirements depend on the product.

Depending on the year in which they were built, these steelworks were either modified retrospectively or were initially equipped with first-generation systems for secondary refining and environmental protection (dust-extraction).[9,22–29]

This kind of classification implies that classical steelworks and works producing large quantities of special constructional steels and mass-produced high-grade steels are operated with scrap as the basic charge-material. On the other hand, mass-production steelworks and works producing high-quality steels, especially those newly built, use sponge-iron as their main charge-material; most even have their own integral sponge-iron production systems.

The latest and newest types of arc furnace steelworks have to meet the most stringent requirements for environmental protection and health and safety at work laid down for modern industrial plants. These differ from other groups only in that the buildings and installations have to conform to all the requirements. This has a very marked effect on investment and operating costs and forces the operation into compromises which sometimes are questionable.

10.3 CONSTRUCTION OF NEW ARC FURNACE STEELWORKS

If we comparatively appraise electric furnace steelworks built in the last ten years on a total-system concept, then, despite differences in local conditions, we find certain recurring schemes, which have proved themselves in practice and can form the basis for new designs. Figures 4–7 summarize the basic models for planning arc furnace steelworks for the production of steels ranging from mass-produced carbon steels to special steels, including the necessary continuous-casting and ingot-production processes.

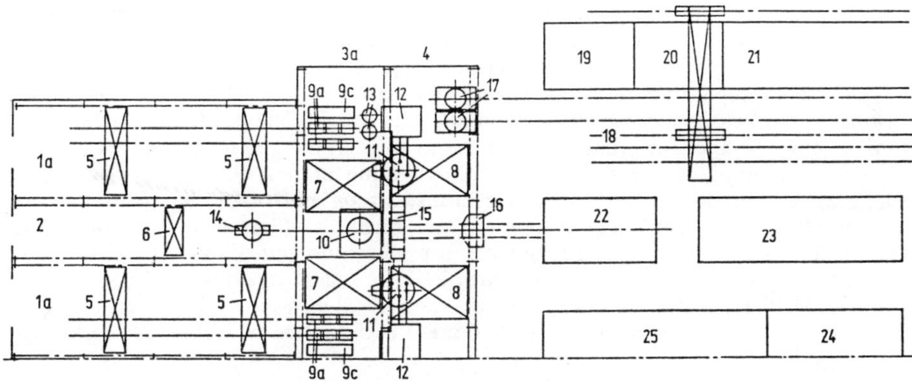

Figure 4. Plan view of an arc furnace steelworks with vacuum-degassing system, roll-ingot casting on a casting bogie, and forging-ingot casting pit; scrap-storage on site. (See page 460 for key)

All these schemes contain two furnaces; this is very appropriate for many reasons, especially in the event of failure of individual components. Moreover, this also enables exchangeable vessels to be used. Extension of the works by adding a furnace or by duplication may be practicable. Experts believe, however, that there should not be more than four furnaces in one shop complex.[30] These planning models, which will be explained in more detail later, can be envisaged as standard schemes in which it is possible and indeed may be sensible to interchange the locations of individual elements, such as scrap-handling and storage or the arrangement of the casting shop, according to local circumstances.

Figures 8 and 9 show the plan and cross-section of one example of a steelworks with integrated sponge-iron production. This is the SIDOR electric furnace steelworks in Venezuela, which produces 2.4 million tonnes of crude steel a year using six 200-t furnaces of 100 MVA supplied power each.[19] This steelworks, which is quite new, also matches the basic schemes to be discussed in the sections that follow.

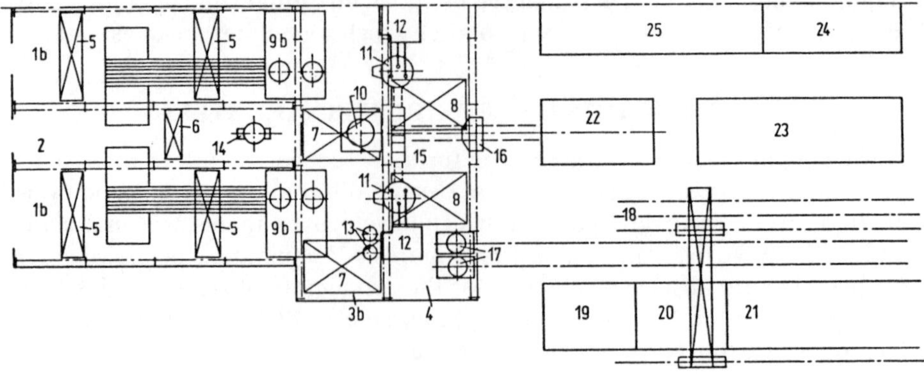

Figure 5. Plan view of an arc furnace steelworks with vacuum-degassing system and continuous-casting system; scrap-storage on site. *Key to figs 4 and 5*: 1a Shops for ingot-casting preparation (casting bogie) and stripping; 1b shops for continuous casting; 2 refractory workshops; 3a tapping and casting shop; 3b tapping shop; 4 furnace bay; 5 cranes for ingot and continuous casting; 6 crane for refractory work; 7 cranes for tapping and casting; 8 furnace-charging cranes; 9a bottom-pouring or single-casting on casting bogie; 9b continuous-casting installation; 9c casting pits for blooms; 10 vacuum-degassing system; 11 arc furnaces; 12 transformer houses; 13 ladle-firing; 14 ladle bogies; 15 bunkers for additions and alloys; 16 control and rest-room; 17 scrap-basket and bogie with built-in scales; 18 railway tracks; 19 alloy store; 20 auxiliary materials store; 21 scrap-storage; 22 dust-extraction filter-house; 23 oil, water, compressed air, oxygen; 24 domestic and office buildings; 25 workshop

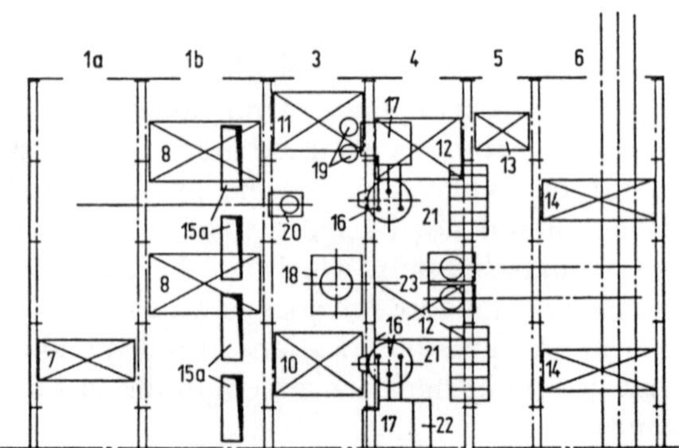

Figure 6. Plan view of arc furnace steelworks with parallel bay for additions and alloys, vacuum degassing systems, and separate casting shop and pit; scrap handling shop in the steelworks. (see page 461 for key)

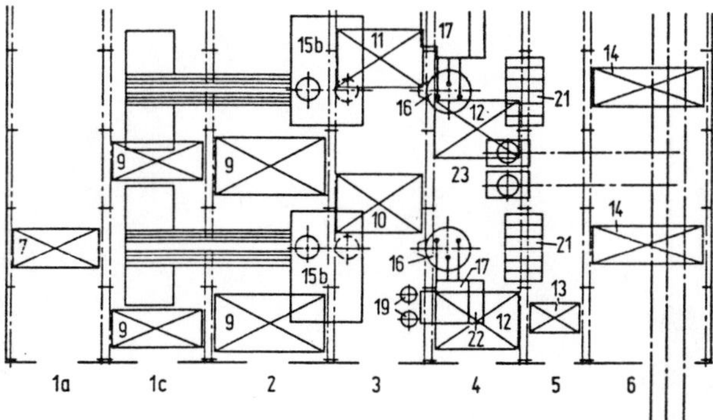

Figure 7. Arc furnace steelworks with parallel bays for additions and alloys as well as continuous casting; scrap-handling shop in the steelworks. *Key to figs 6 and 7* 1a Ingot/continuous-casting despatch shop; 1b casting pit bay for ingots; 1c cooling-bed for continuous casting; 2 Continuous-casting shop; 3 tapping bay; 4 furnace shop; 5 bay for additions and alloys; 6 scrap-handling area; 7 loading crane; 8 casting crane; 9 auxiliary crane in continuous-casting shop; 10 tapping crane; 11 auxiliary crane in tapping bay; 12 furnace-charging crane; 13 auxiliary crane for additions and alloying-materials bay; 14 crane for transferrring scrap into baskets; 15a casting pits; 15b continuous-casting installation; 16 arc furnaces; 17 transformer houses; 18 vacuum-degassing system; 19 ladle-drying; 20 ladle bogie; 21 bunkers for additions and alloys; 22 control and rest-room; 23 scrap-basket and transport bogie with built-in scales

Steelworks also have auxiliaries such as mechanical and electrical workshops and installations for cooling water, compressed air, gas and oil fuels, storage for scrap, alloys, refractories, spares, etc. In addition, there are domestic and office buildings and space for a chemical laboratory. An electricity substation with high-tension switchgear and, perhaps, reactive power compensation, is usually needed nearby to power the arc furnaces. Figures 4 and 5 show the basic arrangement of these auxiliaries.

10.3.1 Scrap-storage and Handling

It always has been preferable to situate scrap-yards close to large steelworks, but this has not created difficulties for small steelworks with low overall production-levels. However, as furnace-sizes increase and production rises, bottlenecks soon arise which can be eliminated by enlarging the scrap-yards. Nowadays, scrap-movement is fully integrated into steelworks process flow. Scrap-stores for steelworks producing high-grade steel differ from those for

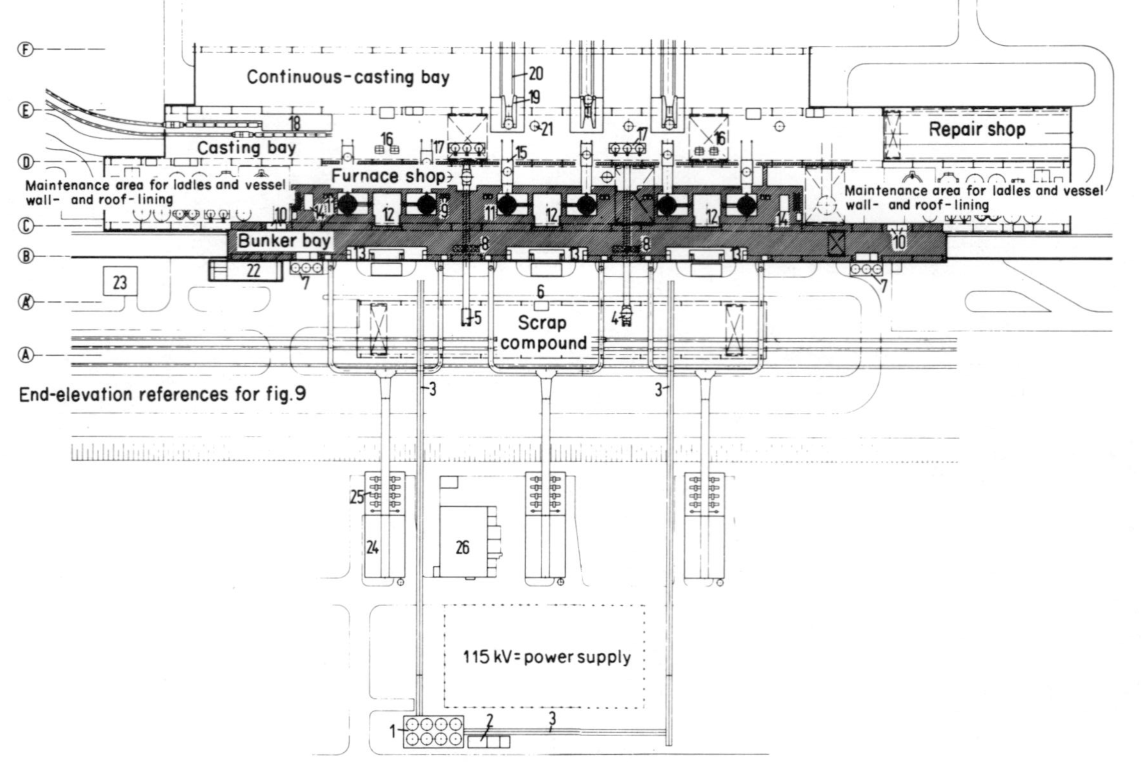

Figure 8. Plan of the Siderurgica del Orinoco (SIDOR) electric furnace steelworks in Venezuela using six 200-t arc furnaces and three continuous-casting units producing slabs[19]. 1 Pellet silos; 2 pellet-feed control; 3 conveyers; 4 scrap-transporter; 5 rail weighbridge; 6 scrap-feed control; 7 silos for lime and coal; 8 ferro-bunker for furnace-charging; 9 ferro-bunker for ladle-additions; 10 dolomite bunker; 11 electric furnace; 12 transformer house; 13 furnace control-centre; 14 electrode-preparation; 15 trolley for steel transfer; 16 slide-exchange area; 17 fires for keeping ladles hot; 18 stand-by casting area; 19 rotary ladle tower; 20 continuous slab-casting system; 21 stand-by casting ladle; 22 central control-room; 23 pump-house; 24 dust-extraction system; 25 blowers; 26 Electricity substation

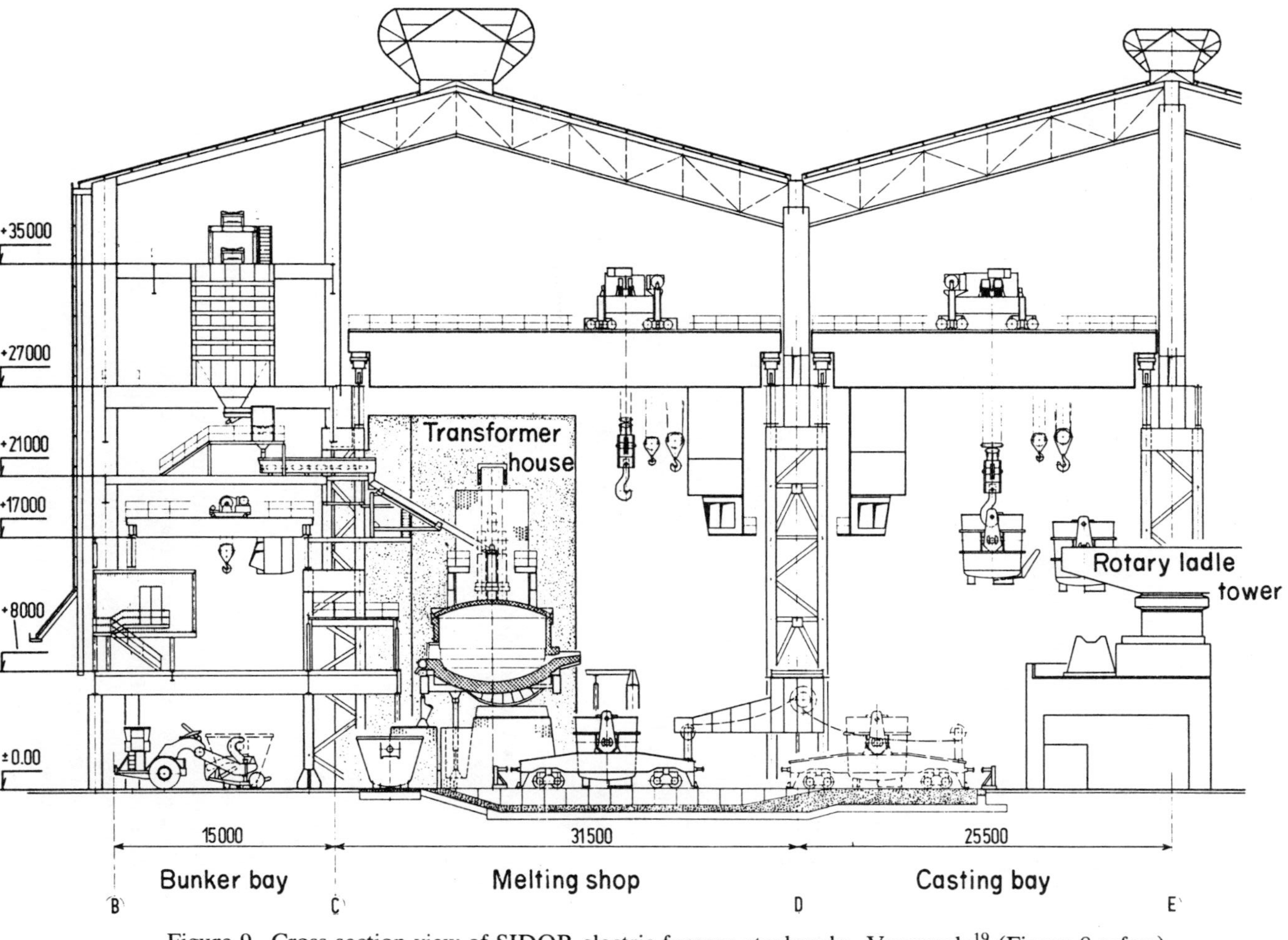

Figure 9. Cross-section view of SIDOR electric furnace steelworks, Venezuela[19] (Figure 8 refers)

high-tonnage steelworks mainly in the number of bunkers, High-grade and special steelworks require a large number of scrap-bunkers of different sizes to ensure segregation of scrap by material type and for preparing accurate charges of recycled and bought-in alloy scrap. The number and size of bunkers depends on the type of programme and the production level. Larger bunkers are used for groups of similarly alloyed scrap, while small bunkers are used to store individual types of special alloyed steel scrap. Steelworks with broad-based programmes for producing special steels require a number of bunkers of different sizes depending on the individual material grades and grade groups and on how much material is recycled (20–30% of production); the bunkers should be able to take at least half a month's arisings. Works mass-producing carbon steels require fewer bunkers, the number of which corresponds to the number of classes of scrap piece-size. Any non-alloy scrap bought-in should be sorted and put into suitable temporary storage.

If the steelworks concerned operates a highly variable programme and a large quantity of scrap has to be stored close to the steelworks, it is preferable to have an integral scrap compound, as shown in Figures 4 and 5. This usually takes the form of a roofless compound spanned by bridge-cranes. The storage capacity is often increased by digging a trough-shaped pit. Incoming scrap is unloaded from trucks and put into the storage compound, unless it can be loaded directly into scrap charging-baskets. The scrap-baskets are carried by self-propelled, diesel tipper-wagons with built-in weighing machines. If the steelworks is located beside water, then the bridge-cranes have outriggers to enable them to unload scrap from barges as well.

Another variant very often found in mass-production steelworks is the scrap-handling area as illustrated in Figures 6 and 7. Scrap is transferred direct into charging-baskets in this shop. Small bunkers between the rails are often used for intermediate storage of particular types of scrap or in order to cater for short-term holdups in scrap-delivery.

Two to four parallel railway tracks are used so that enough wagons can enter the handling area. Here, too, the scrap is weighed continuously during loading; the weight is displayed on a large indicator and may be fed into a computer. The baskets are usually loaded by means of fast cranes of up to 30-t capacity, equipped with powerful single or twin electromagnets.

The more scrap has to be handled in a scrap area, the more carefully the activity and area have to be planned, and it is particularly important that the distances over which scrap is moved by crane should be as short as possible. To expedite scrap-transfer in bays such as those in Figures 6 and 7, the scrap-wagons are guided by special positioning devices into the immediate vicinity of a passing basket. Scrap of uniformly small piece-size can be loaded by dumper wagon directly into the charging-basket, provided the respective sill heights are compatible.[31,32] Under-floor tracks for the scrap-baskets improve flow greatly by eliminating level-crossings.

For tactical reasons, steelworks often incorporate intermediate stockpiles of scrap to meet needs for the next month or more. These should be particularly carefully thought out and located between the point where scrap enters the works and the scrap-processing shop or handling area. Care should be taken to ensure that traffic approaching this intermediate store can move without interference from traffic going the other way, and that stock is regularly turned over, according to demand. Tipper lorries have proved effective for such internal scrap movements; they also save having to use cranes to load the scrap-baskets.

10.3.2 Furnace Shop

Two-bay furnace shops (Figures 6 and 7) are being increasingly built in modern works instead of single-bay shops (Figures 4 and 5). Special production programmes and large furnaces in particular can require such large quantities of additives and alloying materials that a bunker bay for additions becomes indispensable. In principle, this also means that the furnace shop should be laid out longitudinally to improve the traffic flow. It has yet to be seen whether the two-furnace-bay system will prevail or whether the old single-bay system will also be better for environmental protection in future.

It has proved basically worthwhile to mount arc furnaces on stages in larger steelworks, and the application of this principle is being extended. Higher construction costs are compensated by many advantages in operation. Slag work and disposal are helped considerably, and, in addition, more space is made available for movement and storage. Scrap-baskets are lifted up by cranes through an aperture in the stage to feed the furnaces. The height of the stage is determined by the headroom required for the transport.

The way the furnaces are charged has to be taken into account when planning the bunker installation. If there is a steel-degassing system, this should be provided for at the same time. Bunkers between the furnaces enable the furnaces to be fed from a special floor-charger on the stage; this charger should not need to turn during material transfer as is required when the bunkers are situated behind the furnaces. Furnaces which operate mainly on sponge iron and incorporate a fifth hole in the roof may also be fed automatically via this aperature instead of by floor-charger. The fifth hole is also used at the same time for continuously feeding sponge iron to the furnace.

Exhaust-gas removal and noise attenuation are dealt with in section 10.5.2 and in Chapter 13. As shown in Figures 4–7, the melt is tapped into a ladle suspended from a crane in the casting bay. The melt is sometimes tapped into a ladle standing on a casting bogie; this is done for environmental protection and involves special procedures.

10.3.3 Tapping and Casting Shop

Like the furnace shop, the shop where tapping and casting were carried out, and where casting pits were constructed, filled, and emptied, has developed from a single shop into several different ones. Figure 4 illustrates the first stage, in which steel is poured into a casting bogie in the tapping shop. The casting bogie is filled and emptied in the stripping bay which is laid out across the works axis. Adoption of this practice, which is still completely valid today, significantly improved working conditions and increased the capacity of the casting shop. In some cases the stripping bay was separated from the steelworks or transferred to the cogging mill. This was an interim solution, adopted because of space-limitations or heat-conservation, for example; however, it is preferable for the installation to be integral with the steelworks.

The layout shown in Figure 5 arose, of necessity, with the introduction of continuous casting. The continuous-casting system here is served via a rotating tower.

The solution shown in Figure 6 dispenses with the casting bogie; however, heavy-duty cranes are needed in the tapping and casting shop.

The continuous-casting system shown in Figure 7 has several narrow bays, and the work-areas are served effectively in various ways by light cranes.

In conclusion, we can say that the deliberations above and the layouts shown in Figures 4–7 are basic but absolutely characteristic of modern electric furnace steelworks. The considerations and the elements of the layouts outlined above can be used in building-block fashion to help to enlarge old works and to plan new ones to accommodate local conditions and requirements in individual cases.

10.4 OLDER TYPES OF ARC FURNACE STEELWORKS

Figure 10 is a schematic representation of the layout of the Böhler AG steelworks in Düsseldorf, which was built before the First World War to produce special steels. This works has been able to keep up with modern steelworks with the same production range by replacing old equipment and adding a dust-extraction system, vacuum-treatment installation, electro-slag remelting (ESR) system, and other equipment at various times.

The four-furnace electric steelworks at Lukens Steel Company, Coatesville, Pa, shown in Figure 11, exhibits a layout almost identical with that in Figure 7. The scrap area is laid out lengthways; the four rail tracks for the scrap and the bays for additions and alloys are located between the scrap area and the furnace shop. This plant is operated largely as a complete system and is known to comply with all present regulations.

10.5 ENVIRONMENTAL AND HEALTH AND SAFETY ASPECTS

The tightening of environmental protection laws and the requirements for improved health and safety in steelworks have led to very marked differences

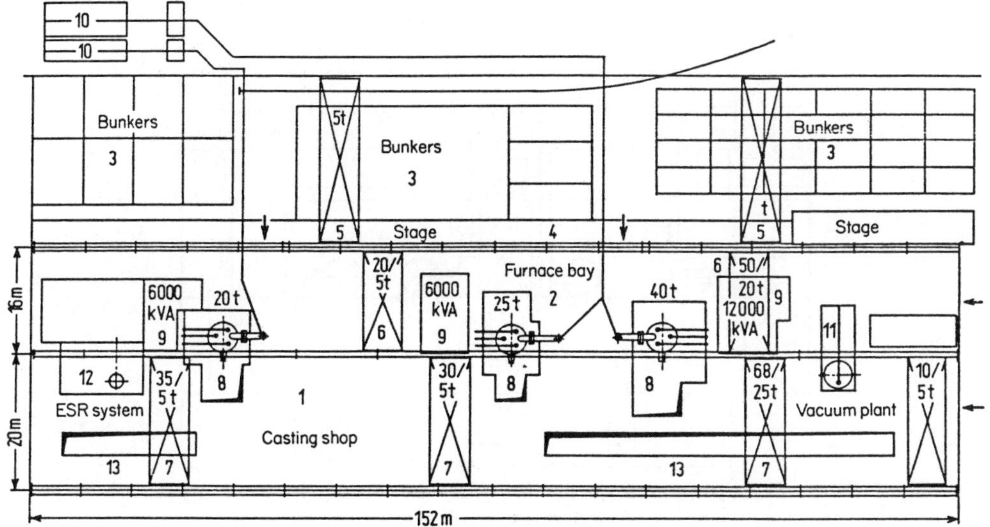

Figure 10. Plan view of the Böhler AG arc furnace steelworks at Düsseldorf–Oberkassel. 1 Casting shop; 2 furnace bay; 3 scrap area with bunkers; 4 stage with bunker-installation; 5 cranes for scrap-loading; 6 cranes for furnace-charging; 7 cranes for casting and auxiliary work; 8 arc furnaces; 9 transformer and switchgear houses; 10 dust-extraction filter houses; 11 vacuum-degassing system; 12 electro-slag remelting system; 13 casting pits

between those arc furnace steelworks just designed and completed in industrialized countries and others built there up to recently. These differences do not have much effect on metallurgical techniques and equipment but greatly influence the steelworks layout, protective buildings, and building costs. The future calls for steelworks to be made 'clean' and 'friendly' to the environment as well as to the worker, ergonomically speaking. The aim is to keep the nuisance and risk from dirt and noise and the stress on workers as low as possible everywhere inside the steelworks, starting with the scrap-yard and extending to outside the works.[33]

10.5.1 Scrap-handling Areas

At least some kind of enclosure will be needed to cut down the noise generated by scrap-handling; however, it will often be necessary to place the scrap shop in a sound-proofed building if there is a residential area nearby. If the latter is required, then it is recommended that the scrap shop should be built on to the

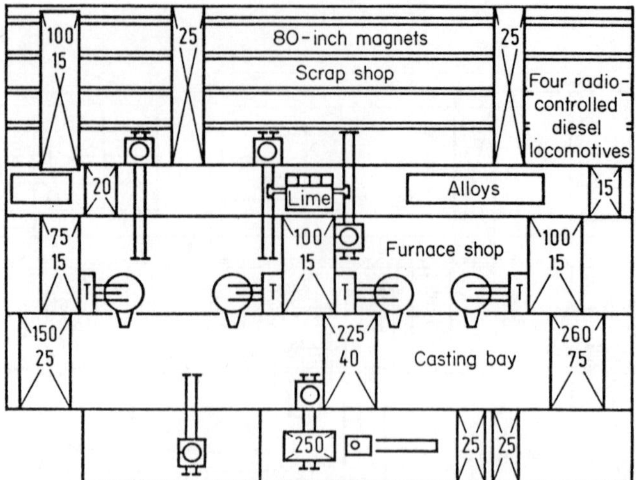

Figure 11. Plan view of the Lukens Steel electric furnace steelworks, Coatesville

steelworks building so that the sound-insulation of one of the side-walls can be simplified.

10.5.2 Steelworks Building and Furnace Bay

Encapsulation of entire steelworks installations carried out earlier protected only the environment outside the works against noise without dealing with noise inside; however, it satisfactorily catered for the capture and scrubbing of enormous quantities of exhaust gases. As furnace-operations are the main source of noise, exhaust gas, and dust, modern protection measures are directed at dealing with the nuisance caused by furnace operations. Simply enclosing the furnace is attractive from the point of view of costs and the effectiveness of the protection provided, but enclosure interferes with operations. All the suggestions and measures adopted so far have been compromises.

Figures 12 and 13 show a cross-section and plan view of an electric furnace steelworks belonging to the Danish firm Det Danske Staalvalsevaerk A/S-(DDS)[34] as an example of a works incorporating the best environmental protection measures at the time of writing. Steel is tapped from the furnaces direct into a ladle standing on a bogie in the furnace shop. Thus the cranes in the furnace shop only need to be used for charging and for odd jobs. The entire furnace shop is constructed as a sound-proofed cell inside the main building. All materials are fed in via sound-proofed doors or charging valves or through sound-absorbent tunnels.

The scrap-charging cranes are free for any tasks arising in the furnace shop.

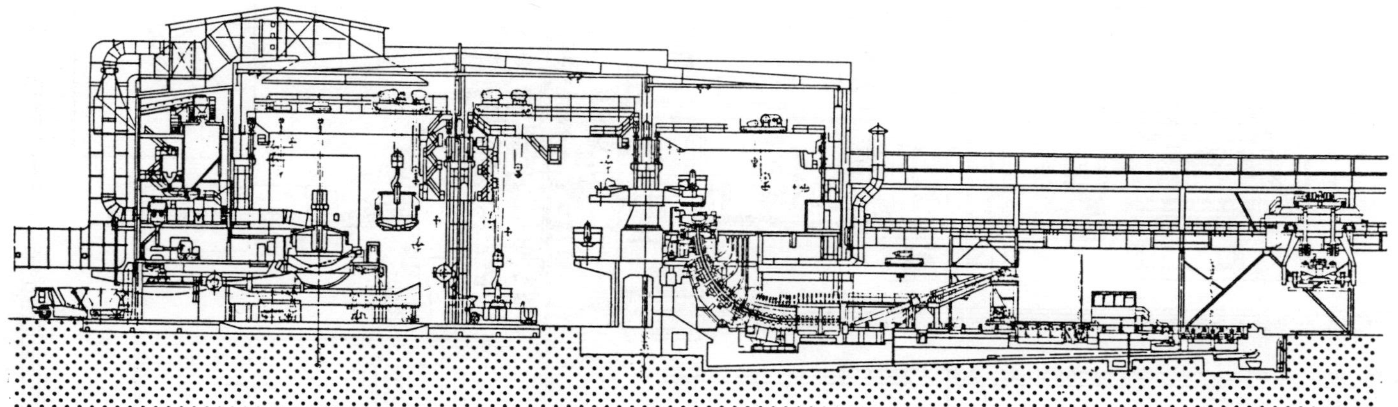

Figure 12. Cross-section of the Det Danske Staalvalsevaerk A/S (DDS) electric furnace steelworks. (Works drawing by DEMAG)

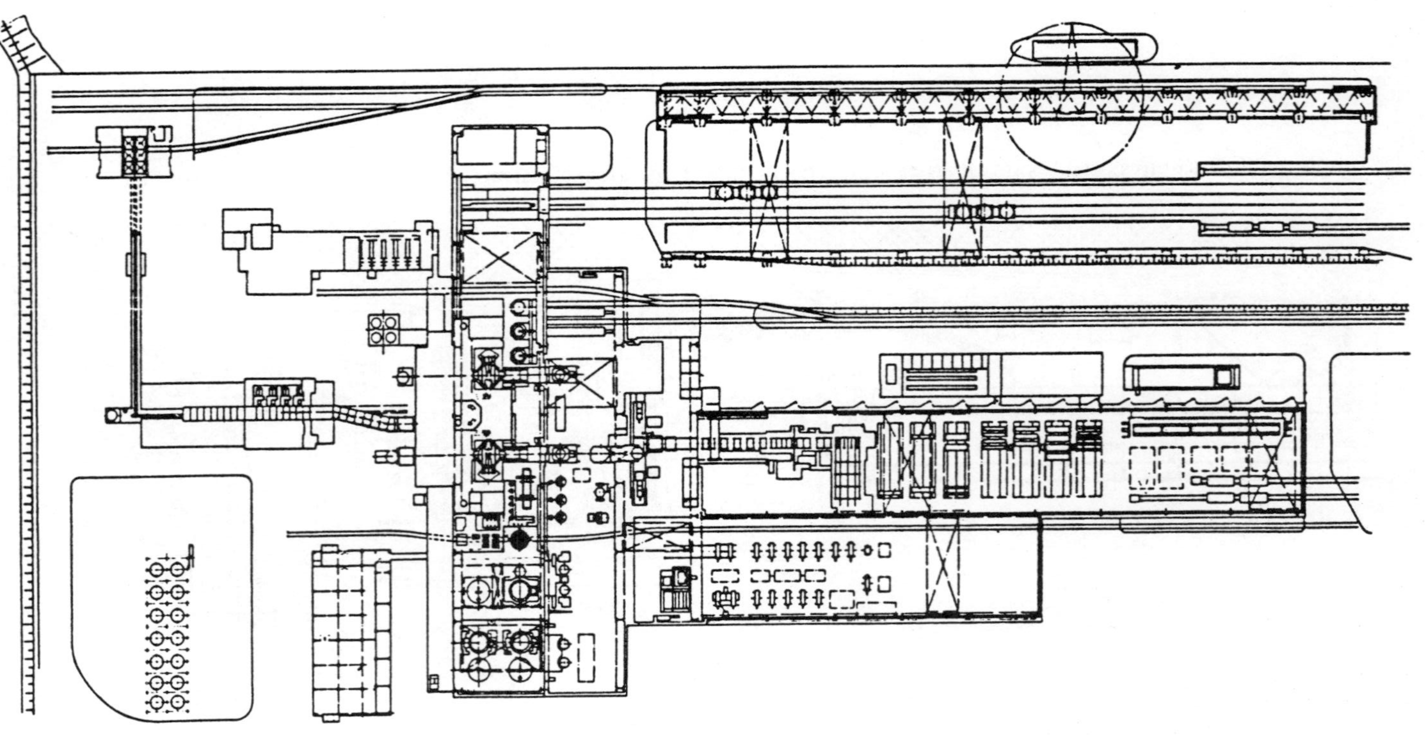

Figure 13. Plan view of the DDS works shown in Figure 12. (Works drawing by DEMAG)

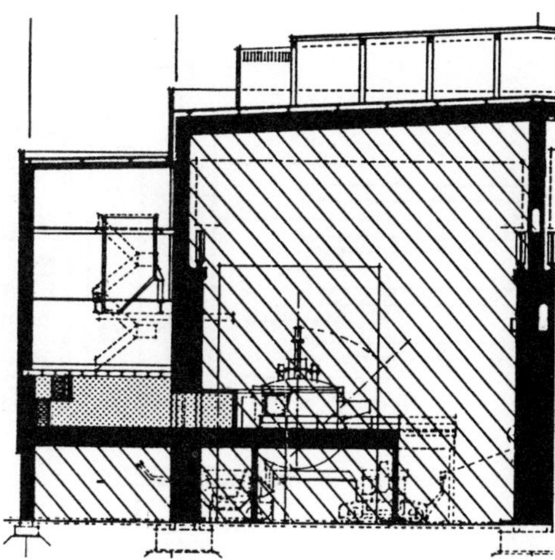

Figure 14. Sound-insulated furnace shop with control cabin. Schematic detail diagram of the DDS works shown in Figure 12. (Works drawing by DEMAG)

The working platform in front of the furnace doors is separate from the furnace shop itself (Figure 14). The entire working stage is roofed over and separated from the furnace shop by a wall. It is actually a room on its own, from which only the furnace doors can be seen and are accessible for work on the furnace. The central control cabin is located on the first upper floor, partly over the furnace shop and partly above the furnace stage space; naturally it is fully sound-proofed and air-conditioned. The furnace is controlled, and the weighing and dosing of additives are monitored automatically from here as are dust-extraction and water-cooling. Replacement electrodes are built up in a completely enclosed, sound-proofed space inside the furnace shop. The furnace vessels are completely exchangeable; re-lining work on these takes place in a separate building which can be isolated from noise and is located opposite the furnace shop. Furnace vessels are exchanged by using two furnace cranes coupled together, where single, very heavy-duty cranes would otherwise have been required. Figure 15 illustrates material flow in the works' finally developed form; this facilitates scrap-basket movement in the furnace bay.

Figures 16 and 17 illustrate the solution developed by Krupp for providing a melting operation producing the least dust and noise possible combined with improved working conditions close to the furnace and in the adjoining parts of the installation.[35] The cranes in the furnace bay now travel across the shop instead of along its axis. This has made it possible to encapsulate the individual furnaces. The cranes are operated by furnace personnel from the shop-floor. In

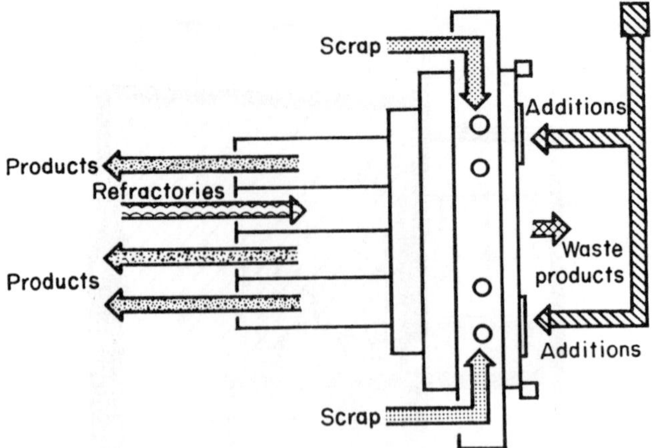

Figure 15. Materials flow in the DDS works, shown in Figure 12, in its finally developed form. (Works drawing by DEMAG)

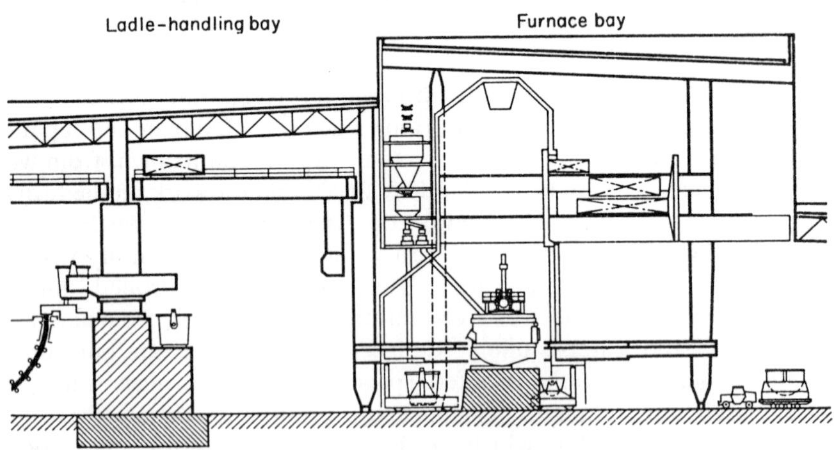

Figure 16. Longitudinal section through an electric furnace steel works with arc furnace specially encapsulated to capture exhaust gases and reduce noise (Krupp system)[35]

this way, a crane can be used for jobs such as charging, electrode-changes, maintenance, and lining work. It is subject to the same amount of movement and heat as a crane travelling along the axis of the shop; however, each crane acts as a separate functional unit for each furnace. This enables a furnace to be compeletly encapsulated; the shrouding is not disturbed while the furnace is operating, and it stays intact during scrap-charging. The charging process, which is shown in Figure 17, is made possible by the use of a moving wall-section attached to the crane and another wall-section of the same size

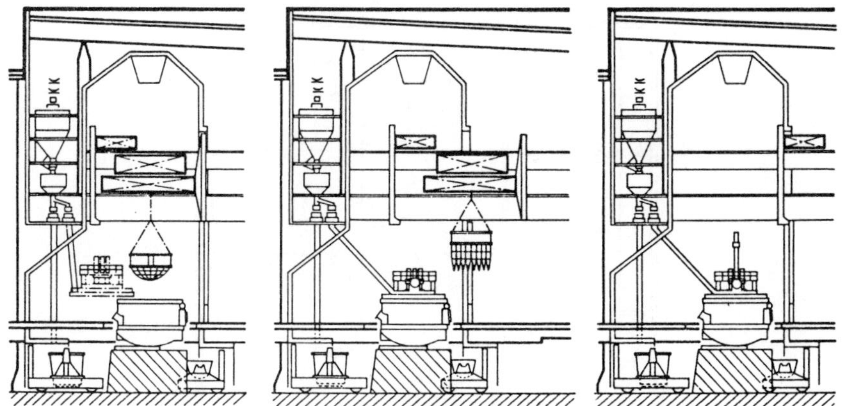

Figure 17. Charging operations with the furnace in figure 16 (Krupp system)

which is self-propelled. Molten steel is tapped off into a ladle standing on a bogie. There is no need for anyone to enter the furnace building during operation, since the whole process can be controlled and monitored by the furnace crew from outside.

One electric furnace steelworks with a high-performance 100-t arc furnace, which came into service in 1978 replacing an old quadruple open-hearth system for making special steel, uses a great deal of the infrastructure that already existed.[36] This indicates consistent and logical development in the construction of the electric furnace steelworks. The 100-t furnace has a 6.8-m diameter vessel, a 75-MVA transformer, electrodes of 610 mm diameter, and three fuel–oxygen burners as well as computer-assisted control. In its first 4 months of service this furnace produced 35 000 t of high-grade steel ingots in an average charge duration of 101 min. In more than a quarter of the first 1000 melts, tap-to-tap times of around 90 min were attained.

Figure 18 is the plan view of a steelworks designed to achieve the greatest possible productivity using a single melting unit. It is operated under optimum production conditions as a functional unit independent of other melting systems and produces steel ready to cast.

The scrap in its charging basket enters the *furnace bay* vertically or horizontally and is fed into the high-performance arc furnace through a hole in the furnace stage which is also used for furnace roof changes. The melt is tapped off into a tiltable ladle mounted on a bogie. It is then moved on the bogie to the steel-degassing installation for post-treatment; the ladle may be arc-heated if necessary; otherwise it may go directly to the casting bay. According to the steel programme and the quality specified, the steel-degassing installation may be replaced by a simple flushing system installed in the casting bay outside the sound-proofed furnace bay; such a flushing system can

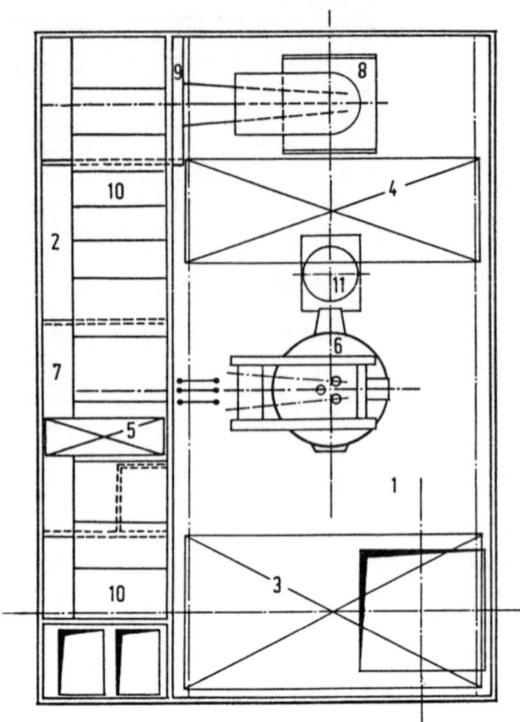

Figure 18. Plan view of a high-performance steelworks using a single-arc furnace in a sound-proofed space. 1 Furnace-bay; 2 side-bay; 3 charging-crane; 4 auxiliary crane; 5 bunker-charging crane; 6 high-performance arc furnace; 7 transformer-house with control-room; 8 steel-degassing system; 9 control-room for steel-degassing system; 10 tall bunkers for additions and alloys; 11 ladle bogie with ladle

serve two furnaces. The furnace cell contains a remote-controlled auxiliary crane in addition to the charging crane. Slag leaves the bay from the front.

The *side bay* contains the transformer house with the control room, which is thoroughly sound-proofed, and the rooms for computers and process control. The free space underneath (on the works floor) can be used to store refractories. There are bunkers on top for additives and alloys which can be fed in either via automatic feed systems or, as shown, by means of a crane. The feed to the furnace, the ladle in the tapping position, and the degassing system are all fully automated and equipped with weighing apparatus, conveyors, and slides. The vacuum pumps, auxiliary equipment, and, in some cases, electrical equipment for heating the ladle in the degassing system are all installed in the side-bay as well.

If there are two arc furnaces and the furnace walls are laid out axisymetrically, a single modified side-bay with the transformer houses in compartments and with a few auxiliaries can be used for both furnaces, especially if the two furnaces are tapped sequentially into a continuous-casting system.

10.6 STORAGE AND TRANSPORTATION OF ADDITIVES

Additives of various kinds are generally stored in tall bunkers or hoppers on the furnace stage or, as already mentioned, in a nearby bay. For mass-production of crude steel, bunkers are required for lime, dolomite, ore, coke, ferromanganese, etc. Special steels require extra bunkers for alloys such as nickel, ferrochrome, etc.

The capacity of the bunkers depends on the programme for the grade of steel being produced. Lime is hygroscopic and can only be stored for a short time before it is used. Bunker-capacities for furnaces of different capacity and throughput are shown in Figure 19; the graph is based[7] on the following consumption rates (kg/t):

Lime	30–36
Dolomite	10
Ore	30
Coke	4
Ferromanganese	8

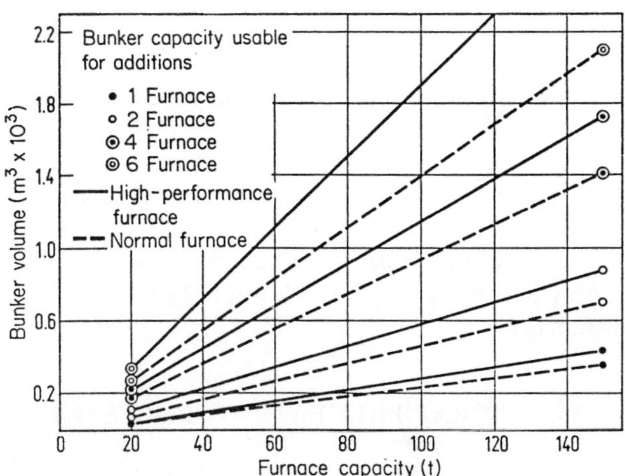

Figure 19. Bunker capacities for furnaces of different size and throughput[7]

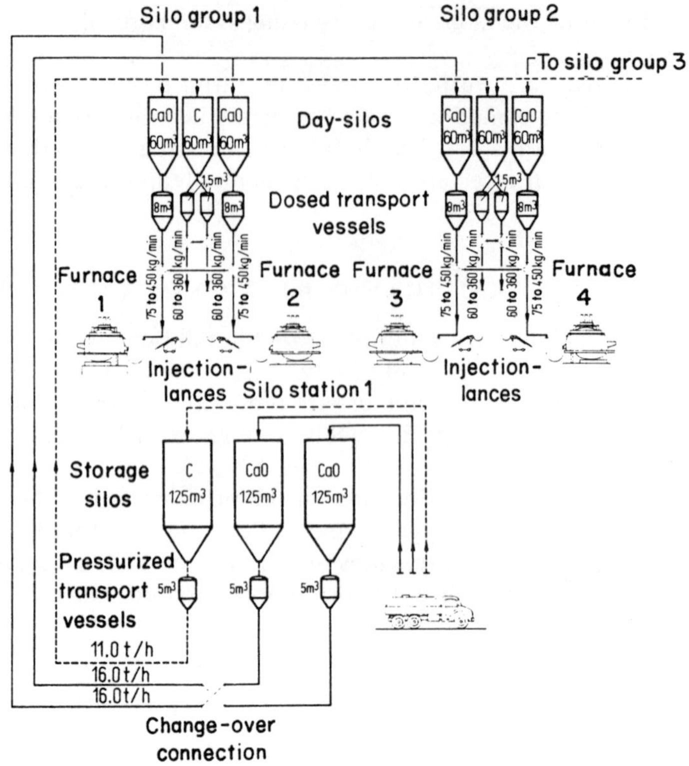

Figure 20. System for storing and pneumatically injecting lime and coal[19]

It is assumed that lime is stored for 3 days and the other materials are stored for 14 days.

The bunkers may be replenished either direct from transporter containers or by conveyor belt with a special feeder-installation. Another possibility is to feed fine lime, fine coal, and breeze pneumatically,[37] in which case special storage silos, capable of being filled pneumatically, must be provided. Figure 20 illustrates a fully automatic pneumatic system capable of moving materials over various distances.

10.7 SCRAP-PREHEATING PROCESSES

Steelworks located in the north have sometimes suffered explosions and serious consequences during post-charging; this was due to moisture in the scrap — mostly snow and ice. In order to avoid such incidents the scrap must be dried. Some arc-furnace operators preheat the charge up to 800°C because they expect this to result in savings in overall energy costs.[38,39] They also expect preheating to result in shorter charge-times, lowered electrode-consumption, increased wall-lining life, and reduced hydrogen contents in the steel produced. Figures 21 and 22 illustrate systems for preheating scrap in baskets. Investment costs can quickly be recovered in locations where primary energy is cheap.[39,40]

The idea of using the hot furnace exhaust gases to preheat scrap led to an installation illustrated in Figure 23.

Preheating scrap to 1000°C requires additional heat using gas or oil. This can be carried out in a revolving cylindrical furnace where the scrap is passed against the flow through the revolving furnace and is then fed continuously to the melting furnace via a hole in the furnace roof.[41] In a further development, shown in Figure 24, the revolving furnace is replaced by a rotating hearth furnace mounted 7–8 m high above the arc furnaces. The scrap takes about 20

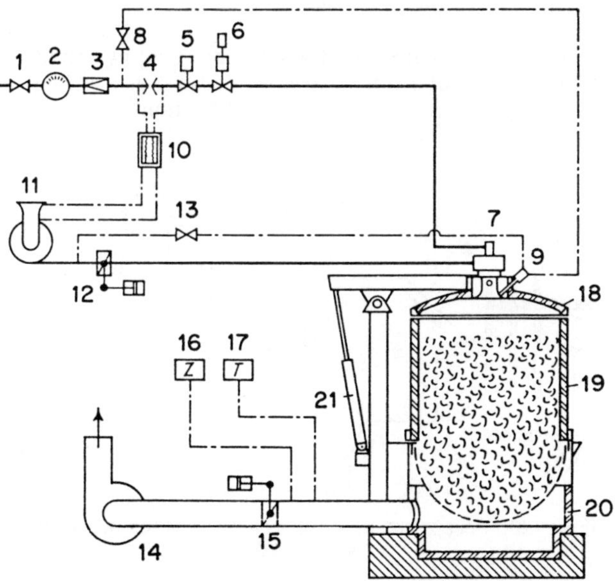

Figure 21. Schematic diagram of a scrap pre-heating system with refractory-lined scrap baskets.[39] 1 Main natural gas valve; 2 gas-meter; 3 reducing-valve; 4 restrictor; 5 fast-acting shut-off valve; 6 regulating valve; 7 burner; 8 ignition shut-off valve; 9 pilot burner; 10 air- and gas-recorder; 11 air-compressor; 12 combuston air throttle; 13 pilot burner air shut-off valve; 14 exhaust-gas fan; 15 exhaust-gas throttle/damper; 16 draught-meter; 17 exhaust-gas thermometer; 18 cover; 19 basket; 20 basket bottom; 21 hydraulic cover lift

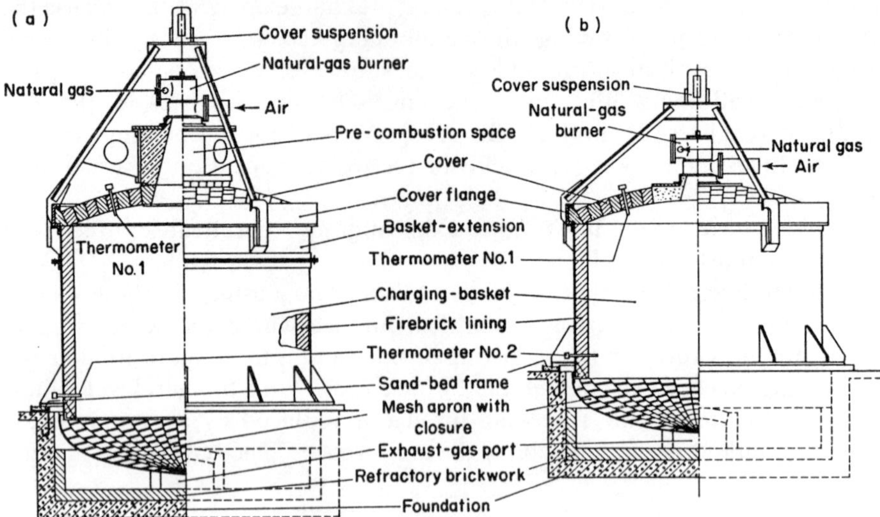

Figure 22. (a) Cross-sectional view of a scrap pre-heating installation with raised scrap basket and pre-combustion space for natural-gas burner. (b) The same installation after modification[40]

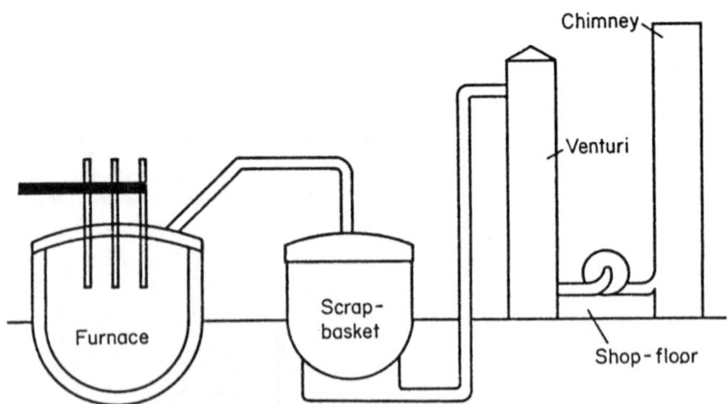

Figure 23. Scrap pre-heating system at Dunswart, Benoni, South Africa

min to pass counter-clockwise through this furnace; during the last 10 min the scrap is heated up to about 1200°C by the arc furnace exhaust-gas and supplementary burners. Oxidation can largely be avoided by heating the scrap up rapidly. The heated scrap is then fed through a tube and a shaker conveyor and drops directly into the arc furnace through a connection in the furnace roof.

The idea of the entire system is to feed scrap continuously through the arc furnace roof and to keep the melting process running with a continuous liquid bath of from 7–15 t. Unlike with the usual system, the vessel of each 80-t arc

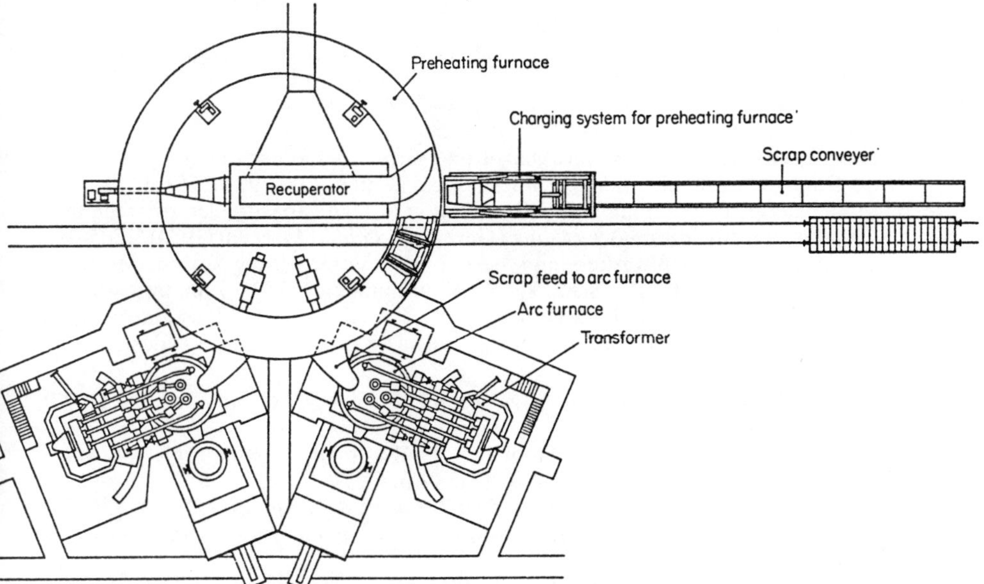

Figure 24. Plan view of the Nuova Scretti S.P.A. scrap pre-heating system at Villadossola, Italy

furnace rotates slowly about its vertical axis throughout the entire melting-down process. The scrap must always drop into a liquid bath; this is ensured by appropriately controlling the supply of energy and of scrap to suit the rotating arc furnace vessel.

The Klöckner–Stahlerzeugung (KS) process goes even further and melts the scrap down ouside the arc furnace.[42] A column of scrap is melted down continuously in a shaft furnace by oil–oxygen or gas–oxygen burners in the lower part of the furnace. The molten metal flows continuously out of the shaft furnace into an arc furnace, where it is post-treated and raised to the required tapping temperature. The development of this process and the technical possibilities it gives rise to are discussed in the literature.[43,44]

A similar principle underlies the direct use of oil–oxygen burners (jet-burners) in an arc furnace (see section 5.6.1.2). The accumulation of successful experience of this process is diminishing the importance of preheating scrap outside the furnace, except where it is necessary because of snow, ice, or oil in the scrap.

10.8 LITERATURE REFERENCES

10.1 The Planning of Arc Furnace Steelworks

1. Lüth, F. Überblick über die bisherige Entwicklung der Ministahlwerke. *Stahl u. Eisen* **92** (1972), 364–8.

2. Blecher, F., Investitions- und Verarbeitungskosten von neuzeitlichen Ministahlwerken. *Klepzig Fachber.* **81** (1973), 430–6.
3. Sperl, H., Das neue Ministahlwerk der Georgetown Texas Steel Corporation in Beaumont. *Stahl u. Eisen* **97** (1977), 686–8.
4. Bauer, M., A method to represent and compute sequences of operations (German). *Stahl u. Eisen* **96** (1976), 1050–5.
5. Dlaska, H., Bestimmung von Anzahl und Große der Elektrolichtbogenofen für ein Edelstahlwerk mit einem Rechnermodell. *Stahl u. Eisen* **95** (1975), 1135–8.
6. Möllenkamp, F. W., Planning and operation of high-capacity electric steelplants with facilities for continuous casting (German). *Stahl u. Eisen* **90** (1970), 1205–14.
7. Baukosten von Stahlwerken und betriebliche Verarbeitungskosten der Stahlwerksverfahren. H. R. Schenk, K. Consemüller, H.-J. Zimmermann, E. Quadflieg, P.-D. Busch together with T. Hinckeldey. Düsseldorf 1970. (Stahleisen — Relationships between technical and economic aspects described.)

McManus, George, J., Profile of a Greenfield steel plant. *Iron Age* **220** (1977), No. 11, 53–65, 68–80.

Hamer, A., and R. P. A. Hoorweg, NKF Staal: a 'mini' mill for 'maxi' requirements. *Wire J.* **10** (1977), No. 9, 138–43.

Britain's private-sector steelmakers. Part 2. *Metals & Mater.* 1977, No. 12, 36–9.

Georgetown Texas — the Korf Group's 'model' plant. *Iron & Steel Intern.* **50** (1977), No. 6, 375, 378–9.

Ministeel works — a discredited concept. *Metal Bull. Monthly* No. 83, 1977, 11–13, 15, 17, 19, 21, 23.

Johnson, Eugene R., and James C. Simmons, CF & J replaces 73 years old open hearth shop with UHP electric arc furnaces. *Iron Steel Eng.* **55** (1978), No. 1, 46–54.

Mini Steel 78. *Metal Producing* **16** (1978), No. 1, 42–9.

Alphasteel unveiled. *Metal Bull. Monthly* No. 87, 1978, 9, 11, 13.

Link, William J., Oregon Steel Mills — An innovator in the steel industry. *Iron Steel Eng.* **55** (1978), No. 7, 29–33.

Collins, Roger, Creating a steelworks overseas. Part 1. *Steel Times* **206** (1978), No. 9, 37–42.

Eketorp, Sven, Decisive factors for the planning of the future steel plants. *Iron & Steelmaker* **5** (1978), No. 12, 37–41.

Good timing blesses J & L's Pittsburgh modernization project. *Metal Producing* **17** (1979), No. 4, 44–8.

Mini Steelworks. *Metal Bull. Monthly* No. 100, 1979, 7–28.

Wheeler, Frank M., and Archie G. W. Lamont, Current trends in electric melt shop design. *Iron & Steelmaker* **6** (1979), No. 1, 36–43.

Mini steelworks — Danieli updates North Americans. *Metal Bull. Monthly* No. 96, 1978, 37, 40–41, 43–5.

Stepper, Gerd, Design and economic construction of electric steel and wire rod and bar mills. *Metallurg. Plant & Technol.* **3** (1980), No. 2, 12, 16, 18–19.

Stepper, Gerd, Design and economic construction of electric steel and wire rod and bar mills. *Wire Ind.* **47** (1980), No. 555, 183–5, 189.

10.2 Principal Types of Arc Furnace Steelworks

8. Leitner, F., Neue Wege für den Bau von Schmelzanlagen eines Edelstahlwerkes, Confidential Report VDEh No. 58, 1944, 7 pp.
9. Thomas, P. M., and A. MacNaughton, Installation and commissioning of an 80-ton

electric arc melting furnace. *J. West Scotl. Iron Steel Inst.* **67** (1959/60), 87–126.
10. Pungartnik, K., Erfahrungen mit den 120-t Lichtbogenöfen des Hüttenwerks Rheinhausen. *Berg- u. hüttenm. Mh.* **106** (1961), 286–99.
11. Kobrin, C. L., The big surge of 'mini' steelplants. *Iron Age* **200** (1967), No. 21, 68–75.
12. Chvoščinskij, A. V., A. P. Kirillov, B. E. Tarasov and Z. A. Zarachova, Elektrostahlwerke mit 100- und 200-t-Öfen. *Stal* **28** (1968), 999–1002; *Stal in Deutsch* **9** (1969), 254–9.
13. McKeever, P. L., A. M. DiGioia and M. A. Czaruk, Unique features of Armco–Butler's electric furnace melt shop. *Iron Steel Eng.* **49** (1972), No. 6, 75–86.
14. Harrison, W. L., Electro-heating plant for the steel industry. *Steel Times* **200** (1972), 535–6.
15. Maas, R., T. Legrand and F. Hauzeur, Das neue Elektrostahlwerk mit Stranggießanlage der S. A. Cockerill-Seraing. *Metallurg. Rep. CRM* No. 38, 1974, 11–21.
16. Thomson, K., and E. Dagnaes-Hansen, Neues Elektrostahlwerk in Frederiksvaerk (Denmark). *Acier Stahl Steel* **40** (1975), No. 6, 206–19.
17. Petocchi, B., V. Picardi and A. Pollarolo, Neues Elektrostahlwerk bei den Dalmine-Stahlwerken. *Boll. tecn. Finsider* No. 352, 1976, 377–88.
18. Reuter, H. J., Das Elektrostahlwerk Lingen der Benteler Gruppe. *Fachber. Hüttenprax. Metallweiterverarb.* **15** (1977), 851, 857–60.
19. Wolters, G., Extension of the steel industry of Venezuela — description of the steelmaking plant of SIDOR (German). *Stahl u. Eisen* **97** (1977), 1145–53.
20. Brusa, U., Angewendete Grundsätze für die Planung und energiemäßige Optimierung eines Elektrostahlwerkes und Walzwerkes zur Herstellung von jährlich 500 000 t Walzstahl. *Metallurg. Ital.* **69** (1977), 1–3.
21. Europas modernstes Elektro-Lichtbogenofen-Stahlwerk in Betrieb. *Fachber. Hüttenpraxis, Metallweiterverarb.* **16** (1978), 406–9.
22. Muhlrad, W., Rauchgasbeseitigung bei elektrometallurgischen Öfen. *Chal. & Ind.* **41** (1960), 237–55.
23. Hohenberger, A., Entstaubung von Lichtbogenöfen. *Stahl u. Eisen* **81** (1961), 1001–5.
24. Erni, E., Konstruktive Besonderheiten der 40-t-Lichtbogenöfen des Stahlwerkes Gerlafingen der Ludw. von Roll'schen Eisenwerke AG. *Berg- u. hüttenm. Mh.* **106** (1961), 280–6.
25. Squires, B. J., Approach to arc-furnace fume control. *Foundry Trade J.* **113** (1962), 499–501.
26. Harms, F., and W. Riemann, Messung der Abgas- und Staubmengen an 70-t-Lichtbogenöfen bei teilweiser Anwendung von Sauerstoff. *Stahl u. Eisen* **82** (1962), 1345–8.
27. Muhlrad, W., Problems concerning dust removal in flue gas from arc furnaces (German). *Stahl u. Eisen* **33** (1963), 921.
28. Kahnwald, H., and O. Etterich, Estimation of flue gas quantity, compositions and temperature and dust evolution in the melting and refining with gaseous oxygen in a 15 ton arc furnace (German). *Stahl u. Eisen* **83** (1963), 1067–70.
29. Bintzer, W. W., Design and operation of a fume and dust-collection system for two 100-ton electric furnaces. *Iron Steel Eng. 41* (1964), No. 2, 115–23.

10.3 Construction of New Arc Furnace Steelworks

30. Eßmann, H., Limits for design and layout of high-powered large arc furnaces (German). *Stahl u. Eisen* **97** (1977), 577–81.

31. Worthington, E., Scrap handling revolution. *Steel & Coal* **187** (1963), 744–6.
32. Bleibe jr., W. W., Automatic scrap handling — Lukens Steel, Coatesville, Pa. *Iron Steel Eng.* **52** (1975), No. 9, 29–34.

10.5 Environmental and Health and Safety Aspects

33. Hausigk, D., and K. H. Giese, Environmental control and ergonomical measures at the construction of a new electric steelmaking plant (German). *Stahl u. Eisen* **97** (1977), 69–74.
34. Scheffler, G., and G. Dikta, Ultra modern electric melting shop reduces impact on the environment (German). *Elektrowärme,* Issue B, **33** (1975), 234–42.
35. Graf. H., F. Josten and G. Meinshausen, Possibilities of efficient and low-pollution steelmaking in the arc furnace (German). *Stahl u. Eisen* **96** (1976), 607–11.
36. Krupp Stahlwerke Südwestfalen AG und Krupp Industrie- und Stahlbau, Planung und Inbetriebnahme eines Hochleistungslichtbogenofens mit Prozeßsteuerung für die Edelstahlerzeugung. *Fachberichte Hüttenpraxis-Metallweiterverarbeitung*, Sprechsaal Verlag Coburg, Issue 10, Oct 1978, pp. 780–8.

Wheeler, Frank M., and Archie G. W. Lamont, Current trends in electric meltshop design. *Proc. Electr. Furn. Conf.* **36** (1978), 139–47.
Baum, Rudolf, Description of the plant and mode of operation of the arc furnace meltshop of Krupp Stahlwerke Südwestfalen AG (German). *Stahl u. Eisen* **100** (1980), No. 10, 517–21 (Stahlwerksaussch. 1038).

10.6 Storage and Transportation of Additives

37. Fenne, M., Pneumatische Zugabe von Kalk und Kohle in Lichtbogenöfen. *Stahl u. Eisen* **93** (1973), 741–2.

Better control of alloying additions. *Iscor-News* **42** (1977), No. 9, 5–6.

10.7 Scrap-preheating Processes

38. Thielker, K. H., Installations for preheating the scrap in electric steelplants (German). *Stahl u. Eisen* **90** (1970), 526–9.
39. Schoenmaker, O. D., II. Operational results of a scrap preheating plant with scrap baskets lined with refractories (German). *Stahl u. Eisen* **90** (1970), 530–4.
40. Schmidt, F., III Operation and results of scrap preheating plants for 10- and 15-ton electric arc furnaces (German). *Stahl u. Eisen* **90** (1970), 534–9.
41. Neumann, F., H. Leu, R. Ptach and U. Brusa, The BBC–Brusa steelmaking process. *Stahl u. Eisen* **95** (1975), 16–23.
42. Klöckner-Youngstown-Stahlerzeugungsverfahren. *Stahl u. Eisen* **96** (1976), 464–5.
43. Langhammer, H.-J., and H. G. Geck, Development of a continuous melt-down process (German). *Stahl u. Eisen* **92** (1972), 501–18.
44. Langhammer, H.-J., Verfahrenstechnische Lösungen von Schrotteinschmelzproblemen bei der Stahlerzeugung. *Arch. Eisenhüttenwes.* **43** (1972), 159–69.

Balbi, M., G. Caironi and G. Tosi, Application of oxy-fuel-burners in arc furnaces for steelmaking (Italian). *Metallurg. Ital.* **71** (1979), No. 4, 140–4, 153.

Electric Furnace Steel Production
Edited by E. Plöckinger and O. Etterich

11 Electric Furnaces in Steel, Grey-cast, and Malleable Cast-iron Foundries

FRANZ NEUMANN, DORTMUND

The output of the foundry industry in the Federal Republic of Germany is about 4–5 million tonnes per annum or 70 kg of cast iron per head of population per year. This represents one-tenth of the total domestic steel-production of about 700 kg per person. These figures appear to be approaching the saturation limit for industrialized nations, since they have changed little in West Germany and the USA over the last few years. This ratio provides a measure of the melting capacity of a foundry, which operates on a lower level than a steelworks in terms of the number and capacity of its furnaces. If we assess the relative apportionment of the production of cast steel, grey-cast, and malleable cast iron in foundries, then we clearly see that grey-cast iron takes up the largest share while steel castings only amount to about 9% and malleable cast iron amounts to about 6% of the grey-cast iron total (including spheroidal cast iron). Because of the economic significance of this relationship, this chapter is slanted accordingly and concentrates on the use of electric furnaces in grey-cast iron production.

11.1 FURNACE SYSTEMS AND PROCESS TECHNOLOGY

11.1.1 Melting Furnaces for Steel and Iron Castings

The basic principles of using electrical energy to melt iron and steel were discussed in detail in Chapter 4. Figure 1 summarizes the four main methods schematically once again; of the four, induction heating and the arc principle are most commonly used in practice. The graphite rod furnace uses resistance heating and belongs to the indirectly heated group (principle C); this is of minor importance in practice, and it is shown enclosed by broken lines in the

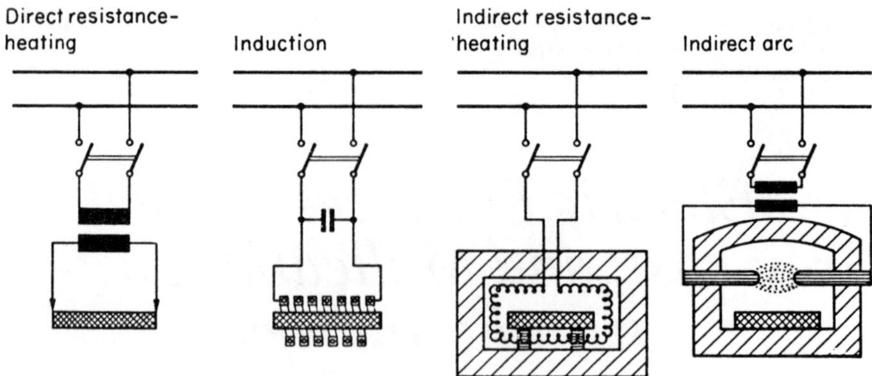

Figure 1. Physical possibilities for electric heating[1]

summary in section 11.1.3. It is used mostly for laboratory work and in special cases for melting small amounts of steel or cast iron. Nevertheless, graphite-rod heating has been used very recently for large holding vessels, with a capacity of up to 30-t, for cast iron,[2] and the method might be developed in the future as an alternative to the induction-channel furnace.

In recent years the oil or gas-fired rotary drum-type furnace[3] has had limited application for melting cast iron but not steel; however, it is confined to smaller installations which are mostly used in place of cold blast furnaces.

The Siemens–Martins open-hearth process, which was formerly used in grey-cast foundries, and the Bessemer process used in steel casting have now lost almost all their former importance. The Bessemer process can only be operated in duplex systems, since the Bessemer converter requires a molten iron charge which is premelted in a cupola furnace and refined in the converter.

Today, foundries mainly use the cupola furnace, the induction furnace, and the arc furnace; the latter predominates in steel foundries with the induction furnace, while grey-cast iron foundries use induction furnaces to an increasing extent as well as cupola furnaces.

11.1.2 Production of Castings

Just as changes have occurred in steel production processes since the Second World War, the melting process for steel-casting production has changed in parallel. Figure 2 illustrates the evolution of the melting process for steel castings in the Federal Republic of Germany since 1950 and the overwhelming significance attained by electric furnace processes in this area. Of the electric processes, arc furnaces take up about 70% of the total and induction crucible furnaces take up the remainder. The open-hearth furnace can no longer provide the flexibility required for the frequently changing rhythm of

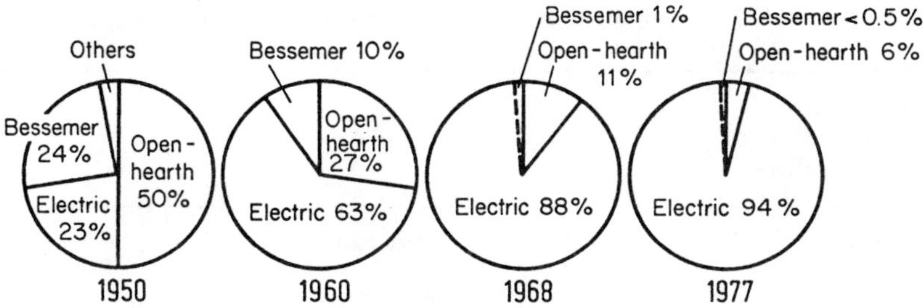

Figure 2. Developments in the distribution of melting processes for steel-casting production in West Germany[4]

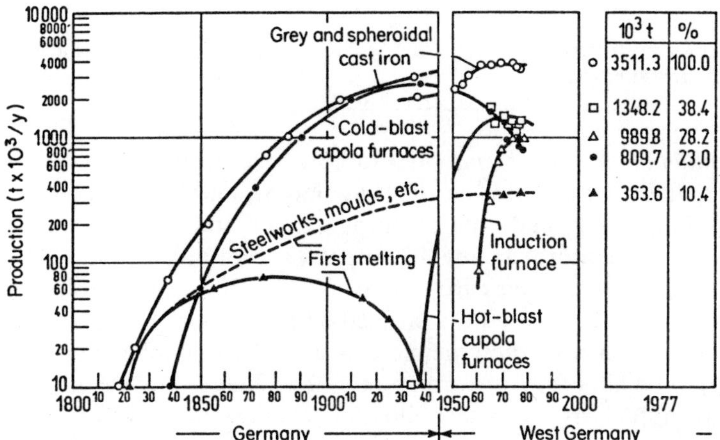

Figure 3. Casting production in Germany and life-curves of the relevant melting processes from the records of the Federal Office of Statistics, Wiesbaden[4]

the modern foundry nor the swift change in analysis and temperatures required, and its rigid operating characteristics make it uneconomic.

Figure 3 illustrates the development of total cast iron production in the Federal Republic of Germany. This also shows the steady increase in the use of electric furnaces and the simultaneous decline in the use of cupola furnaces. Since arc furnaces are only used in a few instances for grey-cast iron in Germany, the electric proportion consists almost entirely of induction crucible furnaces.

11.1.3 Simplex Operation

'Foundry iron' today is largely made in simplex processes; this means that the melt is finished in one vessel, so that it can be poured directly into the mould, even if it is subject to ladle post-treatment. The most important simplex melting processes are listed in order of their present importance in Table 1.

Table 1. Simplex melting processes for producing castings

Grey cast iron Spheroidal cast iron Malleable cast iron	Steel castings
Cupola furnace Hot blast Cold blast	Arc furnace
Induction crucible furnace Coreless, mains frequency Medium frequency (special cases)	Induction crucible furnace Coreless, often medium-frequency since cold start needed
Arc furnace Rotating-drum furnace Graphite rod furnace	Reverberating hearth furnace

The *Induction crucible furnace* has rapidly become established in foundries. Its advantages, which are particularly important for foundry work, are flexibility in quantity and type of output, quick and accurate adjustment of quality and temperature, and environmental acceptibility.

Mains-frequency furnaces are usually used for grey-cast iron foundries, not least for economic reasons; these furnaces must be operated with a liquid initial bath of at least 35% of the crucible capacity. This enables quasi-continuous operation to be adopted, as is frequently desirable in grey-casting but is often impossible in steel casting, where alloys are changed more frequently. For such cases a medium-frequency furnace is used; although this is somewhat more costly to run, it enables single-cast operation to be adopted.

For grey-cast-iron melting, induction crucible furnaces are almost always lined with acid ramming compound. This has particularly good expansion tolerance and crack-resistance in comparison with possible alternatives, but it is sensitive to slags containing FeO and basic ingredients.

For melting steel, acid furnace linings are limited by their low temperature-resistance. Extended operation at temperatures above 1580°C already produces too high a loading which leads to accelerated lining wear, especially during oxidation (see sections 6.1.2.2 and 6.1.6).

In West Germany, the *arc furnace* has become established only in steel foundries but is hardly ever used for grey-cast iron. This is explained by history, since German steel foundries were already operated with arc furnaces before the development of induction furnaces, and so further development of electric melting there proceeded from that basis. On the other hand, the cupola furnace had a secure place for grey-cast iron in Germany. However, the arc furnace is often used for grey-cast iron in the USA and a few other countries;[5a,5b,5c] the reason for this may be traceable to the fact that no development work on induction furnaces was carried on in those countries at first. Altogether, iron production from arc furnaces in the USA increased from 1 million tons in 1970 to 2.9 in 1976 and was expected to reach 4.3 million tons in 1981. The development of the high-performance arc furnace and the good durability of acid linings when melting simpler types of iron both played a part in this expansion. For steel casting the arc furnace offers the advantage of robustness, while its basic lining presents no difficulties; there are few problems with the charge-material, and there are many metallurgical possibilities. In its present form it is nevertheless mostly used in single-cast operation and is considered primarily as a melting machine. Further differences between arc and induction crucible furnaces are dealt with in Chapter 6. In conclusion, we can add that, in comparison with the arc furnace used in a foundry, the induction furnace is more flexible and more environmentally acceptable; these advantages can be decisive in foundry work.[7]

11.1.4 Small Duplex Systems

In *duplex operation* the melting process is divided into two stages carried out in separate units. Melting down takes place in the first stage; the second stage is for post-treatment.[7]

One can only speak of a 'small duplex process' in foundry operation if the second stage is used as a buffer for soaking or holding the melt without any further treatment. This form of operation is used in different ways.

Without doubt, the most common form is the combination of a cupola furnace with a holding furnace. In this way the necessarily continuous operation of the cupola furnace can best be matched to the varying rhythm of foundry operation. The holding furnace separates the actual foundry operation from the melting process; this is naturally also an advantage in smoothing the operation when using an induction or arc furnace. We thus have the following combinations:

Melting	Accumulation Holding
Cupola furnace	Forehearth (gas-heated), channel-type induction furnace
Arc furnace	Channel-type induction furnace
Induction crucible furnace	Channel-type induction furnace

The forehearth used in earlier times is being increasingly replaced by the induction-heated channel furnace, because its temperature can easily be controlled and it can tolerate a certain amount of overheating. Thus casting conditions can be kept constant, even with lengthy interruptions in melting, and this improves the certainty with which particular quality standards can be maintained. However, flexibility in handling quality changes is limited because of the large quantity of stored melt.

A channel-type induction furnace can either be included directly after the cupola furnace, as a substitute for a forehearth, or else placed on the casting line to supply the castings there.

Therefore the holding furnace:

(1) Is used as a storage vessel for equalizing temperature and analysis and carrying out any superheating;
(2) Increases foundry capacity by enabling premelting and replenishing to be carried out outside casting time;
(3) Saves energy costs by using a cheap night-electricity tariff;
(4) Maintains steady casting conditions, even with interruptions in melting operations; and
(5) Improves matching of the melting and casting operations.

Another form of small duplex system involves two or more induction crucible furnaces, powered by a single electricity supply which can be switched from one furnace to another. Here, melting takes place in one furnace while the other acts as a holding and casting furnace. Heat-losses from the latter are restored either by periodic, short-term switchover of the electricity supply normally used for melting or by connecting in an additional holding power supply.

At this point we must also mention induction-heated (channel-furnace) foundry systems,[8,9] which are being introduced in increasing numbers in the course of progress in mechanization and automation. Two systems are in use today. In one type a plug in the bottom of the storage vessel opens and closes to provide a runout for casting; in the other type, the pressure in a tea-pot-shaped vessel is controlled in such a way that a rise in pressure operates a siphon overflow in the outlet.

Both systems are positioned right at the casting line; however, they are not used only for casting but also for storage and holding in order to allow for interruptions in casting. The capacity of such casting installations now in use is between 1 and 10 tonnes.

11.1.5 Large Duplex Systems

This system consists of *two melting units* in tandem; its economics are widely disputed. Technically, the system offers the possibility of exploring the advantages of both units used: for example, economical melting down in a cupola furnace is followed by versatile, quick alloying, and refining in an induction furnace; in such cases the induction furnace is designed with a reduced electrical power supply. The economics of each individual case should be assessed from the point of view of the energy situation, production rate, the type of metal to be produced, etc. In a large foundry it may be economical to use a large central melting unit followed by several small induction furnaces for final preparation of the different melts and their precise adjustment to the required conditions. The following duplex systems have been proved effective in practice:[7]

Melting down	Post-treatment
Cupola furnace	Induction crucible furnace(s)
Arc furnace	Induction crucible furnace(s)
Blast furnace	Induction crucible furnace(s)

The total effect and the advantages and disadvantages of these and other possible combinations can only be assessed in the light of local conditions and intentions.

Use of the arc furnace as a post-treatment and holding unit cannot be recommended for a foundry. Therefore this possibility has not featured in the combinations discussed above.

11.1.6 Energy Consumption

The energy required to melt down metals and their alloys depends on what is called the enthalpy (or heat content) of the substance in question. Figure 4 shows the enthalpy of cast iron, steel, and other materials, expressed in kWh/t. Taking into account the total losses, which can be taken in rough approximation to be 30% for electric melting, the total energy-consumption can be estimated quite well. The energy consumed by a cast iron melt with a heat content of 387 kWh/t at 1500°C can be calculated from the enthalpy to be 552 kWh/t; however, in foundry melting this is sometimes greatly exceeded and may amount to 650 kWh/t or more.

High values like these call for careful checking, since they indicate poor matching of melting and casting operations; the melt may possibly always be ready at the wrong time, and long waiting-times may be evident.

The energy budget for a melting furnace can be examined in summary by means of an energy balance such as the comparative energy flow diagrams for an induction crucible furnace and an arc furnace shown in Figures 5 and 6. It can be seen from the diagrams that the energy consumed by the arc furnace[7,11] is virtually the same as that for the induction crucible furnace[7,12] with the same throughput and in identical conditions; this is also borne out in practice.

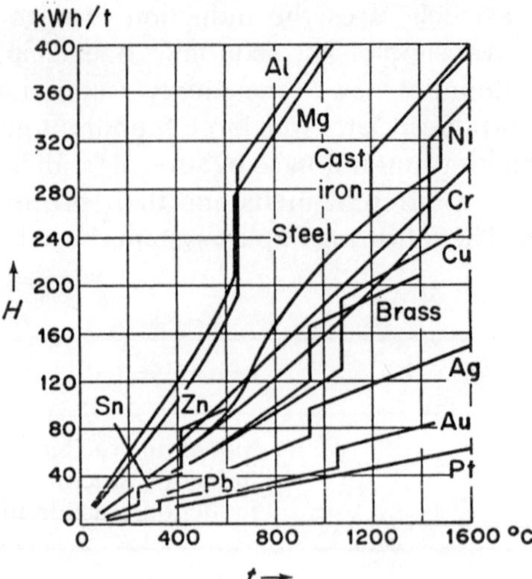

Figure 4. Heat content (enthalpy) H of different metals and of steel, cast iron, and brass[1]

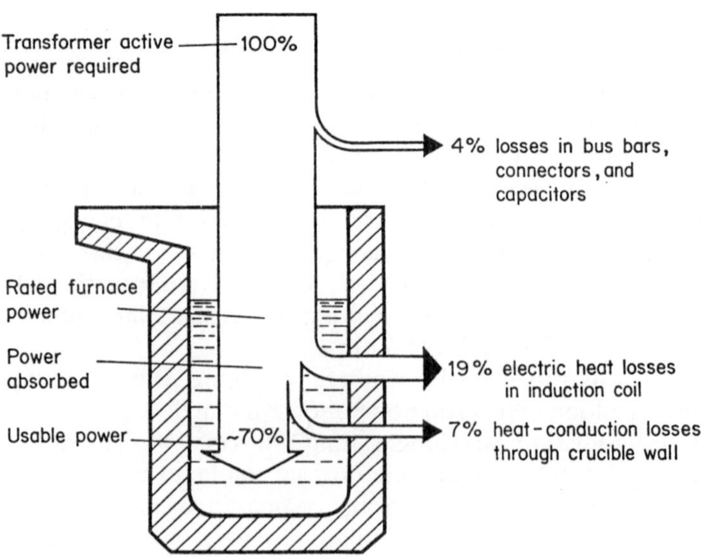

Figure 5. Energy flow diagram for an induction crucible furnace[10]

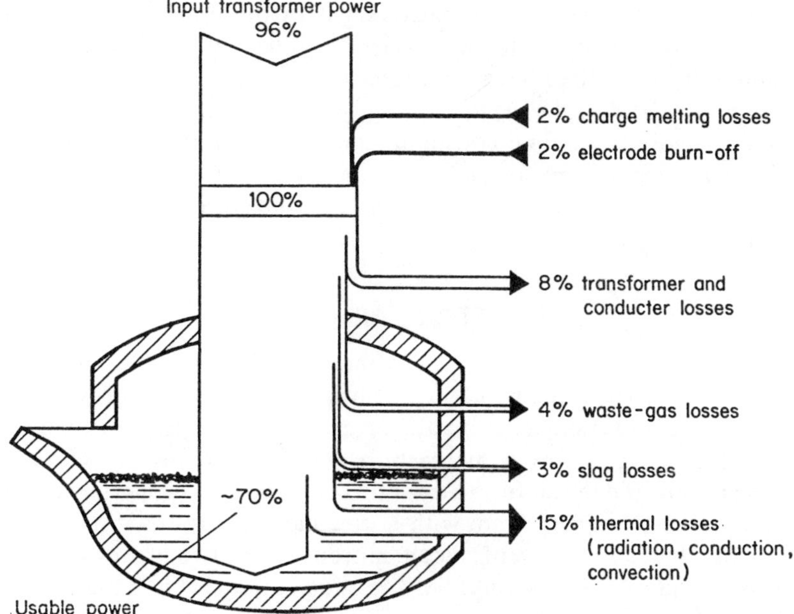

Figure 6. Energy flow diagram for an arc furnace used for remelting

11.2 METALLURGY AND MELTING TECHNIQUE IN FOUNDRIES

In a foundry the electric furnace is used mainly as a remelting unit, so that metallurgical composition melting will not be necessary, except for special cases. This means that a slag reaction is not normally needed, and the furnace can be acid-lined — in so far as the required working temperature allows this.

An arc furnace for melting steel is almost always basic-lined; this enables any possible metallurgical processing to be carried out if needed. The induction crucible furnace is less suitable for this (see Chapter 6).

In a grey-cast iron foundry, melting is almost always carried out with an acid lining because of the advantages already mentioned. In such circumstances a 10-t induction furnace achieves a throughput of 3500 to 4000 tonnes in its life, with a lining wear rate of about 1 to 3 kg/t.[7] The wall and hearth of a 10-t arc furnace can achieve a life of more than 2000 charges, and a silica roof lasts over 130 charges;[5] changing the roof lining to an 80% alumina material increases roof life to 500 charges. However, arc furnace linings wear at a distinctly higher rate than in induction furnaces (6–15 kg/t[7]), and it can be seen that a long life of 2000 charges can only be achieved by regular patching during idle times. Melting spheroidal cast iron results in more attack on the lining, so that the wear rate here is correspondingly increased.

It should be noted here that aggressive elements such as lead, zinc, and magnesium can cause considerable damage to the crucible. It is therefore necessary in certain cases to have slag reactions for particular melts, and attention is drawn to the relevant literature.[12]

In general, metallurgical slag reaction work with acid linings for foundry melting is usually more time-consuming and expensive than preparing clean, sorted scrap; this is true for both arc and induction furnaces. This is why metallurgical work in foundry melting is concentrated on the following processes.

11.2.1 Alloying

Alloying in induction furnaces presents no difficulties because of the intense stirring action; on the other hand, in arc furnaces dissolution and even dispersal throughout the melt take a distinctly longer time, and, for a particular charge quantity (for example, with heavy carburization) the process is relatively more time-consuming and less accurate as well. Temperature is generally important, since the process is speeded up with hotter melts. Carburization takes place more slowly than absorption of silicon; moreover, the two processes accelerate each other, and so silicon should only be added after carburization in order to achieve a better yield.[1,12] Figures 7 and 8 show this clearly.

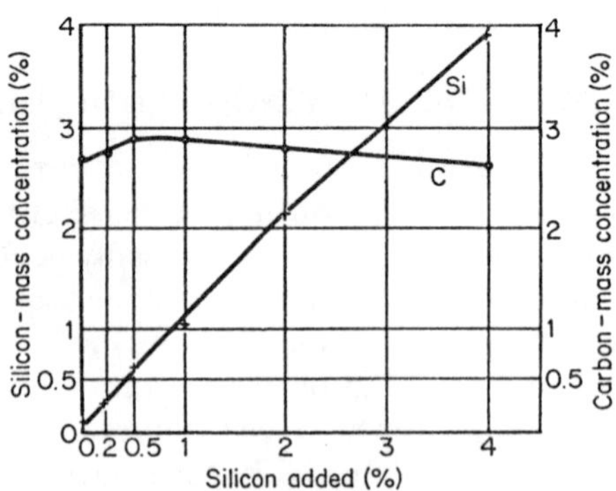

Figure 7. Silicon-absorption in a carbon-rich iron melt[1]

11.2.2 Homogenizing

While normal melting technique is used for producing steel for ingots or continuous casting for subsequent hot deformation, additional metallurgical measures are needed for the production of castings in steel, and also chiefly in grey-cast iron, in order to ensure good castability and optimum cast structure. In this connection homogenizing assumes fundamental importance;[13,14] it is closely associated with oxide skin formation and also nucleation, which has a

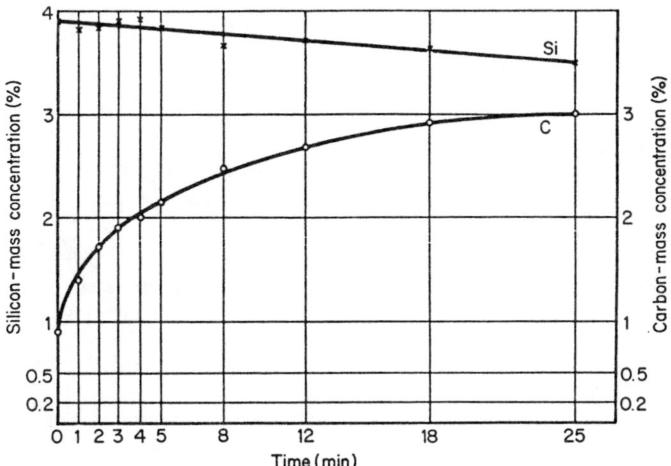

Figure 8. Dependence on time of carbon-absorption and simultaneous loss of silicon in an iron melt.[1] (Initial carbon content 0.94%; 2.5% carbon added)

marked effect on cast structure. These two processes will be dealt with in the two sections after this.

A certain amount of superheating and bath motion are required to homogenize the melt. In the usual remelting processes without metallurgical refining, non-metallic impurities imported with the charge are removed either by coagulation and ascent through the melt or by chemical reduction of their oxides.

It can be shown that certain impurities have an inoculating effect which improves the solidified structure;[15,16] however, their concentration and hence their effect cannot be foreseen. This uncertainty is usually removed these days by homogenizing the melt through moderate superheating before defined and accurate post-treatment.[13] This applies to both steel and grey-cast iron.

In particular, the tensile strength and hardness of castings made from hypereutectic cast iron are improved by homogenizing; this is shown in Figure 9 after research by K. Orths.[14] There is a noticeable jump at a defined level of superheating; this jump depends not only on temperature but also on superheating time. Here the temperature is related to carbon–silicon–oxidation equilibrium, which will be discussed below because of its importance.

Carbon–silicon–oxygen equilibrium and carbon–silicon isotherms also form the important reference basis for oxide-skin formation and nucleation criteria in foundry melting metallurgy, as will be discussed later.

If we begin with the oxidation reactions for carbon and silicon, which are the main components of cast iron in the basic Fe–C–Si system, we can use known thermodynamic data to work out the conditions in terms of temperature and

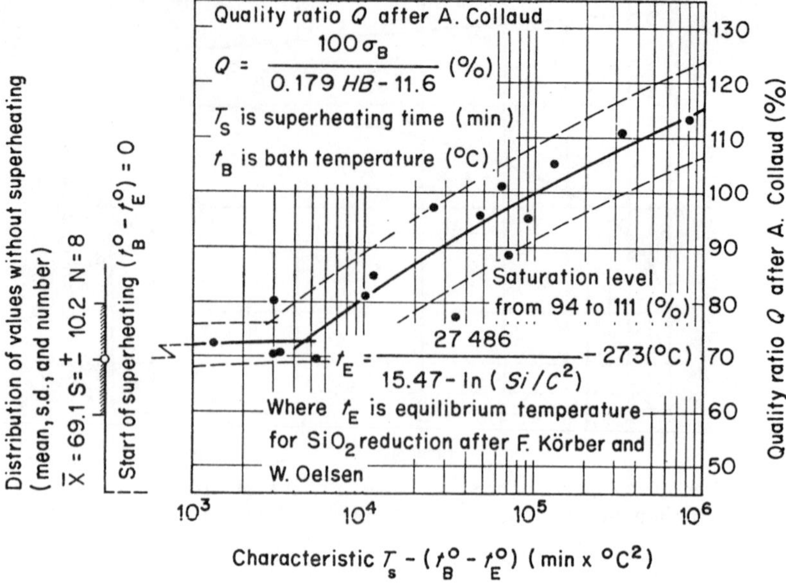

Figure 9. Effect of superheating temperature and time on the quality of eutectic to highly hypereutectic cast iron in first melt[14]

concentration for favouring one of the two oxidation reactions instead of the other.[17] Because the temperature-dependence of the two reactions is opposite (see Figure 10), the characteristics intersect at a particular temperature which indicates the transfer of preference from silicon oxidation to that of carbon. However, it should be noted that, because of the nucleation necessary to form CO, the C–O reaction does not start as soon as equilibrium is surpassed; instead, the melt requires a certain amount of superheat, and then the condition of the crucible surface will have some influence.

The activation energy recognizable in superheating can be expressed in the form of excess pressure for CO evolution p_{co}; thus p_{co} should not be based on 1 at as usually was the practice in earlier representations.[17] Experience shows that superheating by 70–90° is required to start CO-formation;[13,14] this corresponds to a CO overpressure of from 1.5 to 2 at. Therefore a melt only begins to boil in this region, depending on the conditions at the start. These values for steel and cast iron are shown as C–Si isotherms in Figures 11(a) and 11(b). This produces a temperature of 1580 to 1620°C for an unalloyed steel melt containing 0.4% C and 0.3% Si; while for cast iron with 4% C and 3% Si a temperature of 1410 to 1450°C can be estimated.

The CO-formation temperature can be taken as a measure of the superheat required to homogenize the melt. The upper limit for unalloyed steel is 1640°C,

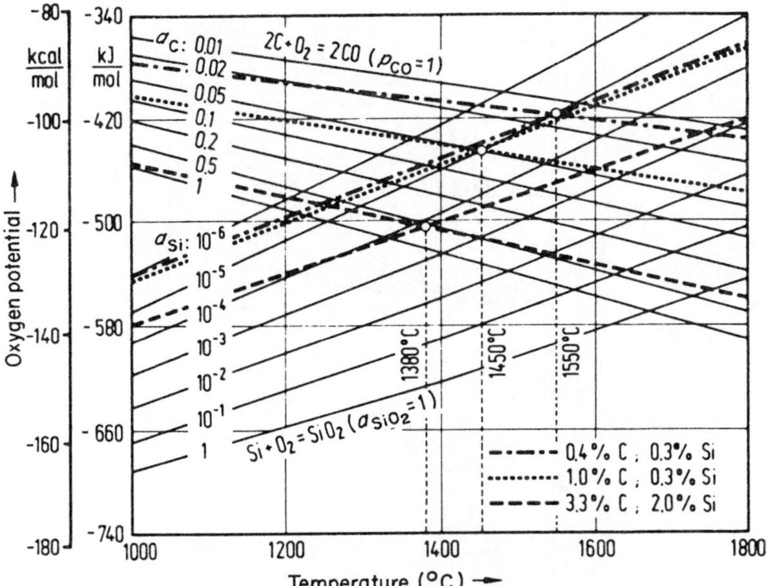

Figure 10. Free reaction enthalpy for carbon and silicon oxidation reactions in relation to temperature for various C and Si activities during melt-down,[17] where p_{co} is excess pressure for CO evolution = 1 at; a_{SiO_2} = 1.

and that for cast iron is about 1470°C; these values should not be appreciably exceeded.

11.2.3 Oxide Skin-formation

Oxide skin forms when highly-fluid SiO_2-based oxides separate out from the melt, mostly in the form of FeO–MnO-silicates, which are responsible for many types of casting defect such as slag-inclusions, frothing, porosity, and others.[13,18] Oxide skin-formation is recognizable as multiple surface activity; this can be illustrated in relation to chemical composition, as in Figure 12.

In foundry practice the oxide skin-formation temperature is kept as low as possible so as to delay separation of silicates as the melt solidifies in the mould and thereby to reduce silicate concentration. The C–Si isotherms show that increasing Si content and reducing C content raise the formation temperature. Sulphur also causes an increase, while manganese lowers the oxide skin-formation temperature.[13,18–20] Besides the effect of the alloying elements, the superheat temperature in particular also has a marked effect. The more the melt is superheated, the more the separation is delayed; i.e. oxide skin-formation will be pushed to lower temperatures.[13,14]

The tendency to oxide skin-formation is recognizable during casting by the

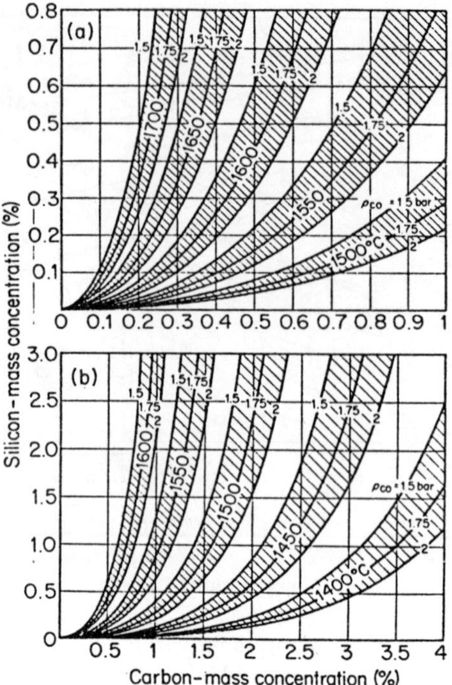

Figure 11. Carbon–silicon isotherms for various p_{co} pressures.[14] (a) Steels up to 1% C and 0.8% Si; (b) cast iron up to 4% C and 3% Si

behaviour of the surface of the melt in the ladle. One objective measure for this is the 'reflectivity number', which results from ascertaining the difference between true temperature and that measured optically.[14]

Casting defects caused by oxide separation are less frequent with electric furnace melts than in iron from cupola furnaces. This is because both the superheat level and the time the melt spends at superheat temperature are usually lower with cupola furnaces.

11.2.4 Nucleation

In recent years the formation of a good, dense, and, above all, fine-grained casting structure has become increasingly important for castings in steel and in grey-cast iron. This calls for a carefully adjusted melting process[21] and optimum nucleation capability.[13]

In steel casting, especially with highly alloyed ferritic and austenitic grades of steel, it is possible to achieve the nucleation required for fine-grained

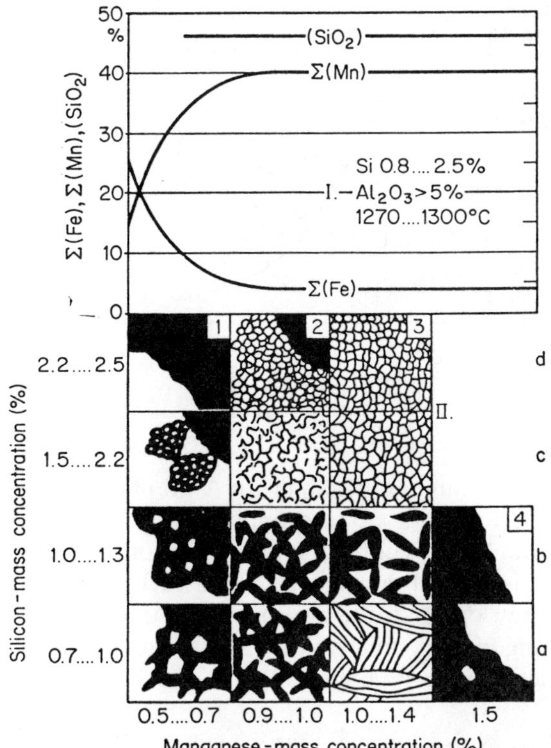

Figure 12. Changes in surface-behaviour figures and slag composition in relation to cast-iron composition[13]

solidification by choosing the right melt superheat temperature and by adding alloying elements to assist nucleation. The theoretical basis for this has not yet been subject to much research; therefore, general processing rules cannot be stated, especially because different inoculation preparations must be used for different steel compositions. Thus, for example, high-melting nitrides, which separate from the melt as it solidifies, have proved particularly effective.

Nucleation during the solidification of cast iron is of even greater importance,[13,21,22] since the characteristic of individual iron castings can vary far more than with steel castings, even if the average composition does not change (see Figure 13). The reason for this lies in the many different structural formations possible when cast iron solidifies; this leads to white or grey solidification and also eutectics in which the number of cells (grain-size) and the graphite formation within the cells can vary widely.

The various nucleation theories give prominence to oxide nucleation theory[13–15] with a certain involvement of sulphur compounds;[24] therefore, in practice melting operations are adjusted accordingly, especially for electric furnaces. The assistance of nucleation is required in order to obtain a fine

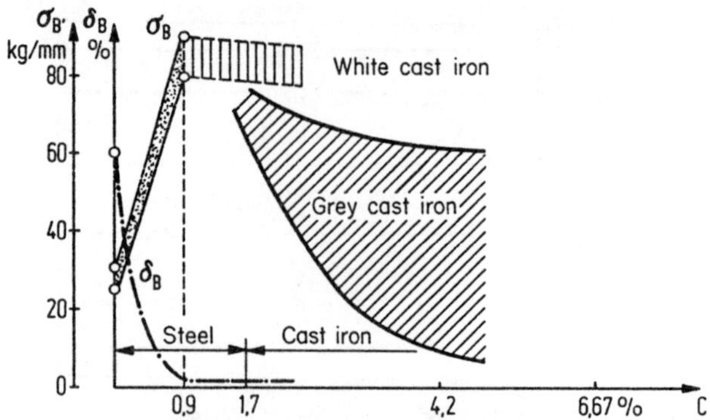

Figure 13. Strength of cast iron and steel in relation to carbon content[23]

grain structure (large number of cells) with A-graphite; this leads to both high strength and low hardness (good workability).

Oxide nucleation theory centres on the formation of oxides on a SiO_2 base, and the graphite is seeded on this. To make such seeds fully effective, it is necessary that they originate in the melt; i.e. they cannot be added from outside. According to oxide nucleation theory, it follows that the oxygen level in the melt at the time when FeSi is added is of decisive importance for oxide precipitation.

K. Orths and K. Weis showed[25] that, because of the relatively high level of nucleation activity during CO bubble-formation, melting steel or cast iron electrically in a relatively narrow temperature band leads to a maximum level of oxygen in the melt. This is indicated by the shaded areas in Figure 14. In this process the CO evolution pressure can rise to 2 at, as was mentioned earlier in the description of C–Si–O isotherms. CO-formation only begins at this stage; it is followed by rapid reduction of the oxygen dissolved in the melt. These relationships are illustrated in Figure 15; they show that, with the oxygen level required, superheating to about 90°C above the equilibrium temperature (C–Si isotherm) can be undertaken before any oxygen is lost.

These relationships must be known so that conditions for nucleation can be optimized. This is illustrated by the 'dead roast' effect,[14] which arises in event of considerable oxygen loss; iron affected in this way will react only weakly to inoculation, if at all. However, such cases occur infrequently and can be corrected, if spotted in time, by raising the oxygen level of the melt during recharging or post-charging or by adding sponge iron.

The best inoculation effect is achieved when superheating in the maximum oxygen area.[13] This type of superheating is also very effective for homogenizing and for retarding oxide skin-formation. However, long soaks in this temperature-region should be avoided, since this type of iron with its elevated

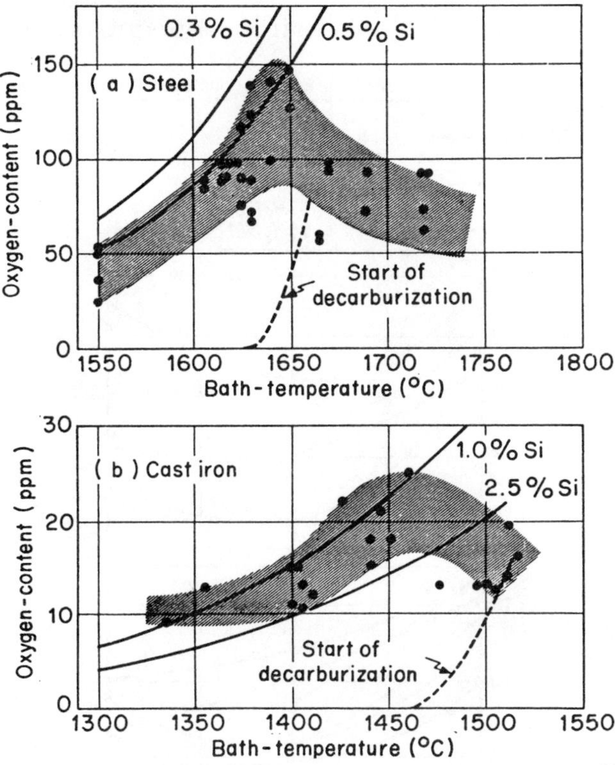

Figure 14. Effect of superheating on oxygen content[25] (a) in steel and (b) in cast iron

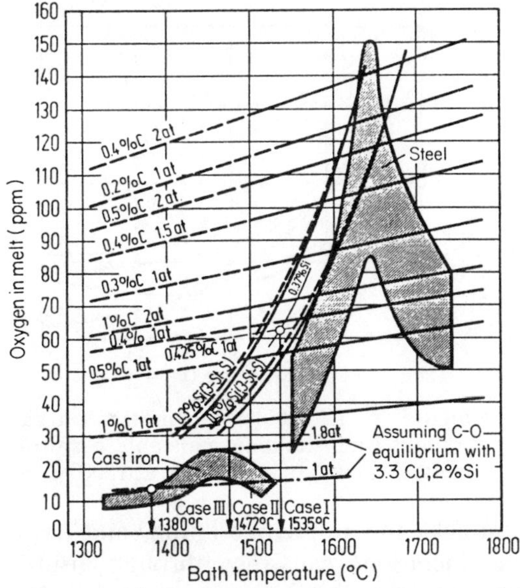

Figure 15. Regulating oxygen-content in melts of cast iron and steel in relation to bath temperature[26]

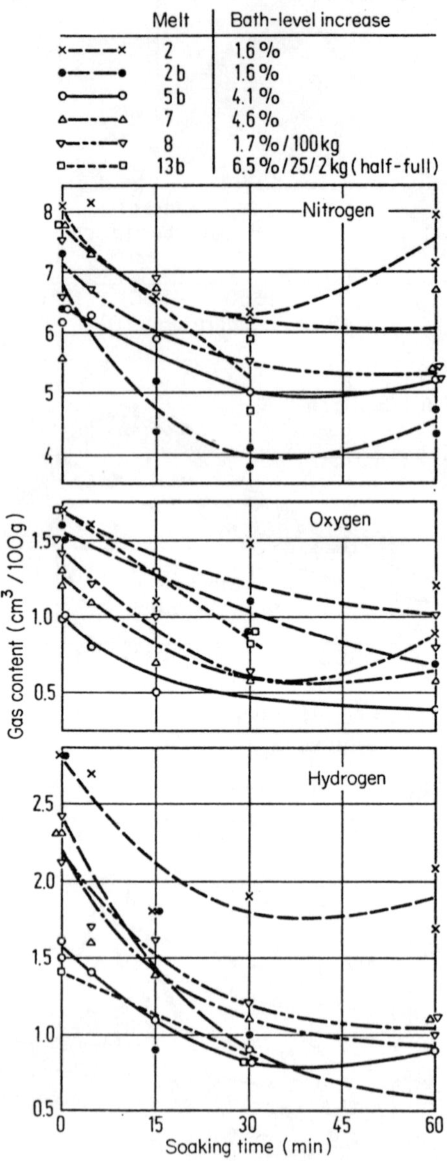

Figure 16. Effect of bath motion on gas content when holding malleable cast-iron melts in an induction furnace at 1500° C[1,12]

oxygen content attacks the refractory lining, and casting defects such as porosity and slag inclusions can arise through insufficient and inexpert inoculation.[14,21]

The tendency to produce such casting defects is still greater if gas such as hydrogen or nitrogen is absorbed. It is true that degassing occurs in induction furnaces. The extent of this depends on the bath motion and the initial gas content imported with the charge; however, the loss can be recovered by gas absorption during extended superheating (see Figure 16). This may be more evident in casting steel, since this metal is more liable to pick up gas because of greater solubility. Atmospheric moisture has some effect on hydrogen pickup, which is more marked with arc furnaces because of dissociation close to the arc.[12]

All the points made in this section on metallurgy and melting technique show how important it is to control both the melting process and also the melt-handling in order to produce satisfactory castings. This is far more important than in a straightforward remelting process. The points are briefly summarized in Table 2 on the next page.

Table 2. Guidelines for melting operations with induction crucible furnaces and their effect on results and casting-characteristics

	Metallurgical melt processing Orientation on C–Si Isotherms		
	Homogenizing	Oxide skin-formation	Nucleation conditions
Definition:	Removal of impurities (suspensions) from the melt and smoothing the temperature and concentration profiles	Separating highly fluid oxides, mostly $FeO–MnO–SiO_2$-based, from the melt during tapping and cooling	Producing a defined melt in respect of inoculability, i.e. response to late inoculation. This is based on knowledge of oxide-nucleation theory in relation to the formation of foreign nuclei (oxides with sulphides) in the melt
Possible defects:	(1) Impairment of casting characteristics because of heterogeneous inclusions (2) Poor control of nucleation because of variable charge quality (inclusions)	(1) Casting defects caused by slag-inclusions partly associated with gas pores (frothiness) (2) Slag build-up in holding vessels and transport and casting systems, causing scars and lining attack	Uncontrolled nucleation in melt leads to: (1) Varying inoculation effects and solidification structures (supercooling), leading to: (2) Large variation in casting properties even with constant macro analysis: (3) 'Dead roast' effect and thus no melt response to inoculation.

Remedies:	(1) Removal of contamination by superheating above C–Si isotherms; associated oxide-reduction (2) Coagulation of suspensions through intense bath motion followed by assumption by slag	Lowering oxide skin-formation temperature; i.e. delaying start of oxide skin formation by: (1) Superheating above C-Si isotherms, causing hysteresis in cooling (2) Raised C content; lowered Si and S content	Controlled melting in respect of melt oxygen behaviour by orientation on C–Si isotherms. Maximum oxygen intake at about 70–90° superheat above C–Si equilibrium temperature at about 1470°C for cast iron containing 3.2% C and 2% Si at about 1550°C for steel containing 1% C and 0.3% Si at about 1630°C for steel containing 0.4% C and 0.3% Si
Aim:	(1) Improving casting characteristics, e.g. by raising Collaud's quality ratio (see Figure 9) (2) Reproducible starting conditions for melt treatment (inoculation)	(1) Reducing the proportion of harmful oxides precipitating (2) Producing an iron which is less liable to oxidize with the possibility of 'long-term killing	There is often an attempt to achieve a fine-grained cast structure with simultaneously good strength and workability (for cast iron this means a large number of cells plus A-graphite)

11.3 LITERATURE REFERENCES

1. Brokmeier, K. H., In *Induktives Schmelzen.* Brown, Boveri & Co. AG. Mannheim. Essen, 1966. pp. 1–2.
2. Schaum, H., Electric resistance heat forehearth new concept in duplexing. *Modern casting* **64** (1974), No. 3, 32–3.
3. Roscrow, W. J., Melting and superheating iron in rotary furnaces. *Brit. Foundrym.* **69** (1966), No. 4, 81–8.
4. Neumann, F., Kriterien und Entwicklungstendenzen der Schmelzverfahren. Paper give before the Fachausschuß Schmelztechnik of the Verein Deutscher Gießereifachleute in Düsseldorf 19 Sept. 1978.

5a. Brown, J. W., A. Bueler and B. P. Courvoisier, Use of arc furnace in foundries (German). *Elektrowärme,* Issue B, **35** (1977), 264–9.

5b. Carmack, Larry G., The control of physical properties in gray iron castings through the use of arc furnaces and the holding furnace. *Iron & Steelmaker* **5** (1978), No. 4, 31–3.

5c. Benyovszky, Moric, *South East Asia Iron & Steel Inst. Quart.* **7** (1978), No. 3, 50–7.

6. Schwabe, W. E., Steelmaking in ultrahigh power electric arc furnaces (German). *Stahl u. Eisen* **89** (1969), 927–37.
7. Neumann, F., Cast iron melting processes compared (German). *Elektrowärme,* Issue B, **32** (1974), 278–82.
8. Wübbenhorst, H., Stand der Mechanisierung und Automatisierung beim Vergießen von Fe–C-Legierung. *Gießerei* **60** (1973), 384–8.
9. Dötsch, E., Anforderungen und Verfahren zum mechanisierten und automatisierten Gießen von Gußeisen. *Gießerei* **64** (1977), 262–7.
10. Pautz, J., Energy consumption analysis in foundries and heat recovery in electric melting shops (German). *Elektrowärme,* Issue B, **35** (1977), 290–4.
11. Neumann, F., H. Leu, R. Ptach and U. Brusa, The BBC–BRUSA steelmaking process (German). *Stahl u. Eisen* **95** (1975), 16–23.
12. Neumann, F., Comparative study of cast iron melting in cupola and coreless induction furnaces (German). *Elektrowärme* **29** (1971), 552–64.
13. Neumann, F., Metallurgische Schmelzführung und ihre Bedeutung für die Treffsicherheit der Gußeigenschaften beim induktiven Schmelzen. Essen, 1972.
14. Orths, K., Zur Kenntnis der Auswirkungen von Schmelzführung und Überhitzung auf einige Fehlererscheinungen und die mech. Eigenschaften von Gußeisen. *Gießerei* **47** (1960), 595–608.
15. Patterson, W., and B. Sigg, Über den Einfluß der Schmelzführung im Induktionsofen und der Schmelzbehandlung auf die Eigenschaften von Gußiesen. *Gießerei, techn.-wiss. Beih.,* No. 25, 1959, 1363–83.
16. Patterson, W., and W. Standtke, Einfluß der Einsatzstoffe der Schmelzführung im Induktionsofen und der Impfbehandlung auf das Gefüge und die mechanischen Eigenschaften von Gußeisen mit Lamellengraphit. *Gießerei techn.-wiss. Beih.,* **15** (1963), 1–24.
17. Neumann, F., and E. Dötsch, Beitrag zur Thermodynamik von Eisen-Kohlenstoff-Silicium-Schmelzen. *Gieß-Forsch.* **27** (1975), 31–8.
18. Dahlmann, A., and K. Löhberg, Schlackenbildung auf Metallschmelzen, insbesondere auf Gußeisenschmelzen und ihre Auswirkungen auf das Gußstück. *Gießerei* **50** (1963), 149–54.
19. Nandori, G., and W. Standtke, Über den Oxidationsverlauf bei Gußeisen aufgrund ungarischer Forschung. *Gießerei* **47** (1960), 535–6.
20. Merz, R., and B. Marincek, Zur Oxidhautbildung verschiedener Roh- und Gußeisenschmelzen. *Stahl u. Eisen* **75** (1955), 196–9.

21. Orths, K., W. Weis and W. Lampic, Verdeckte Fehler bei Gußstücken aus Gußeisen. *Gieß.-Forsch.* **27** (1975), 103–11–.
Einflüsse von Formstoff und Form, Schmelzführung und Desoxidation auf die Entstehung verdeckter Gußfehler bei Gußeisen. *Gieß.-Forsch.* **27** (1975), 113–28.–
Gesetzmäßigkeiten und Zusammenwirken von Regelgrößen bei der Entstehung verdeckter Fehler bei Gußeisen. *Gieß.-Forsch.* **28** (1976), 15–26.
22. Caspers, K. H., Bedeutung der Haltezeit und der Überhitzungstemperatur bei der Erschmelzung von Gußeisen mit Lamellengraphit im NF-Induktionstiegelofen. *Gießerei* **58** (1971), 77–83.
23. Marnicek, B., Lecture given at BBC course Baden/Schweiz 1970.
24. Moore, A., Some factors influencing inoculation and inoculant fade in flake and nodular graphite irons. *AFS Trans.* 73–27, 268–77.
25. Orths, K., and E. Weis, Die Beeinflussung des Kieseläuregehaltes von Gußeisen durch die Schmelztechnik. *Gieß.-Forsch.* **25** (1973), 9–19.
26. Neumann, F., and E. Dötsch, Beitrag zur Thermodynamik von Fe–C–Si-Schmelzen. The metallurgy of cast iron. Lux, Minkhoff, Mollard (Bate. — Geneva), Delta Publishing Com. Vevey 1975.

Gallo, S., Fiat's Crescentino Foundry. *Foundry M & T* (May 1975), 100–16.
Hautmann, M., Betriebsergebnisse mit einer Induktionsgroßanlage (Flüssigverbund) für Gußeisen. *Gießerei* **55** (1968), 456–60.
Leipoldt, Ralf, Entwicklungen zum Schmelzen von Gußeisen und Stahl. Entwicklungen auf dem Gebiet Netzfrequenz-Induktionstiegelöfen I. *Fachber. Hüttenpraxis Metallweiterverarb.* **15** (1977), No. 11.
Wilke, K., Basisch zugestellte Induktionstiegelöfen mit einem Fassungsvermögen von 3,6 t und 32 t zum Schmelzen von Stahl. *Elektrowärme international* **37** (1979), B 5, 244. See also Dötsch E. and B. Schulze, Zumhülsen Netzfrequenz-Induktions-Tiegelofenanlage als Schmelzaggregat in einer Großstück-Stahlgießerei. *Elektrowärme international* **38** (1980), B 1, 19–22.
Company news, Midland gains in rollmaking with new foundry. *Iron Steel Eng.* **55** (1978), No. 10, 73.
Company news, Mack-Hemp Steel Foundry completes $7,75 million expansion. *Iron Steel Eng.* **54** (1977), No. 12, 67–9.

Electric Furnace Steel Production
Edited by E. Plöckinger and O. Etterich

12 *Automation and Process Control*

SIEGFRIED KÖHLE, DÜSSELDORF, AND
GERD WILHELM DREES, DORTMUND

Automation in electric furnace steelworks is becoming increasingly important under the influence of new technical possibilities and the requirement for greater economic efficiency. This can be seen from the large number of articles in technical publications and from conferences held over the years. The vast subject of automation and process control cannot be dealt with in depth here since it is continuously developing. However, the aim of this chapter is to provide a survey of the functions and potential of automation in electric steelworks and foundries, where process-control computers can be very useful, and to elaborate in a few instances and amplify the survey by numerous references to the literature.

12.1 AUTOMATION IN ELECTRIC FURNACE STEELWORKS

Figure 1 shows the functions of automation in electric steelworks, arranged on three levels. The lowest level contains process-control functions which mainly concern individual furnaces. The middle, process co-ordination, level embraces all the furnaces in the steelworks together with the systems they use in common. The top level, works management, contains functions which link the electric steelworks with other areas outside. Literature references 1 to 17 are typical papers about the use of computers in fifteen electric steelworks; however, the list is not intended to be complete. These applications all contain some or most of the functions illustrated in Figure 1.

The heart of the concept is monitoring and control of furnace-charging. In the simplest case the computer tells the melt-operator how to conduct the process.[1] As well as issuing melt-instructions, the computer can control the issue of charge-materials on the scrap-weighing machine.[3] After melt-down, the computer can use the first analysis to select the quality of steel which can be

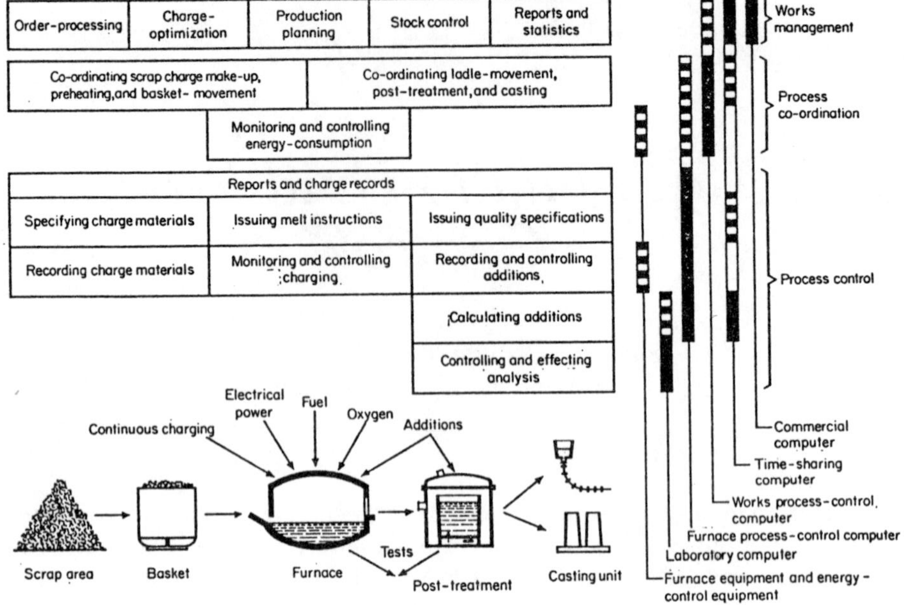

Figure 1. Functions of automation in electric steelworks

produced at minimum cost to meet the requirement.[5] Most computers used in electric steelworks go further than just issuing melt-instructions; they act directly on the process itself by controlling electrical power and the overall energy supply. Where necessary, the computer can also control fuel–oxygen burners and the continuous feed of sponge iron.

As well as control of the thermal state there is also metallurgical control. Laffolie reports on using computers to control the analytical apparatus and for computing the analysis values from the measurements taken by the apparatus.[18] The analysis-values enable the quantity of blowing-oxygen to be calculated as well as the amounts of additions and alloying ingredients. Many electric steelworks use computers in these ways; in a few cases the oxygen supply and the feed of additions are computer-controlled.

While computerizing the alloy additions leads in many cases to precise optimization of costs, furnace control is generally based on empirical procedures. A few attempts at optimizing this procedure can be found in the literature. Dynamic programming is used in order to optimize the charging process throughout the four stages of melting down, oxidizing, reducing, and refining.[19] Melting down has been optimized experimentally by using the maximum principle.[20] Here, only melt-down time and electricity costs are used for the optimization criterion. A more thorough optimization process also takes account of electrode-consumption and refractory lining-wear; here, a dynamic model is used to describe heat-transfer between the area and the electrodes, the charge, the lining, and the atmosphere.[21]

Another task at the process-control level in Figure 1 is compiling accounts and charge-reports. The data collected and processed by the computer can be stored and recalled for statistical evaluation later.

At the process-co-ordination level the most important task is monitoring and controlling energy-consumption; this is done by most computers installed in electric furnace steelworks. Where there are several furnaces, scrap-loading, basket-movement, and, if necessary, scrap-preheating all have to be co-ordinated with each other and with furnace-operation. Another possible task for the computer is co-ordination of furnace-operation, ladle-movement, post-treatment, and casting.

The right-hand side of Figure 1 shows how the various tasks can be allocated to different computers. The hatched areas indicate where the individual computers are only conditionally suitable. Hard-wired control units are adequate for simple furnace-control.[22–5] There are also hard-wired energy control units to monitor and control the consumption of energy.[26,27] Laboratory process-control computers are often used to control analytical apparatus and to process and evaluate the analysis; they are also sometimes used to work out the addition of alloying ingredients. The furnace process-control computer controls one or more furnaces. It also performs the alloying calculations, and the process can then be carried out by the melt operator in the light of actual conditions. The latest analytical data can be fed into the furnace process-control computer by data link from the laboratory computer.

Process-co-ordination tasks can be allocated to a furnace process-control computer; otherwise, if these tasks are too extensive, they can be allocated to a separate process-control computer for the whole operation. This should be linked to the furnace process-control computer so that the two can exchange the necessary information.

These process-control and co-ordination tasks for which no direct link is required between the computer and the process can also be allocated to a time-sharing computer, which can be assessed by steelworks personnel via an electric typewriter or visual display (VDU) terminal. A time-sharing computer or commercial computer is also suitable for works-management tasks, of which Figure 1 shows only requirement-processing, applications-optimization, production-planning, stock control, and reports and statistics.

12.2 ARC-FURNACE CONTROL

Figure 2 illustrates an outline system for controlling an arc furnace. The computer monitors the charging process via signals from various sensors; the most important types are shown in Table 1. For convenience, charge-monitor-

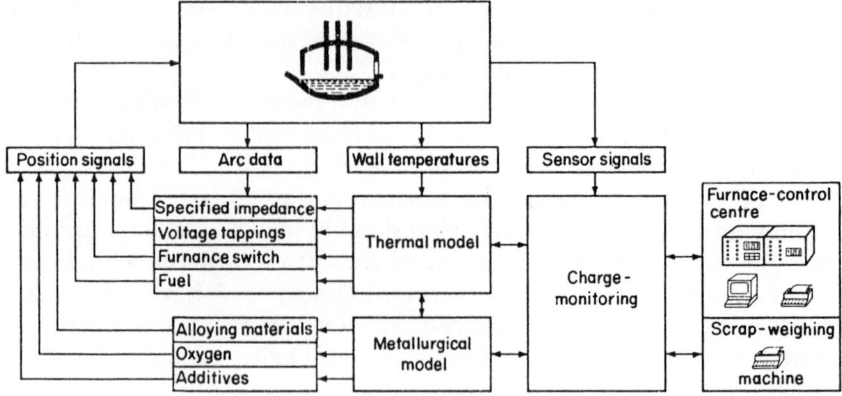

Figure 2. Arc furnace control system

Table 1. The most important sensor-signals for furnace-control

Energy impulse	Electrical power
Furnace switch position	Fuel flow rate
Voltage settings	Oxygen flow rate
Reactor setting	Blowing-oxygen flow rate
Vessel position	Exhaust-gas flow rate
Roof position	Exhaust-gas temperature
Portal position	Cooling-water flow rate
Electrode positions	Cooling-water temperature
Wall temperatures	Phase currents, voltages, and powers
Bath temperature	

ing starts when the furnace vessel is tilted for tapping. Initiation of melt-down is marked when the energy supply is started. Melting down ends when the first plausible bath-temperature is measured; this is followed by the start of the finishing-stage. Post-charging of individual baskets of scrap can be indicated when the computer asks the melt-operator for an appropriate command. The furnace roof is then swung open and the charge is dropped in; a measured amount of energy is then supplied after the roof is closed again.

The melt-operator feeds the necessary information into the computer via an electric typewriter or VDU terminal in the furnace-control centre. Sometimes these dialogues are initiated by the melt-operator and sometimes by the computer. The operator has only to answer questions put to him by the computer, and operator errors are largely eliminated. These dialogues mostly involve input into the computer of charge-number, type of material, specified analysis, and temperature-setting, as well as the types and quantities of alloying materials and additions where these are not input automatically. Charge-material data can be input when a scrap-basket is filled, so that the computer is already primed when the basket is charged into the furnace.

The amount of energy stored and the thermal state of the charge are computed by means of the thermal model. The energy-balance calculation must take account not only of electrical energy but also in particular of heat energy liberated in oxygen-blowing and, if appropriate, the energy introduced via fuel–oxygen burners and the energy-content of preheated scrap. On the energy-loss side it is assumed, in the simplest case, that the loss-rate is constant during the various stages of charge-history or even that it is constant throughout the whole charge. More precise results can be obtained by measuring the flow rates and temperatures in the exhaust-gas and cooling-water and calculating energy-loss. Further energy-losses incurred when removing electrodes and swinging the furnace roof open can also be taken into consideration. Losses in the electrical conductors can be calculated from the currents and resistances in the high-current conductors. The right time to cut the energy supply and, if appropriate, to charge the next basket is worked out by comparing the computed level of energy stored in the charge with the desired value. The desired level of stored energy is obtained from the quantities of the charge-materials and their specific energy-requirements; the latter depend on the chemical composition and physical nature of the materials concerned.

While the individual basket-loads melt down, the computer adjusts the furnace transformer voltage-settings and the impedance target-values in the electrode control-system. In this way, the electrical state of the furnace can be controlled as desired, depending on the energy already supplied and that still required. If the furnace transformer can take unbalanced load, the impedances and voltages can be adjusted differentially in the three phases, so that electrical conditions in the three arcs will be different and asymmetry in the reactances or thermal imbalance in the furnace can be compensated. The fuel–oxygen burners can also be controlled at the same time as the electrical power if necessary.[9]

The system for controlling the arc values can also incorporate a means of using a specified impedance-value as a reference and controlling, say, the power ratios or radiation factors of the three arcs.[4,14,17] Control of the energy-supply according to the stored energy computed by the thermal model can be affected by wall-temperatures.[7,12,14,28,29,30] If the wall-temperature close to one of the electrodes rises prematurely, then the power of the phase concerned can be reduced relative to that of the other phases.

Bath-temperature during finishing is computed by the thermal model.[4,9,10,11] This calculation is performed when the first plausible temperature-measurement is obtained; it is then corrected with every subsequent plausible reading. Here, the cooling effect on the bath of charging alloying materials and other additions must be taken into account. During finishing the electrical power control is used to raise the bath-temperature to a desired level in a specified time.

The analysis-values are transmitted to the furnace computer direct from the laboratory computer via data link;[9,10,12] otherwise they are fed in by the melt-operator. If a new analysis is required, the metallurgical model computes the alloying materials required to produce the specified analysis at minimum cost. Alloy-additions are divided up for charging into the furnace and ladle. Here the temperature to be set for tapping is calculated so that the melt will assume a predetermined temperature after the remaining additions have been fed into the ladle. This tapping temperature governs the way electrical power is controlled via the thermal model. The metallurgical model has the additional job of computing the composition of the melt from the data on the charge-materials while allowing for the melting losses; it then makes corrections from the values obtained by analysis. The required quantities of oxygen and additions have also to be determined. The computer can control the oxygen supply and the feed of alloying materials and additives from their bunkers.

The most important readings and computed values which describe the state of the charge are displayed in digital or analogue form on the melt-operator's desk. Foremost among these are the energy supplied, the energy still to be supplied up to the end of the current stage, the computed and specified bath-temperatures, and the electrical conditions in the three phases. Further information is supplied via the VDU and electric typewriter. All this information and the details of the dialogue between the melt-operator and the computer form a detailed chronological account of the history of the charge, which provides an in-depth insight into the process and is of great use in event of inconsistencies and interruptions. In addition, the most important details are printed out in a summary melt-report at the end of each charge. These data can be stored long term and retrieved for many kinds of statistical evaluation, so that the process-control systems can be analysed and continually improved.

12.3 OPTIMIZING ALLOYING INGREDIENTS AND CHARGE-MATERIALS

When making an alloy from a melt of known weight and composition, it is necessary to determine the types and quantities of alloying additions in order to arrive at an alloy of predetermined composition. The mixture can be worked out manually only as an approximation in the time available. Each addition to the melt, made to adjust the concentration of a particular element, has the effect of increasing the weight of the melt and varying the concentration of other elements which have not yet been considered. Using a computer for this eliminates errors of approximation and rounding as well as the occasional calculating error inherent in manual working.[31–4] This improves the accuracy with which the final composition can be achieved; the concentrations of the different elements can be brought close to their limits, and alloying ingredients can be saved. In addition, the computer requires less time than a manual method for an alloying calculation.

If there are several different alloying materials which contain the same elements in different proportions, there is a range of possible choices of alloying additions. The simplex linear-optimization process can be used to choose the least costly combination of alloying materials. Use of the computer here leads to savings which are only partly attainable by manual means because of the complex relationships involved.

Programs to optimize calculation of the alloying additions are easy to use interactively with the computer, and various operational constraints can be taken into consideration. Composition may be specified in terms of fixed constituent values or as upper and lower limits or ranges for the elements. The relationships between the concentrations of several elements can be taken into account. Melting-losses and the reclamation of alloying elements can be accounted for in the specified analysis or in a separate calculation. Optimization can also be aimed at a predetermined final weight or, where the initial concentration was too high, can specify dilution. Stocks of alloying materials can be controlled and monitored, and the amounts procured can be kept constant or at a minimum. The number of different types of alloying additives can be linked and their quantity can be rounded into specified units.

Charge-materials utilization can be optimized by a method similar to the alloying calculation. Charges of predetermined weight and composition can be made up at minimum cost from the types of scrap and alloying materials available. The charge-materials and proportions can be chosen by using the same considerations as in the alloying calculations. Charges may be optimized individually, or several charges can be planned simultaneously and the materials-allocations can be optimized for several charges in a planning period.[35] If the charge-calculation is related to the different types of scrap on the market, then scrap-purchasing can be considered with minimization of the costs of raw materials generally.

12.4 CONTROLLING ENERGY-CONSUMPTION

Because of the tariff structure incorporated in electricity supply contracts, every works endeavours to keep to a predetermined average power figure during peak periods. Average power is measured by working out the average power over periods of, say, 15 min; it is therefore proportional to the energy consumed during the period. Electricity charges are determined from the maximum demand value of the average power during an accounting period and constitute a significant proportion of total energy costs.

Works electricity-consumption can be controlled through reducing arc power by adjusting the voltage tappings; it can also be cut by withdrawing the electrodes. These actions are additional to the power-control methods for individual furnaces described in section 12.2. The power of each furnace should be controlled in such a way that the average power and with it the energy consumed by the works in each period does not exceed the maximum demand. On the other hand, the predetermined energy-allotment should be used up if possible, provided that it is required by the furnaces. If the works produces any electricity or other energy this should also be controlled by the control system. The energy available for melting should be distributed among the individual furnaces according to priorities at the time. In general, the further a furnace has progressed in processing its charge, the higher will be its priority; however, the charge can also be affected by the way the melting operation is co-ordinated with post-treatment. Finally, the furnace-voltage should be varied as little as possible for controlling electricity-consumption so as to avoid excessive wear in the control system.

There are various methods for monitoring and controlling energy-consumption.[26,27,36–8] In some cases it is not possible to control furnace power by adjusting the voltage-tappings but only by switching off and on again. Only process-control computers can deal adaptively with changing furnace-priorities.[37,38] Other methods, mainly involving equipments incorporating digital modules, can do this only to a limited extent. Most systems for controlling energy-consumption are based on the principle of controlling overall works maximum demand. The desired overall power level is calculated in such a way that the specified amount of energy is used by the end of the maximum demand period if the total power for the period equals the specified value. As this cannot be the case in general, the desired value is continually corrected. Overall power control is aimed at compensating for fluctuations in consumption by individual equipments through changing the power of other furnaces. This is unnecessary, at least at the beginning of the period, and leads to excessive and undesirable intervention. This problem can be abated by switching the control system on only after a specified time or if fluctuations exceed specified limits.

Figure 3 is the block diagram of a system for controlling individual furnace power instead of overall power.[38]. The upper part of the diagram shows how the total works power P_{TOT} comprises the total power P_F of all the furnaces plus the operating power P_W of all non-switchable consumers and, when integrated, provides the total energy W_{TOT} consumed since the start of the period. If the total energy power P_{TOT} fluctuates beyond the tolerance band between the upper and lower limits (P_{TOTU} and P_{TOTB}, respectively), then the furnaces are controlled, according to their priorities, in such a way that the sum P_F of their powers assumes the specified value P_{FS}. P_{FS} is the difference between the specified total power P_{TOTS} and an estimated value P_{WE} of average operating power throughout the remainder of the period. P_{WE} is

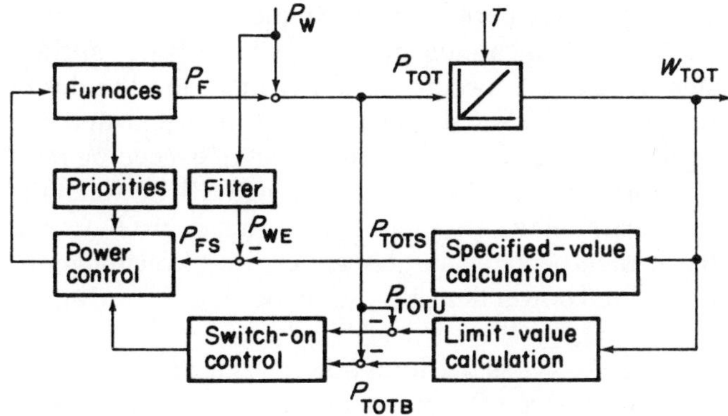

Figure 3. Block diagram of energy-distribution control system[38]

derived from the measured value P_W of other works consumer power via a time-delay element, the time-constant of which decreases with increasing time. In this way fluctuations in operating power P_W have less effect on furnace-power control at the beginning of the period than at the end.

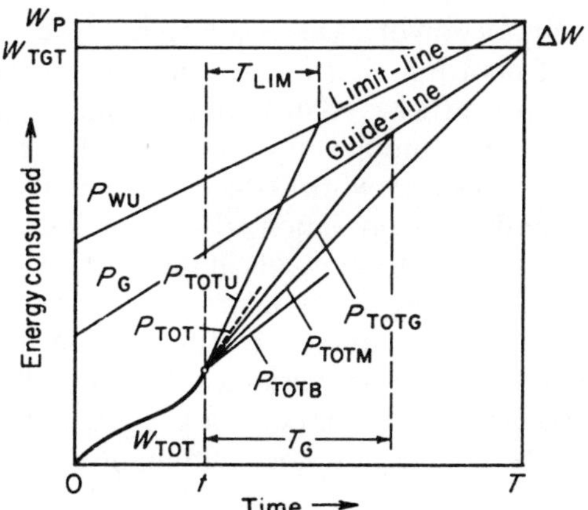

Figure 4. Determining specified and limiting values for total power[38]

Figure 4 shows how the specified and limiting levels of total power are calculated. Curve W_{TOT} indicates the amount of energy consumed from the start of the period up to time t. It must always remain below the limit line, the rise of which is given by the maximum operating power P_{WU}, so that the predetermined total energy allowance W_P is definitely not exceeded by the end of the maximum-demand period. The upper power-limit P_{TOTU} is the level of power which will cause the limit-line to be reached after a time T_{LIM} and no later than at the end of the period. In the simplest case the desired power value is taken to be P_{TOTM}, which indicates energy level W_{TGT} at the target point. Thus the tolerance zone between the upper limit and the desired red value narrows as the energy consumed approaches the limit-line and time progresses towards the end of the period, and so the effect of exceeding the desired level becomes ever more critical. The small difference ΔW between the two targets has the effect of opening the tolerance zone up again just before the end of the period so that it then becomes largely unnecessary to change furnace power again. The lower total power limit P_{TOTB} can be set at a fixed amount below the desired value.

Instead of using P_{TOTM} as the total power value, P_{TOTG} can be taken; this leads to a guide-line of slope P_G after a time T_G and no later than the end of the period. The effect of the guide-line is that the desired value is greater than the preset average power at the beginning of the period, but is less at the end of the period. This more-than-proportional procedure can to some extent avoid the problems of limiting power at the beginning of the period and not using up all the energy-allotment by the end of the period because of insufficient demand.

12.5 AUTOMATION IN THE FOUNDRY

Automation in the foundry involves tasks similar to those in the electric furnace steelworks, and the main function of the computer is co-ordination of the various interlinked processes.[39] Calculating optimum use of charge and alloying materials is less important in the foundry than in the electric steelworks, since it is usual to work with only a small number of different analyses. In quasi-continuous operation with induction furnaces most of the melt with its analysis already set remains in the furnace after tapping, and so only small quantities of alloying additions are needed. The weight of the melt is known precisely where the furnace is fitted with pressure-transducers. Automatic weighing of charge-materials and automatic transportation of materials to the furnaces are both important because of the high frequency of charging operations. Hallot has described automatic charging for six induction furnaces.[40]

The aim of controlling induction furnace electric power is to get the melt to the desired temperature at a specified time. Where a process-control computer is used for this it also has to ensure that heat-losses are minimal and that the relatively thin furnace-lining is stressed as little as possible.[41] The limited supply of electricity during peak periods also necessitates control of energy-consumption in foundries.[42]

Furnace temperature-control must also take account of co-ordination of the supply of casting material from the melting shop with the requirements of the moulding shop. G. W. Drees describes a process-control computer system that solves this problem by monitoring production and repeatedly replanning output.[43] This enables potential holdups in supplies of materials to the moulding shop to be seen in good time and avoided by amending the production plan.

Another paper describes[44] the use of a computer to control and monitor sand-preparation and sand-mould installations. This requires extensive use of sequential control which has often been achieved up to now by means of hard-wired or programmable control units. Other uses of computers in the foundry involve monitoring the flow of materials,[45] including controlling the stocks of raw materials, sand-binder, and castings, as well as checking the quality and dimensions of geometrically complex castings.[39]

12.6 ECONOMIC EFFICIENCY

The literature contains some assessments of the operational and economic advantages of using computers. Furnace-control enables developed processes to be analysed in detail without having to rely on the skill and momentary attention of melt-operators. This results in reductions of about 10% in charge-time[5,10,17] and about 3% in specific energy-consumption,[11,15,24] while lining-life is increased by around 20%.[24,37] Objective and comprehensive documentation (recording) of the process is generally seen to be an advantage since it simplifies accounting and enables process-management to be analysed in detail.

Calculating the alloying additions increases the certainty with which the intended composition can finally be achieved; it is said that the scatter of analysis-values for highly alloyed steels is narrowed by about 50%.[9] This makes it possible to save around one to 25 DM per tonne by choosing optimum-cost alloying additives.[10,11,32]

Controlling energy-consumption increases the amount of energy available during peak periods without affecting electricity-costs; this lowers specific energy-costs. It is said that the utilization of energy allocations can be increased by 4% or 2.5%.[15,38] It is an additional advantage here to exploit furnace-priorities when adjusting furnace power.

The above figures should be taken as approximations; the overall saving in the cost of electric furnace steel-production is 2–3%. It can therefore be expected that the cost of automation will be recovered over a reasonable time because of the advantages that result; indeed, it is reported that investment in automation has been amortized in under a year.[11]

12.7 LITERATURE REFERENCES

1. Gloven, D. O., Computer control for electric furnace steelmaking. *J. Metals* **16** (1964), 963–66.
2. Howes, R. S., Some developments in electric steelmaking at Templeborough. *J. Iron Steel Inst.* **206** (1968), 205–15.

3. Mulcahy, J. A., and G. H. Samuel, Lake Ontario Steel's computer-controlled electric arc meltshop. *Blast Furn. Steel Plant* **58** (1970), 801–5.
4. Riddervold, H. W., M. Mohagen and O. P. Thoresen, Computer controlled steel melting in a 50 tons electric arc furnace. In *Automation I. Internationale Eisenhüttentagung* 1970, Luxembourg, 13–15 April 1970. Verein Deutscher Eisenhüttenleute and Centre National de Recherches Métallurgiques. Liège 1970, pp. 315–23.
5. Stenhouse, J. F., and D. F. Boyd, Application of a digital computer to the control and direction of an electric furnace melt shop. *Iron Steel Eng.* **48** (1971), No. 3, 93–7.
6. Hohendahl, K., Prozeßautomatisierung im Elektrostahlwerk. *Siemens-Z.* **47** (1973), Beih. 'Antriebstechnik und Prozeßautomatisierung in Hütten- und Walzwerken', 35–40.
7. Figoureux, Mathieu and Hohendahl, Planung und Ausführung einer Prozeßführung mit Rechner in einem Elektrostahlwerk für Edelstahl. Paper given at a meeting on automation and process control in electric steelworks and foundries, 14–16 May 1975 in Versailles. Sponsored by Comité Français d'Electrothermie. Ber. II.2.
8. Mommertz, K. H., G. Poklekowski and L. Kütemeyer, Führen des Elektrostahlprozesses durch einen Rechner. In *Internationaler Eisenhüttentechnischer Kongreß*, Düsseldorf, 27–30 May 1974, Vol. II. (Düsseldorf) 1974. Report No. 3.1.2.7. 17 pp.
9. Baum, R., R. Heinke, W. Jung, S. Köhle and G. Poklekowski, Schritte zur Automatisierung des Elektrostahlprozesses. In *Automatisierung in der Eisen- und Stahlindustrie*. Internationaler Eisenhüttentechnischer Kongreß 1976, Brüssel, 17–18 May, Düsseldorf, 20–21 May, Vol. 1b. Centre de Recherches Métallurgiques. Liège 1976. Report No. 4.3.1. 20 pp.
10. Maitre, A., and P. Moreau, Optimisation et programmation par calculateur de process des aciéries éléctriques. *Circ. Inform. techn.* **32** (1975), 599–609.
11. Davene and Renninger, Rechnersteuerung des Elektrostahlwerkes der Firma Creusot-Loire in Le Creusot. See Ref. 7. Report No. II.3.
12. Comes, H., H. Köbke and H. G. Vowinkel, Einsatz eines Prozeßrechners an einem 50 t UHP-Lichtbogenofen und Versuch einer Dynamischen Prozeßsteuerung. See Ref. 9. Vol. 3. Report No. II.4.3.4. 23 pp.
13. Brandt, S., Automatisierung von Elektro-Lichtbogenöfen. *Techn. Mitt. AEG-Telefunken* **66** (1976), 299–302.
14. Bergman, K., Computer control of arc furnaces. Paper for the 8th Internationalen Elektrowärme-Kongresses, 11–15. 10. 1976 in Liège. Report No. IV.4.
15. Yoshino, T., and Y. Mikuni, First computer control system for arc furnaces in Japan. See Ref. 14. Report No. IV.5.
16. Noda, H., and I. Eguchi, On the improvement of UHP arc furnaces for melting. See Ref. 14. Report No. IV.8.
17. Tamao, Y., H. Takabe, N. Fujimori, S. Saito, H. Shimizu, M. Nishio and T. Miyashita, Development of arc furnace computer control system. *Nippon Kokan techn. Rep. Overseas* No. 22, 1976, 51–7.
18. Laffolie, H. de, Der Einsatz von Rechnern im Eisenhüttenlaboratorium. See Ref. 9. Vol. IIc. Report No. 7.1.1. 16 pp.
19. Calanog, E., and G. H. Geiger, Optimization of stainless steel melting practice by means of dynamic programming. *J. Metals* **19** (1967), No. 7, 96–104.
20. Gosiewski, A., and A. Wierzbicki, Dynamic optimization of a steel-melting process in electric arc furnace. *Automatica* **6** (1970), 767–78.

21. Kubena, J., and J. Hlineny, Steuerung der in den Lichtbogenofen zugeführten Leistung mit Verwendung des dynamischen Modells der energetischen Vorgänge. See Ref. 7. Report No. I.2.
22. Schönert, D., Programme-controlled power supply of an electric arc steel-melting furnace (German). *Elektrowärme* **23** (1965), 505–10.
23. Ohmoto, M., A. Kawashiro and M. Aoshika, Newly developed automatic optimum power control equipment in arc furnace. *IHI Eng. Rep.* **4** (1971), No. 2, 46–58.
24. Vinogradov, V., V. Piroznikov, V. Jakovlev, R. Samygin, V. Cukanov and N. Nikokosev, Systeme zur Steuerung des Schmelzprozesses in Lichtbogenöfen und ESU-Anlagen. *Neue Hütte* **20** (1975), 208–12.
25. Leu, H. W., and O. Schläpfer, Hochleistungs-Lichtbogenöfen. *Brown Boveri Mitt.* **63** (1976), 651–4.
26. Severin, R., *Eine Lastkontroll-Anlage*. S.-A. aus Landis & Gyr-Mitteilungen 1/1964.
27. Martin, O., W. Groenewald and G. W. Drees, Economic supply of purchased power through automatic peak limiting (LBA) (German). *Elektrowärme* **30** (1972), 69/B 74.
28. Brachet, J. P., and G. Sartorius, Entwicklung eines Wärmefluß-Meßfühlers und eines Wandstärke-Meßsystems für Lichtbogenöfen. See Ref. 7. Report No. I.5.
29. Bergman, K., Erfahrungen mit Wandtemperaturmessungen in Lichtbogenöfen. See Ref. 7. Report No. I.6.
30. Antoine, J., and J. L. Trevedy, Versuche bei IRSID zur Temperaturüberwachung des Lichtbogenofens mit Hilfe eines Wandwärme-Meßfühlers. See Ref. 7. Report No. I.7.
31. Mommertz, K. H., Berechnung optimaler Legierungs- und Einsatzstoffe für Elektrostahlwerke mit elektronischer Datenverarbeitung. *Stahl u. Eisen* **93** (1973), 933–4.
32. Höpfner, G., and F.-J. Eilert, Anwendung der elektronischen Datenverarbeitung für die Legierungsoptimierung im Stahlwerk einer Stahlgießerei. *Gießerei* **61** (1974), 369–75.
33. Wurth, G., Berechnung optimaler Legierungszusätze im Elektrostahlwerk. *Stahl u. Eisen* **95** (1975), 12–15.
34. Scharf, G., W. Buhr, and R. Börnig, Vorraussetzung und Aufbau einer Legierungseinrechnung für hochlegierte Stahlmarken durch Prozeßrechner. *Neue Hütte* **21** (1976), 351–65.
35. Kellner, S., H. Walter, D. Winkler, D. Kaiser and G. Wurth, Prozeßrechnereinsatz zur optimalen Aufteilung der Einsatzstoffe im Elektrostahlwerk. *Regel.-Techn. u. Prozeß-Datenverarb.* **21** (1973), 313–21.
36. Alexander, K. E., and M. Maczuzak, A power-demand predicting computer control system for Sheffield Div. of Armco Steel. *Iron Steel Eng.* **41** (1964), No. 5., 102–10.
37. Borggrefe, H.-H., K. Hohendahl and G. Zingel, Optimization of the power economy for the operation of an electric steel plant (German). *Stahl u. Eisen* **93** (1973), 1261–5.
38. Jung, W., S. Köhle, R. Lichterbeck and P. Schmidt, Überwachen und Steuern des Energieverbrauches im Elektrostahlwerk. See Ref. 9. Report No. 4.3.2. 21 pp.
39. Grandpierre, M., Data-processing and automation. *Foundry Trade J.* **137** (1974), 317–30.
40. Hallot, L., Automatisierung der Beschickung von 25 t Induktionsöfen in der Gießerei der Francaise de Mécanique in Douvrin. See Ref. 7. Report No. III.2.
41. Maurer, G., Burderus Wetzlar: Prozeßrechner-Lösung Schritt für Schritt. *Online* **13** (1973), 332–3.

42. Drees, G. W., and W. Groenewald, Automatisierungssysteme für Gießereien. See Ref. 7. Report No. III.5.
43. Drees, G. W., and W. Groenewald, Gießereiführung mit einem Prozeß-Datenverarbeitungssytem. *Gießerei* **62** (1975), 147–51.
44. Dalahaye, Copie and Maldidier, Einsatz eines Rechners (T 2000) in einer Formerei für Kästen 1000 × 1000. See Ref. 7. Report No. III.4.
45. Weinberg, H., Datenerfassung und -verarbeitung zur Kontrolle und Steuerung des Stoffflusses in Gießereien. *Gießerei* **58** (1971), 611–18.

Andersson, L., Rechnerunterstützte Betriebsführung in einem schwedischen Stahlwerk. *ASEA-Z.* **23** (1978), 89–92.

Billings, S. A., F. M. Boland, and H. Nicholson, Electric arc furnace modelling and control. *Automatica* **15** (1979), 137–48.

Custer, C. C., The practical use of computer programming in electric furnace product quality control. *Iron & Steelmaker* **5** (1978), No. 5, 28–32.

Esser, F., Rechnersimulation der Einschmelzvorgänge im Elektrolichtbogenofen. *Neue Hütte* **24** (1979), 205–14.

Gill, L. L., Operational aspects of arc-furnace computer control. *Ironmaking & Steelmaking* **5** (1978), No. 5, 217–23.

Guncz, A., Automatisierung von Elektrostahlwerken. *Brown Boveri Mitt.* 1978, 118–24.

Jung, W., S. Köhle, R. Lichterbeck, and P. Schmidt, Monitoring and control of the power consumption in an electric steel plant (German). *Stahl u. Eisen* **98** (1978), 979–84.

Köhle, Siegfried, Use of computers for control of arc furnaces (German). *Stahl u. Eisen* **100** (1980), 522–8.

Lindhoff, Dieter, and Werner Raschke, Central control and supervision by means of a process computer at the SIDOR electric melting shop. *Metallurg. Plant & Technol.* **2** (1979), No. 4, 30–2, 34–5.

Lindig, D., and L. D. Wostinjuk, Ein statistischer Algorithmus zur Kontrolle der Einschmelzperiode im Lichtbogenofen. *Neue Hütte* **24** (1979), 210–14.

Moriya, Atsuo, Shuntaro Ito, Jun Inoue, Tatsuo Murakami and Haruo Yamamura, Development of 80-ton electric arc furnace computer control system. *Sumitomo Search* No. 21, 1979, 8–17.

Mullins, Peter J., Now, arc furnace computer control. *Iron Age Metalworking Intern.* **18** (1979), No. 5, 16MP3/16MP5, 16MP7.

Scheurer, H.-G., and E. Elsner, Use of process computers and microcomputers in electric steel plants (German). *Stahl u. Eisen* **100** (1980), 112–18.

Computerized process control in the steel industry. *Intern. Metals Rev.* **22**, No. (1977), 4, 355–81.

Energy control stops fuel waste. *IAMI* 7/1978, 16MP20, 16MP22.

Segel, J., Approaches to computer control meltshop steelmaking. *Proceedings of the Third International Iron and Steel Congress,* 16–20, May 1978, Chicago, pp. 676–85.

Kröncke, G., R. Baum, R. Heinke, and S. Köhle, New computer controlled ultra high power arc furnace of Krupp Stahlwerke Südwestfalen. *Iron and Steelmaker,* March 1980, 30–5.

Heinke, Rüdiger, Operating results of the computer-controlled, UHP arc furnace of Krupp Stahlwerke Südwestfalen AG (German). *Stahl u. Eisen* **100** (1980), 529–32.

Munsterhjelm, Peter, Development trends in instrumentation and control of electric furnace systems. *Proc. Electr. Furn. Conf.* **36** (1978), 218–28.

Electric Furnace Steel Production
Edited by E. Plöckinger and O. Etterich

13 Electric Steelmaking and the Environment

Klaus Polthier, assisted by Ulrich Haering, Helmut Kahnwald, and Dirk Marchand, Düsseldorf

The last ten years have produced significant developments in waste-gas management and noise-reduction, occasioned by very substantial advances in legislation about health and safety at work and environmental protection in the traditional industrial countries and the development there of a mature environmental consciousness. As well as extraction of exhaust-gases direct from a fourth hole in the arc-furnace roof, special mention should be made of other measures, already in normal service in several works, which involve capture and removal of dust from gases given off during charging and tapping. These measures have made it possible to remove almost all dust from exhaust-gases given off and have led to electric steelworks operations being substantially free from anti-social emissions, although this has incurred significant increases in investment-outlay and operating-costs. The measures adopted in building new arc-furnace works and in retrospective modification of existing installations employ various techniques, mostly involving comprehensive enclosure of the arc-furnace area or of the entire furnace shop or using large hoods with exhaust-gas extractors and dust-removal gear in the works roof.

13.1 LEGAL REQUIREMENTS IN THE FEDERAL REPUBLIC OF GERMANY

Installations which either by their nature or by their operation cause a nuisance or danger to the environment require a permit in the Federal Republic of Germany.[1] Installations requiring permits are defined by law;[2] steel-production and electric furnaces are specifically covered.[2]

Written application is required for the construction and operation of such installations. The form of application, the contents, the addressee details, and particulars about processing are all covered by legislation.[3] Such applications must be accompanied by details of proposed construction and operation and technical reports on the expected levels of emission and nuisance so that an evaluation can be made.

Care must be taken to ensure that the latest techniques are used[1] to protect the environment against danger, disadvantage, or nuisance occasioned by operation of an installation. Standards are laid down in Guideline 3465[4] published by VDI, the Association of German Engineers, which not only covers the direct extraction of dust via a fourth hole in the furnace roof throughout the melting process but also requires that further dust generated during charging, oxygen-blowing, refining, and tapping (see also section 13.2) is collected, as far as possible, and led to a separation plant. Additional instructions are given in clean-air regulations (technical).[5]

General rules about protection against gas, vapour, and dust in working premises are given in legislation on health and safety at work.[6] The maximum permissible levels of concentration in working premises (called MAK values in Germany) are given in a special list.[7] This list, which covered about 500 different substances at the time of writing, is drawn up and updated annually under the auspices of the German Research Association Hazardous Materials Test Committee on behalf of the Federal Minister for Labour and Social Services.

Although these lists are only advisory as published, they assume mandatory character in individual cases when called up in the official permits to construct and operate particular installations. While exceeding one of the maximum permissible levels of concentration close to an electric furnace generally presents no problems if the exhaust-gas system works properly, keeping within permissible noise-levels gives rise to considerable difficulties, particularly with older installations, and can only be achieved by using secondary protection measures such as sound-proofed control cabins and ear-defenders.

Noise at work is subject to factory regulations and noise-induced accident prevention regulations (here termed NAPR for short).[8] The latter are published by the Association of German Industrial Employers' Liability Insurance Companies and apply to all undertakings where employees are exposed to noise. They stipulate that work-places and processes must be arranged so that employees are not subjected to mean noise-levels (measured throughout a full shift) of 90 dB(A) or more. If, despite applying progressive, established measures, noise-level in the working area still exceeds 85 dB(A), then the management must provide personal ear-protection means which should be worn by workers when noise levels reach 90 dB(A) or more, unless the overall risk of accidents is raised by doing so.

The NAPR also contain rules for marking areas in which a noise level of 90 dB(A) is likely to be reached or exceeded. They also lay down the employers' responsibility for providing qualified audiometric testing and precautions for all employees in noisy areas. The results of audiometric tests and any supplementary examinations should be passed to the insurance company concerned.

The problem of hearing-damage caused by noise was already recognized in the early 1960s by the West German iron and steel industry which developed its own detection and monitoring system that is applied with the co-operation of the insurance companies and Industry Control Offices. This provides for noise-levels in work-areas as well as audiometric data and medical results to be recorded and evaluated centrally. This results in documentation which is passed back to the works' medical services and enables judgements to be made on the noise-situation and danger to hearing in any operating area in a steelworks.

Regulations for work-areas, which came into effect with the backing of the German Federal Council on 1 May 1976,[6] lay down defined noise-limits for working areas at 55 dB(A) for work of a predominantly mental nature, 70 dB(A) for simple office work and similar activities, and 85 dB(A) for all other work. The latter figure may be exceeded up to a limit of 90 dB(A) if a level of 85 dB(A) cannot be reached after taking all reasonable steps to reduce the noise.

If it is not possible to keep noise in a proposed new work-area down to these limits, an exception may be made in certain circumstances, provided a suitable case is put to the local government inspector; otherwise the new working-facility must not be built. In this way, these regulations for work-areas represent a tightening up on NAPR and have significant consequences for sound-protection in the construction of new electric steelworks and the modification of old works, since it is not possible today to limit the noise in the working-area around an electric arc furnace to the limiting figure of 85 dB(A).

Electric steelworks can also create considerable noise problems in residential areas, since such works, particularly small ones, are not sited only in industrial areas but are scattered across the country, including in localities with stringent pollution-control regulations.

In places where close proximity is permitted between industrial and residential buildings and where recommended separation distances may be undercut, as in North Rhine Westphalia,[9] it is only possible to keep to the noise-emission guide values, quoted in technical guidelines on noise-protection[10] for purely residential areas and general residential areas, by adopting additional noise-limiting measures such as enclosing workshops completely.

Ways of dealing with noise in electric steelworks are indicated in section 13.3.2.

13.2 EMISSION AND CONTROL OF HARMFUL SUBSTANCES

The production of crude steel in the Federal Republic of Germany has doubled in the last twenty years. At the same time a rapid change occurred in production processes which is best described by the following statistics. In 1955 about 50% of all steel was produced in Siemens–Martin open-hearth furnaces; by 1976 this had dropped to just 15%. In the same period the proportion of steel produced in basic oxygen furnaces grew from zero to around 72%. While this changeover was taking place, electric furnace steel was steadily gaining in importance and in 1976 amounted to 12% of all steel produced in West Germany.

At the same time as this shift in production techniques was taking place it was possible to keep up with the increasing requirements for protection of the environment. These requirements played their part in speeding up the changeover from one process to another. It can be seen today that in the next few decades to come the main method for converting molten pig iron will be the oxygen-refining process while electric furnaces will be used for melting down scrap and other solid-iron carriers such as directly reduced iron ore.

The development of steel-production in the Federal Republic of Germany and of dust-emission for the four main processes in the period 1955-75 are illustrated in Figure 1. The trend towards basic oxygen furnaces and the steady growth of electric furnace steel-production can be clearly seen.

Dust-emission declined steadily in step with the changeover in furnaces, particularly because basic oxygen furnaces replaced Thomas converters which do not have dust-extraction systems. With electric furnaces the reduction in

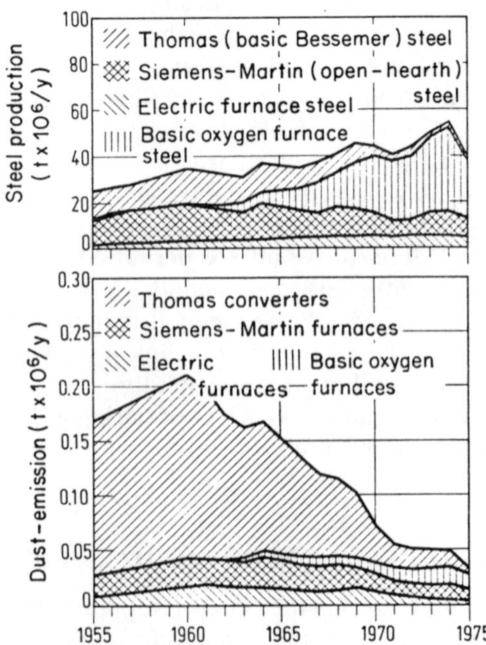

Figure 1. Dust-emission in crude-steel production in the Federal Republic of Germany

dust-emission can be traced increasingly to new furnace-installations equipped with direct exhaust-extraction systems and hoods in the shop roof as well as to the replacement of small furnaces by larger units.

13.2.1 Generation of Dust and Gas

With the *electric arc furnace*, the mass flow, composition, and temperature of gases and entrained solids given off by the furnace vessel depend on various influences, of which the most important are:

(1) The composition of the charge and of impurities such as oil, grease, and sundry combustibles in the charge;
(2) The way the charge is managed, including the supply of energy from electricity, oil–oxygen burners, and oxygen-lances during melt-down;
(3) The amount of gaseous oxygen blown in for refining after the charge has been melted down.

During *scrap-charging* into a hot furnace vessel, gas can only be given off by combustible impurities such as cutting oil, grease, etc. in the charge. The vertical flow above the furnace is mainly caused by heat making the column of air above the furnace rise. This thermal current carries away fine solid particles, such as loose rust rubbed off the scrap during charging. The exhaust-gas flow rate depends mainly on the density of the hot airstream emerging from the open furnace and is increased by burning oil and grease.

During *melt-down* most of the main dust-laden waste gas consists of entrained air and the products of oxidation of the graphite electrodes as well as of combustible matter in the charge that is vaporized in the arcs. Lubricating oil and grease contain varying amounts of sulphur and chlorine, and so the exhaust-gas may contain sulphur and chlorine compounds; however, these can also emanate from plastic waste, such as PVC, PVF, and PTFE-coated plate.

Using oil–oxygen burners to assist melt-down increases the amount of exhaust-gas from the furnace considerably. No nitric oxide is generated in this way; however, it is formed in the arcs and when carbon monoxide burns in air drawn into the furnace vessel.

During the melting period between 5 and 10 kg of dust are generated per tonne of crude steel, depending on how much low-vapour pressure matter there is in the charge and on the nature and rate of energy input. Using oil–oxygen burners to assist melting has the effect of reducing dust-generation by about 50% to specific values of around 2–5 kg per tonne of crude steel.[1]

Gas- and dust-generation are highest during *oxidizing* using gaseous oxygen, so this also affects the design of the dust extraction system. Combustion of carbon from the steel bath generates a quantity of carbon monoxide at a rate that depends on the decarburizing rate, which in turn depends on the rate at which gaseous oxygen is supplied to the melt (Figure 2). As can be seen in Figure 3, the

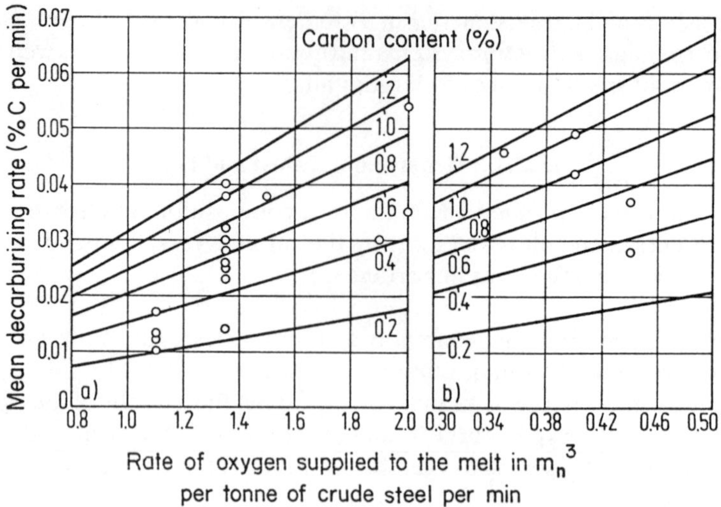

Figure 2. Rates of oxygenation and decarburization possible with (a) high-chromium-alloy melts and (b) non-alloy melts

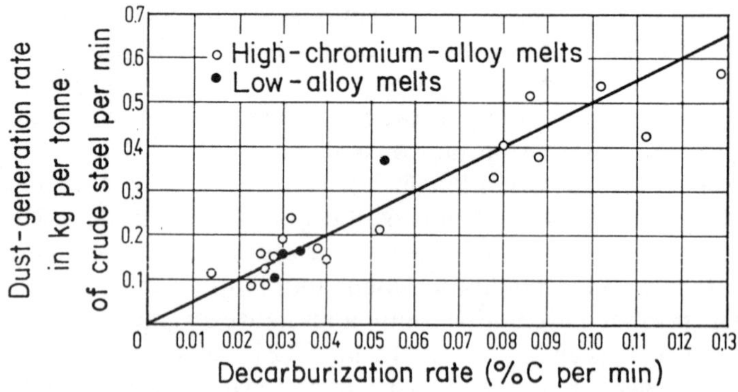

Figure 3. Dust-generation during oxidizing using gaseous oxygen

rate of dust-generation is roughly proportional to the decarburizing rate;[2] the mean value during the oxidizing period can amount to 10 kg per tonne of crude steel overall.

During the *finishing* period dust-generation normally amounts to around 1–2 kg per tonne of crude steel, but if carbon or lime are blown in it can rise to as much as 5 kg per tonne.[1] In addition, if fluorspar is used this can generate gaseous fluorides over a long period.

During *tapping* the specific dust-generation rate may amount to about 0.5 kg per tonne of crude steel,[1] depending on the tapping-temperature, the composition of the steel, the tapping-ladle temperature, and any ladle-processing. The dust is borne up by thermal currents above the ladle.

Taking all furnace-operation phases into account the mean overall specific dust-generation rate over a long period can lie between 14 and 20 kg per tonne of crude steel. This has to be allowed for in the design of the dust-extraction system, including removal of dust from the filters, intermediate bunkerage, pelletizing, and storage. Table 1 shows the scatter range of particle sizes for crude gas dust, averaged over the whole melt.[1,3] The particle-size distribution is almost identical to that for a Siemens–Martin open-hearth furnace operating intensively with scrap.

Table 1. Scatter range of particle sizes of crude gas dust, averaged over the whole melt[1,3]

Particle size (μm)	5	10	20	50	100	200	500
(%)							
Maximum	74	86	93	98	99	100	100
Maximum	30	40	50	65	75	83	90

The nature of the charge is reflected in the composition of the dust. Table 2 is a guide to dust composition; it is based on American data[3] together with data from a few West German works.[1] Scrap charged into electric furnaces is usually relatively clean compared with that for open-hearth furnaces, and this results in comparatively low levels of zinc and lead in the dust. With unfavourable scrap-conditions, e.g. with galvanized or lead-coated scrap, the dust may contain more zinc and lead overall, as shown in columns 1 and 2.

Waste gases do not normally contain sulphur, and this is why the dust absorbs almost no sulphur from the gas phase. The dust therefore contains little sulphur unless large quantities of sulphur-bearing lubricants get into the furnace vessel with the charge.

Levels of zinc, lead, alkaline metals, and alkaline earth metals in the dust from electric furnaces can reach the same kind of levels as in dust from open-hearth furnaces. However, these materials mainly occur as oxides (that are not water-soluble) and not as sulphates and chlorides (see also section 13.4), so that recycling and dumping present fewer problems.

As lime-blowing is inefficient, a large proportion of lime blown in reappears in the dust; this results in calcium levels in the dust of up to 15% or much more in extreme cases. Because the levels of sulphur and chlorine in the dust are low while the level of lime is relatively high, and aqueous dust extract normally has a weak to strong basic pH value.

In one case the dust pH value is moved from the neutral/weak basic area

Table 2. Dust-composition (mean values for entire charge cycle)

	US figures (scatter for each element)	Figures for industrial West German works					
		Non-alloy steels					Alloy steels
		A	B	C	D	E	A
	1	2	3	4	5	6	7
Iron	16.4–38.6	21.6	25.5	31.5	43.6	ND[a]	35.3
Manganese	2.3– 9.3	2.8	4.8	1.4	0.9	ND	2.0
Chromium	0– 8.2	–	0.1	–	–	ND	13.4
Nickel	0– 2.4	–	–	–	–	ND	0.1
Lead	0– 3.7	5.6	1.4	3.1	1.3	ND	0.4
Zinc	0–35.3	26.2	19.8	11.7	5.8	ND	1.4
Silicon	0.9– 4.2	1.0	0.9	1.4	1.7	ND	ND
Aluminium	0.5– 6.9	0.3	0.1	0.7	1.1	ND	ND
Calcium	2.6–15.7	6.6	9.2	14.5	13.2	ND	0.4
Magnesium	1.2– 9.0	1.8	4.5	1.2	1.0	ND	1.2
Alkaline metals	1.0–11.0	ND	ND	ND	ND	ND	0.7
Phosphorus	0– 1.0	0.1	0.5	ND	ND	ND	ND
Sulphur	0– 1.0	0.6	1.1	ND	ND	0.3–0.5	0.1
Chlorine	ND	ND	ND	ND	ND	0.2–0.4	0.4
pH value	ND	ND	12.3	ND	ND	9.6–5.2	ND

[a] ND = no data.

into the neutral/weak acidic region with the introduction of gaseous compounds of sulphur when oil–oxygen burners are used in conjunction with a low level of lime in the dust. This can lead to considerable reduction in the lives of fabric filter-elements made from polyamide fibres; these filters are common and so this effect must be borne in mind when oil–oxygen burners are used.

Table 3 shows how dust-composition changes in the various process phases with a 7-t furnace.[4]

Induction furnaces are usually used only for melting or remelting, and the absence of refining here means that practically no waste gas is produced by the process. The amount of dust generated depends on whether melting is carried out with the furnace roof open or closed. With the roof closed, dust amounts to up to 0.2 kg per tonne of steel; with the roof open it is up to 0.9 kg per tonne.[1] The dust is carried up by the vertical thermal current induced over the furnace vessel.

The *electro-slag remelting* process (ESR) produces no waste gas. The dust and hydrogen fluoride generated are borne upwards by thermal currents above the remelting vessels. The amount of dust produced depends on how much slag evaporates. Hydrogen fluoride is evolved secondarily from the reaction of

Table 3. Composition of dust in arc furnace exhaust-gases at various times in the melting cycle with 34Cr4 steel

Phase	Dust-composition (%)								
	SiO_2	CaO	MgO	Fe_2O_3[a]	Al_2O_3	MnO	Cr_2O_3	SO_3	P_2O_5
Melt-down	9.77	3.39	0.46	56.75	0.31	10.15	1.32	2.08	0.60
Direct reduction	0.76	6.30	0.67	66.00	0.17	5.81	1.32	6.00	0.59
Oxygen-blowing	2.42	3.10	1.83	65.37	0.14	9.17	0.86	1.84	0.76
Refining	Low	35.22	2.72	26.60	0.45	6.70	0.53	7.55	0.55

[a] Iron content was calculated as whole and converted to Fe_2O_3 content.

evaporating slag components with water vapour in the thermally induced air flow.[5] Dust and hydrogen-fluoride generation rates in kg per tonne of remelted steel are proportional to remelted steel output in tonnes per hour, depending on the parametric influences of the ratio of mould surface-area to electrode surface-area as a measure of reaction interface-area and, for hydrogen fluoride-emission, on the amount of fluoride in the slag.

Specific dust-generation rates of between 0.2 and 0.4 kg/t and hydrogen fluoride evolution rates of between 0.15 and 0.25 kg/t have been recorded at a remelted steel output rate of 0.7 t/h in a slag containing 75% fluorspar for an area ratio of between 2.5 and 3.5.[1]

13.2.2 Collection and Cleaning of Exhaust-gases

The dust-laden exhaust-gases generated by electric-furnace processes have to be collected and cleaned in a dust-extraction system. Gases may be collected directly from the furnace vessel or via a hood system in the furnace shop roof above the crane gantry or by total enclosure of the furnace area and collecting gases from the enclosed area. The first two of these are illustrated in Figure 4.

13.2.2.1 Collection of gases from the furnace vessel

There are two main groups of methods for collecting exhaust gases from the furnace vessel:

(1) Direct extraction collection (DEC system) via the fourth hole in the furnace roof;
(2) Indirect local collection (side-draught system).

Exhaust gas collected through the fourth hole in the furnace roof consists of gases given off in the furnace during the process and primary air drawn into the vessel by extraction of the gases. The extraction-system suction is greatest if the exhaust duct is in contact with the fourth-hole elbow and there is no gap for secondary air to enter. However, such an arrangement is not possible for operational reasons because the furnace has to be able to move; it is also undesirable since a certain amount of secondary air is necessary for safety reasons to complete combustion of partly burnt waste gases.

The area of the gap should be as small as possible in relation to the area of the fourth hole, since the area ratio greatly affects the total volumetric flow rate V_A that has to be handled by the dust-extraction system. The area ratio ϕ should always be less than unity and, if possible, should be about 0.5. When ϕ is 0.5, enough secondary air is drawn in to burn off the carbon monoxide completely before the exhaust-gas reaches the dust-extraction plant. Figure 5 shows the volumetric flow rate V_K of the waste gas produced by the furnace and the total

Collection of gases from the furnace itself

Direct extraction via the fourth hole in the furnace roof

Extraction with or without furnace pressure control (DEC system)

Pressure-relief and capture

Indirect local gas-collection (side-draught system)

Collection at the electrodes

All-round collection

Collection of gases above the furnace

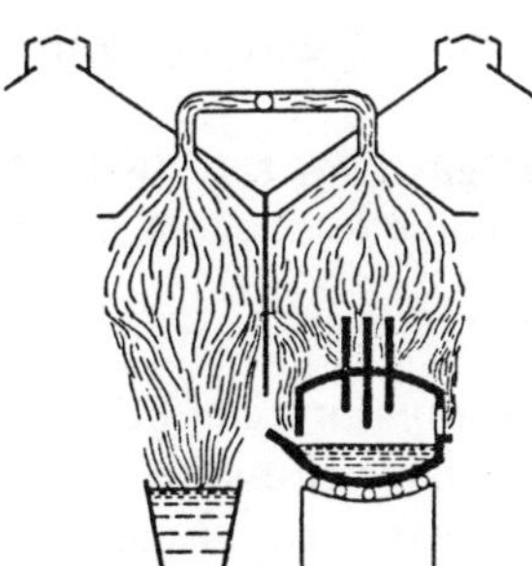

Roof-hood exhaust system

Figure 4. Different ways of collecting exhaust gases from electric arc furnaces

exhaust gas flow rate V_A which has to be handled when mean area ratio ϕ is 0.9; both are plotted against average decarburization rate.[2] The total exhaust-gas volumetric flow rate V_A is made up from the furnace-waste gas flow rate V_k drawn from the fourth-hole elbow and the volumetric flow rate V_s of secondary air drawn in at the gap. Normally V_A lies between about 1000 m^3/h per tonne of crude steel for low-alloy steels and 1300 m^3/h per tonne for high-chromium-alloy steels at high decarburization rates. Various forms of direct extraction system exist:[6–11]

(1) Furnace pressure-control by altering the gap area and the amount of secondary air drawn in (by-pass control);

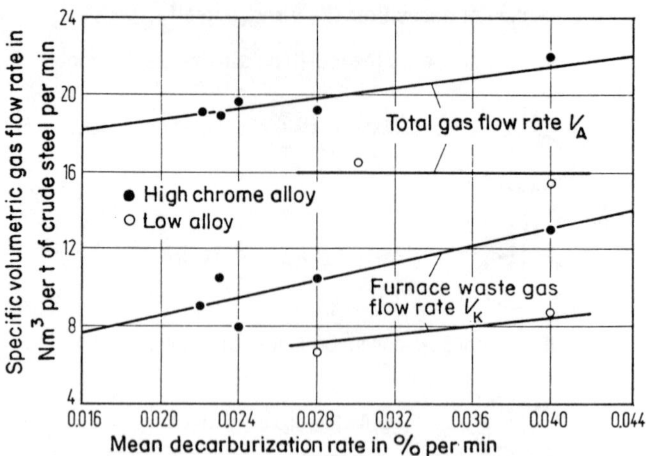

Figure 5. Furnace-generated exhaust gas flow rate (V_K) and total gas flow rate capacity (V_A) required of the dust-extraction system; the ring-gap area ratio ϕ is 0.9

(2) Furnace pressure-control by means of a butterfly valve in the exhaust duct while keeping the gap at the fourth-hole orifice constant;
(3) Furnace pressure-control by using secondary air to gate the fourth-hole aperture;
(4) A process that maintains a constant exhaust-gas flow rate, with flexible curtains in the air gap acting as a non-return valve to prevent overflow in event of excess pressure in the furnace.

In the first two systems mentioned the furnace pressure can be controlled between 0 and 50 Pa; in the third process the furnace pressure can also be controlled in the region of low overpressure. This is particularly necessary when using the two-slag process for the production of alloy steels.

Almost all electric arc furnaces today have direct-extraction systems using one of the processes mentioned. The side-draught system is rarely used, since in most cases the installation and maintenance costs are higher than for a DEC system; however, side-draught systems can handle three to five times more total gas flow rate than direct-extraction systems.

13.2.2.2 Exhaust-gas collection using a hood

During certain stages of the steel-production process, e.g. during charging and tapping, it is not possible to collect the dust by using direct-extraction; however, this can be done by means of a hood system built above the crane gantry.[12–17] In existing steelworks it is often difficult to install a hood system beneath the roof or in the roof itself; even then its efficiency in capturing secondary gas emission is

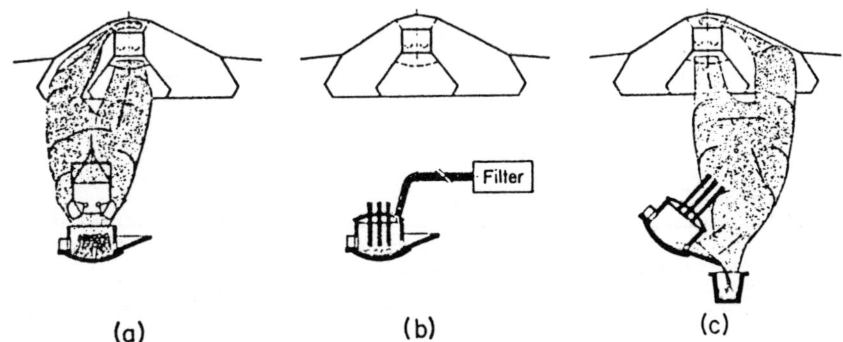

Figure 6. Three-section hood system in roof above an electric arc furnace collecting exhaust gases during specific operating phases. (a) Charging; (b) melting; (c) tapping

often limited. Figure 6 illustrates how a three-section hood system, retrospectively installed above an electric arc furnace, operates in various furnace-process phases. Two hood sections at a time are used during charging and tapping, while direct-extraction is used when the furnace is operated with its roof closed.

Some 85–8% of all dust arising in processing is collected by direct extraction.[18–22] A hood extraction-system retrospectively installed in the roof of an existing steelworks results in an increase of about 10% in the amount collected of the total dust given off. In newly built steelworks almost all dust given off can be collected by using a hood in combination with direct extraction.

An exhaust-gas volumetric flow rate of 700 000 m^3/h should be allowed for each furnace in order to ensure efficient dust-extraction by a hood system in the shop roof; however, this also depends on the height of the furnace bay, the size of the furnace, and the nature of the operation. When several furnaces are operated in one shop it is possible to reduce the flow rate allowance per furnace to 500 000 m^3/h by staggering furnace-operating phases in time. Thus the volumetric flow rate to be handled by a shop roof hood extraction system is about five to ten times greater than that for direct extraction for the same size of furnace.

The capital outlay and operating costs for an exhaust gas collection and filtration plant are accordingly very high. It can be taken as a guide that operating costs increase three- or four-fold when changing over from direct extraction to a combined system, according to the size and number of furnaces.[22,23] Figure 7 is a schematic illustration of a three-section hood system built as a retrospective modification into the roof of a shop containing a 50-t furnace standing on the floor. The three-part hood design has proved effective since this enables more flexible control of exhaust-gas flow rate in the hood inlet to suit conditions during charging and tapping.

In newer steelworks the electric arc furnaces are installed on a furnace

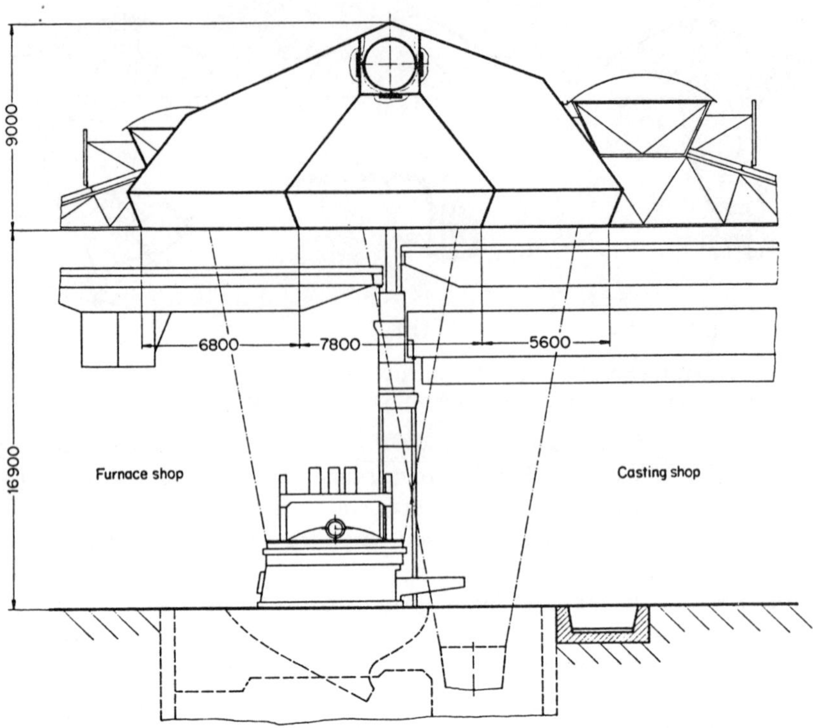

Figure 7. Schematic illustration of a hood-extraction system installed retrospectively above an electric arc furnace (Stahlwerke Südwestfalen AG, arc furnace No. III)

platform so that tapping and slag work can be done on the shop-floor. Figure 8 shows the arrangement of a shop roof-hood system for a 110-t furnace with feed-conveyor installation.

In most cases shop roof-hood extraction systems are used together with direct extraction, and the exhaust-gases are either fed into a single combined dust-extraction installation or to two separate ones. When primary and secondary exhaust-gases are fed together to one filter there is no need for a cooler, as is required for direct-extraction gases alone, since the large volume of air drawn into the hood system cools the exhaust-gases to the temperature necessary for input into the filter.

With smaller furnaces direct extraction can be dispensed with by allowing furnace exhaust gases to convect up to the hood through a vertical water-cooled stack on the furnace roof. An example of this arrangement for an 25-t furnace is shown in Figure 9.

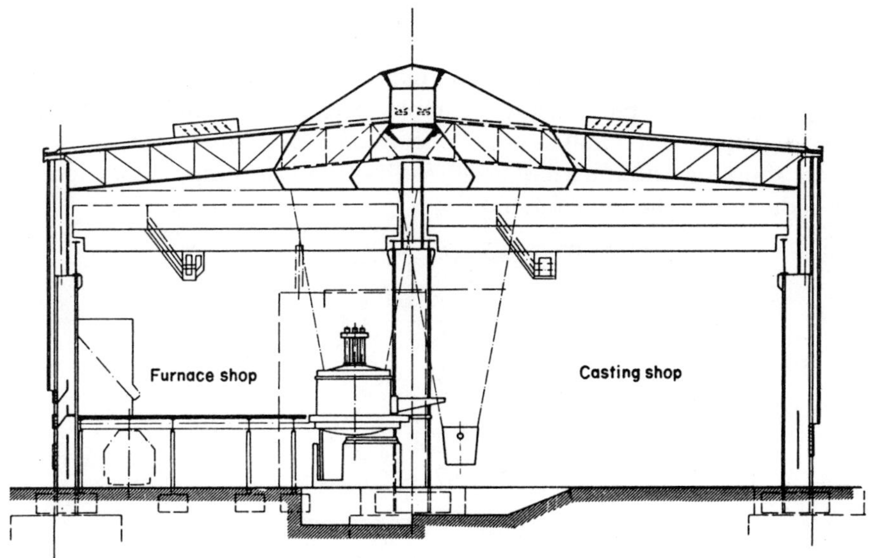

Figure 8. Schematic illustration of hood-extraction system installed retrospectively above an electric arc furnace (Hamburger Stahlwerke GmbH, arc furnace No. 3)

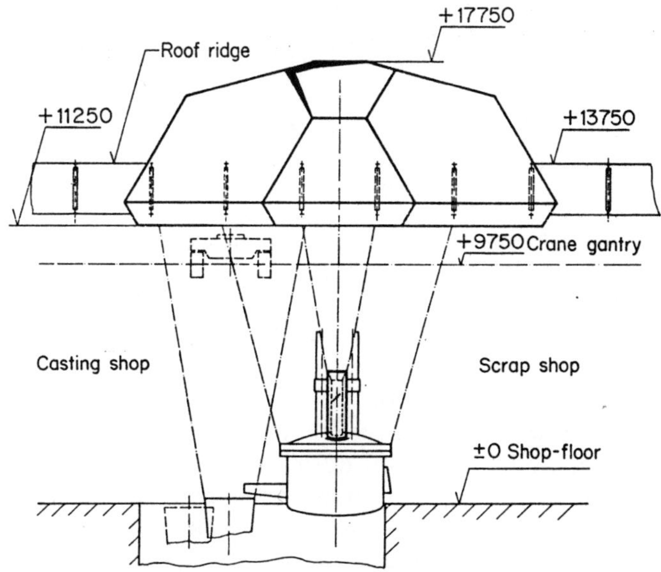

Figure 9. Schematic illustration of an exhaust gas collection-hood used in conjunction with a water-cooled stack on the furnace roof (Marienhütte, Graz)

13.2.2.3 Waste-gas collection in newly built steelworks

The hood systems mentioned so far are built as modifications into the roofs of existing steelworks buildings to comply with legal requirements. New steelworks construction makes use of encapsulation of entire areas such as the furnace shop or the furnace itself for both exhaust-gas collection and noise-reduction at the same time.

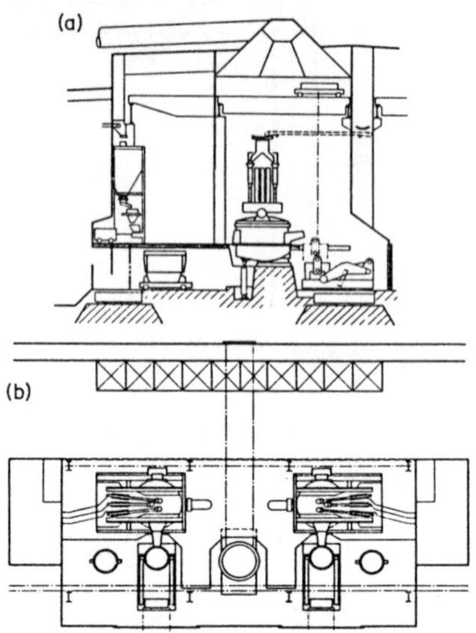

Figure 10. Illustration of the principle of an encapsulated furnace bay. (a) Elevation; (b) plan view

Figure 10 is a schematic illustration in plan and elevation of one way of encapsulating the furnace bay in an electric steelworks with two arc furnaces. This principle is applied in extended form to two 100-t furnaces in a Danish Steelworks (Det Danske Staalvalsevaerk) in which the work-stations on the furnace stage are screened from the furnaces (see Chapter 10, Figures 12–14).

UHP furnaces usually have short tap-to-tap times of 1–2 h because of their high transformer power, and because they use large quantities of oxygen and additional burners during the melt-down phase. This enables one crane per furnace to be utilized effectively running in the material flow direction, unlike in conventional layouts. Such a system eliminates waiting-time caused by cranes obstructing each other and allows any number of electric furnaces to be arranged in a row.

Figure 11 shows section and plan views of a new-built electric furnace shop with two cranes operating one above the other in the material-flow direction. When the furnace roof is in place, waste gases from the furnace are exhausted

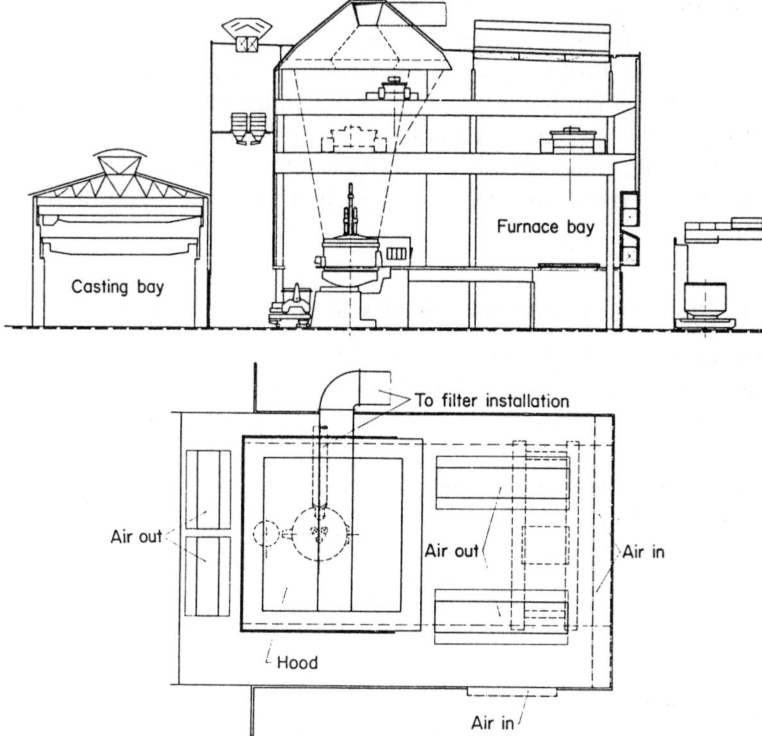

Figure 11. Elevation and plan view of furnace-shop layout for a 110-t UHP furnace (Stahlwerke Südwestfalen AG, arc furnace No. VI)

by direction extraction via the fourth hole in the furnace roof. Gases leaking from the furnace during this phase convect up in the furnace bay, which is enclosed on three sides, and are captured with entrained air in the large hood, measuring 26 × 22 m, installed in the roof. During charging, patching, and tapping the hood-extractor operates at maximum performance and enables the exhaust-gases laden with foreign substances to be collected in the roof and ducted away to the filter-installation.

Complete encapsulation is achieved in a steelworks concept still in planning stage at the time of writing (see Chapter 10, Figures 16 and 17).[24] Dust and furnace gases are collected in the usual way via the fourth hole in the furnace roof. The other dust-laden gases emanating from the furnace are led up inside the enclosed space to the 'roof', from where they are ducted through a special exhaust-system to a dust-extraction plant. As the space served by the exhaust system is smaller than that served by the usual sort of roof-hood extraction system used in electric steelworks, the volumetric flow rate is less and the cost of dust-extraction is lower. The furnace is also enclosed during scrap-charging, thus preventing flame- and gas-ingress into neighbouring bays. The operating procedure is as follows: the crane travels with the scrap-basket to the charging position above the furnace. For this the self-propelled wall moves out of the way towards the end of the crane track. The similar wall mounted on the back of the crane closes as soon as the crane reaches the charging position, and the time for which the enclosure is open is brief. After charging, the 'wall-change' takes place in reverse and the self-propelled 'first wall' resumes its initial position.

13.2.2.4 Planning exhaust-gas collection in existing steelworks

For existing steelworks planning deliberations lead to the concept of confining encapsulation of the electric arc furnace to a small space beneath the crane gantry (Figure 12). This form of encapsulation is also possible with new construction. The furnace is charged after the basket and the two movable roof-sections have been moved into place. Waste gases are exhausted through a duct fixed between the crane-track supports. Steel is tapped into a ladle mounted on a bogie.

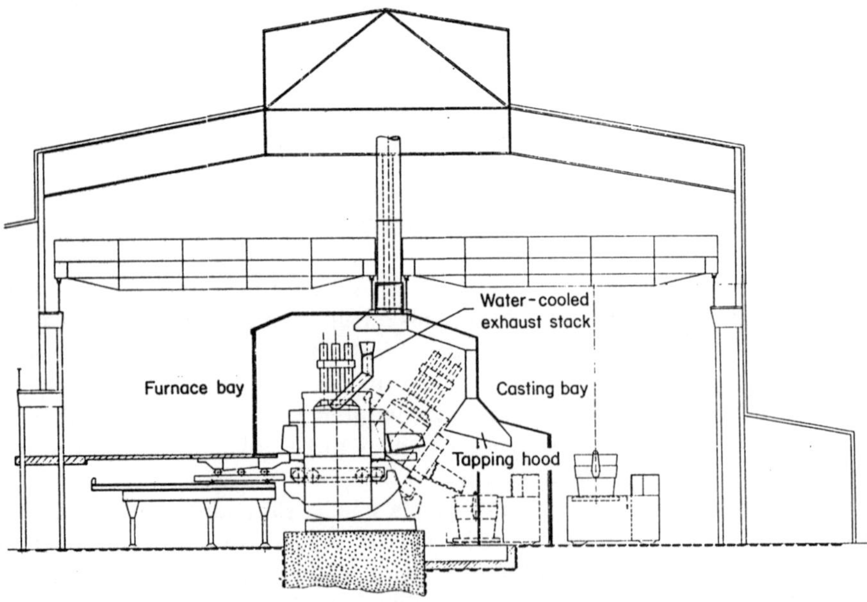

Figure 12. Planning study for encapsulating a 40-t furnace (Norsk Jernverk)

Since exhaust-gases generated in normal furnace operation are removed by direct extraction, it is often sufficient to only provide additional extraction facilities for gases given off during tapping. This is particularly appropriate when a furnace is charged continuously with shredded scrap or sponge iron in pellet or lump form. Two possible arrangements for this form of enclosure for tapping arc furnaces mounted on the shop-floor or stood on a platform are shown in Figures 13 and 14.

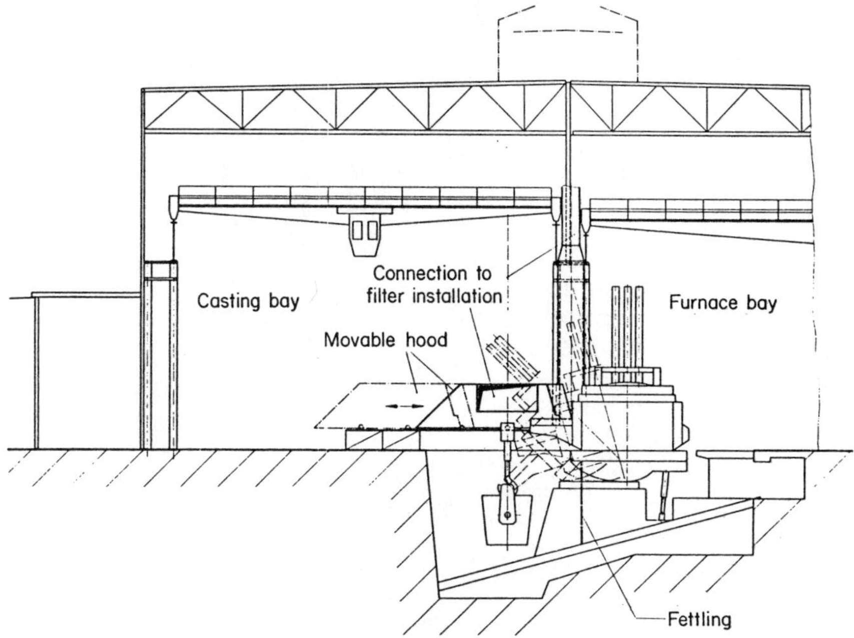

Figure 13. Scheme for a movable tapping hood for a floor-mounted 45-t arc furnace (plan for Zelezarna Ravne)

When a furnace is tapped into a tapping pit the hood shown in Figure 13 can be replaced by a horizontal screen of air with the gas-capture hood mounted at one side.[19,21] Figure 15 illustrates the principle of a pilot-installation of this type. Capture-efficiency can be improved by using compressed-air jets at the side in place of the fan-blowers.[25]

With all three methods of collecting waste gas generated during tapping, the volumetric flow rate to be handled can be reckoned to correspond to the capacity of the direct-extraction system, but it depends on the size of the furnace and the time taken for tapping.

13.2.2.5 Removal of dust from electric furnace fumes

Exhaust-gases collected by one of the methods mentioned in the previous sections have to be cleaned in a dust-extraction installation. In the Federal Republic of Germany, VDI Guideline 3465 stipulates that with the current

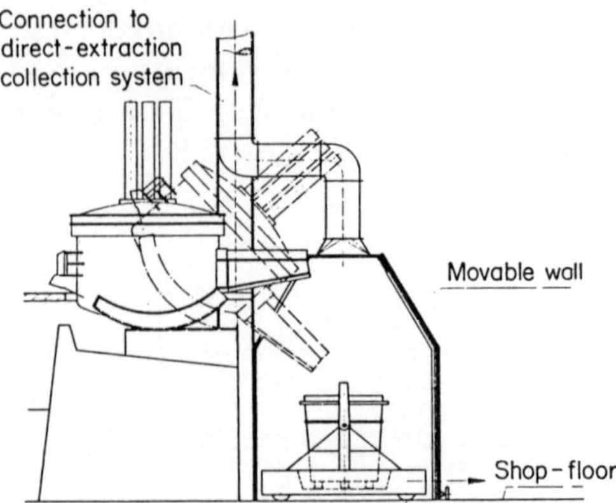

Figure 14. Scheme for encapsulting the tapping process including the tapping-ladle bogie for a 60-t furnace (plan for Firth Brown, Sheffield)

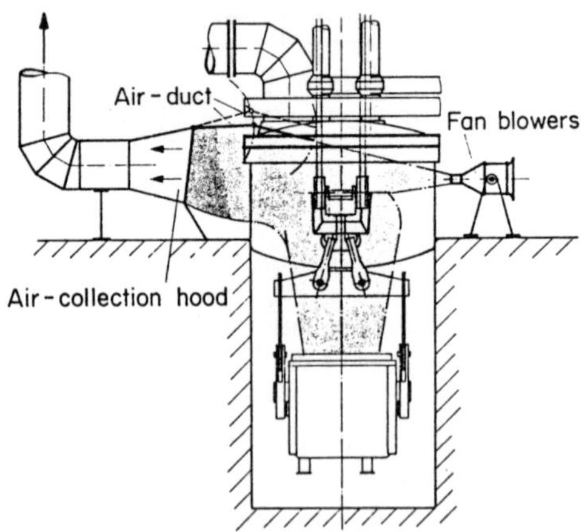

Figure 15. Principle of air-barrier installation (pilot system for 25-, 40-, and 100-t electric furnaces)[19,21]

state of technology, exhaust-gases from new installations in the standard configuration should be purified to a final dust-content of no more than 75 mg/m^3.

In the past electric arc furnaces were equipped with the most varied kinds of dust-extraction systems, such as

Wet electric filters,
Dry electric filters,

Pease-Anthony venturi scrubbers,
Disintegrators, and
Fabric filters.

However, in the last few years there has been a clear trend towards use of fabric-filter installations.[18,26] This development can be traced partly to more stringent clean-air regulations, which place great importance on air-purity standards and their attainment as well as the environmental compatibility of methods of dumping or disposal of materials extracted from exhaust-gases in the course of purification.

The dust is precipitated dry, so that no water-pollution or sludge problems arise. Disposal and dumping can sometimes be simplified by making the dust into pellets immediately after collection.

Where fabric filters are used it is important that the gas-temperature does not exceed the limit for the filter medium. Filter gauze and felt made from polyacrylonitrile such as Redon or Dralon can withstand mildly acid or alkaline fluids and can be used for long periods at temperatures up to 140°C. In recent years, gauze and felt made from polyester or polyamide fibre have become more and more widely accepted. Peak-temperatures arising in the furnace are reduced by using suitable cooling systems and/or mixing in air from the outside or cooler furnace exhaust-gas from the roof-hood system.

Experience shows that gauze made from the above materials can tolerate exhaust-gas flow rates in the region of 0.8–1.1 m^3 per minute per square metre while felt can withstand 1.5–1.8 $m^3/min/m^2$, depending on the operating conditions. If the flow rate is much higher the filter medium will wear out sooner, but if it is lower the capital cost is unnecessarily high.

Figure 16 illustrates an exhaust-gas collection system with cloth filter added on. In the pressure filter shown the dust-laden exhaust-gas is scrubbed in the first three chambers, while the filter tubes are back-flushed in the chamber to clean them.

13.2.2.6 Induction furnace waste-gas removal and scrubbing

Dust-laden waste gases which arise during melting down, preparation, and tapping with induction furnaces are either collected in an adjustable hood above the furnace or sucked into slotted ducts encircling the gap between the crucible and the furnace roof. In normal operation with the furnace roof closed most waste gas is generated during preparation and tapping. Depending on the dust-concentration, exhaust-gases are either cleansed in a filter installation connected to the system or released direct to atmosphere via a chimney.

Dust is generally removed by means of fibre or fabric filters which work dry. Low-pressure scrubbers with or without filter are only used in a few cases. With these kinds of exhaust-gas filter used in appropriate installations, the

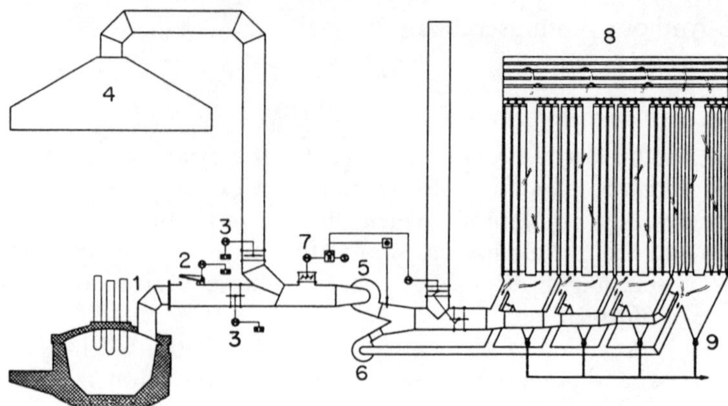

Figure 16. Schematic diagram of a dust-extraction system for an electric arc furnace using both direct extraction and a hood. 1 Furnace-roof exhaust elbow; 2 entrained-air intake flap; 3 changeover flaps; 4 hood; 5 exhaust fan; 6 flushing-air fan; 7 temperature-controller; 8 filter house; 9 dust-removal

concentration of solid matter left in the gas after scrubbing is less than the limiting value of 100 mg/m^3 laid down in clean-air regulations (technical) for foundries.

13.2.2.7 Dealing with exhaust gases from ESR systems[27]

ESR (electro-slag remelting) installations are usually equipped with exhaust-gas extraction systems in order to protect the environment from the gases and dust generated during the remelting process. This is particularly necessary for slag premelting furnaces, since the heaviest dust output can be expected while the slag is being melted, and collecting the gases enables one to prevent maximum permissible concentrations being exceeded in the furnace shop. In addition to having an effective extraction system it is advisable also to use dry materials and a dry atmosphere in order to keep fluorine emission as low as possible.[28]

If necessary, a filter plant can be added to the exhaust system so that the dust is separated out as well as ducted away.

13.2.2.8 Exhaust systems for plasma-melting furnaces[29]

The largest plasma-melting furnaces in operation at the time of writing have nominal capacities of 10 and 30 tonnes.[30,31] With these, as with arc furnaces, exhaust gases are removed via a fourth hole in the furnace roof. The amount of gas and dust generated depends mainly on the metallurgical process used in the furnace.

If the furnace is used only as a melting system for alloy scrap, an atmosphere containing a high proportion of argon is always maintained at a slight overpressure in order to optimize alloy output. With this process the amount of exhaust gas and dust generated is small, and the exhaust-system capacity need not be great.

If refining with gaseous oxygen is also carried out in the plasma furnace, then the capacity of the exhaust- and dust-removal systems should not be the same as for an electric arc furnace.

13.3 NOISE-EMISSION AND ITS REDUCTION

The electric arc furnace is the chief source of noise in an electric steelworks. Higher levels of specific installed electric power demanded because of technological and metallurgical developments lead to shorter charge-cycles and thus to louder and more frequent bursts of noise. This, and a greater critical awareness of industrial noise, have led to a closer examination of the problem in electric steelworks in order to find ways of reducing the noise.

Ultra-high-power furnaces of up to 140-t tapping weight with transformer apparent powers of up to 85 MVA are being planned in the Federal Republic of Germany. These will have tap-to-tap times of up to 1.5 h. These times could be further cut by using oil–oxygen burners to speed up the scrap-melting rate; a Japanese electric steelworks using this technique quotes tap-to-tap times of around 70 min.[1]

Smaller steelworks producing up to around half a million tonnes of crude steel in a year are scattered throughout the country and are not only sited in industrial centres; these works can be a major problem as far as noise is concerned.

13.3.1 Noise Emitted during Electric Steelmaking

The main sources of noise in electric steelmaking are the electric arc furnace and the handling of scrap. Noise is also emitted from heating stands for transport ladles (ladle-firing) as well as by the dust-extraction plant and the fans for water-cooling. In steelworks with roof-mounted extraction-hood installations, the broad external exhaust ducts have a large surface area from which noise is emitted; these must be considered as a further source of noise.

13.3.1.1 Scrap-handling

The noise emitted depends on the quantity of scrap handled in a given time; this is determined by the output of the furnace and the number and size of the scrap-baskets used. The scrapyard often contains scrap-preparation equipment, such as breakers, shears, and crushers, which also contribute to the noise generated.

During scrap-handling, noise levels L_h of up to 106 dB have been measured at ranges of 10 m, depending on the type of scrap and its composition; the typical steady noise level L_s is about 97 dB. The acoustic power level of the noise generated can be calculated to be about 125 dB, having regard to the time involved. As far as the effect of this noise is concerned, it should be borne in mind that the sound is impulsive in character (a series of crashes). For this kind of noise at night, the regulations (see section 13.1, reference 10) stipulate that the guide-value (which is based on the mean noise-level) should never be exceeded by more than 20 dB, even once.

13.3.1.2 Electric furnaces

The noise emitted by an electric arc furnace depends on many factors, of which the most important are the specific installed electric power (normal and UHP furnaces), the type of charge and its composition, the efficacy of the furnace sealing, and the type of process and stage in the process at the time. The noise-level in the workshop is also affected by the geometry of the furnace bay and the way it is built.

When scrap is being melted down the noise is mainly caused by the arcs igniting and changing rapidly at mains frequency and by the scrap moving as it melts. Thus, frequencies of around 100 Hz (twice the mains frequency) are particularly prominent in a typical frequency-spectrum (Figure 17).

The mechanism whereby noise is generated in the furnace is not yet precisely understood, and steps cannot yet be taken to reduce noise at its source in the furnace.

In order to measure the sound emitted from an electric furnace and to obtain values for comparing the various types of furnace, measurements are taken at a defined reference-point situated on the furnace platform 5 m from the vessel and at an angle of less than 45 degrees from the furnace door. This usually produces peak readings of up to 120 dB(A) for a modern UHP furnace during the melting-down phase. A typical noise level plot for two 50-t arc furnaces using sorted scrap is shown in Figure 18.[2] The plot is characterized by high peak values which arise during the melting-down phase shortly after the baskets are charged and the furnace is switched on. The difference in the mean noise levels of 102 and 96 dB(A) is due to the different furnace powers.

With modern arc furnaces operating at powers of over 60 MVA the average noise level during processing of a charge is about 106 dB(A). When the insulating effect of the furnace, which cuts the noise level by 20 dB,[3] is taken into account, the noise-level inside the furnace amounts to about 150 dB(A).

The noise-level in the furnace shop is determined not so much by the sound-deadening effect of the furnace structure as by points, such as the electrode apertures, the fourth hole in the roof, the furnace door, and the roof seal, which are acoustically-weak. This is why improving the sound-insulation

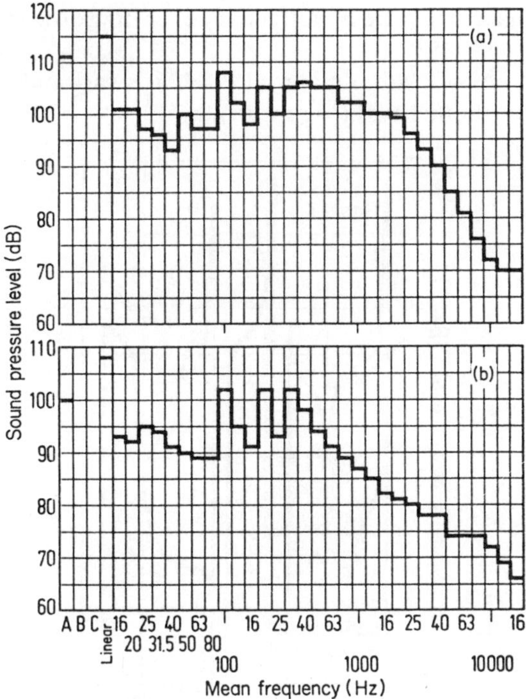

Figure 17. Typical third-octave band-noise spectrum for an arc furnace measured at the same reference point (a) during melt-down, (b) during the refining phase

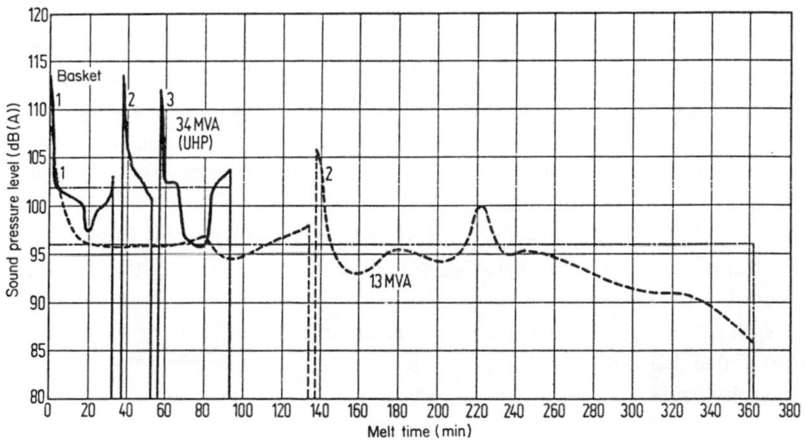

Figure 18. Plot of noise level against time for two 50-t electric arc furnaces; one is operated at normal power, the other is UHP. Both use sorted scrap. Measurements were taken at the reference point

quality of the furnace structure is ineffective for cutting down the noise-level on the shop-floor.

With certain types of charge-material, such as swarf and pellets, the noise level is reduced by about 10 dB (Figure 19).[4,5]

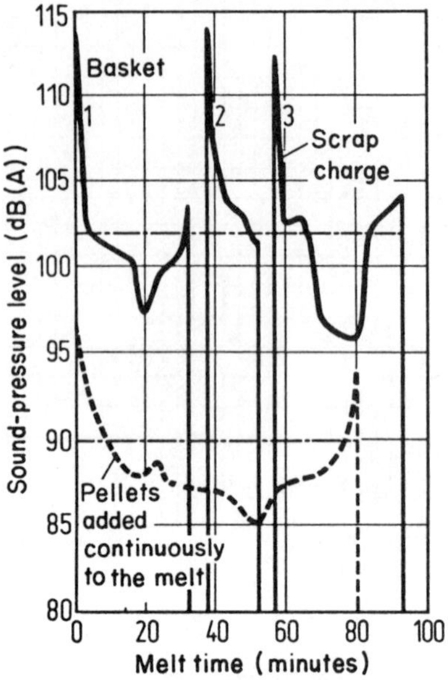

Figure 19. Plot of noise level against time for a UHP furnace using different types of charge, measured at the reference point

With induction furnaces, unlike arc furnaces, the noise generated during melting down scrap does not exceed the background noise-level in the melting shop. The sound-pressure level at 1 or 2 m from the furnace opening is less than 85 dB(A). There are various electrical auxiliaries, depending on the frequency used (mains, medium, or high) which often generate secondary noises that can cause a nuisance.

The noise-level when scrap is melted in a plasma furnace is low.[6] According to the manufacturers, the noise generally does not exceed the background-noise level in the shop.

The electro-slag remelting process produces relatively little noise, and there is almost no difficulty in keeping the noise in the melting-shop to within permissible limits.

13.3.1.3 Other sources of noise

In most electric steelworks the dust-extraction plant and its fans are situated in their own building outside the main steelworks building. The noise emitted is ultimately determined by the power and performance of the plant. It is affected by the volumetric flow rate of the exhaust gases and the pressure-losses in the exhaust system, which in turn depend on:

(1) Furnace size;
(2) Type of extraction system (direct extraction via a fourth hole in the furnace roof and/or roof-mounted extraction hood system);
(3) Type of dust-extraction system (fabric filter, electric filter, wet scrubber);
(4) Type of building housing the dust-extraction system.

The extractor fans are a significant source of noise in a dust-extraction plant; acoustic power can amount to 120 dB(A). Flushing and cooling fans and worm or chain drives for dust-transport can often cause squeals.

For dust-removal systems with secondary extraction where the volumetric flow rate is high, ducts which have a large surface area usually have to be mounted at considerable heights and can thus emit noise without hindrance. Even when the noise-output level at the duct walls is only 85 dB(A), acoustic power levels of about 105 dB can arise which need special attention because of their effect in any neighbouring residential areas.

The water-cooling fans emit noises containing no individual notes and are generally well screened; they therefore are not particularly significant from the noise point of view.

13.3.2 Possibilities for Reducing Noise

Up to now we do not know of any basic means of dealing with the noise emitted by a.c. arc furnaces by carrying out constructional modifications or changing the operating process and reducing the noise at its source in the furnace. For this reason, noise to which furnace personnel are subjected can only be reduced by:

(1) Improving the sound-insulation of the furnace structure;
(2) Acoustic encapsulation of the furnace or furnace bay;
(3) Installing sound-proofed control cabins.

On the other hand, although acoustic barriers around the outside of the furnace shop, such as at air-intakes and -outlets, windows, doors and gateways, do provide protection for the neighbourhood, they have very little, if any, effect on the level of noise on the shop-floor.

13.3.2.1 Steps taken at the furnace

The thickness of the furnace wall, which is determined by production and process considerations, provides quite good noise-insulation which can only be improved by further increasing the wall-thickness. However, this is never effective because the noise emitted by the furnace comes mainly from the openings, which include circular gaps around the electrodes, working apertures, the fourth hole in the furnace roof, and imperfections in the fit of the furnace roof on the vessel.

For this reason, attempts have been made to reduce the gaps between the cooling rings and the electrodes; however, these have a very short life in practice.[3] Even if this were sufficient, the improvement resulting from closing the gaps around the electrodes would only be slight because the noise emitted also depends on the fourth hole in the furnace roof, used for dust-extraction, which has a relatively large cross-section area.

The fit between the furnace vessel and its roof must also be considered; gaps here can be avoided by using suitable sealing methods like a sand-tray. The apertures required for working the furnace should be well sealed by suitable construction, e.g. self-sealing doors.

All the same, it can be said that with the present state of the art the fourth hole in the roof prevents further improvement in noise-reduction through constructional changes to the furnaces.

13.3.2.2 Steps taken around the furnace shop

The noise-reduction measures described in section 13.3.2.1 have little effect, and, since electric arc furnaces are often operated close to residential areas, it is necessary to take steps around the outside of the furnace shop to protect the environment. The nature of these acoustic measures depends on the noise-emission level guide-values which have to be met.[7] Special attention must be paid to acoustic weak points such as doors, gates, and windows as well as air-intakes and -outlets. It is often necessary to build the furnace shop from materials that provide sound-insulation as well as protection against the weather. it may be necessary to double up the doors and gates, provide ventilation-apertures with baffles (coulisses) to cut down noise (Figure 20), and make windows for lighting purposes out of glass bricks.

13.3.2.3 Acoustic isolation of furnaces from other work-areas

According to health and safety regulations (see section 13.1), all work-areas in an electric steelworks must be protected against noise from electric furnaces. In conventionally constructed electric steelworks it has always been possible up to now to partially isolate the furnace shop from other areas because of the flow of materials. In these cases, because of the gaps remaining, there is no point in using building materials with high sound-insulation properties (over 35 dB). If the apertures necessary for transport in and out of the furnace area can be made smaller, for instance by using revolving

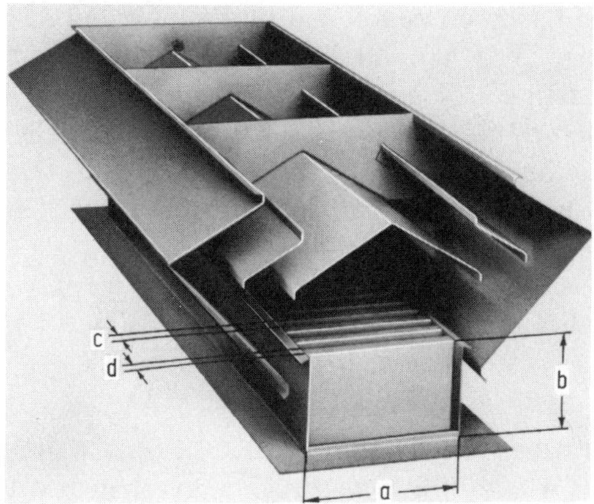

Figure 20. Ventilation outlet with wind cowl and baffles

or roller-doors,[8] the overall noise-reduction achieved will be proportionately greater.

More stringent environmental protection laws and regulations on health and safety at work now require newly built furnace shops to be isolated as completely as possible from neighbouring workshops. This can only be done by repositioning the furnaces compared with conventional layouts. This makes it impossible to use a crane to transfer crude molten steel into the casting bay. In one modern steelworks[9] molten crude steel is transferred from the furnace on a ladle-bogie through a doorway in the acoustic partition, via doors that shut, into the casting shop (see Chapter 10, Figures 12 and 14). This almost total separation of the casting shop from the furnace bay has enabled furnace noise-penetration to be cut by around 30 dB(A), so that the noise-level in the casting shop is no longer affected by the noise from the arc furnace. The furnace has to be positioned away from the building's main support axis because the isolating wall is built along it.

In addition, in these works the working-platform in front of the furnace is acoustically isolated from the two arc furnaces by walls as well as by enclosing the control console.

Large openings are necessary so that materials can be transported into the furnace shop from the scrap-store; this can have a very adverse affect on the acoustic barrier provided by the outer walls. Because of the high frequency of the traffic concerned, it was decided to do without quick-action doors and to use a tunnel to absorb the noise instead.[9]

When sound-proofing is considered together with clean-air requirements then this type of solution is seen to be rather expensive, since large quantities of air have to be exhausted from the large furnace shop and, in addition, the area of sound-absorbent walls required is very great.

Another system has been developed by Krupps, but at the time of writing this had not yet been used in practice.[10] This system is based on the principle of encapsulating the furnace (see Chapter 10, Figures 16 and 17). The furnace is enclosed in a structure like a house. This arrangement has two advantages: it enables all the reaction gases to be collected and shields the surroundings, and particularly the working-positions on the furnace stage, from the noise of the UHP furnace.

It should be possible in this way to cut the noise-level by about 25–40 dB.[10] The total thickness of the acoustic walls is around 400–500 mm.

13.4 DUST-DISPOSAL

Up to the present it has only been possible to utilize a proportion of the dust collected from electric arc furnaces; this means that dust has to be dumped. According to the waste-disposal laws,[1] waste may only be processed, dumped, or disposed of in approved sites or systems. Dust may be dumped at a works' own special dump or an approved dump elsewhere.

Dust from electric furnaces is pelletized in various plants, sometimes with the addition of fine quicklime. This improves transportability and largely prevents the dust being blown away from tips by wind; it also has a particularly beneficial effect on leaching-away behaviour. Tests on lime-stabilized pellets show that compression strengths of up to 800 N per pellet can be obtained.[2]

Extensive testing is required in the approval procedures laid down by the waste disposal laws. In particular, tests must be carried out of the leaching-away behaviour of the material to be dumped. At the time of writing, guidelines were being prepared in North Rhine – Westphalia to lay down the principles for assessing dust in that local government area.

Tests carried out on dust from electric furnaces have shown in the meantime that the leaching-away behaviour of the dust is largely dependent on the sulphate and chloride levels, which, according to Table 2 in section 13.2.1, should be relatively low.

Efforts are being made to find an eventual use for dust from electric furnaces. Already today dust which is low in zinc and lead content (such as in columns 5 and 7 of Table 2, for example) is recharged into electric furnaces in pellet form. Several reports have appeared in recent years[3–6] about possible ways of treating steelworks dust. A few electric steelworks make use of dust treatment facilities in the non-ferrous metal industry. The income from this goes some way to compensate the high costs of other processes.

13.5 LITERATURE REFERENCES

13.1 Legal Requirements in the Federal Republic of Germany

1. Gesetz zum Schutz vor schädlichen Umwelteinwirkungen durch Luftverunreinigungen, Geräusche, Erschütterungen und ähnliche Vorgänge (Bundes-Immis-

sionsschutzgesetz – BImSchG) 15 March 1974. Bundesgesetzbl., T.I, 1974, pp. 721–43 and 1193.
2. Vierte Verordnung zur Durchführung des Bundes-Immissionsschutzgesetzes (Verordnung über genehmigungsbedürftige Anlagen – 4. BImSchV) 14 Febr. 1975. Bundesgesetzbl., T. I, 1975, pp. 499–503 and 727.
3. Neunte Verordnung zur Durchführung des Bundes-Immissionsschutzgesetzes (Grundsätze des Genehmigungsverfahrens – 9. BImSchV) 18 Feb. 1977. Bundesgesetzbl. T. I, 1977, pp. 274–9.
4. VDI-Richtlinie 3465. Auswurfbegrenzung. Stahlwerksbetrieb. Elektrolichtbogenöfen (Draft Oct. 1975). Düsseldorf, 1975.
5. Technische Anleitung zur Reinhaltung der Luft (1. Allgemeine Verwaltungsvorschrift zum Bundes-Immissionsschutzgesetz – TA Luft) 28 August 1974. *Gemeinsames Ministerialbl.* **25** (1974), 426–52, 525.
6. Verordnung über Arbeitsstätten (Arbeitsstättenverordnung – ArbStättV) 20 March 1975. Bundesgesetzbl., T. I, 1975. pp. 729–42.
7. Maximale Arbeitsplatzkonzentrationen 1978. Hrsg. von der Senatskommission zur Prüfung gesundheitsschädlicher Arbeitsstoffe. Mitteilung XIV 16 June 1978. Bonn/Bad Godesberg 1978.
8. Unfallverhütungsvorschrift Lärm (UVV Lärm) des Hauptverbandes der gewerblichen Berufsgenossenschaften. Bonn. Dec. 1974.
9. Schutzabstände zwischen Industrie- und Wohngebieten. Rd. Erl. No. III 12/72 vom 12. 6. 1972. Ministerialbl. Nordrhein-Westfalen, Ausg. C, Gliederungs-No. 2311.
10. Technische Anleitung zum Schutz gegen Lärm (TA Lärm) 16. 7. 1968. Bundesanz. No. 137, 26 July 1968 (Beilage).

Tabler, Shirley K., Federal standards of performance for new stationary sources of air pollution. *J. Air Pollution Control Assoc.* **29** (1979), No. 8, 803–11.
Dear, T. A., Industrial noise regulation and worker protection. *Wire Ind.* **45** (1978), No. 531, 171–2.

13.2 Emission and Control of Harmful Substances

1. Evaluation of unpublished report by the Betriebsforschungsinstitutes des Vereins Deutscher Eisenhüttenleute, Düsseldorf.
2. Petersen, K., and H. Kahnwald, Direct collection of the dust-containing fumes of electric-arc furnaces during the production of high-chromium steels and their de-dusting with the aid of cloth filters (German). *Stahl u. Eisen* **93** (1973), 910–14.
3. Varga, J., and H. W. Lownie, A system analysis study of the integrated iron and steel industry. Battelle Memorial Institute, Columbus Laboratories 505 King Avenue, Columbus, Ohio 43201, 15 May 1969 Reproduced by National Technical Information Service Springfield, Virginia 22151.
4. Harms, F., and W. Riemann, Messung der Abgas- und Staubmengen an 70 t-Lichtbogenöfen bei teilweiser Anwendung von Sauerstoff. *Stahl u. Eisen* **82** (1962), 1345–8.
5. Schlegel, E., Thermische Zersetzung von Flußspat durch Wasserdampf. *Ber. Dt. keram. Ges.* **45** (1968), 520–3.
6. Muhlrad, W., Problems concerning dust removal in flue gas from arc furnace (German). *Stahl u. Eisen* **83** (1963), 921–9.
7. Kahnwald, H., and O. Etterich, Estimation of flue gas quantity, compositions and temperature and dust evolution in the melting and refining with gaseous oxygen in a 15-ton arc furnace (German). *Stahl u. Eisen* **83** (1963), 1067–70.

8. Bintzer, W. W., Design and operation of a fume and dust collection system for two 100-ton electric furnaces. *Iron Steel Eng.* **41** (1964), No. 2, 115–23.
9. Baum, K., F.-A. Hahn, E.-U. Brüderle and G. Urban, Flue gas de-dusting of electric arc furnaces (German). *Stahl u. Eisen* **84** (1964), 1497–1505.
10. Bergmann, P., W. Jesdinsky, G. Reuter and E. Werthmöller, Development and design of a modern dust removing plant for an electric steel plant (German). *Stahl u. Eisen* **87** (1967), 1310–14.
11. Petersen, K., and H. Kahnwald, Direct collection of the dust-containing fumes of electric-arc furnaces during the production of high-chromium steels and their de-dusting with the aid of cloth filters (German). *Stahl u. Eisen* **93** (1973), 910–14.
12. Wilcox, M. S., and R. T. Lewis, A new approach to pollution control in an electric furnace melt shop. *Iron Steel Eng.* **45** (1968), No. 12, 113–20.
13. Venturini, J. L., Operating experience with a large baghouse in an electric arc furnace steelmaking shop. *J. Air Pollution Control Assoc.* **22** (1970), 808–13.
14. Brough, J. R., and W. A. Carter, Air pollution control of an electric arc furnace steelmaking shop. *J. Air Pollution Control Assoc.* **22** (1972), 167–71.
15. Kaercher, L. T., and J. D. Sensenbaugh, Air pollution control for an electric furnace melt shop. *Iron Steel Eng.* **51** (1974), No. 5, 47–51.
16. Vay, K., Totalentstaubung von Lichtbogenöfen. *Fachber. Hüttenprax. Metallweiterverarb.* **13** (1975), 783–85.
17. Baum, J. P., Gas cleaning and air pollution control for iron and steel processes. *Iron Steel Eng.* **53** (1976), No. 6, 25–32.
18. Kahnwald, H., H. W. Kreutzer, and D. Marchand, Staub- und Gasanfall bei der Elektrostahlerzeugung und Möglichkeiten zur Emissionsverminderung durch geeignete Erfassungs- und Entstaubungssysteme. *Proceedings of the 3rd International Clean Air Congress*, Düsseldorf 1973, pp. E3–E7.
19. Marchand, D., Dust and gas evolution in arc furnace steelmaking and possible alternatives for reducing emission with suitable collection and cleaning systems. In Engineering aspects of pollution control in the metals industries. *Proceedings of the Metals Society*, London, 27–29 Nov. 1974. London 1975, pp. 47–57.
20. Flux, J. H., Containment of melting shop roof emissions in electric arc furnace practice. *Ironmaking & Steelmaking* **1** (1974), 121–33.
21. Marchand, D., Possible improvements to dust collection in electric steelplants and summary of all planned and existing collection systems in the Federal Republic of Germany. *Ironmaking & Steelmaking* **3** (1976), 221–9.
22. Kahnwald, H., Staub- und Gasanfall beim Betrieb von Elektrolichtbogenöfen und Maßnahmen zur Verringerung der Emissionen. Association des Ingenieurs Electriciens, Comité Belge de l'Electrothermie et de l'Electrochimie, Liège, Nov. 1973. Fours à arc, pp. 10/3–10/6.
23. Barnes, Th. M., A cost analysis of air pollution controls in the integrated iron and steel industry. Battelle Memorial Institute. Columbus/Ohio, 1969.
24. Graf, G., F. Josten and G. Meinshausen, Possibilities of efficient and low-pollution steelmaking in the arc furnace (German). *Stahl u. Eisen* **96** (1976), 607–11.
25. Fritz, H., and D. Marchand, Design factors in the reduction of stresses caused by dust and heat in metals plant operation and design. The Royal Society of Metals, London, 1977, 56–62.
26. Baum, R., E. Baltes and D. Marchand, The dust-collection of the electric furnace steel plant of Stahlwerke Südwestfalen AG in Siegen-Geisweid (German). *Stahl u. Eisen* **95** (1975), 1286–9.
27. Kubisch, Chr., Personal communication.
28. Schwerdtfeger, K., and K. Klein, Rate of fluorine volatilization from fluoride containing liquid slags. In *Proceedings of the 4th International Symposium on*

Electroslag Remelting Processes. Tokyo, Japan, 7–8 June 1973. Publ. by the Iron and Steel Institute of Japan, Tokyo, 1973, pp. 81–90.
29. Kühnelt, G., Personal communication.
30. Fiedler, H., Zum Stand der Plasmaschmelztechnik in der DDR. *Neue Hütte* **21** (1976), 69–72.
31. Borodacev, A. S., G. N. Okorokov, N. P. Postejev, N. A. Tulin, H. Fiedler, F. Müller, and G. Scharf, Stahlerzeugung in Primärschmelzaggregaten. *Neue Hütte* **22** (1977), 604–7.

Kentucky's Green River Steel acquires sop electric-arc-furnace dust-removal system. *Metal Producing* **15** (1977), No. 15, 46–7.

Drehmel, Dennis, C., Advanced electrostatic collection concepts. *J. Air Pollution Control Assoc.* **27** (1977), No. 11, 1090–2.

Shah, Raymond, Steelmakers use advanced methods for environmental control. *Iron Age Metalworking Intern.* **16** (1977), No. 6, 12MP7, 12MP9, 12MP11, 12MP14, 12MP18–19, 12MP21–24.

Cassettes filter out fumes. *Iron Age Metalworking Intern., Metalprod. Ed.* **17** (1978), No. 8, 80MP25–26, 80MP28.

Goodfellow, Howard D., Safety aspects of electric arc furnace fume systems. *Iron Steel Eng.* **55** (1978), No. 12, 52–7.

Jonsson, John-Eric, Electric arc furnaces. *Steel Times* **207** (1979), No. 2, 129–34.

Smith, C. G., A. Whale, and C. Whitehead, Control of fugitive emissions in the iron and steel industry. *Iron Steel Eng.* **56** (1979), No. 3, 49–56.

Jonsson, John-Eric, Modern techniques in fume control of electric arc furnaces. *Iron Steel Eng.* **55** (1978), No. 4, 42–7.

Dust and fume collection and air pollution systems. *Steel Times* **207** (1979), No. 6, 454–5.

Eisenbarth, Manfred, Secondary dust collection systems in electric and oxygen converter steel plants. *Metallurg. Plant & Technol.* **2** (1979), No. 5, 29, 32, 34–9.

Baum, Peter, and Thomas E. Dixon, Secondary emission control using flatbag roof filters. *Iron Steel Eng.* **56** (1979), No. 11, 31–5.

Marchand, Dirk, and Klaus Polthier, Environmental protection in electric steel plants (German). *Stahl u. Eisen* **100** (1980), No. 11, 569–74 (Betriebsforschungsinst. 98 u. Stahlwerksaussch 1042).

Canfield, Jr, Glenn, Pollution control using a furnace enclosure. *Proc. Electr. Furn. Conf.* **36** (1978), 248–55.

13.3 Noise-emission and its Reduction

1. Hogan, W. T., New economics of steel — more steel with less capacity. *Iron Steel Eng.* **52** (1975), No. 12, 57–60.
2. Schulz, G., and H. U. Haering, Ergebnisse bei Untersuchungen zur Lärmminderung in einem Elektrostahlwerk. In *Moderne Unfallverhütung* No. 18, 1974, 144–6.
3. Thielker, K. H., K. G. Steck, H. U. Haering and V. Irmer, Die Lärmbelästigung an Lichtbogenöfen und die Möglichkeiten der Lärmminderung und des Lärmschutzes. Abschließender wissenschaft. Report to EGKS-Forschung 6242–22/1/061 (unpublished).
4. Schoenmaker, O. D., and J. P. Guldenmundt, use of Purofer sponge iron in a 35-ton electric-arc furnace (German). *Stahl u. Eisen* **94** (1974), 711–28.

5. Ottmar, H., A. Oertler, G. Schmeiduch and D. Ameling, Ausgewählte Versuchsergebnisse über den Einsatz von Eisenschwamm in einem Hochleistungs-Elektrolichtbogenofen. Association des Ingenieurs Electriciens — Comité Belge de l'Electrothermie et de l'Electrochimie, Nov. 1973. Fours à arc.
6. Fiedler, H., Zum Stand der Plasmaschmelztechnik in der DDR. *Neue-Hütte* **21** (1976), 69–72.
7. Habel, F., M. Hausen and G. Schulz, Planning and execution of anti-noise measures (German). *Stahl u. Eisen* **87** (1967), 1077–85.
8. Hausigk, D., and K.-H. Griese, Environmental control and ergonomical measures at the construction of a new electric steelmaking plant (German). *Stahl u. Eisen* **97** (1977), 69–74.
9. Scheffler, G., and G. Dikta, Ultra-modern electric melting shop reduces impact on the environment (German). *Elektrowärme,* Issue B, **33** (1975), 234–42.
10. Graf, G., F. Josten and G. Meinshausen, Possibilities of efficient and low-pollution steelmaking in the arc furnace (German). *Stahl u. Eisen* **96** (1976), 607–11.

Willis, R. R., New emphasis on noise control. A steelmaker's efforts to protect workforce and community. *Iron & Steel Intern.* **51** (1978), No. 1, 41–4, 46–7, 49.

McQueen, Douglas H., Noise from electric arc furnaces. I. General considerations. *Scand. J. Metallurg.* **7** (1978), No. 1, 5–10.

McLarty, Thomas, E., Steel industry noise control. *Iron Steel Eng.* **55** (1978), No. 9, 60–2.

McQueen, Douglas H., Noise from electric arc furnaces. II. Noise generation mechanisms. *Scand. J. Metallurg.* **7** (1978), No. 5, 223–9.

Haering, Hans Ulrich, Klaus Polthier and Jürgen Schmitz, Geräuschemission von Anlagen der Elektrostahlwerke. *Stahl u. Eisen* **99** (1979), No. 11, 562–8 (Betriebsforschungsinst. 89, Stahlwerksaussch. 1022 and Aussch. Umweltschutz 38).

McQueen, Douglas H., Noise from electric arc furnaces. III. Detailed solutions for two models. *Scand. J. Metallurg.* **8** (1979), No. 2, 55–63.

Coping with noise pollution. *Iron Age Metalworking Intern.* **18** (1979), No. 5, 16MP11, 16MP13, 16MP15.

Haering, Hans-Ulrich, Klaus Polthier and Jürgen Schmitz, Noise reduction in electric steel plants and suggestions for planning (German). *Stahl u. Eisen* **100** (1980), No. 2, 53–8.

BS 4142: 1967. Method of rating industrial noise affecting mixed residential and industrial areas.

13.4 Dust-disposal

1. Gesetz über Beseitigung von Abfällen (Abfallbeseitigungsgesetz — AbfG) in the version dated 5 January 1977. Bundesgesetzbl. 1977. Part I, 1977, pp. 42–51.
2. Amelung, E., K. H. Steinbacher, J. Otto and A. Schneider, On the dust collection of scrap-practice open-hearth furnaces (German). *Stahl u. Eisen* **97** (1977), 674–80.
3. Meyer, G., K. H. Vopel and W. Jansen, Investigations of the utilization of dusts and sludges from the waste-gas cleaning systems of blast furnaces and BOP steelmaking plants in the rotary kiln (German). *Stahl u. Eisen* **96** (1976), 1228–33.
4. Maczek, H., H. Rellermeyer, G. Kossek and H. Serbent, Tests for processing metallurgical wastes using the rotary-kiln volatilizing method in an industrial plant (German). *Stahl u. Eisen* **96** (1976), 1233–8.

5. Sugasawa, K., Y. Yamada, Sh. Watanabe, K. Kato, K. Masuda, Y. Sato and T. Kawabata, Direct reduction of metallurgical dusts (German). *Stahl u. Eisen* **96** (1976), 1239–44.
6. Haucke, M., and W. Theobald, Behandlung und Aufbereitung von Stäuben und Schlämmen in der Stahlindustrie. In Behandlung und Beseitigung von Abwasserschlämmen sowie von flüssigen bis festen Sonderabfällen unter Berücksichtigung von neuen Forschungsergebnissen und Erfahrungsberichten installierter Anlagen. Institut für Siedlungswasserwirtschaft der Technischen Hochschule Aachen. Aachen 1976. (Gewässerschutz. Wasser. Abwasser. 21.), 511–41.

Electric Furnace Steel Production
Edited by E. Plöckinger and O. Etterich

14 Chemical Analysis and Temperature Measurement in Steel Production

WILHELM ANTON FISCHER AND WOLFGANG THOMICH, DÜSSELDORF

The control of metallurgical processes in steel production requires constant monitoring of the chemical composition and temperature of the melt. Both of these must be kept within ever narrowing limits in order to meet the technical requirements of the process and the quality demands for the final product as well as the need to run the operation economically. In the course of development of the latter three requirements, the empirical methods used for many years in steelworks were succeeded by exact systems of measurement, and the classical analytical methods have given way to fast, automatic apparatus. The new methods of measurement and analysis also enabled electronic data processing to be used to evaluate the readings taken and the processes to be controlled automatically. These developments, which are important for all phases of the most varied kinds of process in electric steelmaking up to casting, will now be briefly described.

14.1 ANALYSING THE CHEMICAL COMPOSITION OF THE MELT

Wolfgang Thomich, Düsseldorf

In order to enable the chemical composition of the steel melt to be analysed, a sample has to be taken from the molten steel; the sample is then prepared and submitted for testing. The sample must be taken in such a way that the result is representative of the condition of the melt at the time; this requires exact operating rules which, among other things, are laid down in detail in the *Steelworks Laboratory Handbook*[1] published by the VDEh Chemical Committee.

14.1.1 Sampling the Melt

As a rule, samples for chemical analysis are taken from the melting vessel, ladle, or mould. Samples should only be taken from the pouring-stream in exceptional circumstances, and then it is preferred that a special probe should be used for the purpose. Spoon- and mould-samples can only be taken by restricting the pouring stream, so then the possibility of changes in composition cannot be excluded. Attempts to use a spoon or mould to sample the full pouring-stream can lead to accidents.[2] The point at which samples are taken and the number of samples required should depend on the type of melting-vessel and its size, because of possible variations in concentration in the steel bath. The types of element to be measured and the analytical accuracy required have an effect on sampling and the most suitable kind of sample. This is important when concentrations have to be determined of elements having a strong tendency to segregation or when dissolved gases, especially hydrogen, are present in the steel.

It is customary these days to take samples by using a spoon, dipping-mould, or immersion probe, or by suction into an evacuated vessel.[3] All these methods have their advantages and disadvantages, especially bearing in mind undesirable side-reactions involving certain elements. The choice has therefore to be made in the light of the purpose of the sample and the time available.

Special precautions have to be observed when sampling to establish the concentration of hydrogen present in molten steel, since the sample can pick up or lose hydrogen, even when solidified. For these reasons, a sample suitable for measuring other elements cannot be used to determine hydrogen-concentration with sufficient accuracy. Samples are generally taken with a copper mould or by using a vacuum pipette or suction pipette.[4] The choice of method is determined mainly by the purpose of the sample and what possibilities there are for its extraction as well as by the nature of the analytical process used. If particular working rules are observed,[5] very reproducible relative values can be obtained with a copper mould. With the other two methods the total hydrogen content is measured, i.e. even that of hydrogen liberated while the sample solidifies and any that can still diffuse out after the sample has solidified.

14.1.2 Sample-preparation

The sample which has been taken from the steel bath by one of the above methods, and which, moreover, is now in the solid state, must be prepared for analysis. The process required to render the sample ready for analysis depends on the nature of the apparatus to be used and the analytical method involved.

Wet chemical analysis and a number of other conventional analytical processes generally require the sample to be fragmented. Different methods of fragmenting are employed, depending on the type of steel involved.[6–8] Under no circumstances must the process of fragmentation change the nature of the sample, either in the course of preparation itself or by separation. However, it should be pointed out in this connection that conventional analytical methods are hardly ever used nowadays for monitoring the melting process, since they take too long and are labour-intensive.

A sample for spectrometric analysis and emission and X-ray fluorescence analysis must have a flat surface on one side, which is ground and polished. The specimen may have to be sectioned before grinding or may require only to be ground and polished without sectioning; this depends on the method by which the sample is taken.[9] The type of surface required and the equipment and media used for grinding and polishing all depend on the purpose of the analysis and the type of analytical process used. For instance, when the aluminium content of a sample has to be determined, then grinding must not involve the use of conventional corundum abrasives. Samples used for emission and X-ray fluorescence analysis also require different types of surface-finish. In general, there has to be a compromise between the best techniques possible and what is defensible in terms of time and material-consumption.

14.1.3 Analytical Processes

As the melting and refining processes used in steel-production are being continually speeded up, it is always necessary to fit the analytical processes to the same requirements. The conventional method needs time for its chemical reactions to create a solution and then separate out the various elements or to melt the sample and remove the elements in the gas phase. Generally speaking, the more chemically resistant a steel is, the longer the analysis takes. The time required for analysis can only be shortened by applying new methods such as fast modern processes based on spectrometry and melting at high temperatures. Some of these processes are so new that there is little coverage in the specialist literature, and therefore they will be discussed rather more extensively here than conventional processes.

14.1.3.1 Conventional analytical methods

The processes used in laboratories today are described in the form of detailed work-instructions in the *Steelworks Laboratory Handbook*, volumes 2 and 5.[10,11] Each time one of these processes is extended or otherwise revised it is always described in up-to-date form in a collection of leaflets. In addition to classical methods such as volumetric, quantitative, and gravimetric analysis, we now already include photometry, potentiometry, complexometry, coulometry, conductometry, chromatography, and polarography — to mention only a few — among the conventional methods, and we already include atomic absorption among these processes as well. For oxygen, nitrogen, and hydrogen elements we additionally count extraction by melting or heating and hot-precipitation as conventional processes.

14.1.3.2 Fast modern methods

The time required for analysis and the accuracy and reproducibility of results can only be improved simultaneously by using automatic analysers, which are partially or wholly controlled by a computer which also evaluates the signals received. The more complex the structure of the steel is, the more corrections are required and the more involved are the computations to obtain precise results. Emission spectrometry, X-ray fluorescence analysis, infra-red analysis, and also thermal conductivity analysis have all largely proved themselves here.

14.1.3.2.1 Emission spectrometry

The basic principles of emission spectrometry have already been described in detail several times.[12,13] and so only the latest state of development at the time of writing will be indicated here. A micro-processor is used to control the spark and count the number of hits; frequencies up to 50 Hz are used both to shorten the time required for analysis and to improve its reproducibility. The computer also monitors all the necessary parameters and converts incoming signals into percentages so that all interference effects from the matrix and from other elements are corrected by computation.

Apparatus in use today consists of multi-channel devices which operate in vacuum under computer control and which can analyse all the elements present in the steel. However, this equipment suffers from excessive scatter at very high concentrations and is limited by inhomogeneity that can occur in the test samples.

14.1.3.2.2 X-ray fluorescence spectrometry

For reasons of time, equipment in use in steelworks laboratories almost always has more than twenty channels operating simultaneously, enabling all elements having an atomic number greater than 9 to be measured under vacuum. This means that elements such as carbon and boron cannot be quantified. A further drawback is the inferior detection-threshold for some elements compared with emission analysis. On the other hand, X-ray fluorescence analysis has the advantage of better reproducibility (area analysis instead of spot analysis), and it enables non-metallic substances to be measured. This equipment is also controlled and monitored by computer. The results can be fed back into the process-control computer direct. The principles of X-ray fluorescence analysis are described in the literature.[14]

14.1.3.2.3 Measurement of carbon and sulphur contents

Methods for measuring carbon and sulphur contents were for a long time among the fastest techniques used in the analysis of steel. The introduction of spectrometric techniques enabled infra-red analysers to be used and the time taken to be further shortened by measuring both elements simultaneously, especially by the integration of micro-computers and electronic weighing into the process. The introduction of higher-powered high-frequency furnaces and also the measurement of carbon monoxide have both contributed to improving analytical methods considerably. Modern systems have two measurement ranges to optimize readings of the concentrations of individual elements. The spectrometer can be connected direct to the computer so that the analysis can be integrated with the overall result.

14.1.3.2.4 Measurement of levels of oxygen, nitrogen, and hydrogen

Oxygen, nitrogen, and hydrogen levels are also determined very quickly today by using very high-performance systems almost exclusively. Extraction is carried out under vacuum or inert gas in a graphite crucible at very high temperature (2000°), using a thermal conductivity cell or infra-red cell for indication. Values are given in percentages and can be fed straight into the spectrometer computer. The equipment is designed for hydrogen-measurement in such a way that diffusible hydrogen levels can be measured as well as total hydrogen content. The concentrations of all three elements are analysed automatically and the process is controlled by the micro-processor.

14.1.4 Time Required for Analysis and Propogation of Results

The time required for a complete analytical operation depends on the time taken to obtain the sample and then to transport and analyse it and feed back the results. The analysis itself generally takes the least time in the sequence. The possibilities for transportation and the time taken to move samples are therefore of decisive importance for positioning the analytical apparatus. It is worth trying to centralize the laboratory for the sake of economy as long as the time taken to move samples is not longer than the time required to obtain and analyse them. Otherwise, the analytical apparatus should be decentralized and located close to the melting-unit.

The results are either transmitted direct from the laboratory computer to a works process-control computer or are sent to the works by teleprinter or a television system. Results are only very rarely transmitted by telephone. Samples and records of results are stored in the laboratory.

14.2 USING IMMERSION-LANCES FOR TEMPERATURE-MEASUREMENT, THERMAL ANALYSIS, AND OXYGEN-MONITORING

Wilhelm Anton Fischer, Düsseldorf

Immersion-probes enable measurements to be taken at source in the furnace or ladle or in intermediate vessels. The immersion pyrometer is usually used nowadays to obtain liquidus temperature and other probes have been developed to determine carbon and oxygen contents via thermal analysis as well as for monitoring oxygen activity by means of EMF measurement. All these methods are based on the system developed for temperature-measurement by immersion pyrometry.

14.2.1 Immersion Pyrometry

Steel-bath temperature is nowadays always measured thermoelectrically. Despite the higher first cost and running expenses, thermoelectric immersion pyrometry has completely replaced the optical pyrometer method.[1]

The immersion thermocouple consists of a sensor head, which can only be used once, and a lance designed for long life. The sensor head consists of a base made from refractory material and a thermocouple in a U-shaped quartz tube.[2]

The thermocouples are made from wire of not more than 0.10-mm diameter and consist of the following alloys: PtRh 10% with Pt,[3] PtRh 13% with Pt,[2] and PtRh 30% with PtRh 6%.[2] These can operate at up to maximum temperatures of 1600, 1700, and 1800°C, respectively. The ceramic sensor head is fitted into a cardboard tube 600–1200 mm long to form the sensor probe; this is inserted in turn into the steel probe lance (Figure 1). Correct mating and electrical contact of the thermocouple probe with the connections in the lance are signalled electrically together with an indication that the device is ready for use.

The temperature-reading can be recorded in analogue form with a compensating recorder (quality at least class 5, settling time 2 s, paper advance rate at least 4 mm/s) or digitally on to magnetic tape. A digital instrument system has advantages over an analogue system in that readings are taken objectively, strictly correlated with sensor locations, and can be stored; its maintenance costs are lower, too.[4]

An average time of 3–4 min is required to fit the sensor head to the lance, move the lance to where it is needed, and immerse it in the molten steel. The reading is indicated 5–6 s after immersing the sensor. Readings in the steel-bath temperature range are reproducible around ±2 to ±4°C.[4] The reliability of readings is markedly affected by interference in the system, such as stray voltages from outside or electrolytic voltages caused by humidity in the lance; however, it also depends greatly on correct handling and timing during immersion.[2,5]

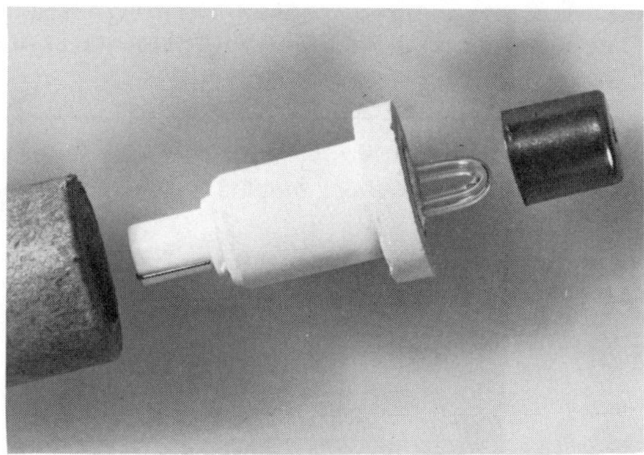

Figure 1. Immersion pyrometer sensor for taking steel-bath temperature

14.2.2 Solidification Analysis for Determining the Liquidus Temperature of a Steel Melt

Heat of crystallization liberated during solidification of a steel melt causes a kink (or for undercooled melts, a hump) in the time-temperature curve derived from thermo-electric sensing. The temperature at which the kink or hump occurs is the liquidus-temperature. It has been found that the liquidus-temperatures of plain and alloy steels determined in this way agree well with figures obtained by summing the reductions in the melting-point of pure iron due to individual elements taken from the relevant phase diagrams.[6]

Table 1 shows drops in melting point in degrees Centigrade per percentage mass of the appropriate companion or alloying element related to the liquidus-temperature, 1539°C, of pure iron.[7] Steel melt liquidus-temperatures obtained from the data in the table form the basis from which the best casting temperature is chosen for the particular circumstances.[8]

14.2.3 Thermal Analysis for Determining Carbon Content

Figure 2 illustrates an instrument probe for determining the carbon content of steel by thermal analysis using the immersion-probe method. We start here with a standard commercial immersion thermocouple.[10] The steel flows in from the side and is first killed (with aluminium in a separate chamber) before solidification in the measuring crucible is recorded using a temperature-recorder or by digital means from temperature-readings sensed by a Pt/PtRh thermocouple. In contrast with the older process involving casting a sample from the steel bath in a ceramic crucible,[11,12] the immersion probe method is less liable to interference, enables readings to be taken with a steel bath which is only slightly superheated, and leads to a quicker reading with the possibility of measuring temperature at the same time.[10,13]

Table 1. Drops in the freezing temperature of molten iron caused by impurities or common alloying elements for steel melts shown with percentages of the different alloying elements in the melt[7]

Element	Drop in freezing point per percentage weight (°C)	Percentage of element present in the melt
Hydrogen (H_2)	1300 (calculated)	0–?
Nitrogen (N_2)	90 (calculated)	0– 0.03
Oxygen (O_2)	80 (calculated)	0– 0.03
Carbon (C)	65 at 0%	0– 3.8
	70 at 1%	
	75 at 2%	
	80 at 2.5%	
	85 at 3%	
	91 at 3.5%	
	100 at 4%	
Phosphorus (P)	30	0– 0.7
Sulphur (S)	25	0– 0.08
Arsenic (As)	14	0– 0.5
Tin (Sn)	10	0– 0.03
Silicon (Si)	8	0– 3.0
Manganese (Mn)	5	0– 1.5
Copper (Cu)	5	0– 0.3
Nickel (Ni)	4	0– 9.0
Molybdenum (Mo)	2	0– 0.3
Vanadium (V)	2	0– 1.0
Chromium (Cr)	1.5	0–18.0
Aluminium (Al)	0	0– 1.0
Tungsten (W)	1	18%W with 0.66%C

The starting point of the calculation is the freezing temperature of 1539° for pure iron.

Example of calculation:

2.5%C	2.5 × 80° =	200°
0.6%Mn	0.6 × 5° =	3°
0.033%P	0.033 × 30° =	1°
0.04%S	0.04 × 25° =	1°
8.0%Ni	8.0 × 4° =	32°
		237°

A melt with this composition will freeze at 1539 minus 1302°C. The best casting temperature is around 80 to 100° higher, ie 1380 to 1400°C.

For non-alloy steel melts containing 0.02–1.0% carbon, 89.3% of all carbon level readings are accurate to within ±0.02%.[12] The measured liquidus-temperature must be corrected when determining the carbon content of alloy steel melts.[7]

When the differential cooling curve is recorded at the same time, its behaviour enables conclusions to be drawn about the purity of non-alloy and low-alloy casting steel melts with regard to oxygen, sulphur, and phosphorus.[14]

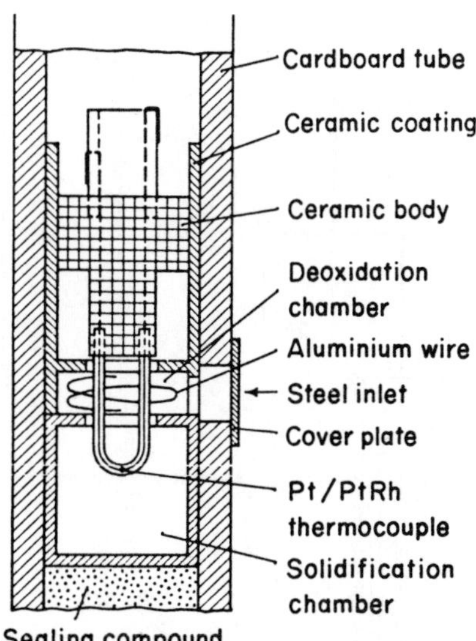

Figure 2. Sensor probe for measuring steel carbon content by thermal analysis[9]

The amount of oxygen dissolved in the steel bath can be determined by carrying out thermal anaysis before and after deoxidation with aluminium and observing the difference in liquidus-temperature.[15]

14.2.4 Electrochemical Determination of Oxygen Content

Unlike analytical evaluation of total oxygen-content, electrochemical determination only applies to oxygen dissolved in the steel bath, the amount of which is of interest in the course of the process.[16,17] The technique embraces measurement of oxygen-content over a range of from 1 to 1500 ppm. The difference between the total oxygen-level determined analytically and that of dissolved oxygen found electrochemically can provide a valuable indication of the amount of oxide in suspension in the melt.

14.2.4.1 Solid electrolyte cells for quick-dip measurement

Solid electrolyte for quick-dip measurements is nowadays usually put into sintered tubelets (35 × 6 × 4 mm), made from ZrO_2 partially stabilized with MgO, enclosed and gas-tight on one side. These tubelets can withstand quick-dipping into molten steel without cracking. Inside the tubelets the reference electrodes are made from a mixture of Cr/Cr_2O_3 or Mo/MoO_2 and are fitted with molybdenum wire connectors. As the oxygen potential of the reference electrodes is heavily dependent on temperature, these cells are

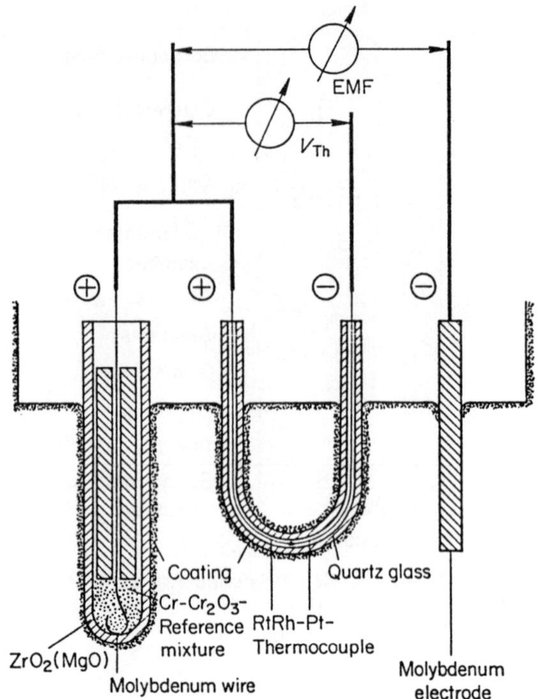

Figure 3. EMF probe sensor for electrochemical determination of oxygen content in a steel bath (schematic)[18]

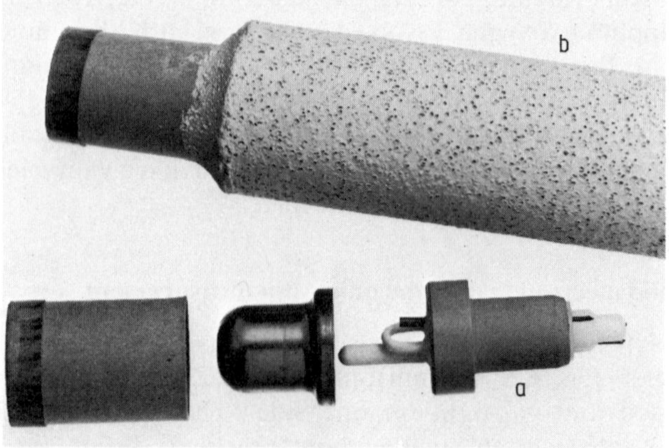

Figure 4. Arrangement for measuring EMF and temperature by quick-dip.[18] (a) Sensor with protective casings; (b) sensor probe

also incorporated into the immersion sensors already mentioned so that temperatures can be measured at the same time. Figure 3 is a schematic illustration of a longitudinal section of a sensor of this kind. Figure 4 shows the sensor with its cardboard or metal casing and the assembled sensor probe in its ceramically coated cardboard tube. The servicability of all the important parts is indicated by lamps.

The measurement signals are fed via special screened leads either to a compensating recorder or a digital indicator. Readings can be taken in 10–15 s; they are reproducible to within about ±5%.[19] Digital indicators incorporate microprocessors which convert stored EMF values into oxygen activity levels.[16] Any semiconductive properties of the solid electrolyte can also be taken into account.[17] It is additionally possible to input the effect on dissolved oxygen activity of companion and alloying elements into the computer.[16,20–22] This can provide an immediate digital indication of dissolved oxygen content and hence, for instance, of how much deoxidant is required. The data can also in principle be recalled from the computer.

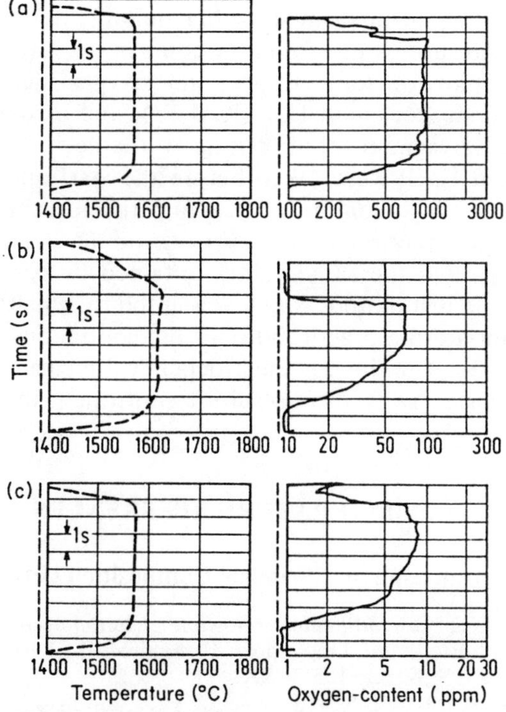

Figure 5. Plots obtained with oxygen-sensors by quick-dip measurements of steel melts (a) during oxygen-blowing in a 25-t arc furnace, (b) shortly before tapping, (c) after deoxidation with aluminium in the casting ladle. (After B. Hanas *et al.*[23])

Figure 5 shows plots, obtained with this kind of probe, of temperature and oxygen-content in a steel bath during oxygen-blowing and just before tapping from a 25-t arc furnace as well as after Al deoxidation in the ladle.[23]

The activity levels and concentrations of elements with an affinity for oxygen can also be ascertained indirectly from the EMF figures via the equilibrium relationships of the dissolved oxygen. This applies, for example, to the final carbon content achieved in oxidizing[24] and to aluminium in precipitation deoxidation.[25]

14.2.4.2 Prolonged measurements with solid electrolyte cells

In contrast to the very comprehensive literature on results obtained with quick-dip measurements in steelworks using the cells described, published findings from prolonged measurements using cells built as fixtures into furnaces, ladles, and special-treatment units are quite scarce. The only exception is in regard to laboratory tests, where building in the cells and their connectors presents far fewer difficulties than in industrial practice. On the other hand, laboratory results can show that suitably designed solid electrolyte cells can be used successfully for many hours in steel-melts.[16]

Figure 6 is a plot of temperature and oxygen activity level for a chromium–aluminium steel melt (X10CrAl 13), and of oxygen potential in the waste gas during vacuum-refining in an 80-t ladle.[22] Immersion-readings of temperature and EMF were taken before the start and after the end of refining in order to check continuous readings taken using electrolytic cells built into the equipment. There is very good agreement between the immersion-readings and the same values obtained by continuous measurement over the whole test period, which amounted to 2 h altogether. This can be seen to substantiate the practicability of the continuous sensor cell under the conditions tested. Continuous electrochemical measurement of oxygen-potential in the waste gas is also useful for controlling the vacuum-refining process.[22,26]

14.3 LITERATURE REFERENCES

14.1 Analysing the Chemical Composition of the Melt

1. *Handbuch für das Eisenhüttenlaboratorium,* Chemikerausschuß des Vereins Deutscher Eisenhüttenleute, Düsseldorf, 1974, Sheet B 5.3-1.
2. See Ref. 1, Sheet B 5.3-2 and B 5.3-14.
3. See Ref 1, Sheet B 5.3-14–B 5.3-16.
4. See Ref. 1, Sheet B 5.3-16–B 5.3-22.
5. See Ref. 1, Sheet B 5.3-18.
6. *Handbuch für das Eisenhüttenlaboratorium.* Vol. 3, 2nd edition, Chemikerausschuß of the Vereins Deutscher Eisenhüttenleute, Düsseldorf 1966, pp. 24–47, 128–36.

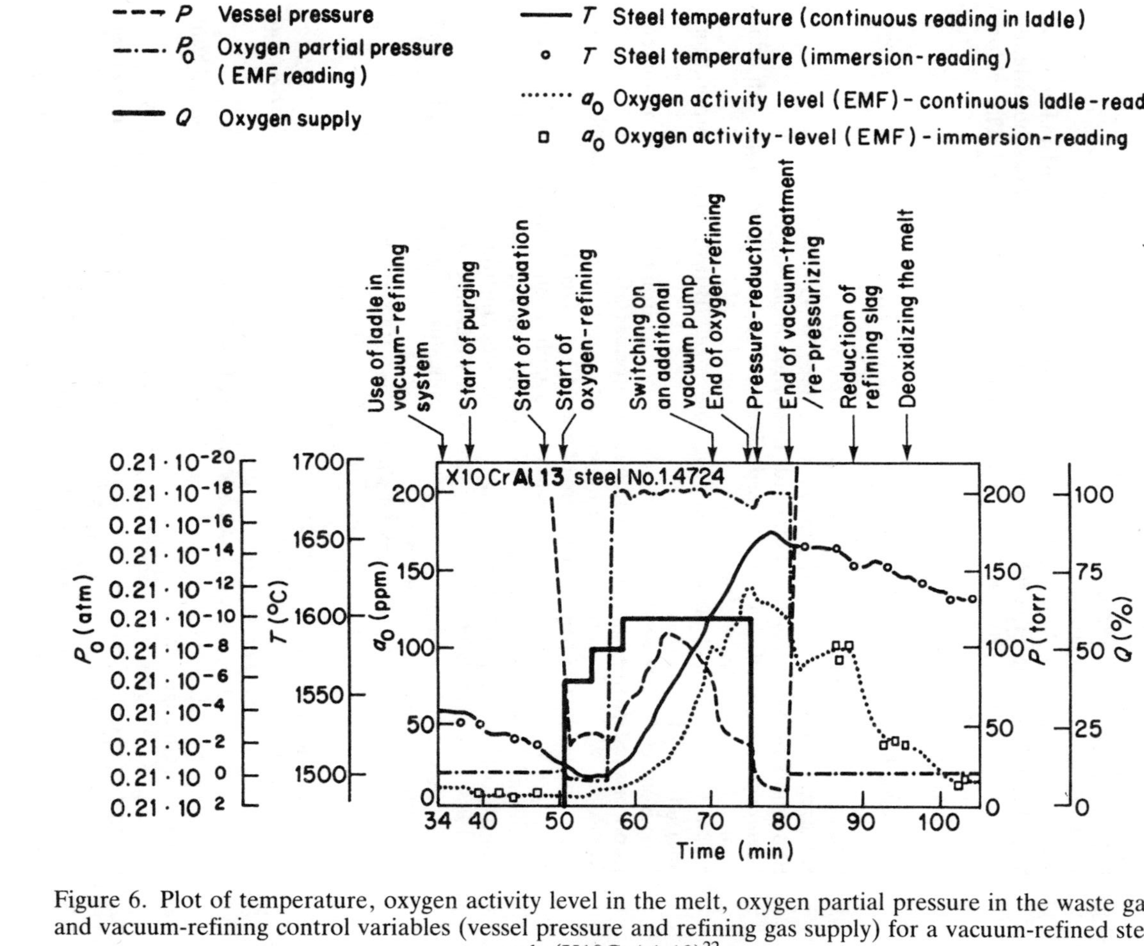

Figure 6. Plot of temperature, oxygen activity level in the melt, oxygen partial pressure in the waste gas, and vacuum-refining control variables (vessel pressure and refining gas supply) for a vacuum-refined steel melt (X10CrA1 13)[22]

7. See Ref. 1, Sheet B 5.3-2.
8. *Analyse der Metalle*. Vol. 3, Probenahme, 2nd edition, Chemikerausschuß der Gesellschaft Deutscher Metallhütten- und Bergleute e. V. Berlin/Heidelberg/New York and Düsseldorf, 1975, pp. 56–74, 187–92, 195.
9. See Ref. 1, Sheet B 5.3-4.
10. *Handbuch für das Eisenhüttenlaboratorium*. Vol. 2, 2nd edition, Chemikerausschuß of the Vereins Deutscher Eisenhüttenleute, Düsseldorf, 1966, pp. 14–164.
11. See Ref. 1, Sheet C 1.1–C 9.1-2.
12. See Ref. 10, pp. 285–310.
13. *Analyse der Metalle,* Vol. 2, Part 2 Selen-Zirkonium. Physikalisch-chemische Verfahren. 2nd edition, Chemikerausschuß der Gesellschaft Deutscher Metallhütten- und Bergleute e. V. Berlin/Heidelberg/New York and Düsseldorf 1961, pp. 1348 ff.
14. See Ref. 10, pp. 311–46.

Harrison, T. S., *Proc. Anal. Div. Chem. Soc.* **12** (1975), 229.
Harrison, T. S., and J. Borrowdale, *Proc. 28th BSC Chem. Conf.* 1975, p. 9.
White, G., *Proc. 29th BISRA Chem. Conf.* 1976, p. 71.
Kidman, L., *Proc. 22nd BISRA Chem. Conf.* 1969, p. 43.
BS Handbook No. 19, British Standards Institution, London, 1970.
Standard Methods of Anaylsis of Iron, Steel and Associated Materials, 5th edn, United Steel Cos Ltd, Sheffield, 1961.
Methods of Chemical Analysis of Iron and Steel, British Steel Corporation, Sheffield, 1974.
Davidson, G. A., The sampling and analysis of cast iron using the quantometer, Report issued by the Applied Research Laboratories (GB) Ltd.
Ramsden, W. R., Recent advances in automatic spectrometers for metal analysis, Report issued by the Applied Research Laboratories (GB) Ltd.
Ramsden, W. R., The ARL Quantovac and its application to the analysis of carbon and low alloy steels, Report issued by the Applied Research Laboratory (GB) Ltd.
Control of composition in steelmaking, Iron and Steel Institute Special Report, No. 99, 1967. (See also Padget, G., *Proc. 20th BISRA Chem. Conf.* 1967, p. 9).
Determination of chemical composition: its application in process control, Iron and Steel Institute Special Report, No. 131, 1971.
Waasdorp, A. A., *Proc. 21st BISRA Chem. Conf.* 1968, p. 76.
Gale, P., *Proc. 22nd BISRA Chem. Conf.* 1969, p. 9.
Statham, R. F., *Proc. 22nd BISRA Chem. Conf.* 1969, p. 25.
BSC Open Report, MG/CC/577/72.
Bareham, F. R., and J. G. M. Fox, X-ray fluorescence analysis, *BISRA and BNFMRA Symposium on Spectroscopy*, April 1959.
Jenkins, R., and J. K. de Vries, *Metals Mater.* **5** (1971), No. 8, 125.
Compton, A. H., and J. K. Allison, *X-rays in Theory and Experiment,* Van Nostrand, Chicago, 1935.
Birks, L. S., *X-ray Spectrochemical Analysis,* Interscience, New York, 1959.
Liebhafsky, H. A., H. Pfieffer, H. Winslow, and P. Zemany, *X-ray Absorption and Emission in Analytical Chemistry,* Wiley, New York, 1960.
Ambrose, A. D., *Proc. 18th BISRA Chem. Conf.* 1965, p. 71.
Bagshawe, B., and P. H. Scholes, BSC Open Report, CDL/CAC/23/74, 1974.
Jones, R. J., P. Gale, P. Hopkins, and L. H. Powell, *Analyst* **90** (1965), 623; Ref. (3), Carbon Method (iv).
Bruch, J., *Iron Steel Inst. Proc. Intern. Conf. on Determination of Chemical*

Composition, Brighton, 28–30 Sept. 1970, p. 64.
Speight, G., and R. M. Cook, *J. Iron Steel Inst.* **160** (1948), 398.
Barraclough, K. C., *Murex Review* **1** (1954), 318.
Cook, R. M., and J. D. Hobson, *J. Iron and Steel Inst.* **169** (1951), 24.
Wells, J. E., and K. C. Barraclough, *J. Iron Steel Inst.* **155** (1947), 27.
Hobson, J. D., *Proc. 12th BISRA Chem. Conf.* Harrogate, October 1958, pp. 12–22.
Swinburn, D. G., *J. Iron Steel Inst.* **209** (1971), 620.
Sloman, H. A., C. A. Harvey, and O. Kubaschewski, *J. Inst. Metals Bull. Metall. Rev.* **80** (1952), 391.
Cook, R. M. and G. E. Speight, *J. Iron Steel Inst.* **176** (1954), 252.
Shanahan, C. E. A. and F. Cooke, *J. Iron Steel Inst.* **188** (1958), 138.
Shanahan, C. E. A., *Iron Steel Inst. (London) Spec. Rept.* 1960, No. 68, 75.

14.2 Using Immersion-lances for Temperature-measurement, Thermal Analysis, and Oxygen-monitoring

1. Pepperhoff, W., Temperaturstrahlung. Darmstadt 1956. (Wissenschaftliche Forschungsberichte. Naturwissenschaftliche Reihe. Vol. 65.)
2. Burton, J., Betrachtung zur Temperaturmessung in Stahlbädern mit Hilfe von Thermoelementen mit auswechselbaren Köpfen. Circ. Inform. techn. **26** (1969), 2493–2517.
3. DIN 43710. Issued Sept. 1977.
4. Schiefer, P., and M. Bretz, Digital temperature measurement by immersion thermocouples (German). *Stahl u. Eisen* **87** (1967), 383–9.
5. Barth, G., Neuere Untersuchung über die Genauigkeit der Tauchthermoelementmessung. *Stahl u. Eisen* **84** (1964), 953–7.
6. Roeser, Wm., and H. T. Wensel, Erstarrungstemperatur von sehr reinem Eisen und einigen Stählen. *J. Res. nat. Bur. Stand.* **26** (1941), 273–87.
7. Erniedrigung der Erstarrungstemperatur von Stahlschmelzen durch Verunreinigungen oder übliche Legierungselemente. In *Stahleisen-Kalender* 1951. Verein Deutscher Eisenhüttenleute. Düsseldorf 1950. p. 138.
8. Guthmann, K., Günstige Gießtemperatur im Vergleich zum Erstarrungspunkt von Eisen- und Stahlschmelzen. *Stahl u. Eisen* **71** (1951), 399–402.
9. Hagen, K., and P.-G. Oberhäuser, Kohlenstoffbestimmung und Temperaturmessung während des Frischens im Sauerstoff-Aufblaskonverter. *Stahl u. Eisen* **89** (1969), 1315–19; see especially p. 1317.
10. Dastur, P. N., C. B. Griffith and G. W. Perbix, Development of a carbon and temperature probe for BOF computer control. *Iron Steel Eng.* **45** (1968), No. 3, 77–83.
11. Tectip type S expendable thermocouple for carbon content determination of steel. (Ed.) Leeds and Northrup Co.
12. Bardenheuer, F., and K. Hagen, Schnellbestimmung des Kohlenstoffs im Stahl mit Hilfe der thermischen Analyse. *Stahl u. Eisen* **87** (1967), 395–6.
13. Kern, D. W., and P. D. Stelts, The lance that measures the BOP. *Iron Age* **201** (1968), No. 24, 79–80.
14. Ableidinger, K., and P. Strizik, Betriebskontrolle der Schmelztechnik für unlegierten und legierten Stahlguß am basischen Lichtbogenofen. *Gießerei Rdsch.* **23** (1976), No. 9, 96–104.
15. Mahn, G., and B. C. Mahanty, Rapid determination of oxygen in molten steel by means of the thermal analysis (German). *Stahl u. Eisen* **89** (1969), 1314–15.

16. Fischer, W. A., and D. Janke, *Metallurgische Elektrochemie*. Düsseldorf/Berlin/Heidelberg/New York, 1975, pp. 203, 206–7, 266, 318–23, 347.
17. Schoop, J., H. G. Fleige, G. Zielinski, B. Olthoff, U. Kersting and F. Schruff, Adjustment of the deoxidation degree of various steel grades with the aid of EMF measurements (German). *Stahl u. Eisen* **98** (1978), 1088–92.
18. Ferrotron, Elektronik GmbH, 4030 Ratingen 4 (Lintorf). Prospekt der Sauerstoffmeßsonde für flüssige Metallbäder. No. 1 — Stand Jan. 1976.
19. Morello, B., Forschungen über Methoden der physikalischen Analyse in der Eisen- und Stahlindustrie. In *Europäische Informationstagung über die Anwendung von Messungen, Kontrollen und Analysen in der Eisen- und Stahlindustrie*, Luxembourg, 2–5 April 1974. Luxembourg 1974, pp. 2–12.
20. Čerkasov, L. A., and W. A. Fischer, Der gemeinsame Einfluß mehrerer Legierungselemente auf die Sauerstoffaktivität von Metallschmelzen. *Arch. Eisenhüttenwes.* **42** (1971), 699–702.
21. Fischer, W. A., and D. Janke, Einfluß von Kohlenstoff, Silicium, Aluminium oder Titan auf die Sauerstoffaktivität in legierten Stahlschmelzen. *Arch. Eisenhüttenwes.* **47** (1976), 589–94.
22. Meierling, P., Steuerung des metallurgischen Prozesses beim Erschmelzen von Chromstählen durch Messung des Sauerstoffverlaufes über kurz- und langzeitige EMK-Messungen in der Stahlschmelze. (Dr-Ing.-Diss., Techn. Hochsch. Aachen) Aachen, 1975.
23. Hanas, B., M. Widell, O. von Krusenstierna and K. Torsell, Development of an oxygen probe for the control of metallurgical processes. *Trans. Iron Steel Inst. Japan* **11** (1971), Suppl. I, 367–71.
24. Hautecler, M., F. Anselin and M. Dutrieux, Einsatz der Sauerstoffaktivitätsmessung im LD-AC-Werk der Société Métallurgique Hainaut-Sambre. *Metallurg. Rep. CRM* No. 39, 1974, 3–9.
25. Hagen, K., P. Hammerschmid, O. P. J. Curé and T. P. C. Bollens, Determination of the aluminium contents in killed steel with the aid of EMF measurements (German). *Stahl u. Eisen* **95** (1975), 398–402.
26. Otto, J., Die Konzentrationsbewegung hochchromhaltiger Stähl beim Vakuumfrischen und die Prozeßüberwachung durch kontinuierliche Sauerstoffpartialdruckmessung des Abgases. Aachen 1975. (Dr-Ing.-Diss. Techn. Hochsch. Aachen.)

Janke, Dieter, Klaus Schwerdtfeger, Johannes Mach and Gerd Bamberg, Test performance of the needle sensor for oxygen measurements under steelmaking conditions (German). *Stahl u. Eisen* **99** (1979), No. 22, 1211–15.

Electric Furnace Steel Production
Edited by E. Plöckinger and O. Etterich

15 Principles of Costing and Calculating Economic Efficiency

KARL PITTEL, COLOGNE

15.1 THE PRINCIPLE OF ECONOMIC EFFICIENCY

According to the principle of economic efficiency, an undertaking should try to achieve its selected aims in the cheapest and most expedient manner, i.e. to attain the highest degree of success with the least resources or to do the best possible with the means available.

An undertaking may operate economically but not profitably if, for example, the proceeds leave no surplus. On the other hand, an enterprise can be profitable without meeting the standards of the principle of economic efficiency in regard to timing, for example.

The principle of economic efficiency for a complete undertaking applies equally to all functional areas of the undertaking, i.e. to purchasing, production, and sales, and to the production areas of each main and subsidiary element as well as to all main and subsidiary cost centres in the enterprise.

15.2 PRODUCTIVITY CHARACTERISTICS IN ASSESSING ECONOMIC EFFICIENCY

While the economic efficiency and profitability of an undertaking enable the total activity of the undertaking to be evaluated in economic terms, technically quantifiable values are assessed in terms of productivity.

15.2.1 Overall Productivity

Continuous measurement of productivity should reveal any changes which are not influenced by prices. Productivity is generally defined by a formula published by the OEEC:[1]

$$\text{Productivity} = \frac{\text{Production output or quantity produced}}{\begin{array}{c}\text{Production factors and input quantities}\\ \text{(capital, equipment, working time,}\\ \text{labour, raw materials, energy)}\end{array}}$$

However, overall productivity and any improvement in productivity are difficult to assess from this formula, since the denominator contains a mixture of input factors such as time, weight, etc. and money values (capital), which all have different dimensions.

In empirical economic research it is usual to work out the results of production by means of econometric production functions; this requires the use of expensive computer programs. For this reason, production elasticity values are worked out for the input factors used in the production functions. These elasticity values show how much output increases when one output factor is increased by 1% while the others remain constant; they have to be reassessed frequently. This method of measuring productivity is not particularly appropriate for assessing economic efficiency, since the problems of estimation are even greater at factory level than for a branch or a complete undertaking.

15.2.2 Partial Productivity

Partial productivity values are therefore often calculated for the capital, labour, and materials areas, so that a technically quantified assessment can be made.[2] Material productivity or yield have been accurately observed in steelworks for a long time.

However, in considering partial productivity values it should not be forgotten that production factors are partly interchangeable. For example, installing a continuous-casting system may increase labour and materials productivity values but, at the same time, reduce capital productivity.

15.2.3 Costing Data and Systems

While productivity is used for technically quantified assessments, any economically quantified assessment of the production process should be based on the economic efficiency criterion. In this way the relevant production factors are evaluated in money terms which can be added together, and the result indicates the expenditure or costs.

In industrial practice, economic efficiency is monitored by means of a costing system. Simply calculating actual costs has the disadvantage that the total costs incurred during an accounting month in main and subsidiary cost centres are split up among cost units in using the relevant designated accounting values. This kind of accounting system is adequate if a plant is run at almost constant loading times when capacity is limited. It is thus only possible to check costs by making comparisons over a period of time and equating actual costs for the previous accounting period. However, as the different factors which govern cost change continually, particularly because of changes in the level of activity, a comparison of actual costs cannot provide a clear measure of the economic efficiency of an undertaking. For this reason a system of accounting using predetermined guide costs was introduced in the iron and steel industry in the 1950s to compensate for these disadvantages.[3] The guide cost system uses actual costs together with costs that would arise from the actual production quantities and composition according to the predetermined guide costs. The guide costs for input and processing are calculated from the following equation:

$$\text{Guide cost} = \text{production} \times \text{specific consumption} \times \text{guide price}$$

A breakdown according to quantity and price components allows for differences between guide costs and actual costs; the most important causes of such differences are variations in activity level, consumption rate, and price.

The guide cost accounting systems described later suffer from the disadvantage of having been created to suit particular installations such as steelworks, rolling mills, and transportation systems.[4–7] The next developments were therefore necessarily directed at extending the guide cost system to apply to the entire production area,[8] in order to enable financial direction and operational control to work together.[9–10] However, most of the effort in research was always clearly directed towards the improvement of results by influencing costs.

15.3 THE NEW CONCEPT OF WORKS ACCOUNTING SYSTEMS

There have been increasing improvements not only in the planning of overall costs but also in methods of medium- and short-term planning, e.g. for investment, provision of raw materials, and budgeting for the costs of administration.

With increasing pressure from competition, management needs comprehensive economic data on all the main areas of activity such as marketing, production and procurement. These are required for formulating short-, medium-, and long-term decisions, directing and controlling activities in all areas of the undertaking, and accounting for any changes made. The data should all be provided by the works accounting system.

The aims of works accounting systems had to be fundamentally reconsidered and reorientated so that the various problems posed by the above considerations could be solved. Many years of co-operation between experts from works belonging to the Betriebswissenschaftlichen Institut der Eisenhüttenindustrie (Institute of Steel Industry Organization and Management) and the Verein Deutscher Eisenhüttenleute (German Steel Industry Association) resulted in the publication of 'Guidelines' for works accounting.[11,12] These provide the basis for the establishment and operation of a results-oriented works-information scheme, the individual parts of which comprise a complete closed-loop system with assessment rules and computing procedures oriented on the objective. The system has a formal uniform structure and uses consistent concepts throughout.

Accordingly, the total system described in the guidelines has three main elements:

Planning calculations,
Documentation accounting,
Check calculations.

The main aim is to quantify costs, returns, and achievement for each area.

15.3.1 Planning Calculations

Planning calculations help in the systematic formulation of decisions concerning practical alternative actions under the conditions expected. This part of the system is forward-looking; it provides carefully chosen technical and economic guide values enabling procurement, production, and sales activities to be defined and quantified.

15.3.2 Documentation Accounting

Documentation accounting records data on quantities, timing, and prices, etc. for actual processes in all areas of the undertaking. The assessment of economic efficiency is devised from this data base, which also is the source from which cost and return figures are obtained.

15.3.3 Check Calculations

Check calculations bring planning calculations and documentation accounting together and serve to highlight deviations between guide values obtained from planning calculations and actual values recorded from documentation accounting. The data are also analysed in order to reveal the causes for any such deviations. Check calculations are thus concerned solely with values arising at particular times under normal circumstances and are not burdened with the details of what occurred in previous periods of time which used to hinder the old check system with its comparison of different times.

Planning calculations and documentation accounting are concerned with determining the economic results for particular periods of time (year, quarter, or month) and indicate the economic effects on markets of actions that have been planned and carried out on individual operations or systems or on the undertaking as a whole. In the marketing area, sales programmes are intended to be carried out in particular periods of time. In the production area programmes are planned and executed to match the sales programme. These two elements of the accounting system are particularly well suited for calculating economic efficiency by comparing periodic receipts from the sales programme with costs for the corresponding production programme.

In the following section we will deal with the source data for assessing results, i.e. receipts from sales and corresponding production costs; we will describe only cost accounting and charging within the framework of planning calculations, documentation accounting, and check calculations.

15.4 USING GUIDE-VALUE FUNCTIONS IN ACCOUNTING

The multitude of technical production variables have to be incorporated into practical equations. The accounting system meets this requirement by means of guide-value functions. This is an important innovation in works accounting.

These guide-value functions encompass the technical structure of the undertaking as well as the relationship of the business to the market.

The general mathematical formula for a guide-value function is:

$$y_i = b_{ij} \cdot x_j$$

where y_i is the target value, and $i = 1 \ldots m$,
x_j is the influence value, and $j = 1 \ldots n$,
b_{ij} is called the Standard of the guide-value function.

The function describes the linear dependence of a target value on one or more influence values.

In the production area, guide-value functions describe the technically conditioned relationships between the consumption of materials and the most important influence values which determine cost, e.g. in the execution of a production programme. These relationships are expressed numerically through consumption and output Standards, and, when used in conjunction with evaluation factors, lead to a cost plan. The Standards are expressed as money values for each influence-value unit where material consumption is of lesser importance or for the service of capital where consumption can be ignored.

Guide-value functions also provide the elementary framework for production-related calculations and for determining product clearing prices and intra-plant output for use in documentation accounting.

In documentation accounting, the use of clearing prices which are calculated in advance as part of the annual planning enables the former practice of calculating costs for all stages of production to be abandoned. This speeds up the monthly accounting process and facilitates check calculations from which useful conclusions can be drawn. This highlights deviations between actual and calculated clearing costs (quantity produced × clearing price) for each cost centre and every stage of calculation and production.

Guide-value functions are also used in check calculations. With this means, guide values are calculated for costs under actual production conditions in the period in question. Any deviations between these values and those derived from documentation accounting are analysed so that there causes can be identified. This enables useful scales to be derived for assessing the economic efficiency of the processes in the plant.

15.4.1 Influence Values

The overall costs in an undertaking arise from the manufacture of products or the provision of services. The main factor influencing materials consumption emanates from the production programme. However, in costing by cost type, not all types of cost can be said to be directly related to the production programme. The main influence value is the production programme and may be configured according to product characteristics (quality or size groups). However, it is also necessary to use other operational influence values, if possible, closely related to the main influence value. A whole range of costs are not related to operational influence values in the short term. In such cases we have to use calendar time or length of period, which are influence values independent of the operation. Guide-value functions for costing normally contain the following influence values:

Primary influence values
Quantities produced, arranged by cost unit,
Length of period.
Derived influence values
Input quantities,
Operating times,
Utilization times.

Derived influence values should always be used where there is no direct relationship between materials consumption and primary influence values. These derived influence values should first be determined as target values in relation to quantities produced (primary influence value); they can then be used as influence values for direct measurement of materials consumption costs.

15.4.2 Types of Guide-value Function

The following guide-value functions are used in costing:

Materials-consumption functions, including materials cost functions and processing costs functions.
Output functions.

15.4.2.1 Materials-consumption functions

A guide-value function is formed for each type of cost if possible and is used as an active, established influence value for costing. Every individual type of cost is dependent on at least one influence value.

15.4.2.1.1 Consumption functions for materials costs

The consumption of charge (input) materials and the amount of material wasted are both, as a rule, proportional to the quantity produced. The materials and waste Standards represent specific rates for input-materials consumption and waste accretion per unit of production under given conditions for input and operation. For example:

$$\frac{\text{Scrap fed in (kg)}}{\text{Crude steel produced (t)}}$$

or

$$\frac{\text{Slag yield (kg)}}{\text{Crude steel produced (t)}}$$

15.4.2.1.2 Consumption functions for processing costs

Materials consumption is not normally determined direct from the quantities produced. In these cases consumption is measured via an influence value (e.g. input quantity or process time) derived from the quantity produced. Specific consumption for a particular cost source per unit of active influence value (e.g. electricity consumed per utilization hour) constitutes a processing cost Standard. Some Standards are expressed as money values per unit of influence value; these are used for costs which cannot be derived separately from consumption and evaluation factors (e.g. for service of capital and operational taxes) or where consumption is not evaluated for economic reasons (e.g. tools, auxiliaries, and working materials).

The most important influence values, to which processing cost Standards can be related, are:

Times: calendar time, operation time, utilization time, man-hours,
Quantities: production; if necessary subdivided into cost unit features; and input, if need be, subdivided into qualities and sizes.

15.4.2.2 Output functions

Output functions describe the relationship between quantities produced and process times. Two types of Standard are used for determining quantities normally associated with production for different types of time. These are:

Output Standards,
Time Standards.

Output Standards represent the specific requirement for a particular type of time associated with one unit of production under normal conditions of operation, e.g.

$$\frac{\text{Utilization time (h)}}{\text{Quantity produced (t)}}$$

Since output Standards normally relate to a particular type of time, idle time can be determined via time Standards related to the main type of time, e.g.

$$\frac{\text{Interruption time (h)}}{\text{Utilization time (h)}}$$

It is reasonable to assume that the relationship between these types of time is proportional.

15.5 COST CENTRES

The constitution of cost centres and their arrangement according to functional area and technical accounting unit are very precisely covered in the 'General Guidelines'.

Cost centres should be constituted and delineated according to the following criteria:

(1) The cost centre should comprise complete working processes.
(2) The limits of the cost centre should match those of the responsibility of the manager concerned.
(3) Materials consumption and the relevant influence values should be directly ascertainable for each cost centre.

Primary cost centres are those which have a direct effect on the manufacture of the product and other operations. Auxiliary cost centres are ones which only have an indirect effect on manufacture of the product and other processes by feeding other cost centres. Administration cost centres are those concerned with the administration of the undertaking.

A number of cost centres may be combined to form a single calculation unit or stage if the quantities of product emanating from cost centres working together are variable. This applies particularly if:

(1) There are appreciable changes in the quanitity of material between input and output:

(2) Part of the material is processed in special cost centres, or is passed to other production lines or processes,
(3) Some material is fed in from another production line for further processing.

Cost centres and calculation units for steelworks are laid down in the 'Special Guidelines'.

15.6 TYPES OF COST

Annex 2 to the 'General Guidelines' illustrates the arrangement and grouping of all the types of cost. In particular, it describes the content of each individual type of cost and lays down rules for evaluating feed and waste materials for documentation accounting and for estimating the values of processing costs for the works and administration. It also describes how to record and compute costs for works and administration cost centres or for limiting accounts or calculation units.

15.7 OPERATING THE ACCOUNTING SYSTEM WITH A MODEL

A complete description of the acccounting system for an electric furnace steelworks would greatly exceed the scope of this chapter. We can therefore merely take a brief look at the accounting system and try to give the reader an impression by describing part of it.

15.7.1 Description of Model Steelworks

The model steelworks has two 100-t ultra-high-power arc furnaces and produces two types of steel. For steel type A the furnace is operated as a melting machine, and the crude steel is degassed and alloyed after tapping into a vacuum installation. Steel type B is a low-alloy steel which is melted, alloyed, and finished in the furnace. Each type of steel is cast into 5- or 6-tonne ingots in the casting pit. The time available for operation in this electric steelworks amounts to 626 h per planning period.

15.7.2 Performing the Planning Calculation

15.7.2.1 Determining planned production quantities and times

Starting from a sales plan oriented on market possibilities for the rolling mill attached to the works, the lead material requirement is calculated with the help of the agreed input material Standard, according to the type of steel and the weight of ingot to be produced. For example, Table 1 shows how quantities for

Table 1. Survey of input and output Standards and the planned production quantities and times derived from them for the model electrical steelworks

Line	Accounting plan	Input and output Standards and planned production quantities and times derived					Plan-month					
	Calculation units type of steel/ingot weight	Unit	Input Standards and planned production quantities				Output Standards and planned times					
			Av 5 t	Av 6 t	B 5 t	B 6 t	Type of steel	Unit	Av 5 t	Av 6 t	B 5 t	B 6 t
Line	1	2	3		4		5	6	7		8	
	Casting											
1	Ingot production (requirement for semi-finished product)	t	15 200	14 020	11 400	7 680						
2	Input Standard, liquid crude steel	t/t	1.0526	1.0414	1.0526	1.0416	Output Standard	hOT/t	0.0114	0.0126	0.0114	0.0126
3	Crude steel requirement	t	16 000	14 600	12 000	8 000	Operating time	h	183	184	137	101
4	Waste Standard scrap	t/t	0.0526	0.0414	0.0526	0.0416						
5	Scrap-yield	t	800	580	600	320						
	Vacuum installation		AV								AV	
6	Crude-steel production	t	30 600		–		Output Standard	hOT/t	0.6198		–	
7	Input Standard, liquid crude steel	t/t	0.980		–		Operating time	h	605		–	

8	Input Standard, alloys	t/t	0.030	–					
9	Waste Standard, melting losses	t/t	0.010	–					
10	Crude-steel requirement	t	30 000	–					
	Furnace operation		A	B			A	B	
11	Crude-steel production	t	30 000	20 000	Output Standard	hUT/t	0.0200	0.0250	
12	Input Standard, metallic charge	t/t	1.050	1.065	Utilization time	h	600	500	
13	Input Standard, additives	t/t	0.057	0.069	Time Standard:				
					Operating time / Utilization time	h / h	1.1	1.1	
14	Waste Standard, slag	t/t	0.090	0.100	Operating time	h	660	550	
15	Requirement for metallic charge materials	t	31 500	21 300					
16	Requirement for additives	t	1 710	1 380					
17	Slag-yield	t	2 700	2 000					
	Distribution of type Programme		Production Quality				Output Standard hOT/t	Plan-operating Time	Hours
18	Furnace 1 Type A	t	27 500				0.0220		605
19	Furnace 2 Type A	t	2 500				0.0220	55	
20	Furnace 2 Type B	t	20 000				0.0275	550	605

Key: OT = Operating time.
UT = Utilization time.

electric steelworks production per cost unit (types A and B) can be calculated in the various calculation units from the sales plan and the requirement for semi-finished products for the works rolling mill in opposition to the materials flow. In addition, the planned amounts of waste or left-over material can be worked out for each calculation stage by using the waste material Standard.

The production plan quantities worked out in this way must still be checked at each calculation stage to see whether they can be attained in terms of capacity. The utilization times required to execute the production programme (Table 1, columns 7 and 8, line 12) are obtained by multiplying the planned production quantities for each type of steel by the output Standard:

$$\frac{\text{Utilization time (h)}}{\text{Production (t)}}$$

The operating time required can be worked out with the aid of the time Standard (line 13):

$$\frac{\text{Operating time (h)}}{\text{Utilization time(h)}}$$

This should then be compared with operating time available.

This calculation should be repeated for each calculation unit or stage in order to ascertain whether any equipment in the electric steelworks could constitute a production bottleneck. Table 1 shows that the operating time for type A steel is greater than the time available, while for type B steel it is less. It can be shown that the operating times requested for furnaces 1 and 2 can be brought to within the time available by careful allocation of production of the two types of steel to the two furnaces.

15.7.2.2 Influence- and accounting-values for melting operations

The influence values used for individual cost centres and the accounting values chosen for calculating the performance of a cost centre are illustrated in Table 2 for the main cost centres in the melting calculation stage. Influence values chosen as accounting values should be selected as being dominant in terms of cost and as those which best represent cost centres in production. Finally, the table also shows the utilization and output Standards related to the cost units; these Standards are calculated from the accounting price per unit of accounting value for each cost unit (type A and B).

15.7.3 Determining Cost per Unit of Influence Value and Price Charged for the Accounting Value for the Furnace Cost Centre

Consumption figures for the furnace operation cost centre for the individual types of cost in connection with the relevant influence values (namely length of

Table 2. Influence and accounting values for the melting operation cost-centre calculation stage together with consumption and output Standards for calculating cost units

Planning calculation	Cost centres, influence-, and accounting-values for melting calculation unit Planning calculation				Calculation	
Calculation unit	Cost centre	Influence value	Accounting value	Unit	Consumption (t/t) and output (h/t) Standard per cost unit Type A	Type B
Melting operation	Scrap area	Period length				
		Solid charge	Solid charge	t/t	1.050	1.065
		Operating time				
	Additions	Period length				
		Additions used	Additions used	t/t	0.057	0.069
		Operating time				
	Furnace operation	Period length				
		Production				
		Operating time				
		Utilization time	Utilization time	h/t	0.0200	0.0250
	Ladle operation	Period length				
		Liquid crude steel plus slag removed	Liquid crude steel plus slag removed	t/t	1.090	1.100
	Common operating costs	Period length	Operating time derived from period length	h/t	0.0220	0.0275

period, production, operating time, and utilization time) are expressed as consumption Standards in Table 3. The sets of costs per type of cost and influence value are calculated by evaluating the specific materials consumption for each influence value. The set of costs for any influence value is obtained by adding together all the sets of costs for each type of cost for each influence value.

A differentiated clearing price should be computed for both types of steel (A and B) for calculating the output of this cost centre for the settling account for the melting operation cost unit (Table 4). Here the utilization time influence value is selected as the accounting value used as the calculation unit for cost centre output; this is because it is cost-wise dominant and because it best represents the demands of furnace operation. The planning costs per unit of utilization time to the extent of 2435.86 DM/h shown in Table 3 can be used directly for the accounting value. The costs which depend on other influence values must be expressed as Standards related to the utilization hour accounting value.

The costs which depend on operating time are added to the utilization value via the time standard:

$$\frac{\text{Operating time (h)}}{\text{Utilization time (h)}} = 0.55$$

Production-dependent costs are added to the accounting value with the set of costs for each unit of production influence value multiplied by the quantity produced per utilization hour.

Period-dependent costs, which cannot be allocated according to cause, are first related to operating time and then added to the accounting value together with the time Standard used for accounting for operation time.

From this the price for the utilization hour accounting value is 3684.09 DM/h for type A and 3610.69 DM/h for type B.

Correlation of the accounting value to the type A and B cost units, respectively, is done via the output Standard per utilization hour. From this the processing costs for the furnace operation cost centre amount to:

3684.09 DM/h × 0.0200 h/t = 73.68 DM/t for type A steel
and 3610.69 DM/h × 0.0250 h/t = 90.27 DM/t for type B.

The processing costs for type A steel, taken in relation to the individual influence values, are made up as follows (DM/t):

Period-dependent	12.90
Production-dependent	7.34
Operating time-dependent	4.72
Utilization time-dependent	48.72
Total	73.68

15.7.4 Applying the Accounting System to Auxiliary Plants Generally

The rules for planning calculations, documentation accounting, and check calculations also apply to auxiliary operations in general. Planned outputs for auxiliary operations are derived from the planned consumption figures for the main process. Planned costs for each individual auxiliary operation are determined with the aid of the influence-value calculation, and the clearing price is worked out per unit of accounting value for each auxiliary operation. Three separate clearing prices are established for charging for the output of the auxiliary operations in primary cost centres:

(1) The first clearing price contains input and operating process costs without service of capital;
(2) The second clearing price adds on the service of capital for the auxiliary itself as well as service of capital arising from charging for the output of the auxiliary operation;
(3) The third clearing price includes costs for administration centres distant from the operation which are nevertheless part of the combined operating cost centre for the auxiliary operation.

Table 5 shows the three-part clearing prices calculated for the accounting values for the various auxiliary operations.

Thus, for example, the water cost for the furnace operation cost centre is charged under the first clearing price at the rate of 32.-DM per accounting quantity of 1000 m^3 of water, which is evaluated from the consumption Standard (0.3500 per unit of operating time influence value, i.e. hour).

Planned consumption is worked out under the second clearing price at the level of 6.20 DM per unit of accounting value, and the cost set is shown in Table 3, column 14, line 17 under service of capital from the output account.

The costs for planned consumption evaluated under the third charging calculation are added to the calculation stage for the total operation cost centre.

15.7.5 Calculation of Cost Units

15.7.5.1 Calculating materials costs

The first part of Table 6 shows materials costs per tonne for steel type A, derived with the aid of material cost functions. These material costs per unit of cost unit are obtained by adding the consumption Standards for the individual input materials to the waste materials Standards and then multiplying by the planned price per unit of input/waste material.

Table 3. Determination of planning costs for each type of cost and unit of influence value for the furnace cost centre as well as planning process costs

	Planning calculation Calculation unit: melting operation Cost centre: furnace operation			Determination of planned processing costs per type of cost							
	Type of cost	Influence value (IV)		1st influence value: period length Plan quantity: 1 month				2nd influence value: liquid crude steel production Plan quantity: 50 000 t			
No	Designation 1	Unit 2	Plan price per unit 3	Consumption Standard 4	Plan consumption 5 4 × plan qty IV 1	Plan costs 6 3 × 5	Plan costs per unit of 1st IV 7 7 × plan qty IV 1	Consumption Standard 8	Plan consumption 9 8 × plan qty IV 2	Plan costs 10 3 × 9	Plan costs per unit of 2nd IV 11 10 ÷ plan qty IV 2
1	Personnel costs	h	11.25	3 042	3 042	34 222	34 222				
2	Oxygen	$Nm^3 \times 10^3$	80.–								
3	Melting power consumed	kWh x 10^3	55.–								
4	Water	$m^3 \times 10^3$	32.–								
5	Operation media and materials	DM	1.–	42 200	42 200	42 200	42 200				
6	Electrodes	kg	3.52								
7	Refractories	DM	1.–								
8	Aggregate operational costs	DM	1.–	125 250	125 250	125 250	125 250				
9	Charge for works-associated admin. centres	DM	1.–	26 230	26 230	26 230	26 230				
10	Charge for operational admin. and maintenance	DM	1.–	25 233	25 233	25 233	25 233				
11	mechanical workshop	man-h	22.30	1 930	1 930	43 039	43 039	0.1273	6 365	141 940	2.84
12	electrical workshop	man-h	27.20	215	215	5 848	5 848	0.0064	320	8 704	0.17
13	building works	man-h	19.70	2 370	2 730	46 689	46 689	0.1958	9 790	192 863	3.86
14	Repair materials	DM	1.–	27 500	27 500	27 500	27 500				
15	Furnace lining	DM	1.–	148 750	148 750	148 750	148 750				
16	Service of capital	DM	1.–	177 820	177 820	177 820	177 820				
17	Service of capital from output charge	DM	1.–	6 936	6 936	6 936	6 936	0.4687	23 438	23 438	0.47
18	Planned cost for cost centre	DM				709 717	709 71			366 945	7.34

								Planning month Planned process costs		
3rd influence value: operating time Plan quantity: 605 h				4th influence value: utilization time Plan quantity: 1 100 h				Plan costs		
Consumption Standard	Plan consumption	Plan costs	Plan costs per unit of 3rd IV	Consumption Standard	Plan consumption	Plan costs	Plan costs per unit of 4th IV	Plan consumption	DM mth	No.
12	13 12 × plan qty IV 3	14 (3 × 13)	15 14 ÷ plan qty IV 3	16	17 16 × plan qty IV 4	18 (3 × 17)	19 18 ÷ plan qty IV 4	20 (5 + 9 + 13 + 17)	21 (6 + 10 + 14 + 18)	
36.9587	22 360	251 550	415.79					25 402	285 772	1
				0.2136	235	18 797	17.09	235	18 797	2
				26.3636	29 000	1 595 000	1450.–	29 000	1 595 000	3
0.3500	212	6 776	11.20					212	6 776	4
								42 200	42 200	5
				222.7273	245 000	862 400	784.–	245 000	862 400	6
				99.0909	109 000	109 000	99.09	109 000	109 000	7
								125 250	125 250	8
								26 230	26 230	9
								25 233	25 233	10
								8 295	184 979	11
								535	14 552	12
								12 160	239 552	13
								27 500	27 500	14
								148 750	148 750	15
								177 820	177 820	16
2.1719	1 314	1 314	2.17	85.6818	94 250	94 250	85.68	125 938	125 938	17
		259 640	429.16			2 679 447	2435.86		4 015 749	18

Table 4. Calculation of the clearing price per unit of utilization-hour accounting value for the furnace cost centre

Planning calculation — Calculation of process costs per unit of accounting value — Planning month

Calculation stage: melting operation — Accounting value: utilization time
Cost centre: furnace operation — Unit of accounting value: 1100 h

Cost unit	Period length			Production			Operating time			Utilization time			
	Period-dependent costs per hour of operating time	Operating hours per unit of accounting value	Period-dependent costs per unit of accounting value	Production per unit of accounting value	Cost per unit of account-value	Influence value dependent costs per unit of accounting value	Influence value quantity per unit of accounting value (time Standard)	Cost per unit of influence value	Influence value-dependent costs per unit of accounting value	Influence value quantity per unit of accounting value	Cost per unit of influence value	Influence value-dependent costs per unit of accounting value	Price per unit of accounting value
1	2 (from Table 3)	3 (from Table 1)	4 (2 × 3)	5	6	7 (5 × 6)	8	9	10 (8 × 9)	11	12	13 (11 × 12)	14 (4 + 6 + 9 + 12)
Type A	1173.08	0.55	645.19	50	7.34	367.–	0.55	429.16	236.04	1.0	2435.86	2435.86	3684.09
Type B	1173.08	0.55	645.19	40	7.34	293.60	0.55	429.16	236.04	1.0	2435.86	2435.86	3610.69

Table 5. First, second, and third clearing prices per unit of accounting value for auxiliary operations

Planning calculation	Example for charging for auxiliary operations			Plan-month
Auxiliary operation	Units	Clearing price 1	Clearing price 2	Clearing price 3
1	2	3	4	5
Electricity for lighting	kWh $\times 10^3$	64.25	3.25	1.20
Electricity for melting	kWh $\times 10^3$	55.–	3.25	1.20
Water	$m^3 \times 10^3$	32.–	6.20	2.50
Mechanical workshop	man-hours	22.30	2.50	5.80
Electrical workshop	man-hours	27.20	2.10	5.60
Building works	man-hours	19.70	0.70	4.50
Transportation	tonnes	1.20	0.20	0.15

Table 6. Part I: Planned materials costs per tonne for steel type A. Part II: Charging cost centre performance using clearing prices for the various accounting values

PART I

Planning calculation	Determination of total planned input quantities and costs and calculation of planned costs per unit of cost unit	Plan month
Calculation stage: melting operation	Cost unit: material costs for Type A Steel Planned production: 30 000 t liquid crude steel	

Line	Cost type	Unit	Plan price per unit of costs (DM/unit)	Consump-Standard (t/t)	Planned input (t)	Planned total costs (DM)	Planned cost per unit of cost unit (DM/t)
	1	2	3	4	5	6	7
	Input costs						
1	Own scrap	t	190.–	0.310	9 300	1 767 000	58.90
2	Bought-in scrap	t	170.–	0.600	18 000	3 060 000	102.–
3	Alloyed scrap	t	250.–	0.140	4 200	1 050 000	35.–
4	FeMn	t	750.–	–	–	–	–
5	FeSi	t	900.–	–	–	–	–
6	Aluminium	t	2 100.–	–	–	–	–
7	Chromium	t	1 500.–	–	–	–	–
8	Molybdenum	t	25 000.–	–	–	–	–
9	Total metallic charge	t		1.050	31 500	5 877 000	195.90
10	Lime	t	110.–	0.050	1 500	165 000	5.50
11	Other additions	t	180.–	0.007	210	37 800	1.26
12	Total charge costs					6 079 800	202.66
	waste material (credit)						
13	Slag	t	30.–	0.090	−2 700	−81 000	−2.70
14	Scrap	t	190.–	–	–	–	–
15	Total residue credit	t			−2 700	−81 000	−2.70
16	Material costs					5 998 800	199.96

15.7.5.2 Charging cost-centre costs

The second part of Table 6 shows how cost-centre costs are charged with the help of the relevant accounting values. Processing costs, totalled for steel type A from the clearing price per unit of utilization time accounting value from

Table 6 cont.

PART II

Calculation stage: melting operation					Processing costs	
Cost centre	Accounting value AV	Plan costs per unit of accounting value DM/AV	Consumption Standard (S) t/t Output Standard (O) h/t	Planned input quantities (t) planned hours (h)	Planned process costs total (DM)	Planned process costs per unit of cost unit (DM/t)
1	2	3	4	5	6	7
Scrap area	Solid charge					
Additions	Additions charged					
Furnace operation	Utilization time	3684.09	(0) 0.0200	(h) 600	2 210 454	73.68
Ladle operation	Liquid crude steel + slag removed					
Total operating costs	Operating time					
Total process costs					2 962 250	98.74
Planned cost						
Unit cost					8 961 050	298.70

Table 4, multiplied by the output Standard for steel type A, are included under the furnace operation cost centre.

15.8 DETERMINING ACTUAL COSTS AND STATEMENT OF TOTAL DEVIATION

15.8.1 Determination of Actual Costs

Documentation accounting collects data on material quantities, times, and prices for actual processes.

Table 7 shows the actual costs for the furnace operation cost centre for a month under actual conditions. Actual production amounted to 45 500 tonnes, of which type A was 28 000 t and type B was 17 500 t. Operating time was 591 h, and utilization time was 1060 h. By agreement, utilization time is also the accounting value.

Table 7. Furnace operation cost centre: actual production, actual quantities of influence values, actual costs

Documentation accounting Processing costs Actual month

Calculation stage: melting operation
Cost centre: furnace operation

Influence values (IV):	Production	Operating time	Utilization time
Actual quantities of IVs:	45 000 t	591 h	1 060 h
Accounting value:	utilization time		

Line	Cost type			Actual costs	
	Description	Unit	Price per unit	Actual consumption	DM/month
	1	2	3	4	5 (3 × 4)
1	Personnel costs	h	11.25	25 200	283 500
2	Oxygen	$Nm^3 \times 10^3$	80.–	200	16 000
3	Electricity for melting	$kWh \times 10^3$	55.–	28 200	1 551 100
4	Water	$m^3 \times 10^3$	32.–	200	6 400
5	Equipment and materials	DM	1.–	42 200	42 200
6	Electrodes	kg	3.52	238 000	837 760
7	Refractories	DM	1.–	103 000	103 000
8	Total operating costs	DM	1.–	125 250	125 520
9	Charge for associiated administration centres	DM	1.–	26 230	26 230
10	Operational administration	DM	1.–	25 233	25 233
11	Maintenance Mechanical workshop	man-hours	22.30	7 900	176 170
12	Electrical workshop	man-hours	27.20	560	15 232
13	Building	man-hours	19.70	10 500	206 850
14	Repair materials	DM	1.–	35 000	35 000
15	Furnace lining	DM	1.–	148 750	148 750
16	Service of capital	DM	1.–	177 820	177 820
17	Service of capital from output charge	DM	1.–	121 166	121 166
18	Cost centre actual costs	DM			3 897 561

15.8.2 Charging in Documentation Accounting

According to the accounting system, the cost centre outputs are charged in the calculation stage at the charge prices for the various accounting values determined in the planning calculation. Charge costs are therefore obtained by multiplying the accounting value charge price by the specified amount of the accounting value which may occur in actual production.

These specified amounts, multiplied by the respective output Standards for steel types A and B from Table 1, should be 560 utilization hours for type A and 438 utilization hours for type B.

If we multiply the calculated specified utilization hours for each type by the appropriate price for the accounting value and type (Table 4), then we obtain the following charges (DM):

Type A:	560 × 3684.09 =	2 063 090
Type B:	438 × 3610.69 =	1 581 482
	Total	3 644 572

15.8.3 Cost Centre Total Deviation

The costs to be charged amount to	3 644 572 DM
The actual costs from Table 7 were	3 897 561 DM
Difference	252 989 DM

The difference between the charged costs and the actual costs is the total deviation.

15.9 CHECK CALCULATION

15.9.1 Use of Guide-value Functions

Guide-value functions are also used in check calculations. They help the determination of guide values for consumption, output, and costs under actual production conditions over a period when costs determined in documentation accounting are measured. Guide consumption figures for materials processed by the cost centre are established thereby, and the actual quantities for the primary influence value (quantity produced) and the specified quantities of the derived influence values (e.g. operating time, utilization time) are used in the guide-value functions. Quantity deviations are derived by comparing guide consumption figures with actual consumption. Since consumption functions and evaluation factors are separate parts of the accounting system, quantity deviations between guide costs and actual costs, as well as deviations conditioned by price, can both be determined.

15.9.2 Resolving Total Deviation According to Cause

The total deviation for a cost centre, which results from the difference between actual costs and charged costs, is resolved with the aid of the guide value functions into a series of deviations related to influence values. The most important types of deviation are described below.

15.9.2.1 Price deviation

The effects on costs of the influence of the supply market are determined in the price deviation. Here actual consumption figures are evaluated using the difference between planned prices and actual prices.

Price deviations occur only with primary cost types, namely materials and processing, which can be enumerated as physical quantities. Consumption figures for secondary types of cost are evaluated through clearing prices in documentation accounting. There is thus no price deviation for these types of cost in the cost centres concerned.

Price deviation (PD) can be calculated from the following formula:

$$PD = C_A(P_P - P_A)$$

where C_A is actual consumption for the period
P_P is the planned price per unit of input material
P_A is the actual price per unit of input material.

15.9.2.2 Deviation in materials selection

Materials-selection deviation is the result of altered quantity ratios arising from the substitution of input materials and a change from the specified standard composition. Materials-selection deviation can arise with both materials costs and processing costs. It should be judged together with price deviation, as it can be related back to adaptation of materials selection to a change in price.

Materials selection deviation (MD) is obtained from the following formula:

$$MD = C_A(P_{SM} - P_{AM})$$

where C_A is actual consumption for the period
P_{SM} is the mixture price per unit of standard composition
P_{AM} is the mixture price per unit of actual composition.

Actual consumption C_A is related to the basic unit of quantity for the materials which can be substituted, e.g. kcal for fuel or tonnes for charge materials. The mixture prices for standard and actual compositions result from the planned prices for the substitutable materials.

15.9.2.3 Consumption deviation

Consumption deviation indicates excess or shortfall of consumption of any kind of consumable material or process cost compared with the guide consumption figures. Guide consumption is calculated with the aid of the guide value function for the quantity of influence value and is evaluated with the planned price for the relevant type of cost.

Consumption deviation (CD) is ascertained from:

$$CD = (C_G - C_A) \cdot P_P$$

For substitutable materials the actual consumption C_A and guide consumption C_G are related to the basic unit of quantity, e.g. kcal for fuels. The planned price P_P then corresponds to the mixture price P_{SM} for the specified composition.

15.9.2.4 Output- and Yield-conditioned deviations

This type of deviation includes increased or decreased consumption of materials in processing, which results from the difference between the guide quantity of a derived influence value and that actually obtained.

If production is the preset influence value in the guide-value function for a cost centre, and the charge and operating/utilization times are influence values derived from it, then, from the actual production figures for the period concerned, we can calculate the guide value for the charge by using the material Standard, and we can work out the guide values for the operating and utilization times with the aid of the output Standard. The effect on processing costs of the deviation between the guide and actual charge quantities is called the yield-conditioned deviation. The deviation relating to time values (operating utilization time) is the output-conditioned deviation.

It is advantageous to calculate both output-conditioned and yield-conditioned deviations summarily as a single value for each influence value.

The yield-conditioned and output-conditioned deviations (influence-value deviation) ID, represented in this example by the output-conditioned deviation, are calculated with the following formula:

$$ID = (C_{GG} - C_G) \cdot P_P$$

While the guide consumption C_G is determined from the actual operating time, the other guide consumption C_{GG} is derived from the guide-value function using the guide value for operating time. P_P is the planned price for the type of cost concerned.

15.9.2.5 Activity-level deviation

When calculating clearing prices for the accounting values of a cost centre after completion of the plan for the year, period-dependent costs (K_{per}) are added to the accounting values on the basis of normal activity levels. If the actual activity level is different from normal, then period-dependent costs will be over or undercharged by an amount called activity-level deviation. This is not determined separately for every type of cost but is worked out in summary for each cost centre.

If the accounting value is subdivided into several characteristics, then activity-level deviation is generally determined from the operating time. But first the period-dependent costs (K_{per}) have to be related to the unit of planned quantities of the accounting value (e.g. normal operating time OT_N).

The period-dependent costs charged for the cost centre are obtained by multiplying this quotient by the guide value for operating time OT_G.

The guide value for operating time is the sum of the actual values for cost-centre activity, charge, or production multiplied by the output Standard. We obtain the activity level deviation (AD) by comparing this result with the absolute period-dependent costs K_{per}:

$$AD = \frac{OT_G - OT_N}{OT_N} \cdot K_{per}$$

If a cost centre charges its output in the terms of production (PR), then the activity-level deviation can be obtained in the following way:

$$AD = \frac{K_{per} \cdot PR_A}{PR_N} - K_{per}$$

where PR_N is production at normal activity level
PR_A is production at actual activity level.

15.9.3 Showing Consumption and Price Deviation for Materials Costs

The guide consumption for each type of input and waste material is determined for steel types A and B in Table 8. The actual value of the production quantity influence value is then obtained by using the guide consumption function. The actual input quantities are compared with the guide consumption figures. The difference in quantity, when evaluated in the price plan, indicates the consumption deviation in DM per month.

The differences between actual and planned prices multiplied by actual values for input and waste for the various materials, lead to the price deviation in DM per month.

15.10 CALCULATING ECONOMIC EFFICIENCY

Examination of the economic efficiency of investment is closely allied to the determination of the costs of an industrial operation. This involves examining the economic efficiency of existing production installations, new installations planned, and changes in production processes. The steady growth in plant capacity, the high capital expenditure involved for each installation, the close

Table 8. Consumption and price deviations for materials

Check calculation		Determination of consumption and price deviation in materials accounting									Actual month		
Line	Material	Type A Actual production 28 000 t		Type B Actual production 17 500t		Guide Input Quantity Total	Actual Input Quantity Total	Difference	Plan Price	Actual Price	Difference	Consumption Deviation	Price Deviation
		Input Standard	Guide input quantity	Input Standard	Guide input quantity								
		(t/t)	(t)	(t/t)	(t)	(t)	(t)	(t)	(DM/t)	(DM/t)	(DM/t)	(DM)	(DM)
1		2	3	4	5	6 (3 + 5)	7	8 (6 − 7)	9	10	11 (9 − 10)	12 (8 × 9)	13 (7 × 11)
1	Own recycled scrap	0.310	8 680	0.320	5 600	14 280	14 394	−114	190	185.–	+5.–	−21 660	+71 970
2	Bought-in scrap	0.600	16 800	0.560	9 800	26 600	26 853	−253	170	168.–	+2.–	−43 010	+53 706
3	Alloyed scrap	0.140	3 920	0.160	2 800	6 720	6 660	+60	250	265.–	−15.–	+15 000	−99 900
4	FeMn	–	–	0.011	193	193	184	+9	750	732.–	+18.–	+6 750	+3 312
5	FeSi	–	–	0.005	88	88	96	−8	900	1010.–	−110.–	−7 200	−10 560
6	Chromium	–	–	0.009	157	157	152	+5	1 500	1 650.–	−150.–	+7 500	−22 800
7	Lime	0.050	1 400	0.060	1 050	2 450	2 695	−245	110	105.–	+5.–	−26 950	+13 475
8	Other additions	0.007	196	0.009	158	354	347	+7	180	178.–	+2.–	+1 260	+694
9	Slag	0.090	2 520	0.100	1 750	4 270	4 480	−210	30	30.–	0	+6 300	0
10	Total											−62 010	+9 897

association of utilization and production capacity, together with the employment of manpower and progress in mechanization and automation, all call for the economic efficiency of improved or new production installations to be checked continuously.

The process used can vary according to what is being studied, i.e. whether the size of the plant, the equipment, the production process, or the product is to be subject to an investigation of economic efficiency. These all have a predominantly mathematical character and use the same quantities such as investment capital, utilization time, interest payment, production output, loading, and operating costs. The following procedures apply according to which quantities are known or assumed and which quantity has to be calculated.

15.10.1 Investment Calculation Procedures

(1) Capital limit value calculation. This is used to estimate the maximum permissible investment levels for cost savings and increasing production.
(2) The limiting quantity and limiting loading calculation. This includes the minimum quantities to be produced or the utilization as an economic measurement.
(3) Amortization calculation. This works out the time in which investment capital will be redeemed together with interest paid.
(4) The profitability calculation. This works out the difference between outlay and income on investment capital and hence derives the ret urn on the investment.

For the assessment of return on investment it is recommended that, out of the various accounting methods available, the method of internal rate of interest[13] should be used.

Internal rate of interest is defined as interest with which surplus income is debited in order to be equivalent to the outlay. Management often lays down a minimum interest rate for investment which must not be undercut by the internal interest. Waldschmidt[14] has demonstrated a method for calculating minimum return on capital based on laying down a minimum rate of interest.

It should be borne in mind that none of these procedures can produce reliable results unless initial values can be determined with sufficient accuracy.

15.10.2 Determining Steelworks Build-costs and Effect of Investment on other Areas

Schenck and his associates have published a report on determining the costs of building and operational processes for steelworks.[15]

When investigating possible investments or changes in steelworks processes, the effects of a planned change on preceding or follow-on process areas are often not recognized or considered in calculating economic efficiency. An incomplete economic efficiency calculation which does not take all economic effects into account can lead to wrong decisions.

The Verein Deutscher Eisenhüttenleute (VDEh-German Steel Industry Association) economic committee have produced check lists[16] for working out the above effects in calculating economic efficiency in investment. Table 9 shows an extract from such a check list for an electric furnace steelworks.

An accountant working on an investment calculation can use these check lists for every cost centre in a steelworks and its most important installations (there are cross-references to other leaflets in the check list) to examine the effects on the function areas concerned, e.g. purchasing or other production areas and especially auxiliaries.

The examination begins qualitatively, with yes/no decisions. When effects are anticipated or known, the requirements should be annotated in the remarks column and on a special comment sheet where they may also be quantified.

Leaflet 7.0.0 on works transportation is subdivided into sheets 7.2.1 on works railways and 7.3.1 on non-rail transport. References in the leaflet indicate that transport volume should be classified and quantified according to charge and waste materials, special consumables (refractories), and removal of products to other production areas. The works transport system should then be analysed to see whether it can handle the volume of traffic.

Using this information, the works transport management decides which means of transportation and which routes to use for the various loads. The foremost consideration is whether the volume of traffic can be handled by existing means or whether it will be necessary to invest in further transportation equipment.

Despite all their advantages, these check lists still do not exclude the possibility that the most economical solution will not always be obvious when there are several alternatives. With the present state of knowledge, it hardly seems possible to devise a systematic method for analysing economic efficiency which can overcome this problem with any degree of certainty.

The two Institutes for works operational research and applied research of the VDEh (German Steel Industry Association) can be approached for advice on organizing and setting up an accounting system for planning calculations, documentation accounting, and check calculations.

15.11 LITERATURE REFERENCES

1. Magnus Radke *Die große betriebswirtschaftliche Formelsammlung*, Munich, 1966.
2. Richtlinien für die Berechnung der Produktivität in der Hüttenindustrie. Arbeitsausschuß Produktivität in the Fachverband der Bergleute und Eisen erzeugenden Industrie. Vienna, 1974.

Table 9. Extract from economic efficiency calculation check list

		Steelworks area			4
		Electric furnace steelworks			3
		Furnace operation			
	Proposal No.	Processing			2
		Sheet			1

	Operation Area	Yes	No	Sheet No	Remarks
1.	Common works installations				
	Office buildings				
	Staff/manpower arrangements				
	Central instrumentation (including process control computers)				
	Laboratories				
	Refractory storage				
	Other common installations				
3.	Scrap-processing			7.2.1	
	Shop, yard				
	Scrap-storage and preparation				
	Scrap-transport to furnace, e.g. cranes, wagons, trays, baskets,			7.2.2	
	Weighing machines				
6.	Furnace operation				
	Shop				
	Cranes				
	Ground-level transportation gear			7.3.1	
	Furnace vessel and roof				
	Electrodes and electrode control			6.4.1	
	Transformer station with reactor and oil circuit				
	Oxygen supply			6.6.1	
	Water supply			6.1.1	
	Instrumentation and control				
	Furnace lining and repair with auxiliaries			1.1.0	
	Ladle transport system				
7.	Dust-extraction system				
	Indirect gas-cooling (boilers)				
	Boiler water-supply			6.1.1	
	Steam-generation			6.3.1	
	Direct gas-cooling and dust-removal				
	Power supply for electrical filters			6.4.1	
	Sludge and dust-removal			7.2.1	

3. Glaszinski, H., Richtkostenrechnung in einem Eisenhüttenwerke. *Z. handelwiss. Forschung* **4** (1952), 447–63.
4. Hoffstadt, J., Die Kosten-Einflußgrößenberechnung, dargestellt am Beispiel eines Siemens-Martin-Stahlwerkes. *Stahl u. Eisen* **77** (1957), 1488–96.
5. Steffen, M., and V. Steinecke, Einflußgrößenrechnung zur Kostenplanung eines kontinuierlichen Feinstahlwalzwerkes mit Matrizen. *Stahl u. Eisen* **82** (1962), 206–11.
6. Flegelskamp, N., and H. Röhler, Control and account of the cost of works railway transport with due consideration of the different factors of influence (German). *Stahl u. Eisen* **89** (1969), 877–85.
7. Waldschmidt, J., Cost of a works railway and factors of influence on it (German). *Stahl u. Eisen* **92** (1972), 809–816.
8. Waldschmidt, J., Characteristic features of a standard cost accounting system designed to control costs (German). *Stahl u. Eisen* **87** (1967), 737–45.
9. Scholler, W., H.-G. Eckhold, Economic plant management with the aid of standard cost accounting (German). *Stahl u. Eisen* **86** (1966), 8–16.
10. Distler, J., L. Gorius, H.-P. Hoffman and K. H. Sandhöfer, The cost system of the Stahlwerke Röchling–Burbach as aid to the management with special consideration of the standard cost system (German). *Stahl u. Eisen* **97** (1977), 342.
11. Richtlinien für das Betriebliche Rechnungswesen der Eisen- und Stahlindustrie. Vol. 1: Allgemeine Richtlinien. Vol. 2: Besondere Richtlinien. Sonderausgabe eines Rechenbeispiels mit Textteil und Tabellenteil. Betriebswirtschaftliches Institut der Eisenhüttenindustrie. Düsseldorf 1976.
12. Kilz, K. Die neuen Richtlinien für das Betriebliche Rechnungswesen in der Eisen-und Stahlindustie. *Stahl u. Eisen* **95** (1975), No. 15, 705–09.
13. *Planung, Durchfürung und Prüfung von Investitionen.* Schriftenreihe der Wirtschaftsvereinigung Eisen- und Stahlindustrie, Issue No. 3, Düsseldorf 1968.
14. Waldschmidt, J., Investment rating in a binational enterprise (German). *Stahl u. Eisen* **98** (1978), 1405–11.
15. Schenck, H. R., K. Consemüller, H.-J. Zimmermann, E. Quadflieg, P.-D. Busch together with Th. Hinkeldey, Baukosten von Stahlwerken und Betrieblichen Verarbeitungskosten der Stahlwerksverfahren. Stahl-eisen-Einzeldarstellungen technisch-betriebswirtschaftlicher Zusammenhänge. Düsseldorf 1970.
16. Check-Listen für Wirtschaftlichkeitsrechnungen Fachausschußbericht No. 7008 des Vereins Deutscher Eisenhüttenleute. Düsseldorf 1970.

Morgan, J. I., and A. I. Johnson, A modular approach to computerized material balance calculations. *Computers Operations Res.* **1** (1974), No. 2 (Aug.), 263–74.

Fabian, T., A linear programming model of integrated iron and steel production. *Managem. Sci.* **4** (1958), 415–49.

Lawrence, I. R., and A. D. I. Flowerdew, Economic models for production planning. *Operational Res. Quart.* **14** (1962), No. 1, 11–29.

Steinecke, V., and M. Steffen, Factor analysis of cost accounting for a continuous small section rolling mill — intra firm input/output model. In *Proceedings of the 3rd International Conference on Operational Research,* Oslo 1963; Paris and London 1964, pp. 353–63. (Mostly identical to Ref. 5.)

Cuninghame-Green, R. A., Process synchronization in a steelwork — a problem of feasibility. In *Proceedings of the 2nd International Conference on Operational Research,* London 1960, pp. 323–32.

Tuckett, R. F., Combined cost and linear programming models of industrial complexes. *Operational Res. Quart.* **20** (1969), No. 2, 233–7.

Franke, R., A process model for costing. *Managem. Accounting* **56** (1975), No. 7, 45–7.

McCulloch, G. A., and R. Bandyopadhyay, Application of operational research in production problems in the steel industry. *Internat. J. Prod. Res.* **10** (1972), No. 1, 77–91.

Cost Planning and Accounting using Matrices-costmat. Guide for Application Programming. Pt. 1: Executive Guide. Pt. 2: Application Guide. IBM United Kingdom Ltd, Peterlee (County Durham) 1975–6, Order No. CE 15-6077 and CE 15-6078.

Electric Furnace Steel Production
Edited by E. Plöckinger and O. Etterich

16 Units and Conversion Tables

In the Federal Republic of Germany the law of 2 July 1969 about units used in measurement systems and instructions dated 26 June 1970 for implementation of the law have been in force since 5 July 1970.[1–3] This law is based on the SI system: the Système Internationale d'Unité. The transition period in West Germany ran out on 31 December 1977, and, from 1978, the old units were replaced in publications there by the new legal units.

Table 1. SI base units

Base quantity	Base unit	
	Name	Symbol
Length	metre	m
Mass	kilogram	kg
Time	second	s
Electric current	ampere	A
Temperature	kelvin[a]	K
Amount of substance	mol	mol
Luminous intensity	candela	cd

[a] Where Celsius temperatures are indicated, the expression degree Celsius (symbol °C) is used in place of kelvin.

Table 1 shows the seven SI base units which are defined in the Annex to DIN 1301 part 1.[4] Table 2 shows the internationally agreed prefixes denoting decimal multiples and submultiples (SI and metric). Units derived from the basic SI units are shown in Table 3. Quantities frequently used in this book originated from DIN standards[4–22] and are shown in Table 4.

16.1 LITERATURE REFERENCES

1. Anwendung der SI-Einheiten in *Stahl u. Eisen* and in *Archiv für das Eisenhüttenwesen. Arch. Eisenhüttenwes.* **49** (1978), 43–5.

Table 2. Internationally agreed prefixes

Factor by which the unit is multiplied	Prefix	Symbol
10^{-18}	atto	a
10^{-15}	femto	f
10^{-12}	pico	p
10^{-9}	nano	n
10^{-6}	micro	μ
10^{-3}	milli	m
10^{-2}	centi	c
10^{-1}	deci	d
10^{1}	deca	da
10^{2}	hecto	h
10^{3}	kilo	k
10^{6}	mega	M
10^{9}	giga	G
10^{12}	tera	T
10^{15}	peta	P
10^{18}	exa	E

Note: When using unit symbols that are similar to prefix symbols, care must be taken to avoid confusion: e.g. torque m·N; millinewton mN.

Table 3. Derived SI units with special names and symbols

Quantity	SI unit Name	Symbol	Relationship
Planar angle	radian	rad	1 rad = 1 m/m
Solid angle	steradian	sr	1 sr = 1 m^2/m^2
Periodic frequency	hertz	Hz	1 Hz = 1 s^{-1}
Activity of radioactive substance	becquerel	Bq	1 Bq = 1 s^{-1}
Force	newton	N	1 N = 1 kgm/s^2
Pressure, mechanical stress	pascal	Pa	1 Pa = 1 N/m^2
Energy, work, quantity of heat	joule	J	1 J = 1 N·m = 1 W·s
Power, thermal current	watt	W	1 W = 1 J/s
Energy dose	gray	Gy	1 Gy = 1 J/kg
Electrical charge, quantity of electricity	coulomb	C	1 C = 1 A·s
Electrical potential, electric voltage	volt	V	1 V = 1 J/C
Electrical capacitance	farad	F	1 F = 1 C/V
Electrical resistance	ohm	Ω	1 Ω = 1 V/A
Electrical conductivity	siemens	S	1 S = 1 Ω^{-1}
Magnetic flux	weber	Wb	1 Wb = 1 V·s
Magnetic flux density, magnetic induction	tesla	T	1 T = 1 Wb/m^2
Inductance	henry	H	1 H = 1 Wb/A
Celsius temperature	degree Celcius	°C	1°C = 1K
Luminous flux	lumen	lm	1 lm = 1 cd·sr
Luminous intensity	lux	lx	1 lx = 1 lm/m^2

Table 4. Units of Measure, dimensions, and conversions in international SI units[4]

Quantity	Symbol	SI units		Definition and remarks	Units outside SI system	Conversion, remarks
		Unit name	Symbol (dimension)			
Geometric quantities						
Length	l	metre	m	Base unit		
Area	A, S	square metre	m^2			
Volume	V	cubic metre	m^3		litre (l)	1 litre = 1 dm^3 = 0.001 m^3
Standard volume	V_n	cubic metre	m^3			
Time and space quantities						
Time	t	second	s	Base unit	minute (min) hour (h) day (d)	
Velocity	c, u, v, w		m/s		km/h	c = light and sound velocities u = peripheral velocity w = relative velocity
Acceleration	a		m/s^2		cm/s^2	
Mass flow	m		kg/s		t/s; t/h	
Volumetric flow rate	V		m^3/s		l/s; l/h	
Frequency	f	hertz	Hz	1 Hz = 1/s		
Speed of rotation	n		s^{-1}		min^{-1}	$1 min^{-1} = s^{-1}/60$
Quantity of matter						
Quantity of matter	n	mol	mol	Base unit		References 14, 16

Mechanics (mass and force)						
Mass, weight	m	kilogram	kg	Base unit	g; Mg = tonne (t)	Mass units are also used for weight — the result of a weighing (Reference 11)
Density	ρ		kg/m^3		t/m^3; g/cm^3; kg/l	The expression 'specific gravity' should not be used (Reference 12)
Specific volume	v		m^3/kg			
Force	F	newton	N	$1\ N = 1\ kgm/s^2 = 1\ W{\cdot}s/m$		Force = mass × acceleration. The units p, kp, and dyn became obsolete in West Germany from 1 Jan 1978 (Reference 11) 1 kp = 9.81 N
Torque	M	newton-metre	N·m	1N·m = 1J = 1 W·s		
Pressure	p	pascal	Pa	$1\ Pa = 1\ N/m^2$	$bar = 10^5\ Pa$	kp/m^2, mmWs, torr, and atm became obsolete in West Germany on 1 Jan. 78. 1 bar = 0.987 atm = 1.02 at = 750 torr 1 torr = 0.00133 bar = 133 Pa
Mechanical stress			N/m^2, Pa			$1\ MPa = 1\ N/mm^2$ $1\ kp/cm^2 = 9.81\ N/cm^2 = 0.981\ bar$
Surface tension			N/m			
Energy quantities						
Work and energy (mechanical and electrical)	E, A, W	joule	J	1 J = 1 N·m = 1 W·s	W.h	1 W·h = 3.6 kJ (Reference 14) 1 kJ = 0.278 W·h

Table 4. (cont.)

Quantity	Symbol	SI units		Definition and remarks	Units outside SI system	Conversion, remarks
		Unit name	Symbol (dimension)			
Quantity of heat	Q	joule	J	1 J = 1 N·m = 1 W·s		cal and kcal became obsolete in West Germany on 1 Jan. 78 (Reference 14) 1 cal = 4.1868 J 1 kJ = 0.239 kcal
Thermodynamics and heat transfer						
Temperature	T	kelvin	K	Base unit		The unit grd became obsolete in West Germany on 1 Jan. 75 (Reference 14)
Temperature difference			K			
Celsius temperature	t	degree Celsius	°C			1°C = 1 K
Entropy	S		J/K			0°C ≙ 273.15 K 1 kcal/°C = 4.19 kJ/K
Enthalpy	H	joule	J			1 J = 1 N·m = 1 W·s 1 kcal = 4.19 kJ
Thermal conductivity	λ		W/m·K		J/m·s·K	1 kcal/m·h·grd = 1.16 W/m·K (Reference 15)
Heat-transfer coefficient	α		W/m^2·K		J/m·s^2·K	1 kcal/m^2·h·grd = 1.16 W/m^2·K (Reference 15)
Electrical quantities						
Current	I	ampere	A	Base unit		References 7, 17

Voltage	*V* or *U*	volt	V	1 V = 1 W/A		In West Germany the standard symbol for voltage is *U*
Resistance	*R*	ohm	Ω	1 Ω = 1 V/A		
Quantity of electricity, charge	*Q*	coulomb	C	1 C = 1 A·s	A·h, A·s	3600 C = 1 A·h (Reference 7)
Capacitance	*C*	farad	F	1 F = 1 C/V	μF	
Field strength	*E*	volt/metre	V/m			
Current density	*J* (*S* in West Germany)	ampere/square metre	A/m^2			In West Germany the standard symbol for current density is *S*
Alternating current and mains quantities						
Apparent power	*S*	watt	W		GVA, MVA, kVA, VA	Reference 10
Active power	*P*	watt	W		GW, MW, kW, W	Reference 10
Reactive power	*Q*	watt	W		Mvar, kvar, var	Reference 10
Impedance	*Z*	ohm	Ω			Reference 10
Resistance	*R*	ohm	Ω			Reference 10
Reactance	*X*	ohm	Ω			Reference 10
Magnetic quantities						
Magnetic flux	ϕ	weber	Wb	1 Wb = 1 V·s		Reference 8
Flux density, magnetic induction	*B*	tesla	T	$1 T = 1 Wb/m^2$		Reference 8: 1 Gauss (G) = 10^{-4}T
Magnetic field strength	*H*	ampere/metre	A/m		1 Oe = 79 557 A/m	Reference 8: the oersted is not an SI unit
Inductance	*L*	henry	H	1 H = 1 Wb/A		Reference 8
Permeability	μ	henry/metre	H/m	1 H/m = 1 Wb/(A·m)		Reference 8
Acoustic quantities						
Sound pressure			Pa	$Pa = N/m^2$		Reference 18
Acoustic power			W			
Sound intensity			W/m^2			
Noise level, volume					phon	References 19, 20
Noise level					dB (decibel)	References 21, 22

2. VIII. Internationale Einheitensystem (SI). Umrechnungstafeln und fachliche Tafeln. In *Taschenbuch für die Stahlindustrie.* 1978. VDEh, the Association of German Iron & Steel Engineers. Düsseldorf 1977, pp. 176–7.
3. Haeder, W., and E. Gärtner *Die gesetzlichen Einheiten in der Technik.* 4th edition. Deutschen Normenausschuß (DNA). Berlin 1974.
4. DIN 1301. Part 1: Units; names, symbols. Issued Oct. 1978.
5. DIN 1301. Part 1, Supplement 1: Units; names and symbols analogous to units. Draft issued Jan. 1979.
6. DIN 1301. Part 2: Units; submultiples and multiples for general use. Issued Feb. 1978.
7. DIN 1324: Elektrisches Feld, Begriffe. Issued Jan. 1972.
8. DIN 1325: Magnetisches Feld, Begriffe. Issued Jan. 1972.
9. DIN 1357: Einheiten elektrischer Größen. Issued Nov. 1971.
10. DIN 40110: Wechselstromgrößen. Issued Oct. 1975.
11. DIN 1305: Mass, force, weightforce, load; definitions. Issued May 1977.
12. DIN 1306: Density; definitions. Issued Dec. 1971.
13. DIN 1314: Pressure; basic definitions, units. Issued Feb. 1977.
14. DIN 1345: Thermodynamik, Formelzeichen, Einheiten. Issued Sep. 1975.
15. DIN 1341: Wärmeübertragung, Grundbegriffe, Einheiten, Kenngrößen. Issued Nov. 1971.
16. DIN 1310. Draft: Zusammensetzung von Mischphasen (Gasgemischen, Lösungen, Mischkristalle). Issued Aug. 1977.
17. DIN 4897: Elektrische Energieversorgung: Begriffe. Issued Dec. 1973.
18. DIN 1320: Akustik; Grundbegriffe. Issued Oct. 1969.
19. DIN 1318: Lautstärkepegel; Begriffe, Meßverfahren. Issued Sep. 1970.
20. DIN 45630. Part 1: Grundlagen der Schallmessung; physikalische und subjektive Größen von Schall. Issued Dec. 1971.
21. DIN 5493: Logarithmierte Größenverhältnisse; Pegel, Maße. Issued Aug. 1972.
22. DIN 40148. Part I: Übertragungssysteme und Vierpole; Begriffe und Größen. Issued Sep. 1966.

Kverneland, Knut O., *World Metric Standards for Engineering,* Industrial Press, New York, 1978. 760 pp. Many illustrations and tables.
British Standard 3763: 1976. The International System of units.
BS 5775: 1979–82. Specification for quantities, units, and symbols.
BS 4727: 1971–83. Electrical engineering glossary.
BS 661: 1969. Glossary of acoustical terms.

Index